# Innovative Analyses
## of
# Human Movement

## Nicholas Stergiou
**University of Nebraska at Omaha**

**Editor**

**Human Kinetics**

**Library of Congress Cataloging-in-Publication Data**

Innovative analyses of human movement / Nicholas Stergiou, editor.
  p. ; cm.
Includes bibliographical references and index.
  ISBN 0-7360-4467-1 (hard cover : alk. paper)
  1. Human locomotion--Mathematical models.  2. Human
locomotion--Statistical methods.  3. Human mechanics--Methematical
models.  4. Human mechanics--Statistical methods.
  [DNLM: 1. Movement--physiology.  2. Biomechanics.  3. Mathematical
Computing.  4. Models, Statistical.  WE 103 I525 2004]  I. Stergiou,
Nicholas.
  QP301.I35 2004
  612.7'6--dc21

                            2004004100

ISBN: 0-7360-4467-1

The Web addresses cited in this text were current as of May 9, 2003, unless otherwise noted.

**Acquisitions Editor:** Loarn D. Robertson, PhD; **Developmental Editor:** Anne Rogers; **Assistant Editor:** Amanda S. Ewing; **Copyeditor:** Joyce Sexton; **Proofreader:** Sue Fetters; **Indexer:** Robert Howerton; **Permission Manager:** Dalene Reeder; **Graphic Designer:** Fred Starbird; **Graphic Artist:** Denise Lowry; **Photo Manager:** Kareema McLendon; **Cover Designer:** Keith Blomberg, **Cover Image:** Fractal image courtesy of Paul Carlson, www.mbfractals.com; **Photographer (interior):** © Marco Gundt/Jump; **Art Manager:** Kelly Hendren; **Illustrator:** All art is author provided, except for figures 4.1, 8.1, and 8.2, which are by Accurate Art; **Printer:** Sheridan Books

Printed in the United States of America          10  9  8  7  6  5  4  3  2  1

Human Kinetics
Web site: www.HumanKinetics.com

*United States:* Human Kinetics, P.O. Box 5076, Champaign, IL 61825-5076
800-747-4457
e-mail: humank@hkusa.com

*Canada:* Human Kinetics, 475 Devonshire Road Unit 100, Windsor, ON N8Y 2L5
800-465-7301 (in Canada only)
e-mail: orders@hkcanada.com

*Europe:* Human Kinetics, 107 Bradford Road, Stanningley, Leeds LS28 6AT, United Kingdom
+44 (0) 113 255 5665
e-mail: hk@hkeurope.com

*Australia:* Human Kinetics, 57A Price Avenue, Lower Mitcham, South Australia 5062
08 8277 1555
e-mail: liahka@senet.com.au

*New Zealand:* Human Kinetics, P.O. Box 105-231, Auckland Central
09-523-3462
e-mail: hkp@ihug.co.nz

# Contents

## Part I

## Methods to Examine Variability in Human Movement   1

### Chapter 1   Single-Subject Analysis   3

### Chapter 2   Considerations of Movement Variability in Biomechanics Research   29

### Chapter 3   Nonlinear Tools in Human Movement   63

# Part II
# Methods to Examine Coordination and Stability in Human Movement                                              91

## Chapter 4   Applied Dynamic Systems Theory for the Analysis of Movement                           93

## Chapter 5   Directional Statistics                                              121

# Part III
# Advanced Methods for Data Analysis in Human Movement   187

## Chapter 9    Power Spectrum Analysis and Filtering          223

# List of Contributors

Barry T. Bates, PhD
Professor Emeritus, University of Oregon

Ugo H. Buzzi, MS, PhD candidate
University of Michigan

Timothy R. Derrick, PhD
Assistant Professor, Iowa State University

Janet S. Dufek, PhD
Research Scientist, Human Performance & Wellness, Inc.

Giannis Giakas, PhD
University of Ioannina

Jack Heidel, PhD
Professor and Chair of the Department of Mathematics,
University of Nebraska at Omaha

C. Roger James, PhD
Associate Professor, Texas Tech University

Max J. Kurz, MS, PhD candidate
Rhoden Biological Research Fellow, University of Nebraska at Omaha

Robert H. Schor, PhD
Associate Professor, University of Pittsburgh

Patrick J. Sparto, PhD, PT
Assistant Professor, University of Pittsburgh

Joshua M. Thomas, MS, PhD candidate
Iowa State University

Songning Zhang, PhD
Associate Professor, The University of Tennessee at Knoxville

Weimo Zhu, PhD
Associate Professor, University of Illinois at Urbana-Champaign

# Preface

In recent years, the electronic lists where the majority of human movement scientists belong have been flooded with e-mails asking for help with data analysis. Usually the questions pertain to the proper usage of new mathematical and statistical methods. The recent advances in computing power have also generated a multitude of possibilities in handling large data sets and properly analyzing them. Scientific societies have recognized the need to answer such questions and have organized workshops and roundtables. The American Society of Biomechanics has conducted workshops on single-subject analysis and data smoothing. The American College of Sports Medicine has organized workshops related to movement variability in biomechanical research and dynamical systems methods for the assessment of movement coordination. The Gait and Clinical Movement Analysis Society has also conducted workshops to address topics related to curve comparisons. While these workshops have had great success and the e-lists have been very helpful in solving technical questions, there has been a great need for a book that addresses frequently asked statistical and technical questions. This book was prepared to fill this void and to answer questions that have arisen as a consequence of the rapid developments in the technical and scientific world.

Therefore, this book provides an in-depth explanation of new and innovative ways to analyze data of human movement. The goal in editing this book was to offer students, as well as faculty members, a resource presenting some of the most recent advanced methods for data analysis in a tutorial fashion. Each chapter contains a large number of examples, suggested readings, and a carefully developed glossary that further contributes to this endeavor. The majority of the chapters also contain work problems for immediate application of the methods described. In addition, each chapter in this book creates new avenues for examining data and thus answering a multitude of research questions.

Many human movement scientists in their research have been very creative in using tools from mathematics and engineering. This book continues in this direction by presenting ways to incorporate mathematical tools such as nonlinear dynamics, directional statistics, and bootstrapping. From that perspective, this book can also contribute significantly to the future development of human movement sciences.

The book is organized in three sections with nine chapters. Part I presents various techniques to evaluate variability. These techniques are both statistical, as in chapters 1 and 2 (Single-Subject Analysis; Considerations of Movement Variability in Biomechanics Research), and mathematical as in chapter 3 (Nonlinear Tools in Human Movement).

In part II, the focus is on tools used to evaluate coordination and variability as described by the Dynamic Systems Theory. First, chapter 4 introduces the tenets and tools of this theory (Applied Dynamic Systems Theory for the Analysis of Movement). Chapter 5 then discusses the special statistical needs that such an approach requires (Directional Statistics), and chapter 6 presents alternative mathematical techniques to address questions related to coordination and variability (Mathematical Measures of Coordination and Variability in Gait Patterns).

Part III presents methods for analyzing complex data sets using the recent advances in computing power. These techniques are statistical, as in chapters 7 and

8 (Time Series Analysis: The Cross-Correlation Function; Principles and Applications of Bootstrapping Statistical Analysis), and mathematical as in chapter 9 (Power Spectrum Analysis and Filtering).

I hope the reader will benefit from the detailed descriptions of the methods presented by the authors and will be able to immediately apply these methods with ease. I envision this book as an invaluable toolbox for upper-level undergraduates, graduate students, and researchers in exercise science, biomechanics, biomedical engineering, and motor control. As such, I anticipate that this book will enhance the research in human movement sciences and especially in the fields of biomechanics, motor learning and control, rehabilitation medicine, biomedical engineering, and ergonomics.

# Acknowledgments

I would like first to thank all the contributing authors of this book. They worked extremely hard and with diligence, providing drafts on time and addressing all comments and revisions asked from them. I am indebted to my chair, Dr. Daniel Blanke, and my doctoral student, Max Kurz, for assisting me with various aspects of the editing process. I would also like to thank my parents, Jesus and Baia, my brother, Dimitris, and my wonderful wife, Ann, for providing love, support, and constant encouragement during this endeavor. Finally, this work would have never materialized without the continuous advice and direction of Anne Rogers and Loarn Robertson of Human Kinetics.

# PART I

# Methods
# to Examine Variability
# in Human Movement

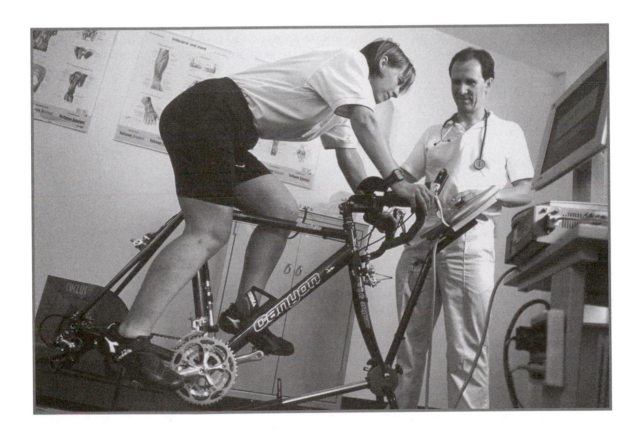

This first section presents statistical and mathematical techniques to evaluate human movement variability, a topic that has received increased attention in recent years.

In chapter 1 the focus is on within-subject variability, mostly from a statistical standpoint. First, a brief historical perspective is presented to explain the development of modern-day group statistical techniques and the relationship to the individual. Then an alternative statistical experimental methodology, the single-subject analysis, is introduced. The authors also discuss how single-subject analysis can form the basis for many research topics such as movement control, individual performance, and injury mechanisms and present the rationale for the single-subject design.

Chapter 2 examines the role of movement variability in human movement research. After a brief introduction, the author discusses theoretical foundations of the different types of variability and their sources, providing insight into the consequences of movement variability and illustrating how the characteristics of variability can influence the research design. This chapter also discusses the functional consequences of variability relative to their implications for both performance enhancement and injury causation. Several unique and very interesting theoretical models for both performance and injury are presented. Common (i.e., standard deviation, coefficient of variation) methods for quantifying variability are examined using examples from selected lower extremity movement tasks (i.e., running, jumping, landing). A final section explores possible future directions in considering variability relative to human movement research, including methodological and functional consequences, and quantification of variability.

Chapter 3 examines variability from a completely different view. The authors consider human movement variability from a nonlinear standpoint. They discuss the differences between traditional linear tools (i.e., standard deviation) and nonlinear methods (i.e., Lyapunov exponent, correlation dimension, approximate entropy) to study human movement variability, providing several examples to illustrate the application of these tools. The authors emphasize that nonlinear analysis can provide a unique approach to research questions and assist in understanding the control process of human movement and movement pathology.

# CHAPTER 1

# Single-Subject Analysis

Barry T. Bates, PhD, C. Roger James, PhD, and Janet S. Dufek, PhD

There exists a variety of analytical techniques for the study and interpretation of human movement responses. One such technique that has met with some contemporary resistance in human movement evaluation is **single-subject (SS) analysis.** This approach (SS analysis) is not a novel concept. The intensive study of individual human behavior was introduced in the mid-1800s in such areas as psychology, physiology, and psychiatry (Barlow and Hersen, 1984). In these studies, methodology often consisted of making repeated measurements of responses (trials) to different stimuli (experimental conditions); the results from these types of individual studies, with replication, often provided insight to allow for the formulation of generalizable outcome statements.

The study of the individual organism reached an early peak in Pavlov's work on the principles of association and learning (Pavlov, 1928), which Skinner later cited as influential on some of his work (Skinner, 1966a). Skinner (1966b) noted that "instead of studying a thousand rats for one hour each, or a hundred rats for ten hours each, the investigator is likely to study one rat for a thousand hours" (p. 21), clearly emphasizing the value of making repeated measurements on a single subject over a period of time. Skinner's approach, however, was a notable exception to an increasing emphasis on group statistical analyses.

Group statistical analysis can be traced back to the investigations of Adolphe Quetelet (1796-1874), who observed that individual traits followed a **normal distribution** grouped around a mean or average that he believed represented the ideal of the observed trait (Stilson, 1966). Quetelet viewed the "average man" as a desirable goal rather than as a descriptive entity and accordingly viewed variations in traits as errors or failures of nature. Quetelet's viewpoints did not gain favor; however, the study of individual differences did flourish during the early to mid-1900s (Barlow and Hersen, 1984) and was expanded with the development of additional descriptive statistical techniques. It was during this time that Pearson, assisted by Galton, developed the notion of correlation and emphasized the use of quantification in biological and psychological research (Barlow and Hersen, 1984). Pearson further promoted the concept that inaccurate data could produce accurate conclusions if one used the proper statistic (Boring, 1950).

Group statistics had the advantage, as perceived by its advocates, in that error as well as variability among individuals could be averaged out of the results if the group being examined was large enough. This notion of individual differences about the mean led to the need to make comparisons among individuals and to compare groups of individuals. In turn, this led to the application of "average" results to "average" individuals, while some researchers with conflicting opinions continued to emphasize the idea that there were no "average" individuals (Dunlap, 1932; Sidman, 1960).

Development of inferential statistical techniques in the early 1900s was the final stage in the evolution of analysis procedures during this phase of history. Most of the sophisticated inferential procedures used today were developed by R.A. Fisher. Fisher (1925) also documented the properties of statistical tests that allow for the generalizability of results from an empirical sample to the population from which the sample was drawn (i.e., $t$ and $F$ tables). It is important to understand the origin of this work since its underlying philosophy lies clearly within the "average" person framework. As a mathematician turned agronomist, Fisher had the objective of producing the greatest crop yield from any given plot of land. Variables including plant variety and soil type were manipulated and/or controlled to determine which plot "on average" produced the greatest output (Barlow and Hersen, 1984). Development and use of appropriate statistical tests to evaluate plot production and variable manipulation allowed Fisher to address his objective of maximum plot production. One shortcoming in this approach was that the fate of the individual plant (or subject) was irrelevant since the focus was on average plant or total plot output. This view of the fate of the individual (plant or human subject) is obviously not always appropriate or applicable to many modern-day questions of interest in the human movement sciences.

Fisher's group analysis techniques grew in popularity in the 20th century because of the desire to achieve generalizability of results, apply results to practical situations (i.e., crop production), and control experimental error by controlling the effects of extraneous variables. The common assumption is that group designs provide the means to generalize results obtained from a sample to the population from which the sample was drawn. In contrast, many experts today believe that SS analysis approaches limit the generalizability, and in turn the validity of results.

Group statistical analysis techniques were well established by the late 1960s. By that time biomechanics and other multidisciplinary areas of investigation in human movement science were emerging and consequently began to dominate research methodology. However, an examination of more recent history (1960s-1990s) as reported by James and Bates (1997) indicates that there has been a mild resurgence of interest in SS analysis and use of the individual as the basic unit of investigation (Bates, 1996; Bergin and Strupp, 1970; Bouffard, 1993; Conners and Wells, 1982; Dufek and Zhang, 1996; Edgington, 1987; Kratochwill, 1992; Sidman, 1960).

# Expanding Experimental Design Horizons

As previously indicated, SS analysis has met with some resistance in human movement evaluation. A difficult task for most of us is changing the way we think. We feel more secure processing information in our traditional ways. We must be willing to try alternative approaches, or at least let others try them. The advancement of science is often dependent on alternative measurement procedures. It is important to remember that measurement is not neutral and that all measurement schemes lose more information than they gather no matter how thorough and well conceived they may be. In addition, we all have biases—we cannot "study anything separate from ourselves. Our acts of observation are part of the process that brings forth the manifestation of what we are observing" (Wheatley, 1992: p. 36). We choose what we study and how we study it, including the experimental methods, the controls, the independent and dependent variables, and the analysis techniques. Since all outcomes are dependent on the state of the organism interacting with the environment at a specified moment in time, much information is lost forever. Furthermore, the data cannot be duplicated even through study replication. How then can we retrieve the data we lost when we went looking for the data we found? One way is to encourage as many research approaches as possible. In this way we will maximize our chances of discovering variations of those lost data that might further our understanding of the underlying mechanisms and processes controlling human movement.

Researchers using traditional group methodology are seeking "generalizability" and believe that this goal can be achieved as soon as all of the variables can be accounted for. Some system components and processes are undoubtedly generalizable within limited definable constraints, but as a total, self-actuated system, "man" is seldom predictable (Stapp, 1971), making it unlikely that this complex "human" system can be captured in a single generalized model. After all, the human system continuously feeds back on itself through experiences and the resulting changing perceptions. This fact—that we are complex, continuously changing interactive systems, each possessing "uniqueness"—not only justifies but sometimes also dictates the use of SS evaluations to better understand individual movement patterns.

Therefore, it is the intent of this chapter to elucidate the assumptions, applications, and limitations of SS analysis while keeping within the framework of human movement science applications. It is important to note that SS analysis does not imply "case study" investigation. Rather, it is an experimental technique that invokes an in-depth examination of individuals in order to better understand what unique movement characteristics, if any, they have in common. In keeping with the chapter objectives, we hope to lend insight to investigators who might assist in the design and analysis techniques used in subsequent research studies.

# Human Movement Characteristics

In order to understand the nature of human movement and thus develop an appropriate research methodology, it is important to first understand some of the more important factors that influence and affect behavioral observations. The following discussion addresses four important factors: (1) movement constraints, (2) human variability, (3) response patterns, and (4) aggregation.

## Movement Constraints

Bernstein (1967) and later Higgins (1977) were the pioneering proponents of the constraints that exist and that influence human movement. The constraints they identified (biomechanical, morphological, environmental), as well as the interactions among the constraints, affect in direct and indirect ways all human movement outcomes. Biomechanical constraints (Higgins, 1977) are defined as limitations imposed on the human system by physical laws (i.e., gravity, friction, etc.). Morphological or anatomical constraints are those limitations imposed on the system as a result of the physical structure and psychological makeup of the individual. Environmental constraints are the result of extraneous factors that affect performance, including personal arousal, crowd response, lighting, temperature, and the like. These are factors that are external to the organism but that affect the intrinsic responses (Higgins, 1977). Operating as an umbrella over all of these constraints, however, is the objective of the movement, termed the task constraint (Higgins, 1985). It is this constraint, in conjunction with experiences and memory, that most directly dictates the responses of the individual. That is, the task constraint refers specifically to the goal of the movement. If one is trying to run fast, then speed is optimized, versus running endurance. If one is trying to carry a very heavy load, then mechanics and leveraging principles will be optimized (focused on) versus movement aesthetics, for example. In short, the constraining nature of the task to be performed dictates the contributions of the remaining three constraints and subsequently produces a movement pattern that can be repetitive or variable, given the experience and prior knowledge of the performer. It is important to note that all constraints act interdependently and in combination with previous experiences to formulate the future movement outcome. In addition, the system is functionally pliable; changes (e.g., volition, learning, perception, growth and development) are possible within the bounds of the imposed constraints, allowing for a seemingly infinite number of movement outcomes or solutions to any movement task (James and Bates, 1997). Bates (1996) suggested that although the system has a considerable number of degrees of freedom, the number of functional degrees of freedom (choices) is "seemingly infinite."

## Human Variability

The human neuromusculoskeletal (n-m-s) system is extremely complex. This complexity was inferred by Colonel John Stapp (1971) during his closing remarks at the 1970 Symposium for Biodynamics Models and Their Applications when he described man generically as an obstinate and irregular object.

> This fifty liter rawhide bag of gas, juices, jellies, gristle, and threads movably suspended on more than 200 bones presided over by a cranium, seldom predictable and worst of all living, presents a challenge to discourage a computer into incoherence. (p. 115, cited in Miller, 1979)

The result of this structural complexity is an even more complex functional system that is inherently variable. Newell and Corcos (1993) stated that variability is inherent within and between all biological systems and is the result of interactions among the structural and functional characteristics of the system and the constraints imposed on motion. Because variability is an inherent component of movement both within and among individuals, one simply cannot replicate a movement pattern exactly given the various physical and environmental constraints surrounding the performance. An understanding of variability is important, therefore, in the design of both SS and group experiments.

Intra- or within-subject variability, "randomness" observed as variations of the same response for a specific set of conditions, affects the reliability of individual measurements; one can accommodate this variability statistically by obtaining multiple measurements (trials). In so doing, one can greatly enhance the power of the statistical test for both group and SS designs. For specific details on the effect of the number of measurements (trials) on statistical power for both group and SS designs, see Bates and colleagues (1992). Inter- or between-subject variability is illustrated as greater or lesser variations among individuals performing the same motor task. This form of variability is typically treated as error and can be accommodated statistically by increasing the number of subjects (or trials) in the experiment provided that the **homogeneity of variance** assumption has not been violated. One of the critical underpinnings in the rationale for SS analysis, however, lies in the observation that between-subject variability can also be the result of unique strategies or response patterns employed by the various subjects due to each individual's unique makeup. This form of variability violates the homogeneity of variance assumption, compromises the validity of the data, and cannot be accommodated statistically in group designs.

## Response Patterns

**Strategy** is defined as a unique n-m-s solution for the performance of a motor task (Bates, 1996). Selection of a particular strategy can be voluntary or involuntary, but the execution of a strategy (to accomplish a task) results in a unique pattern of movement with its own pattern of intra-individual variability. Patterns of movement are constrained to respond to three peripheral sources of variation (constraints): mechanical or biomechanical, morphological, and environmental (Bernstein, 1967; Higgins, 1977). The intrasubject variability observed, as well as actual or real between-subject differences observed in human movement, is caused by these constraints. One critically important aspect of this constraint domain that influences strategy selection is related to the perceptions and experiences of the performer. For example, in response to the addition of ankle weights, a gymnast and a mountaineer would likely perceive the need to modify landing performance differently based on their past experiences relative to weight of performance footwear; that is, slippers versus boots. Empirical evidence in support of individual strategies can be found in the research literature. Here we discuss selected examples across a variety of movement tasks and situations for illustrative purposes.

Bates and colleagues (1979) identified unique performance characteristics among five elite runners that were initially masked with use of a group descriptive analysis approach. The "average" runner, as depicted by the mean data, did not resemble any of the individuals in the group. In a more recent study of running performance, Lees and Bouracier (1994) evaluated experienced and inexperienced runners longitudinally, across days, in search of commonalities among selected kinematic variables describing various running patterns. The authors anticipated within-day variability in order to support their hypothesis of a "movement pattern fixation." They stated this conceptual hypothesis as follows:

> . . . in order to solve the movement problem associated with running, the subject may be able to select a method of running (a solution) from a pool of solutions which suit the particular requirements on the particular day of testing. (p. 198)

Dufek and Zhang (1996) examined the landing performance of elite volleyball players following execution of a blocking maneuver across the competitive season. The authors went on to create prediction models from kinematic variables during execution of the block that would predict landing impact forces. Results of their work, using both a group and an SS approach, produced group prediction models for forefoot and rearfoot landing force that were not representative of any of the seven individual volleyball players evaluated. The general (group) model was that of an "average" performer not representative of any single athlete participating in the study.

Another area in which strategies can influence conclusions is that of learning and/or motor skill acquisition experiments. Schmidt and colleagues (1992) stated that acquisition of motor skill involves the successive reparameterization of a dynamical control structure in the direction of increasing stability. The authors went on to indicate that the intentional process of reparameterization is itself dynamical. The process of motor skill acquisition is both dynamic and variable; therefore it should be apparent that performance options exist for individual subjects and that when combined in a group analysis, these individual learning effects could be obscured. In their work, Schmidt and colleagues (1992) did identify two unique strategies for the acquisition of the interlimb movement pattern under investigation.

Worringham (1993) also examined motor skill acquisition via the evaluation of spatial variability associated with complex arm movement tasks. Worringham made a priori awareness of subject differences clear, and in an attempt to gain statistical power in the experiment, the importance of avoiding any subject grouping was stressed. The author implied that the need to understand acquisition of the skill was possibly more important than incorporating traditional statistical treatment techniques during data analysis.

Benzecri and Benzecri (in Loslever, 1993) further emphasized Worringham's point when they stressed the importance of developing one's model to fit the data rather than forcing one's data into a model. Loslever (1993) further suggested that it is often unreasonable to expect a single model to fit all subjects in a given study because of excessive between-subject variability that could result from individuals' use of unique strategies. Thus the author suggested that the model should be adjusted for each class of behavior identified, which in practice would require examination of subjects at a deeper level than as members of the group.

In a final example that strongly supports the need for SS analysis, Schlaug and colleagues (1994) investigated intersubject variability of cerebral activation patterns during the acquisition of a complex sequential finger movement pattern. Results of group mean activation images illustrated some consistent task-specific activation sites in the brain; however, there was little or no spatial overlap of these patterns among subjects. The observed individual changes suggested a prominent inter- and intra-individual plasticity of cerebral activation patterns.

These exemplar studies, along with the previously discussed theoretical bases, provide strong support for the existence of independent performance strategies among individuals. A *strategy* was defined earlier as a selected n-m-s solution for the performance of a given motor task, within the existing constraint domain. In the case in which the constraint domain is manipulated within the experimental setting, causing a perturbation to the system (as is typical in human movement research), the system response or lack of response will depend on the individual's recognition or perception of the perturbation, which in turn will be a function of past experiences. The possible response patterns are identified in a model developed by James and colleagues (2003) (figure 1.1). The resulting response patterns can be purely mechanical or "Newtonian," meaning that the perturbation is completely ignored (the original motor program is used) and the outcome is predicted based solely on the principles of physics; or they can be fully accommodating or "neuromuscular," meaning that the system adjusts to produce an identical response on the dependent variable by modifying other characteristics of the performance; or they can be a combination of these two responses (biomechanical, overaccommodating, underaccommodating). The extent of the fully **accommodating response** will be dependent on the magnitude of the applied stressor relative to the physical capabilities of the individual at a given point in time. On the basis of this model, it should be apparent that each performer has myriad heterogeneous performance options that make homogenous grouping difficult, at best. Examples of these different types of strategies or responses have been frequently observed for impact forces during running (Nachbauer and Nigg, 1992; Nigg et al., 1987; Nigg and Segesser, 1992; Stergiou et al., 1999) and landing (Dufek and Bates, 1990; Dufek et al., 1995b; Hreljac, 1998).

As previously indicated, individual performance strategies can result in increased intersubject variability that not only reduces the statistical power in a group design, but also often leads to false support for the null hypothesis, depending on the distribution of subjects along the strategies continuum. At any point or over any constrained area within the performance domain, subjects will form a relatively homogeneous group (representing membership in a defined population), and their responses can be evaluated using traditional group statistical designs. For example, if all individuals perform along one of the lines (Newtonian or fully accommodating) within a specific region (biomechanical, overaccommodating, underaccommodating), they can be evaluated using group traditional statistical methods. When subjects performing in different regions are combined to form a group however, group statistical methods will provide invaluable data for all individuals or at least a subgroup of individuals. On the other hand, the per-

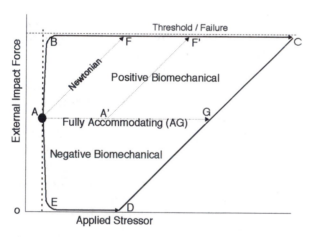

**Figure 1.1** Strategy model for impact accommodation. The boundaries of polygon ABCDE define the limits for all possible subject performances for accommodating impact force characteristics (ordinate) in response to changes in an applied stressor (abscissa). An example for a landing activity impact force characteristic is peak ground reaction force (GRF), and an applied stressor is landing height (LH). Point A would represent the initial value of GRF at some initial LH. Subsequent performances (values of GRF) would be plotted at different values of LH, and all performances would fall with the region defined by the polygon. The pattern and location of the responses would define the strategy, either positive or negative biomechanical, as depicted in the model.

From James, Bates, and Dufek, 2003.

formance of any given individual regardless of response can be evaluated using an SS approach provided that sufficient data are obtained and that the response remains stable during the observation/evaluation interval.

## *Aggregation*

An increasing number of researchers, especially those working in areas of applied research, are becoming more dissatisfied with traditional group statistical approaches (Barlow and Hersen, 1984). Their dissatisfaction stems from the inability to strongly associate experimental results to observed behaviors as well as limitations placed on their ability to treat or predict individual behavior (Barlow and Hersen, 1984; Bergin and Strupp, 1970). For example, Dufek and colleagues (1995b) predicted impact forces for individual subjects running and landing as well as for the respective groups. Results indicated that there were few common impact force predictor variables for individual subjects relative to the "group" prediction models. The same phenomenon has also been reported in the psychology literature. Payton (1994) reported that psychotherapy made no difference to the mental health of patients when evaluated on a group basis; individual-subject data, however, revealed that some mental health patients improved while other patients worsened. These and other studies (Bates, 1996; Dufek and Zhang, 1996) are examples of a mathematical (statistical) cancellation effect due to aggregation of the data when subjects execute the same task using different performance strategies. Aggregation masks individual performance strategies across a group of subjects and often results in false support for the null hypothesis, along with a loss of information relative to the manipulated dependent variables and a general lack of interest in the original research question (James and Bates, 1997).

Traditional group statistical techniques are typically invoked in order to examine differences among unique classes of data such as treatment conditions or subject groups (Bouffard, 1993). Bouffard explains that the class mean or aggregate is used to represent the performance of the entire class and that conclusions are drawn based on differences between the averages. These expressed differences are then presumed to be true for each and every member of the class. Bakan (1955) emphasized failure to distinguish between two types of propositions that are central to the ability to generalize results: general-type and aggregate-type propositions. A general-type proposition "asserts something which is presumably true of each and every member of a designable class" while an aggregate-type proposition "asserts something which is presumably true of the class considered as an aggregate" (p. 211).

The aim of science is to develop general principles or laws (Bouffard, 1993). The two types of propositions can therefore be viewed as the bases for making "lawful relationships about averages" (aggregate type) and "lawful relationships about people" (general type) because in order for a law to be universal, it must produce or predict the same or a similar result for every person (Taylor, 1958). Therefore, a researcher interested in the behavior of one or more individuals should incorporate methodologies that utilize general-type propositions such as SS analysis.

Differences in individual and aggregate-type data have been acknowledged, yet it has been customary in research incorporating group designs to assure that the functional form of a mean data curve reflects the form of the individual-subject mean data (Estes, 1956). James and Bates (1997) pointed out that the rationale behind arguments about individual versus aggregate data lies in the mathematical relationships that exist among the complex functions representing individual behavior. Many authors have shown that individual-subject data, when averaged, can produce an aggregate curve unlike any of the individual component curves (Baloff and Becker, 1967; Bouffard, 1993; Estes, 1956; Zhang et al., 2000). As explained by Baloff and Becker (1967), this phenomenon occurs because the differential effects cancel when averaged whereas similar components can be accentuated. The use of aggregate data to predict individual-subject behavior may therefore produce little or no meaningful information about the individual or may entirely misrepresent the types of individual behavior from which the aggregate was created (Baloff and Becker, 1967; Bouffard, 1993; Estes, 1956), and inductive inference from the mean to the individual is therefore impossible (Estes, 1956).

In response to the issues raised and discussed here, some human movement researchers have presented empirical results from both individual and aggregate data perspectives (Bates and Stergiou, 1996; Caster and Bates, 1995; Dufek et al., 1995b). In all of these studies, the researchers carried out both group and SS analyses in order to ascertain how the selected methodology might influence conclusions drawn. Although these studies employed different group techniques, the conclusion in each case was that the group results were not representative of the individual-subject results.

# Issues Relative to Data Analysis and Evaluation

Many of the statistical techniques commonly used for group analysis are equally applicable for SS evaluations. The criteria for acceptance as a legitimate test in the case of SS designs are the same as for group designs—for example, one must satisfy the assumptions underlying the selected procedures (Bates, 1996).

Criticisms of the SS approach have typically centered on the possible violation of statistical assumptions, as well as lack of generalizability of results. It is commonly held that measures taken from different individual subjects must be independent and that if subjects are assigned and/or selected to a treatment condition in accordance with the statistical rules for random assignment to groups, the results of the study will be generalizable to the population that the sample represents. It has also been generally assumed, although not previously supported by data, that multiple measures obtained from the same individual must be dependent (not independent) since they are generated by the same biological system. In addition, since the total number of subjects investigated in any given SS experiment is typically small and subjects are not randomly selected, it is assumed that the results of the study have very limited value relative to generalizability of results.

## Violation of Statistical Assumptions

Normality of the distribution of data, as well as **independence** of data sets, have been raised as statistical assumption violations when an SS approach is invoked. **Normality** of data in group experiments is typically assumed by most researchers, with the best test of normality being a visual inspection of the data distribution (Guildford, 1936; Stevens, 1986). In addition, statistical textbook authors typically indicate that most statistical tests are robust relative to violations of normality (Guildford, 1936; Stevens, 1986). The observation that there is relatively little importance placed on data normality in group experiments makes it difficult to understand objections to any possible violation of normality in SS designs. However, since the issue has been raised, analyses addressing this point have been completed and are included in this discussion.

Data to address the normality assumption are presented for a theoretical model as well as for an actual empirical data set. The theoretical model used Monte Carlo procedures in conjunction with a normal distribution random model generator (Bates et al., 1992). Random samples of various sizes representative of individual-subject performances were generated from a normally distributed data set and evaluated for normality using the Shapiro-Wilk (W) test for normality (SAS Institute, 1993; Shapiro and Wilk, 1965) and a correlation test between the generated data values and empirical values from a normal distribution (Ryan and Joiner, 1974) as suggested by Hildebrand (1986). The mean values are presented in table 1.1 and are the results of 1,000 replications.

**Table 1.1  Normality Results for 1,000 Replications of Normal Distribution Model**

| Sample size | $r$* | W** |
|---|---|---|
| 10 | .962 | .478 |
| 20 | .970 | .476 |
| 25 | .973 | .495 |
| 30 | .973 | .502 |
| 50 | **.975**[a] | .478 |

*Mean correlation coefficient (Ryan and Joiner, 1974) for empirical data versus randomly generated data (see text for full description); **Mean Shapiro-Wilk (W) test values for normality of the randomly generated data sets (e.g., $p <$ W = .478 indicates that data are normally distributed at $p < .05$ for n= 10); [a]**bold** = non-normal at $p < .05$ (see text for explanation).

The results of the model given in table 1.1 were dependent only on sample size as shown by the data. The probability of non-normality for all samples was approximately 5% as would be expected at the $\alpha$ = .05 level of significance. The mean probability level for the Shapiro-Wilk test was around .05, again as would be expected. The mean correlation coefficient increased as sample size increased (from .962 to .975), showing a greater congruity (i.e., greater explained variance) between the sample and normal curve. Even though explained variance continued to increase with sample size, it is interesting to note that the value of .975 for $n$ = 50 was non-normal ($p < .05$) due to the increased statistical power of the test. The same tendency can be observed in the mean W values as probability values increase through $n$ = 30 followed by a decrease between $n$ = 30 and $n$ = 50 indicating a decrease in normality (statistically).

The real data analyses were conducted on various sets of trials performed by individual subjects within a single experimental setting. The analyses were carried out on three running data sets and three landing data sets with individual subjects performing between 20 and 30 trials per experimental condition. For the three running data sets, a total of 17 subjects performed under two experimental conditions. Eighteen subjects performed in three landing studies in which the number of conditions by day varied from two to six. The six combined data sets represent performances by 35 subjects for an average of 2.4 experimental conditions per test session totaling 552 subject-variable sets. Specific instrumentation and data collection protocols are described elsewhere (Bates and Stergiou, 1996; Caster et al., 1992; James et al., 1992; SAS Institute, 1993; Schot, 1991). The Shapiro-Wilk (W) test was used to evaluate normality of these data sets.

A summary of the real data SS normality test results is given in table 1.2. These values all exceed the expected value of 5% (for $\alpha$ = .05), indicating a violation of the normality assumption with extreme results observed for early temporal events (T1, Tmin). Many of the measures evaluated are constrained physically or temporally in one direction, forcing some skewness of the data in the opposite direction. For example, impact forces in running typically occur in about 15 to 35 ms following contact (Bates et al., 1981), while the two impact forces in landing are observed around 12 and 50 ms following foot contact for the forefoot and rearfoot, respectively (Dufek and Bates, 1990). These events are all constrained in one direction by an absolute temporal limit (0 ms at contact), forcing variations in the positive direction. The strong negative relationship ($r = -.82$) between mean impact time and percent non-normality for the data sets used in this evaluation, shown in figure 1.2, illustrates the effect of the 0-ms constraint on the relationship.

Other temporal variables such as time to joint position can also be constrained, but usually to a lesser extent, by joint limits (full extension) and directional adjustments (one can only flex from full extension). Events or measurements such as impact force magnitudes (F1, F2) that are less physically constrained are more likely to approach normality, as the data indicate. The constraints identified are certainly not intrasubject dependent; they apply to intersubject samples as well.

Several variables from two group analyses (Dufek and Zhang, 1996; James and Zhang, 1994) were also evaluated for normality. One group consisted of 16 volleyball players performing block jumps (six trials); the second group was made up of 12 runners performing on a treadmill (five trials). Because of the limited number of data sets (two), a bootstrapping technique (Wasserman and Bockenholt, 1989) was used to generate normality probability data. The results for 1,000 replications are presented

**Table 1.2 Normality and Trial Independence Results for Single-Subject Data Analyses of Three Running (R) and Three Landing (L) Data Sets**

| Variable | Non-normal* | Lag 1* | Lag 2* | Lag 3* | N** |
|---|---|---|---|---|---|
| **FORCE** | | | | | |
| F1 | 11.5 | 11.5 | 9.8 | 3.3 | 61 |
| Fmin | 25.4 | 4.8 | 7.9 | 6.3 | 63 |
| F2 | 5.5 | 6.8 | 4.1 | 2.7 | 73 |
| Impulse | 15.0 | 5.0 | 0.0 | 0.0 | 20 |
| **TIME** | | | | | |
| T1 | 72.6 | 5.5 | 0.0 | 1.4 | 73 |
| Tmin | 69.8 | 7.9 | 0.0 | 1.6 | 63 |
| T2 | 24.0 | 5.3 | 2.7 | 0.0 | 75 |
| Tp | 41.7 | 4.2 | 4.2 | 0.0 | 24 |
| Tk | 20.6 | 5.9 | 5.9 | 5.9 | 34 |
| **KINEMATIC** | | | | | |
| $\theta$ | 33.3 | 7.6 | 6.1 | 3.0 | 66 |
| ROM | 20.0 | 6.7 | 10.0 | 6.7 | 30 |
| V / A | 36.4 | 13.6 | 4.5 | 0.0 | 22 |
| **KINETIC** | | | | | |
| JM | 20.0 | 5.0 | 10.0 | 5.0 | 20 |
| JP | 35.0 | 0.0 | 5.0 | 0.0 | 20 |
| **MUSCULAR** | | | | | |
| iEMG | 42.9 | 11.4 | 7.1 | 7.1 | 70 |
| Mean | 33.0 | 7.1 | 4.8 | 3.0 | 46 |

* = percent of non-normal and $r$-values significantly different from zero ($p < .05$); ** = number of data sets evaluated; F1 = first maximum vertical ground reaction force (vGRF) [R, L]; Fmin = relative minimum vGRF [R, L]; F2 = second maximum vGRF [L]; Impulse = vGRF impulse [R, L]; T1= time to F1 [R, L]; Tmin = time to Fmin [R, L]; T2 = time to T2 [L]; Tp = time to maximum pronation [R]; Tk = time to first maximum knee flexion [R]; $\theta$ = angular joint displacement for hip, knee, and ankle joints [L]; ROM = range of motion; V / A = velocity and acceleration for hip, knee, and ankle joints [L]; JM = joint moment for hip, knee, and ankle joints [L]; JP = joint power for hip, knee, and ankle joints [L]; iEMG = integrated EMG activity for selected lower extremity muscles [L].

in table 1.3. Similar to the situation with the SS results, the percent of non-normal distributions exceeded the expected value of 5% for all variables, suggesting non-normality of the data sets. The results overall, however, are better than the SS values (18.6% vs. 33.0%) but still 3.7 times greater than the expected value. These results along with the SS values suggest exercising some caution especially with the use of procedures sensitive to non-normality for both SS and group statistical analyses. Perhaps there is a need for all researchers to test for normality on a regular basis for both SS and group data sets. In a practical sense, when data sets do violate the normality assumption, whether SS or group, many statisticians suggest that one should evaluate the possible minimal effect on the $F$-distribution, acknowledge its limited effect, and move on (Keppel, 1982; Stevens, 1986). Others (Rebousin and Morgan, 1996) suggest more strongly that it is the responsibility of the analyst to weigh the effects of non-normal distributions and select the most appropriate analysis tool for the given data set.

A second and probably more important assumption that has been challenged in SS designs is trial independence. Subjects in a group are typically assumed to produce independent responses; however, when the same subject performs numerous trials within an SS experiment, the trial responses are assumed by many opponents of SS analyses to be dependent. Regarding the nature of intrasubject variability, the litera-

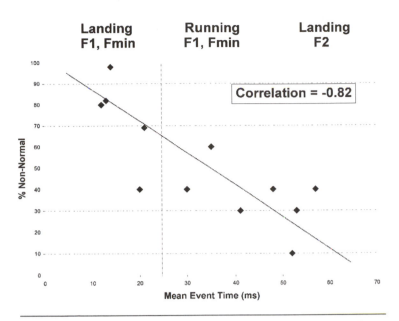

**Figure 1.2**  Normality rate for time of occurrence of impact force values.

ture (Bates, 1996; James and Bates, 1997) suggests that there should be no more dependence for any single variable among trials produced by a given individual than for trials generated by different subjects. Individual expectations and assumptions suggest that some dependence among variables across trials might exist. Dependence for a single variable viewed in isolation should appear negligible, however, because of the complexity of the human organism and the multiple functional degrees of freedom associated within the system. It should also be noted that in most SS experiments, trials are not generated in a consecutive manner, although they can be for various activities including consecutive swim strokes or foot strikes as observed on the treadmill. Data collection procedures typically impose a temporal constraint between samples during which time the subject might be mentally rehearsing, resting, or performing repetitions of the task. In the case of locomotion over the force platform, for example, the performer must return to the starting point, which involves additional steps that could be relatively similar or different. In the case of more discrete exemplar tasks such as landing from a drop or jump, the subject typically returns to the initial point upon volition. These events would certainly affect dependence if it did exist.

Trial dependence/independence was evaluated for the 552 subject-variable data sets previously identified using autocorrelation (SAS Institute, 1993). **Autocorrelation** is a procedure used to determine the relationship, if any, between various trial sequences in a data set using correlation techniques (Bates et al., 1996). In this technique, one begins with consecutive trials (1-2, 2-3, 3-4, etc.: lag 1) followed by trial sequences with one or more intermediate trials (1-3, 2-4, 3-5, etc.: lag 2; 1-4, 2-5, 3-6, etc.: lag 3) with a limit of $n - 8$ lags ($n$ = total number of trials), based on a significant Pearson $r$ criterion of .707 ($R^2$ = .500). The first three **lag correlations** using this procedure were computed, representing the relationship/dependence between successive trials (lag 1), trials with an intervening trial (lag 2), and trials with two intervening trials (lag 3).

Although more lag coefficients could have been computed, it did not seem logical that trials further removed could be meaningfully related. A summary of the results from the correlation analyses is given in table 1.2. The values in the table represent the percent of correlation coefficients significantly different from zero ($p < .05$). The average value of 7.1% for lag 1 indicates a slightly greater than chance value of 5.0% for dependence, while the other two

**Table 1.3  Normality Results for Group Data: Analysis of One Running [R] and One Landing [L] Data Set**

| Variable | Non-normal* | N** |
|---|---|---|
| F1(g) | 27.9 | 12 |
| F1 | 9.8 | 16 |
| F2 | 9.9 | 16 |
| T1 | 15.7 | 16 |
| T2 | 6.0 | 16 |
| θr | 31.3 | 12 |
| θl | 17.6 | 32 |
| Mean | 18.6 | 14.9 |

* = percent of non-normal values significantly different from zero ($p$ < .05); ** = number of data sets evaluated; F1(g) = impact force measured in gs [R]; F1 = first maximum vertical ground reaction force [R, L]; F2 = second maximum vertical ground reaction force [L]; T1 = time to F1 [R, L]; T2 = time to T2 [L]; θr = maximum knee joint angle [R]; θl = maximum knee joint angle [L].

lag values (4.8% and 3.0%) were less than the expected chance values. The mean explained variance for those variables that were significantly different from zero was only about 18%. Even though the analysis identified limited dependence among trials within selected data sets, the overall mean result of 4.97% supports the assumption of independence.

Given no dependence within a data set, and provided that treatment and conditions are independent, then it logically follows that there is independence between data sets. In other words, the arrangement of trials is arbitrary within conditions, and therefore computing a correlation coefficient between condition trials is just as meaningless. When multiple responses across conditions in a group design are examined, statistical adjustments are made for repeated or correlated measures. Attempting to use repeated-measures techniques in an SS design is definitely inappropriate, since these approaches assume dependence or a correlation among mean trial or individual values between conditions that were shown to be independent. For SS analyses, the more appropriate approach is to assume the trial values to be independent measures and use the corresponding independent test procedure (Bates et al., 1992).

In summary, the data presented further support the model proposed by Bates and colleagues (1992) that represents the subject as a normal distribution random generator, producing responses to stimuli (conditions) that are statistically independent both within and between subject conditions. The model presented by Bates and colleagues (1992) allowed for the manipulation of several variables including number of replications of the experiment, the number of empirical trials obtained per subject-condition, the differences between condition means (effect size), and the variability associated with the mean condition distributions both within and between subjects. This model has been used in several studies (Bates et al., 1992; Bates and Stergiou, 1996; Dufek et al., 1995a) to provide a better understanding of human performance.

## Generalizability of Results

Another primary criticism directed toward SS experiments and analyses is the lack of generalizability of results. Proponents of group designs profess that a large number of subjects is needed to control for intersubject variability and to provide adequate statistical power to generate significant results that are generalizable to the greater population from which the subjects were selected. It is important to remember that generality is based on replication of the study and not sample size (Bates, 1996). A group design, regardless of sample size, tests the theory only once. Sidman (1960) stated that the replication of an experiment with two subjects establishes greater generality for the data among individuals than does replication using two groups of subjects whose data have been combined. Denenberg (1982) showed that statistically, the probability of drawing from a large population, at random, three subjects who produce the same response is quite small, less than $p = .05$.

In order to allow one to infer from the sample to the population, the sample must be representative of the population; this in turn requires a more homogenous group that then limits generalizability of results. How restrictive this becomes depends on individual-subject complexities, selective movement patterns relative to the task, or both. Because voluntary movement (behavior) is continuously influenced by environmental and system constraints (Higgins, 1977; Schmidt et al., 1992), the likelihood of a homogenous response—all subjects in the group responding in the same fashion—appears limited in many situations. In any event, subjects who should be grouped for statistical treatment are those that produce the desired outcome (dependent variable response) using the same set of independent variables in approximately the same proportions, and then only for the period of time during which these performance characteristics remain relatively constant (Bates, 1996).

Differential response patterns from a relatively fixed set of performance characteristics within a group statistical design threaten the external validity of the experiment and often result in no observed group differences. It would seem that in many situations, differential response patterns (observed as intersubject variability) are likely to be the rule rather than the exception because of the complexity of the human n-m-s system and its numerous associated functional degrees of freedom, along with the different experiences, perceptions, and expectations each subject brings to the experimental setting. The consequences of this influence are

*easily averted by insisting that all experimental effects be clearly demonstrated on the behavior of one organism at a time with its single environmental history. The question of the representativeness of the effect for other subjects in the population (subject generality) can then be easily and effectively approached by succeeding investigations with other subjects. (Johnston and Pennypacker, 1980, p. 256)*

## Theoretical and Ethical Underpinnings of Single-Subject Design

Single-subject experiments can be used quite effectively to generate new hypotheses or cast doubt on existing assumptions or theories. Replication of SS experiments on other individuals can provide information to support or refute previous individual-subject findings. Sufficient support (in number of individuals examined) can eventually lead to modifications of the hypothesis or elucidate the conditions under which the hypothesis does or does not remain tenable, and these modifications can then be tested on other individuals. As previously indicated, an advantage of an SS design is that it is not threatened by intersubject variability. Another advantage is its efficiency (cost and time) for evaluating the effects of a number of alternative treatment conditions, especially during the early stages of formulating or modifying hypotheses.

Kerlinger (1979) has suggested that scientists "are not and cannot be concerned with the individual case. They seek laws, systematic relations, explanations of phenomena. And their results are always statistical" (p. 270). Kerlinger surmises that predictions made by scientists apply to an "abstract" or "average" individual. In direct contrast, Bates and his colleagues (1992) suggested that it is often the individual that is of primary importance and not some mythical "average" person that is of concern experimentally, especially in areas such as high-level motor performance, injury, rehabilitation therapy, and learning.

A critical shortcoming of group statistical designs can become evident when differences between condition means are identified and the treatments are operationalized to "similar groups." Since 50% of the individuals in each treatment group fall below the mean value, it is possible that members residing in this end of the distribution did not respond favorably to the intervention, and it is also possible that some might have even responded unfavorably. If the group is of primary concern and the individual members forming the group are "irrelevant," then this outcome is perfectly acceptable. However, in many instances, an awareness of who was affected, in what way(s), and the magnitude of the effect(s) is important as previously indicated, especially in clinical applications, elite performance, and learning. In instances such as these in which the response of the individual is of interest and is important, this situation could result in an ethical dilemma and should force us to evaluate some of the ways in which we look at research problems.

# Experimental Design

While empirical examples are somewhat limited at present, the most common statistical applications found in the literature include mean difference tests to identify between-condition differences (Bates and Stergiou, 1996; Dufek et al., 1991; Schot et al., 1994), simple correlation and regression techniques to predict and determine relationships (Dufek and James, 1994; Stergiou and Bates, 1994), and **multiple regression** analyses to explain relationships and observe interactions among variables (Caster and Bates, 1995; Dufek and Bates, 1992; Dufek and Zhang, 1996). This section includes selected studies using these techniques as representative examples. Another technique that is not currently prevalent in the literature but can potentially benefit both SS and group analyses is a procedure known as bootstrapping (Wasserman and Bockenholt, 1989) or randomized SS experimentation (Edgington, 1987). It is not a statistical procedure per se but a Monte Carlo procedure for approximating the entire sampling distribution (population) from a small sample in order to make more meaningful statistical inferences. In addition, nonparametric tests have been used due to the inherent assumption of nonrandom sampling.

## Analysis Techniques

A wide range of analysis techniques and tools is available for incorporation in SS analyses. Just as there are numerous parametric and nonparametric statistical tools from which to choose for use in data analysis of group design experiments, with choices based on the research question asked, the nature of the empirical data obtained, and the design of the study, the same limitations and delimiters exist when one is selecting SS analysis tools. The following is a brief overview of procedures that have appeared in the literature, along with explanations, when appropriate, on how to carry out the procedures.

A number of possible established SS analysis techniques are at the disposal of researchers presently. These include (1) nonparametric techniques, (2) **"bootstrap"** and/or randomization techniques, (3) Model Statistic procedure, (4) "Fisherian" techniques including analysis of variance, and (5) multiple regression techniques.

**Nonparametric tests** are appropriate for SS analysis since one of the assumptions of this less powerful technique is the lack of random sampling (measurements are not assumed to be obtained from a normal distribution). The **Mann-Whitney U Test** is one specific technique that has few statistical assumptions. Its assumptions include at least an ordinal scale of measure and measurement of the dependent variable on a continuous scale (McCall, 1994). In addition, there is no hypothesis being tested on specific parameters. In this test, the null hypothesis ($H_o$) assumes that the populations are similar whereas the alternative hypothesis ($H_A$) assumes different populations. Relative to SS analyses, the interpretation of these assumptions would be that $H_A$ (experimental condition) responses are different from $H_o$ (control condition) responses in any given subject. The specific procedures for computing a Mann-Whitney $U$ Test are given in example 1.1 and can be found in numerous elementary statistics textbooks (e.g., McCall, 1994).

## Example 1.1

### DESCRIPTION OF DATA SET

Data are from a single male subject (age 24 years; height 1.70 m; mass 75.23 kg) performing bilateral drop landings onto a force platform from a height of 0.60 m. Vertical ground reaction force (GRF) measurements were obtained (1,000 Hz) from the left foot before (Condition A) and after (Condition B) lower extremity fatigue. The dependent variable used in this example represents the first GRF force peak (F1) associated with forefoot impact. A complete methodological description can be found in James (1991).

| Data in order collected | | | Data sorted in ascending order of F1 | | | |
|---|---|---|---|---|---|---|
| Condition | Trial | F1 (N/kg) | Rank | Condition | Trial | F1 (N/kg) |
| A | 1 | 19.07 | 1 | A | 2 | 15.57 |
| A | 2 | 15.57 | 2 | A | 5 | 16.92 |
| A | 3 | 17.56 | 3 | A | 10 | 17.17 |
| A | 4 | 18.59 | 4 | A | 3 | 17.56 |
| A | 5 | 16.92 | 5 | A | 8 | 18.27 |
| A | 6 | 20.34 | 6 | A | 7 | 18.50 |
| A | 7 | 18.50 | 7 | A | 4 | 18.59 |
| A | 8 | 18.27 | 8 | B | 9 | 18.79 |
| A | 9 | 18.81 | 9 | A | 9 | 18.81 |
| A | 10 | 17.17 | 10 | A | 1 | 19.07 |
| B | 1 | 20.14 | 11 | B | 5 | 19.52 |
| B | 2 | 22.03 | 12 | B | 10 | 20.01 |
| B | 3 | 24.04 | 13 | B | 1 | 20.14 |
| B | 4 | 21.95 | 14 | B | 7 | 20.20 |
| B | 5 | 19.52 | 15 | A | 6 | 20.34 |
| B | 6 | 20.85 | 16 | B | 8 | 20.83 |
| B | 7 | 20.20 | 17 | B | 6 | 20.85 |
| B | 8 | 20.83 | 18 | B | 4 | 21.95 |
| B | 9 | 18.79 | 19 | B | 2 | 22.03 |
| B | 10 | 20.01 | 20 | B | 3 | 24.04 |

## STEPS FOR CALCULATING $U_{OBS}$

1. Sum the ranks for Condition A. $T_A = 1 + 2 + 3 + 4 + 5 + 6 + 7 + 9 + 10 + 15 = 62$.

2. Compute $U_{obs}$ using: $U_{obs} = n_A n_B + [n_A(n_A + 1) / 2] - T_A$ where $n_A$ and $n_B$ are the number of observations in Conditions A and B, respectively. $U_{obs} = 10 \times 10 + [10(10 + 1) / 2] - 62 = 93$.

3. Compare $U_{obs}$ to a critical value of U from a decision table (e.g., McCall, 1994, pp. 420-423). For example, for a two-tailed test at $\alpha = .05$, $U_{crit} = 77$. Since $U_{obs} > U_{crit}$ ($p \leq .05$), the decision is to reject the null hypothesis.

*Note:* This method is valid only when *n* in either group is less than 20. When *n* is greater than 20, the observed value of *U* approaches a normal distribution and a Z-transformation is required (McCall, 1994).

Another technique one can incorporate in SS analysis is the bootstrap or **"randomization"** procedure (Edgington, 1987). In this procedure, one directly addresses the question of temporal effects on the experimental outcome, that is, the effects of practice, learning, accommodation, and expectations and experience on performance.

To address the temporal effects of the measured outcomes, one first computes all combinations of empirical data obtained (McCall, 1994, p. 321) in order to create a randomness among the data. Next, one computes the mean differences (test statistic) for the newly generated data sets. The next step is to determine the number of test statistics that are as great as or greater than the mean difference observed empirically. The final step is to determine the probability level by dividing the number of observed test statistics satisfying the previous step by the number of data combinations. This procedure is shown schematically in figure 1.3 and is illustrated numerically in example 1.2.

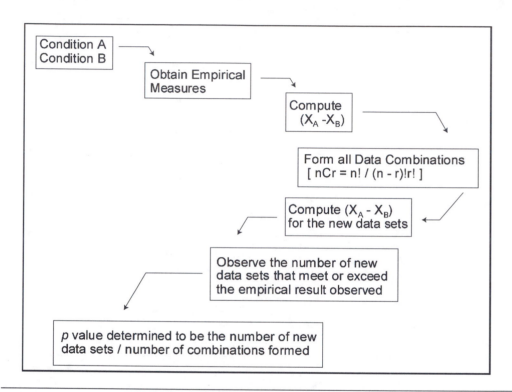

**Figure 1.3** Flowchart illustrating procedure for randomization technique.
Adapted from Edgington, 1987.

## Example 1.2

### DESCRIPTION OF DATA SET

Data were obtained from one male subject running over the force platform (1,000 Hz) at his preferred velocity (±5%) wearing each of two types of shoes (Shoe 1 = Condition 1; Shoe 2 = Condition 2) (Edgington, 1987). The dependent variable represented is first maximum vertical GRF (N/kg). Numerous data trials were obtained; however, only the first three trials per condition are provided.

| Subject 1 | Condition 1 (C1) | Condition 2 (C2) |
|---|---|---|
| Trial 1 | 19.1 | 16.8 |
| Trial 2 | 18.2 | 15.9 |
| Trial 3 | 17.8 | 17.2 |

1. Compute $\bar{X}_{C1} - \bar{X}_{C2}$:

$$\bar{X}_{C1}: (19.1 + 18.2 + 17.8) / 3 = 18.37$$

$$\bar{X}_{C2}: (16.8 + 15.9 + 17.2) / 3 = 16.63$$

$$18.37 - 16.63 = 1.73$$

2. Form all data combinations using: [nCr = n! / (n – r)! r!] where $n$ = total number of observations (6) and $r$ = number of trials per condition (3):

$$6! / ((6 - 3)!\ 3!) = 6 \times 5 \times 4 \times 3 \times 2 \times 1 / (3 \times 2 \times 1) \times (3 \times 2 \times 1) = 20$$

Combinations are formed by logically creating groups of numbers representing any given three trials. The 20 combinations for three trials (t) would be:

|         |         |         |         |
|---------|---------|---------|---------|
| t1 t2 t3 | t2 t3 t4 | t3 t4 t5 | t4 t5 t6 |
| t1 t2 t4 | t2 t3 t5 | t3 t4 t6 |         |
| t1 t2 t5 | t2 t3 t6 | t3 t5 t6 |         |
| t1 t2 t6 | t2 t4 t5 |         |         |
| t1 t3 t4 | t2 t4 t6 |         |         |
| t1 t3 t5 | t2 t5 t6 |         |         |
| t1 t3 t6 |         |         |         |
| t1 t4 t5 |         |         |         |
| t1 t4 t6 |         |         |         |
| t1 t5 t6 |         |         |         |

Substituting the data provided into this matrix would produce the following complete data sets:

|      |       |       |       |       |       |       |       |       |       |       |
|------|-------|-------|-------|-------|-------|-------|-------|-------|-------|-------|
|      | **19.1** | 19.1 | 19.1 | 19.1 | 19.1 | 19.1 | 19.1 | 19.1 | 19.1 | 19.1 |
|      | **18.2** | 18.2 | 18.2 | 18.2 | 17.8 | 17.8 | 17.8 | 16.8 | 16.8 | 15.9 |
|      | **17.8** | 16.8 | 15.9 | 17.2 | 16.8 | 15.9 | 17.2 | 15.9 | 17.2 | 17.2 |
| Mean | **18.37** | 18.03 | 17.73 | 18.17 | 17.90 | 17.60 | 18.03 | 17.27 | 17.70 | 17.40 |
|      | **16.8** | 17.8 | 17.8 | 17.8 | 18.2 | 18.2 | 18.2 | 18.2 | 18.2 | 18.2 |
|      | **15.9** | 15.9 | 16.8 | 16.8 | 15.9 | 16.8 | 16.8 | 17.8 | 17.8 | 17.8 |
|      | **17.2** | 17.2 | 17.2 | 15.9 | 17.2 | 17.2 | 15.9 | 17.2 | 15.9 | 16.8 |
| Mean | **16.63** | 16.97 | 17.27 | 16.83 | 17.10 | 17.40 | 16.97 | 17.73 | 17.30 | 17.60 |
| Diff | **1.73** | 1.07 | 0.47 | 1.33 | 0.80 | 0.20 | 1.07 | –0.47 | 0.40 | –0.20 |
|      | 18.2 | 18.2 | 18.2 | 18.2 | 18.2 | 18.2 | 17.8 | 17.8 | 17.8 | 16.8 |
|      | 17.8 | 17.8 | 17.8 | 16.8 | 16.8 | 15.9 | 16.8 | 16.8 | 15.9 | 15.9 |
|      | 16.8 | 15.9 | 17.2 | 15.9 | 17.2 | 17.2 | 15.9 | 17.2 | 17.2 | 17.2 |
| Mean | 17.60 | 17.30 | 17.73 | 16.97 | 17.40 | 17.10 | 16.83 | 17.27 | 16.97 | 16.63 |
|      | 19.1 | 19.1 | 19.1 | 19.1 | 19.1 | 19.1 | 19.1 | 19.1 | 19.1 | 19.1 |
|      | 15.9 | 16.8 | 16.8 | 17.8 | 17.8 | 17.8 | 18.2 | 18.2 | 18.2 | 18.2 |
|      | 17.2 | 17.2 | 15.9 | 17.2 | 15.9 | 16.8 | 17.2 | 15.9 | 16.8 | 17.8 |
| Mean | 17.40 | 17.70 | 17.27 | 18.03 | 17.60 | 17.90 | 18.17 | 17.73 | 18.03 | 18.37 |
| Diff | 0.20 | –0.40 | 0.47 | –1.07 | –0.20 | –0.80 | –1.33 | –0.47 | –1.07 | *–1.73* |

Authors' original data and computations, following the technique reported by Edgington, 1987.

*(continued)*

*Note:* Illustration of 20 data sets (original data in bold) formed from step 2 along with the respective mean and difference values computed in step 3. Difference values in *italics* are those subgroups that are greater than or equal to the empirically observed difference of 1.73.

3. Compute $\bar{X}_{C1} - \bar{X}_{C2}$ for the 20 new data sets formed in step 2.
4. Observe the number of data sets that are greater than or equal to the empirically observed difference: 1 (see preceding chart).
5. Probability of true differences between C1 and C2 is determined to be the number of new data sets meeting the criteria in step 4 (1) / number of data combinations in step 2 (20):

$$p = 1 / 20 \text{ or } p = .05$$

Another uniquely SS method that has appeared in the literature is the **Model Statistic** (Bates et al., 1992) procedure. This analysis might be viewed as the SS "*t*-test" approach.

The purpose of the technique is to take advantage of the repeated-measures concept of the within-subject design rather than using an independent technique that lacks comparison sensitivity or violates statistical assumptions associated with between-subject repeated-measures techniques (Bates et al., 1992). The Model Statistic procedure was developed to evaluate SS differences between conditions in an SS experiment. The subject is assumed to be a response generator producing responses to a particular stimulus or perturbation (condition) that are statistically independent and distributed about a central tendency describing an input population distribution. It is further assumed that no appreciable learning takes place during the hypothesized data-collecting session.

A computer program was used to generate individual test statistics (table 1.4) for selected sample sizes (3 to 50 trials) and $\alpha$ levels (.10, .05, .01). The test statistics were generated by randomly selecting two samples of the appropriate size from a normally distributed population with a mean of 0.0 and standard deviation of 1.0. The absolute differences between the sample means were computed and stored for 5,000 paired samples, resulting in a normalized distribution of mean absolute differences. The 90th, 95th, and 99th percentile values from these sampling distributions were identified as the test statistics for $\alpha$ levels of .10, .05, and .01, respectively.

To statistically evaluate ($\alpha$ = .05) a mean difference between two conditions, the appropriate test statistic is multiplied by

**Table 1.4 Critical Values (Test Statistic) for Model Statistics: Normal Population**

| Sample size | $\alpha = .10$ | $\alpha = .05$ | $\alpha = .01$ |
|---|---|---|---|
| 3 | 1.3733 | 1.6533 | 2.2133 |
| 4 | 1.2643 | 1.5058 | 1.9867 |
| 5 | 1.1597 | 1.3662 | 1.7788 |
| 6 | 1.0629 | 1.2408 | 1.6044 |
| 7 | 0.9751 | 1.1306 | 1.4623 |
| 8 | 0.0896 | 1.0351 | 1.3473 |
| 9 | 0.8270 | 0.9536 | 1.2542 |
| 10 | 0.7673 | 0.8857 | 1.1776 |
| 11 | 0.7172 | 0.8307 | 1.1129 |
| 12 | 0.6757 | 0.7867 | 1.0581 |
| 13 | 0.6415 | 0.7516 | 1.0117 |
| 14 | 0.6132 | 0.7234 | 0.9720 |
| 15 | 0.5896 | 0.7001 | 0.9375 |
| 16 | 0.5695 | 0.6798 | 0.9070 |
| 17 | 0.5522 | 0.6618 | 0.8796 |
| 18 | 0.5371 | 0.6458 | 0.8548 |
| 19 | 0.5237 | 0.6311 | 0.8318 |
| 20 | 0.5114 | 0.6175 | 0.8102 |
| 25 | 0.4592 | 0.5572 | 0.7145 |
| 30 | 0.4194 | 0.5097 | 0.6437 |
| 35 | 0.3896 | 0.4729 | 0.5949 |
| 40 | 0.3673 | 0.4442 | 0.5626 |
| 45 | 0.3500 | 0.4207 | 0.5414 |
| 50 | 0.3352 | 0.4000 | 0.5256 |

Data from Bates, Dufek, and Davis, 1992.

the mean standard deviation, typically for the two sample standard deviations $(SD = [(SD_1^2 + (SD_2^2) / 2]^{1/2})$, to produce a critical difference. In some instances the best estimate of the mean standard deviation may be computed from more than two data sets. The interpretation of a mean absolute difference greater than the critical difference is that there is only a 5% chance that the difference is due to sampling error. Therefore, given the possibility of a Type I error occurring 5% of the time, the observed difference suggests that the two samples come from two distinct populations. Since all comparisons are planned, no adjustments for family-wise Type I errors are made. Several $\alpha$ levels are given in table 1.4 for consideration. Using the critical values given in table 1.4 for $\alpha$ = .05 and 25 trials of per condition data, the mean difference is statistically significant if $|\bar{X}_1 - \bar{X}_2| > 0.5572 \times [(SD_1^2) + (SD_2^2) / 2]^{1/2}$. A thorough numerical example of the Model Statistic procedure is given in example 1.3.

## Example 1.3

## DESCRIPTION OF DATA SET

Frontal plane kinematic data (200 Hz) were obtained from one subject running at his preferred velocity (±5%) on a treadmill in two shoe conditions (Shoe 1 = Condition 1; Shoe 2 = Condition 2). The maximum pronation angle (dependent variable) was calculated for numerous footfalls (trials).

| Subject | Condition | Trial | Max pronation | Subject | Condition | Trial | Max pronation |
|---------|-----------|-------|---------------|---------|-----------|-------|---------------|
| 1 | 1 | 1 | –5.83 | 1 | 2 | 1 | –8.50 |
| 1 | 1 | 2 | –6.18 | 1 | 2 | 2 | –6.61 |
| 1 | 1 | 3 | –4.97 | 1 | 2 | 3 | –6.07 |
| 1 | 1 | 4 | –4.75 | 1 | 2 | 4 | –8.33 |
| 1 | 1 | 5 | –7.03 | 1 | 2 | 5 | –7.22 |
| 1 | 1 | 6 | –5.20 | 1 | 2 | 6 | –6.36 |
| 1 | 1 | 7 | –5.11 | 1 | 2 | 7 | –8.77 |
| 1 | 1 | 8 | –7.19 | 1 | 2 | 8 | –7.81 |
| 1 | 1 | 9 | –6.12 | 1 | 2 | 9 | –7.97 |
| 1 | 1 | 10 | –7.16 | 1 | 2 | 10 | –8.34 |
| **Mean** | | | **–5.955** | | | | **–7.596** |
| **SD** | | | **0.890** | | | | **0.919** |

The basic premise of the Model Statistic is to compare the empirically observed condition difference to a probabilistic critical difference. If the observed condition difference is greater than or equal to the probabilistic critical difference, the conditions are statistically significant.

1. Compute the mean and standard deviation for each condition. Using spreadsheet tools:

$$\bar{X}_1 = -5.955, SD_1 = 0.890; \bar{X}_2 = -7.5996, SD_2 = 0.919$$

*(continued)*

2. Compute the empirically observed condition difference:

$$|\bar{X}_1 - \bar{X}_2| = |(-5.955) - (-7.596)| = 1.641$$

3. Compute the probabilistic critical difference:

critical difference = critical value (table 1.4) $\times$ mean standard deviation

4. Critical value (table 1.4) for 10 trials for $\alpha = .05$ is 0.8857.
5. Mean standard deviation = $[(SD_1^2 + SD_2^2) / 2]^{1/2}$.

$$\text{mean standard deviation} = [(0.890)^2 + (0.919)^2 / 2]^{1/2} =$$
$$[(0.7921) + (0.8446) / 2]^{1/2} = (0.8183)^2 = 0.9046$$

6. Critical difference = $0.8857 \times 0.9046 = 0.8012$.
7. Compare observed condition difference to critical difference:

$$1.641 > 0.8012; \text{ conclude conditions are different at } \alpha = .05.$$

Use of traditional analysis of variance techniques for SS analyses requires only a brief comment, as the computational formulas are not modified. Of note is the fact that rather than subject mean values, individual trial values are used. Therefore, the "within" error term in the generalized analysis of variance model is the difference between the trial value and the mean of all trials. The "between" error term is the difference between the condition mean and the total mean deviation, and the "total" error is the difference between each individual trial and the total mean deviation.

Multiple regression can also be used as a viable SS analysis tool (Dufek et al., 1995b; Dufek and Zhang, 1996). Again, the computational procedures do not change from those used in group designs. However, one must take care to preserve the independent variable (IV)-to-dependent variable ratios. One way of dealing with this possible problem **(multicollinearity)** is the use of correlation matrices to eliminate highly correlated IVs. Using this technique, one would first examine all cross-correlations sharing greater than 50% common variance ($r > .707$) and eliminate one of the related variables. The decision of which IV to eliminate should be based on what is known about the activity, what has been shown in the literature, and, of primary importance, the research question being asked. After reducing multicollinearity among the predictor variables, one can proceed to compute prediction models using multiple regression techniques. If one uses a typical stepwise procedure, a more liberal $\alpha$ level may be set for initial variable entry (i.e., .15) while the typically accepted $\alpha = .05$ overall model significance level is retained. In this way, the best combination of IVs may be identified.

## Examples From the Literature

We now review two more complex preplanned studies from the research literature to further elaborate on ways to integrate SS and group analysis techniques. The primary purpose of a study by Bates and Stergiou (1996) was to evaluate the interactive effects between performer perceptions/strategies and response patterns. The group experimental design incorporated an **analysis of variance** (ANOVA) [shoe conditions ($n = 3$) by (subjects ($n = 18$) nested in shoe hardness ($n = 2$))]. Six subjects were each nested in two shoe hardnesses within each of three shoe conditions (soft, medium, hard). So that both group and SS analyses could be performed and adequate statistical power maintained, subjects performed 25 running trials in each of the two hardness conditions. The dependent variables of interest were impact force and maximum knee joint angle (deg of flexion). Since the SS analyses resulted in both significant and nonsignificant responses to the shoe hardnesses independent of shoe conditions, the subjects were regrouped (post hoc) based on the SS impact force results for further analyses using a one-factor (shoe hardness) repeated-measures ANOVA. Finally, a series of Pearson product-moment correlation coefficients were computed between the two dependent variables.

The group ANOVAs provided rather limited insight into the research question, indicating only significant impact force differences for the soft shoe condition and knee joint angle differences for the hard shoe condition. Individual-subject analyses using a Model Statistic approach (Bates et al., 1992; Dufek et al., 1991) identified nonsignificant ($p < .05$) impact force differences for 8 subjects while 10 subjects exhibited significant differences distributed among the soft ($n = 5$), medium ($n = 2$), and hard ($n = 3$) shoe conditions. The knee joint angle analyses resulted in nine significant differences with no apparent trend relative to the impact force results. A statistically significant correlation coefficient of $r = -.59$ between impact force and knee joint angle, however, indicated that some accommodation did take place on average but that the extent varied among subjects, with some demonstrating a greater Newtonian or mechanical component and others showing greater neuromuscular accommodation. The results support the response strategy continuum concept discussed previously, with individual subjects performing differently due to the availability of numerous functional degrees of freedom in conjunction with their previous experiences and resulting perceptions. The combined approach (group and SS analyses) used in this investigation allowed for a more complete and meaningful assessment of the research question than would have been possible using either one or the other experimental protocol alone.

The second study, by Dufek and colleagues (1995b), used a multiple regression approach. In this research, two parallel experiments were undertaken with the primary purpose to examine the effects of subject expectations and perceptions resulting from previous experiences on lower extremity function during impact activities. Similar experimental methodologies were incorporated for the two studies using different subjects performing either landing or running activities. The objective outcome measure, impact force, was evaluated using regression modeling techniques. The study was exploratory and not designed to "find an answer but, to maximize the amount of information made available" (Walker and Catrambone, 1993, p. 409) to help elucidate the complexities of human performance. Subjects in the landing model substudy ($n = 6$) performed 30 trials for a normal, self-selected landing technique and an additional 10 trials for comparatively "soft" and "stiff" landings, as defined by knee joint angle. The six running model substudy participants performed 30 trials at a normal, self-selected running pace along with 20 trials for each of two perturbed conditions (understride and overstride). In both cases, the data were evaluated to ensure that all performances represented similar skills. In the landing study, for example, the vertical GRF curves were tested to assure that all trials belonged to the family of landing curves (homogeneity), yet the non-normal conditions provided heterogeneity among conditions. For each experiment, the group prediction models for impact force were strong, accounting for 86.7% and 66.6% explained variance for landing and running, respectively. The individual-subject models, however, were all unique; and none were similar in their predictor elements to the group models. Independent of the movement examined, results led to a similar conclusion—group models formed to predict impact forces created an "average" performance profile that was not representative of the individual subjects. The individual-subject impact force prediction models created profiles indicative of the different "strategies" invoked by each performer. Evaluation of the individual profiles across subjects within each activity illustrated both similarities and differences among subjects that were different from the group model results.

Further examination of landing impact force values elicited for the group and subject conditions in this study demonstrated predominantly positive biomechanical responses with a strong Newtonian component to the imposed perturbations, that is, increased impact force values corresponding to increased task demands of lower extremity stiffness. This result taken in isolation would lead to the conclusion that all subjects responded similarly and could therefore be "grouped," which would produce generalizable results. The error in this thinking was demonstrated through an examination of the unique predictor models generated for the individual subjects. One might hypothesize that the subjects in this experiment responded at different levels along the Newtonian-neuromuscular continuum and found unique *internal* solutions to the imposed task constraints. The "average performance" group model solution did little to explain the internal mechanisms of lower extremity function during impact activities for the individual subjects. This observation was made possible only as a result of the availability of individual-subject information.

These two examples emphasize the importance of SS analyses to confirm or reject, in total or in part, the findings from a group evaluation. In these examples, the SS results definitely provided additional and important insight for better understanding of human movement. Other research examples could have been selected and discussed in which the two outcomes were comparable and the group design would have been adequate and sufficient. Unfortunately, one often faces the predicament of not knowing how to approach a specific problem a priori because of a lack of information, in which case an SS or combined approach can be useful in developing hypotheses. Because of the different outcomes and conclusions that can result from a group analysis independent of SS evaluations, it may be necessary to modify the way we approach the study of some human movement problems.

The appropriate experimental design for any research problem is based on the research question being asked and what is known about performance characteristics. The advantage of group designs, if appropriate and properly executed, is that they give the researcher the ability to subsequently generalize results to a similar population. A shortcoming in many research studies, however, lies in the assumptions regarding the performance characteristics of the pool of included subjects. In many instances these assumptions are incorrect because of the many performance options (strategies) available to each individual performing a given motor task. Individual subjects within an experiment can perform a gross task (such as running or landing) differently (figure 1.1); this can result in inaccurate or false research conclusions. Because of the functional complexity of the human n-m-s system and the imposed constraints, the only

> *individuals that should be grouped for statistical treatment are those that produce the desired outcome (dependent variables) using the same set of independent variables in approximately the same proportion and then only for the period of time that those performance characteristics remain relatively constant. (Bates, 1996, p. 635)*

One possible approach to address this predicament, as demonstrated by the previous examples, is to combine group and SS designs with the intent of gaining additional insight into the problem(s) of interest when the research question makes this appropriate. Developing the option to statistically evaluate the group as a whole as well as individual subjects, however, requires a more comprehensive research design, as previously discussed. This combined approach can be used in several ways. First, a priori, the researcher can decide to conduct a group and an SS analysis to identify consistencies or a lack of consistency in performance among groups and individuals. As part of this methodology, the SS analysis can be used to regroup the subjects post hoc according to their performance strategies. Evaluating these post hoc groups simply identifies the magnitude of differences between the post hoc groups and allows for comparative evaluation of the a priori and post hoc groups. Second, following a group analysis resulting in no significant differences, the investigator can look for conflicting subject performances post hoc. As in the previous example, post hoc groups could be formed and evaluated. Which is the most appropriate approach is probably dependent on the research question being investigated.

# Summary

The primary objective of this chapter was to provide historical, theoretical, and practical arguments in support of an SS analysis design as an acceptable and appropriate means of analysis for some research questions as opposed to group analysis designs by default. Both SS and group designs have been used since the mid-1800s. Group designs have become dominant with the development of inferential statistical techniques and their ability to generalize results to populations representative of the sample. Single-subject techniques have brought into question, in a number of instances, the homogeneity of the individuals making up these samples and therefore the validity of generalizability. Single-subject analyses also often refute the value of mean or "average" results and the concept of the "average" individual. Criticisms of the SS approach have typically centered on the possible violation of statistical assumptions including data normality and trial independence and the lack of generalizability of results. These objections were discussed and data provided to refute these criticisms.

The nature of human movement was discussed with an emphasis on movement constraints, variability, response patterns or strategies, and the effects of data aggregation. Movement constraints result in intrasubject variability and between-subject differences consequent to strategy selection. The individual is a complex, continuously changing interactive system that feeds back on itself through experiences and the resulting changing perceptions. Myriad response patterns can result along a continuum from purely mechanical or "Newtonian" to fully accommodating or "neuromuscular." Practically all are some combination of these two responses, that is, "biomechanical."

A secondary objective was to present analysis methods and techniques that can be used to complete a scientifically rigorous SS experiment. Common statistical applications discussed included mean difference tests to identify between-condition differences, simple correlation and regression techniques to explain relationships, and multiple regression analyses to explain relationships and observe interactions among variables. Examples from the literature were presented to demonstrate the use of these different statistical applications.

# Suggested Readings and Other Resources

## Historical and Theoretical Information

Barlow, D.H., and Hersen, M. 1984. *Single case experimental designs: Strategies for studying behavioral change*. New York: Pergamon Press.

Bates, B.T. 1996. Single-subject methodology: An alternative approach. *Medicine and Science in Sports and Exercise,* 28(5): 631-638.

Schlaug, G., Knorr, U., and Seitz, R.J. 1994. Inter-subject variability of cerebral activations in acquiring a motor skill: A study with positron emission tomography. *Experimental Brain Research,* 98: 523-534.

## Statistical Background Information

Bates, B.T., Dufek, J.S., and Davis, H.P. 1992. The effect of trial size on statistical power. *Medicine and Science in Sports and Exercise,* 24(9): 1059-1068.

Dufek, J.S., Bates, B.T., and Davis, H.P. 1995. The effect of trial size and variability on statistical power. *Medicine and Science in Sports and Exercise,* 27(2): 288-295.

James, C.R., and Bates, B.T. 1997. Experimental and statistical design issues in human movement research. *Measurement in Physical Education and Exercise Science,* 1: 55-69.

## Full Papers for Examples Discussed

Bates, B.T., and Stergiou, N. 1996. Performance accommodation to midsole hardness during running. *Journal of Human Movement Studies,* 31: 189-210.

Dufek, J.S., Bates, B.T., Stergiou, N., and James, C.R. 1995. Interactive effects of group and single-subject response patterns. *Human Movement Science,* 14: 301-323.

# References

Bakan, D. 1955. The general and the aggregate: A methodological distinction. *Perceptual and Motor Skills,* 5: 211-212.

Baloff, N., and Becker, S.W. 1967. On the futility of aggregating individual learning curves. *Psychological Reports,* 20: 183-191.

Barlow, D.H., and Hersen, M. 1984. *Single case experimental designs: Strategies for studying behavioral change.* New York: Pergamon Press, pp. 1-31.

Bates, B.T. 1996. Single-subject methodology: An alternative approach. *Medicine and Science in Sports and Exercise,* 28(5): 631-638.

Bates, B.T., Dufek, J.S., and Davis, H.P. 1992. The effect of trial size on statistical power. *Medicine and Science in Sports and Exercise,* 24(9): 1059-1068.

Bates, B.T., James, S.L., Osternig, L.R., and Sawhill, J.A. 1981. Effects of running shoes on ground reaction forces. In: *Biomechanics VII-B,* eds. A. Morecki, K. Fidelus, and K. Kedzior, 226-233. Champaign, IL: Human Kinetics.

Bates, B.T., Osternig, L.R., and Mason, B.R. 1979. *Variations in velocity within the support phase of running.* Del Mar, CA: Academic Press, pp. 55-59.

Bates, B.T., and Stergiou, N. 1996. Performance accommodation to midsole hardness during running. *Journal of Human Movement Studies,* 31: 189-210.

Bates, B.T., Zhang, S., Dufek, J.S., and Chen, F.C. 1996. The effects of sample size and variability on the correlation coefficient. *Medicine and Science in Sports and Exercise,* 28(3): 386-391.

Bergin, A.E., and Strupp, H.H. 1970. New direction in psychotherapy research. *Journal of Abnormal Psychology,* 76: 13-26.

Bernstein, N. 1967. *Coordination and regulation of movement.* Oxford: Pergamon Press.

Boring, E.G. 1950. *A history of experimental psychology.* New York: Appleton-Century-Crofts.

Bouffard, M. 1993. The perils of averaging data in adapted physical activity research. *Adapted Physical Education Quarterly,* 10: 371-391.

Caster, B.L., and Bates, B.T. 1995. Assessment of mechanical and neuromuscular response strategies during landing. *Medicine and Science in Sports and Exercise,* 27(5): 736-744.

Caster, B.L., Bates, B.T., and Dufek, J.S. 1992. A multi-dimensional assessment of the functionality of cross training athletic footwear. *Proceedings of the Second North American Congress on Biomechanics,* 275-276. Chicago: Organizing Committee.

Conners, C.K., and Wells, K.C. 1982. Single-case designs in psychopharmacology. In: *New directions for methodology of social and behavioral sciences: Single case research designs,* eds. A.E. Kazdin and A.H. Tumas, 61-76. San Francisco: Jossey-Bass.

Denenberg, V.H. 1982. Comparative psychology and single-subject research. In: *New directions for methodology of social and behavioral sciences: Single case research designs,* eds. A.E. Kazdin and A.H. Tumas, 19-31. San Francisco: Jossey-Bass.

Dufek, J.S., and Bates, B.T. 1990. The evaluation and prediction of impact forces during landings. *Medicine and Science in Sports and Exercise,* 22(3): 370-377.

Dufek, J.S., and Bates, B.T. 1992. Lower extremity performance models for landing. *Human Movement Science,* 11: 299-318.

Dufek, J.S., Bates, B.T., and Davis, H.P. 1995a. The effect of trial size and variability on statistical power. *Medicine and Science in Sports and Exercise,* 27(2): 288-295.

Dufek, J.S., Bates, B.T., Davis, H.P., and Malone, L.A. 1991. Dynamic performance assessment of selected sport shoes on impact forces. *Medicine and Science in Sports and Exercise,* 23(9): 1062-1067.

Dufek, J.S., Bates, B.T., Stergiou, N., and James, C.R. 1995b. Interactive effects of group and single-subject response patterns. *Human Movement Science,* 14: 301-323.

Dufek, J.S., and James, C.R. 1994. Modeling landing impacts. *Proceedings of the Canadian Society of Biomechanics,* 176-177. Calgary: Organizing Committee.

Dufek, J.S., and Zhang, S. 1996. Landing models for volleyball players: A longitudinal evaluation. *Journal of Sports Medicine and Physical Fitness,* 36: 35-42.

Dunlap, K. 1932. *Habits: Their making and unmaking.* New York: Liveright.

Edgington, E.S. 1987. Randomized single-subject experiments and statistical tests. *Journal of Counseling Psychology,* 34(4): 437-442.

Estes, W.K. 1956. The problem of inference from curves based upon the group. *Psychological Bulletin,* 53: 134-140.

Fisher, R.A. 1925. On the mathematical foundations of the theory of statistics. In: *Theory of statistical estimation.* Cambridge, England: Cambridge Philosophical Society.

Guildford, J.P. 1936. *Psychometric methods.* New York: McGraw-Hill.

Higgins, J.R. 1977. *Human movement: An integrated approach.* St. Louis: Mosby.

Higgins, S. 1985. Movement as an emergent form: Its structural limits. *Human Movement Science,* 4: 119-148.

Hildebrand, D.K. 1986. *Statistical thinking for behavioral scientists.* Boston: PWS.

Hreljac, A. 1998. Individual effects on biomechanical variables during landing in tennis shoes with varying midsole density. *Journal of Sports Sciences,* 16: 531-537.

James, C.R. 1991. *Effects of fatigue on mechanical and muscular components of performance during drop landings.* MS thesis, University of Oregon.

James, C.R., and Bates, B.T. 1997. Experimental and statistical design issues in human movement research. *Measurement in Physical Education and Exercise Science,* 1: 55-69.

James, C.R., Bates, B.T., and Dufek, J.S. 2003. Classification and comparison of biomechanical response strategies for absorbing landing impact. *Journal of Applied Biomechanics* 19: 106-188.

James, C.R., Dufek, J.S., and Bates, B.T. 1992. Effects of fatigue on mechanical and muscular components of performance during drop landings. *Proceedings of the Second North American Congress on Biomechanics,* 553-554. Chicago: Organizing Committee.

James, C.R., and Zhang, S. 1994. Impact force characteristics of treadmill beds. Unpublished laboratory report.

Johnston, J., and Pennypacker, H. 1980. *Strategies and tactics of human behavioral research.* Hillsdale, NJ: Erlbaum.

Keppel, G. 1982. *Design and analysis: A researcher's handbook.* Englewood Cliffs, NJ: Prentice Hall.

Kerlinger, F.N. 1979. *Behavioral research: A conceptual approach.* New York: Holt, Rinehart and Winston.

Kratochwill, T.R. 1992. Single-case research design and analysis: An overview. In: *Single-case design and analysis: New directions for psychology and education,* eds. T.R. Kratochwill and J.R. Levin, 1-14. Hillsdale, NJ: Erlbaum.

Lees, A., and Bouracier, J. 1994. The longitudinal variability of ground reaction forces in experienced and inexperienced runners. *Ergonomics,* 37: 197-204.

Loslever, P. 1993. Error and data coding in the multi-dimensional analysis of human movement signals. *Proceedings of the Institute of Mechanical Engineers,* 207(2): 103-110.

McCall, R.B. 1994. *Fundamental statistics for behavioral sciences.* Fort Worth, TX: Harcourt Brace.

Miller, D.I. 1979. Modelling in biomechanics: An overview. *Medicine and Science in Sports and Exercise,* 11(2): 115-122.

Nachbauer, W., and Nigg, B.M. 1992. Effects of arch height of the foot on ground reaction forces in running. *Medicine and Science in Sports and Exercise,* 24: 1264-1269.

Newell, K.M., and Corcos, D.M., eds. 1993. *Variability and motor control.* Champaign, IL: Human Kinetics.

Nigg, B.M., Bahlsen, H.A., Luethi, S.M., and Stokes, S. 1987. The influence of running velocity and midsole hardness on external impact forces in heel-toe running. *Journal of Biomechanics,* 3: 951-960.

Nigg, B.M., and Segesser, B. 1992. Biomechanics and orthopedic concepts in sport shoe construction. *Medicine and Science in Sports and Exercise,* 24: 595-602.

Pavlov, I.P. 1928. *Lectures on conditional reflexes* (W.H. Grantt, trans.). New York: International.

Payton, O.D. 1994. *Research: The validation of clinical practice.* Philadelphia: Davis.

Reboussin, D.M., and Morgan, T.M. 1996. Statistical considerations in the use and analysis of single-subject designs. *Medicine and Science in Sports and Exercise,* 28(5): 639-644.

Ryan, T.A., and Joiner, B.L. 1974. *Normal probability plots and test for normality.* Unpublished technical report, Department of Statistics, Pennsylvania State University.

SAS Institute. 1993. *Version 6.0 user's guide.* Cary, NC: SAS Institute.

Schlaug, G., Knorr, U., and Seitz, R.J. 1994. Inter-subject variability of cerebral activations in acquiring a motor skill: A study with positron emission tomography. *Experimental Brain Research,* 98: 523-534.

Schmidt, R.C., Treffner, P.J., Shaw, B.K., and Turvey, B.T. 1992. Dynamical aspects of learning an interlimb rhythmic movement pattern. *Journal of Motor Behavior,* 24(1): 67-83.

Schot, P.K. 1991. *Analysis of bilateral symmetry of lower extremity function during landing.* PhD dissertation, University of Oregon.

Schot, P.K., Bates, B.T., and Dufek, J.S. 1994. Bilateral performance symmetry during drop landing: A kinetic analysis. *Medicine and Science in Sports and Exercise,* 26(9): 1153-1159.

Shapiro, S.S., and Wilk, M.S. 1965. An analysis of variance test for normality (complete samples). *Biometrika,* 52(3-4): 591-611.

Sidman, M. 1960. *Tactics of scientific research: Evaluating experimental data in psychology.* New York: Basic Book.

Skinner, B.F. 1966a. Invited address to the Pavlovian Society of America.

Skinner, B.F. 1966b. Operant behavior. In: *Areas of research and application,* ed. W.K. Honig. New York: Appleton-Century-Crofts.

Stapp, J.P. 1971. Closing remarks: The future. *Symposium of biodynamic models and their applications, AMRL_TR_71_29.* Wright-Patterson Air Force Base, Ohio.

Stergiou, N., and Bates, B.T. 1994. Running impact force modeling. *Proceedings of the Canadian Society of Biomechanics,* 178-179. Calgary: Organizing Committee.

Stergiou, N., Bates, B.T., and James, S.L. 1999. Lower extremity asynchrony during running. *Medicine and Science in Sports and Exercise,* 31(11): 1645-1655.

Stevens, J. 1986. *Applied multivariate statistics for the social sciences.* Hillsdale, NJ: Erlbaum.

Stilson, D.W. 1966. *Probability and statistics in psychological research and theory.* San Francisco: Holden-Day.

Taylor, J.G. 1958. Experimental design: A cloak for intellectual sterility. *British Journal of Psychology,* 49: 106-116.

Walker, N., and Catrambone, R. 1993. Aggregation bias and the use of regression in evaluating models of human performance. *Human Factors,* 35(3): 397-411.

Wasserman, S., and Bockenholt, U. 1989. Bootstrapping: Applications to psychology. *Psychophysiology,* 208-220.

Wheatley, M.J. 1992. *Leadership and the new science: Learning about organization from an orderly universe.* San Francisco: Berrett-Koehler.

Worringham, C.J. 1993. Predicting motor performance from variability measures. In: *Variability and motor control,* eds. K.M. Newell and D.M. Corcos, 53-63. Champaign, IL: Human Kinetics.

Zhang, S., Bates, B.T., and Dufek, J.S. 2000. Contributions of lower extremity joints to energy dissipation during landings. *Medicine and Science in Sports and Exercise,* 32(4): 812-819.

# CHAPTER 2

# Considerations of Movement Variability in Biomechanics Research

C. Roger James, PhD

**H**uman movement is variable. Human movement variability has been the focus of numerous works across multiple disciplines within the movement sciences. Many excellent books, chapters, review papers, symposium papers, and empirical studies have been published on the topic of intra-individual variability from a variety of theoretical perspectives. It is not the aim of this chapter to support or refute any of those perspectives. Philosophical discourse about the origins and nature of variability is left to those scientists who are invested in this area of research. Newell and colleagues (1993, 1998) have provided outstanding theoretical accounts of variability, and the reader is directed to these and other works for additional information. Instead, the aim of this chapter is to contribute to the variability discussion from an applied perspective in which the consequences of variability, rather than its nature, are the primary focus. With use of this approach, understanding the factors that affect kinematic and kinetic outcomes is relevant for reasons relating to both biomechanical function (e.g., performance, injury mechanisms) and methodology. The specific goals of this chapter are to (1) briefly recount some past and contemporary views about the nature of intra-individual movement variability in order to provide a framework within which readers may more fully appreciate discussion of the consequences of variability, (2) discuss the potential consequences of variability relative to biological health, (3) review several methodological considerations for the presence of variability in biomechanics research, and (4) discuss some traditional methods for quantifying variability.

## The Nature of Intra-Individual Movement Variability

Variability is pervasive throughout the multiple levels of movement organization and occurs both within and between individuals (Newell and Corcos, 1993). Variability exists because of the many complex systems and constraints that must interact in order to produce movement and is a direct result of the degrees-of-freedom coordination problem expressed by Bernstein (1967; Whiting, 1984). Variation in the structure or function of biological systems within an individual, interacting with the constraints provided by the task, the environment, and the individual's psychological state at the time of movement execution, contributes to movement variability (Higgins, 1977; James and Bates, 1997). Control of the vast number of degrees of freedom of the neuromuscular and musculoskeletal systems is central to the variability discussion, and the number of degrees of freedom involved increases from macro to microscopic levels of organization (Newell and Corcos, 1993).

Movement variability has been viewed as both detrimental and beneficial to skilled coordinated movement. One view of variability is that it represents error in movement planning, execution, and outcome. This perspective, which originated from engineering and cybernetic-control concepts applied to the human, helped form the basis for information-processing models in motor control (Gallistel, 1981; Schmidt, Zelaznik, Hawkins, Frank, and Quinn, 1979; van Emmerik and van Wegen, 2000). Schmidt and colleagues (1979) proposed three sources of impulse variability (error) in a speed-accuracy trade-off information-processing model:

1. Error in selecting the appropriate motor program for the given task (central command error)

2. Scaling errors in setting the parameters responsible for executing the program (central error, peripheral error, or both)

3. Random noise in the system as the program is executed (peripheral error)

Alternatively, some have perceived variability as beneficial to movement organization and execution. This view emerged from the study of the behavior of chaotic nonlinear dynamical systems applied to human movement. From this perspective, variability is believed to be an emergent property of the self-organizing behavior of the nonlinear dynamical properties within the neuromotor system (Bingham, Schmidt, Turvey, and Rosenblum, 1991; Kelso and Tuller, 1984; Scholz, 1990; Schoner and Kelso, 1988; Turvey, 1990). The variability is thought to arise from two primary sources: (1) stochastic (random) fluctuations and (2) deterministically chaotic fluctuations (mathematically predictable when the initial conditions are known, but qualitatively appearing random; Kelso and Ding, 1993). From the dynamical systems perspective, at least four benefits of variability have been suggested. First, variability determines the stability of a movement pattern around an attractor. Large amounts of variability suggest unstable movement patterns, while small amounts of variability indicate stable patterns. Second, variability allows flexibility within the neuromotor system to permit the learning of a new movement pattern through adjusting the appropriate parameters. Third, variability allows flexibility to select or change to new, previously learned movement patterns by rescaling parameters so that different attractors can be accessed. Fourth, variability provides stochastic perturbations that allow constant sampling of different movement patterns (i.e., exploratory behavior) so that the most appropriate pattern can be selected (Kelso and Ding, 1993; Kelso, Scholz, and Schoner, 1986; Newell and Corcos, 1993; van Emmerik and van Wegen, 2000; Zanone and Kelso, 1992).

Several different structures of variability have been identified in physical and biological systems, and these different structures suggest merits of both views of variability. Stochastic "white" noise is common, but variant properties other than white noise also exist (Newell and Slifkin, 1998; Schroeder, 1991). Many biological signals have been observed to exhibit frequency profiles that follow inverse power laws of the form $1/f^n$, where $f$ is the frequency. Random white noise is $1/f^0$, which indicates no structure within the signal. Other frequency profiles of noise are labeled by different colors, such as pink ($1/f$), brown ($1/f^2$), and black ($1/f^2$; Newell and Slifkin, 1998). Each successively greater inverse power represents greater structure within the signal. The presence of a variance profile other than white noise within a biological movement signal indicates that error and variability are not synonymous (Newell and Slifkin, 1998). Logically, if variability is not error, then it may not be detrimental to the temporal evolution of the movement pattern or the outcome. However, it is not easy to establish beneficial effects of variability, and it is possible that the presence of structured variability exists simply as a characteristic of the movement and does not play any positive or negative role.

Newell and Slifkin (1998) suggested that knowledge of the structure of a signal is important for determining its underlying nature and that the structure should be examined in addition to the more commonly quantified magnitude of variability (e.g., standard deviation or coefficient of variation). The structure of variability (i.e., white, pink, brown, or black noise) may change over time during movement execution and for different task requirements, suggesting that organizational properties of the neuromotor system may also be changing. For example, Newell and Slifkin (1998) reported that the variability in the time series profile of a continuous isometric force production task was different for moderate force magnitudes (near 50%) than for force magnitudes of low and high extremes. This conclusion was supported by two very different variability quantities, the coefficient of variation and an approximate entropy variable (see chapter 3, pp. 76, for discussion of the approximate entropy variable). The coefficient of variation (standard deviation normalized to the mean of the score distribution) suggested a u-shaped function, with lesser variability near the midrange. Conversely, the approximate entropy variable (a measure of the degree of regularity of the time domain evolutionary properties of the force) suggested an inverted u-shaped function, with lesser regularity occurring at the midrange values of the force production. An important observation from these findings was that the force level exhibiting the lowest signal-to-noise ratio (i.e., least coefficient of variation magnitude) was approximately the force level exhibiting the most complex structure of variability (i.e., greatest approximate entropy; Newell and Slifkin, 1998). Independent interpretation of the results from each variable could lead to different conclusions about the variability associated with the given task. Evaluation of only the coefficient of variation variable would have suggested low variability at the midrange force, whereas independent examination of the approximate entropy variable would have led to the opposite conclusion. Two lessons from these results are that (1) the choice of variables is critical for making conclusions about variability and (2) traditional variability variables, such as standard deviation and coefficient of variation, do not demonstrate the underlying complexity of the system.

The choice of the variables used to quantify movement variability depends on the specific application. Investigators seeking to understand the nature and complexity of the neuromotor or other biological system should consider using a collection of different types of variability measures, including traditional quantities such as standard deviation or coefficient of variation, along with other measures such as the power frequency structure of the signal, approximate entropy, and dimensionality (estimation of the number of degrees of freedom regulating the output; Newell and Slifkin, 1998; see chapter 3, pp. 73, for a discussion of dimensionality of the signal). The approach of choice should be determined by the theoretical foundations driving the research and the specific research questions asked. Recognition of the presence of different types and characteristics of variability permits the formulation of a simple conceptual model of the **total variability** associated with movement. The model can be represented by the following equation:

$$V_T = V_n + V_e \tag{2.1}$$

where $V_T$ is the total variability that exists within a system and is observable during movement, $V_n$ is the variability due to the nonlinear dynamical processes within the system, and $V_e$ is the variability due to error. $V_e$ can be partitioned further to reveal additional components, as shown in equation 2.2:

$$V_e = V_{eb} + V_{em} + V_{ee} \tag{2.2}$$

where $V_{eb}$ is the biological error or noise that is present within the neuromotor system, $V_{em}$ is the error contributed by methodological processes (e.g., measurement error, data reduction and processing error), and $V_{ee}$ is the **error variability** due to all other sources external to the organism (e.g., environmental changes, variations in task requirements). From equation 2.1, the methods used to study the nature of variability (e.g., Buzzi, Stergiou, Giakas, Dierks, and Georgoulis, 2001; Stergiou, Buzzi, Hageman, and Heidel, 2000) are assumed to isolate $V_n$ and permit the assessment of the inherent control processes within the neuromotor system. Traditional quantities provide an estimation of the total of all sources of variability ($V_T$) observable in the movement. Certainly both approaches have merit for the study of variability, each revealing unique information. However, the consequences of variability in some applications may be best characterized by the total variance involved. Some of these applications are the focus of this chapter, along with several traditional methods for quantifying variability. Additional methods for quantifying variability are discussed in chapters 3 and 6.

# Variability and Biological Health

Variability has been linked to the health of biological systems in a variety of contexts (Ivry and Corcos, 1993; Glass and Mackey, 1988; Goldberger and West, 1987; Babloyantz and Destexhe, 1986; van Emmerik and van Wegen, 2000). Many biological rhythms necessary for life (e.g., heartbeat, respiration, menstrual cycle, sleep-wake cycle, and gait, among numerous others) are affected by variability in their functional processes. The dynamics of health were once thought to be ordered and regular, while variation was considered an indicator of disease (Goldberger and West, 1987). However, since the emergence of chaos dynamics and its application to biological systems, many healthy systems have been observed to exhibit greater variability than those that are diseased. Thus, a current prevailing viewpoint is that biological variability in the correct amount is essential for health. However, variation of biological rhythms outside of normal limits, either too great or too small, may lead to a class of diseases called "dynamical diseases" (Glass and Mackey, 1988), which are characterized by abnormal temporal organization. **Dynamical diseases** are typified by changes in the dynamics of some relevant variable, including changes from small- to large-amplitude variations, the appearance of new rhythms or periodicities, and the disappearance of variation accompanied by the appearance of more constant dynamics (Glass and Mackey, 1988).

## Examples of Variability Associated With Biological Health

Several examples of variation in biological processes that are associated with health or disease have been observed in cardiac physiology, brain pathology, and other neurological impairments, as well as in the movement sciences. Goldberger and colleagues reported different variability patterns in heart rhythms among healthy and diseased patients (Goldberger and West, 1987; Goldberger, Rigney, and West, 1990). Ventricular fibrillation, a cause of sudden death, was once believed to exhibit turbulent cardiac dynamics; but more recent evidence has suggested that fibrillation exhibits a narrow-band spectral representation characteristic of less variability (Goldberger and West, 1987; Goldberger, Bhargava, West, and Mandell, 1986). Similarly, extremely periodic dynamics of the electrocardiogram signal is a characteristic of torsades de pointes, another cardiac arrhythmia associated with sudden death (Goldberger and West, 1987). Tsuji and associates (1994) demonstrated a relationship between decreased heart rate variability and lesser survival rate in elderly individuals. The current evidence suggests that healthy heart rhythms are associated with an information-rich broadband frequency spectrum ($1/f$-like distribution; i.e., pink noise) that is associated with greater complexity and often greater variability (Goldberger and West, 1987).

A similar relationship between variability and health has been reported in brain pathology and other neurological impairments (Goldberger and West, 1987; Ivry and Corcos, 1993; Newell, van Emmerik, and Sprague, 1993; van Emmerik and van Wegen, 2000), although the results appear to be dependent on the parameters evaluated. Babloyantz and Destexhe (1986) reported a reduction in the dimensionality (i.e., complexity) in electroencephalogram (EEG) data during epilepsy versus normal function. Similar results have been observed as a function of sleep stage and coma (Gallez and Babloyantz, 1991), with greater dimensionality observed in the more alert states. Stam and colleagues (1994) reported a reduction in EEG complexity in Parkinson's patients compared to a healthy control group.

Using more traditional measures (e.g., standard deviation of finger tapping intervals; standard deviation of a force-control task) to quantify variability, Ivry and Corcos (1993) reported greater variation in the timing of tapping of individuals with lesions of the cerebellum and frontal cerebral cortex compared to healthy control and Parkinson's patients. The increased variability was attributed to a disruption in the internal clock timing process believed to be related to the cerebellar and cortical lesions (Ivry and Corcos, 1993). Parkinson's patients were not different from controls on the described tapping task (Ivry and Corcos, 1993), even though evidence from EEG studies (e.g., Stam et al., 1994) suggests variability differences between Parkinson's and control groups. Additionally, children with soft signs of cerebellar dysfunction have been observed to exhibit greater variability on a tapping task compared to children with soft basal ganglia signs or compared to controls (Lundy-Ekman et al., 1991 as reported in Ivry and Corcos, 1993). Conversely, the children with soft basal ganglia signs exhibited greater variability on a force-control task than the children in either of the other two groups (Lundy-Ekman, Ivry, Keele, and Woollacott, 1991; Ivry and Corcos, 1993). The research reported by Ivry and Corcos (1993), Keele, Davidson, and Hayes (1998), and others provides evidence for the importance of the cerebellum and basal ganglia in the regulation of force and timing control in motor activities. Impairments to these areas of the brain seem to alter the movement consistency-variability relationship, thus illustrating the role of variability (either too great or too small) in characterizing these conditions.

Further evidence for the association of variability and biological health comes from research on individuals with tardive dyskinesia, a syndrome that may arise from long-term neuroleptic treatment for individuals undergoing antipsychotic drug therapy (Newell and Slifkin, 1998). Persons with tardive dyskinesia commonly exhibit stereotypic (repetitive and predictable) movement behaviors such as body rocking, hand waving, facial gestures, and tongue protrusions (Newell et al., 1993). Newell and colleagues (1993) reported evidence suggesting that facial stereotypic movement patterns in persons with tardive dyskinesia, although repetitive and qualitatively appearing invariant, may exhibit greater variability than similar repetitive movement patterns performed by healthy control subjects. Van Emmerik, Sprague, and Newell (1993) studied the postural sway characteristics of patients with tardive dyskinesia compared to healthy control subjects and found more consistency in the rhythm of center of pressure movements in the patients, but with a greater overall magnitude of movement. The authors suggested that the patients may have utilized movement patterns consisting of less complexity compared to the control individuals, which may have constrained their ability for postural adaptation.

The literature contains numerous other observations of the presence or role of variability in the function and health of biological systems. Collins and colleagues (1996) observed an improvement in tactile sensation of cutaneous mechanoreceptors when stochastic noise was added to the input stimulus. Greater variability in posture control (center of pressure movement) during quiet upright stance has been observed in young healthy individuals compared to elderly individuals deemed at greater risk for falling. Additionally, the elderly persons appeared to be less able to control variability as the postural stance shifted so that the center of pressure path approached the stability boundaries (van Emmerik and van Wegen, 2000). In both the tactile and postural sway examples, variability is considered beneficial because it permits exploratory behavior and optimization of the involved functional processes.

Finally, clinical physiological data have suggested a possible link between certain characteristics of chaotic oscillations in the regulation of hematopoesis and the presence of some blood disorders (Glass and Mackey, 1988). Specifically, it has been observed that premature cell death during the proliferation phase in stem cell populations results in oscillations in the dynamics of the cell population (Glass and Mackey, 1988).

Collectively, these examples suggest that variability may distinguish healthy from diseased biological systems. However, the numerous examples presented fail to definitively elucidate whether more or less variability is advantageous in all cases, or whether variability characteristics for involuntary physiological processes are similar to those for voluntary movement. The evidence suggests that the amount of variability desired is dependent (minimally) on the biological system involved and the variable under examination. In many examples, diseased states were characterized by too little variability, while in other cases healthy systems exhibited less variation. In none of the examples is the increase or decrease in variability thought to be the cause of the pathology; instead it is a consequence of the altered system dynamics. The next section presents a hypothesis for the role of reduced variability in contributing to some types of musculoskeletal injuries.

## Variability and Overuse Injury: A Musculoskeletal Loading Hypothesis

Although numerous examples of association between variability and health have been observed, no direct connection has been made between total movement variability and musculoskeletal injury. During exercise, sport, and other common but demanding movement activities, functional injury to otherwise healthy tissue results from a failure of the musculoskeletal system to adapt to mechanical stress. Acute traumatic injuries result when stress acutely exceeds tissue tolerance. Overuse injuries develop when stress occurs repetitively and with sufficient magnitude and frequency to outpace the physiological adaptive process.

Williams (1993) presented a model for overuse injuries that illustrated the importance of biological adaptation (musculoskeletal remodeling) in maintaining tissue health (figure 2.1). Two stress-adaptation pathways that determine the condition of the involved tissues were presented: (1) rate of remodeling exceeds the rate of tissue damage, which leads to stronger tissue; and (2) rate of remodeling is delayed compared to the rate of tissue damage, which leads to overuse injury (Williams, 1993). For example, remodeling in bone occurs as a result of a coupling between the deposition and resorption processes (Zernicke and Loitz, 1992). Resorption, the initial response to the adaptation stimulus, occurs first and approximately one week prior to deposition (Zernicke and Loitz, 1992). When mechanical loading occurs at a rate greater than some critical frequency, resorption outpaces deposition (Whiting and Zernicke, 1998). A chronic imbalance of the resorption-deposition coupling process results in weakened tissue and may lead directly to tissue damage (e.g., stress fracture) or may increase the susceptibility of the tissue to a future traumatic event. Similar processes lead to overuse injuries in cartilage, ligament, tendon, and skeletal muscle, but with physiological mechanisms specific to those tissues (Nordin and Frankel, 2001). Consequently, factors that influence the rate of remodeling, rate of damage, or both, affect the health of the tissue. Since stress is associated with both

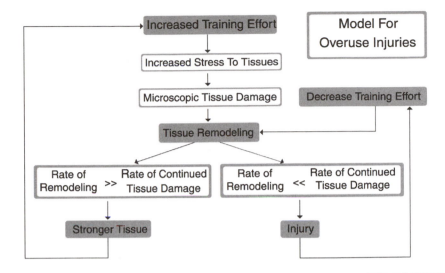

**Figure 2.1**   Model for the relationship between loading, remodeling, and overuse injury.

Adapted, by permission, from K.R. Williams, 1992, Biomechanics of distance running. In *Current issues in biomechanics* edited by M.D. Grabiner (Champaign, IL: Human Kinetics), 21.

remodeling and damage, the stress-adaptation pathway followed is determined by factors that alter tissue stress characteristics.

Anecdotal evidence has suggested that variability might play a positive role in preventing overuse injury by providing a broader distribution of stresses among different tissues or altering the stress magnitude, direction, rate, or frequency within the same tissue, thus providing a longer adaptation time between loading events. For example, in order to prevent overtraining, coaches and athletes commonly use training techniques like cross-training and periodization to increase both mechanical and physiological stress variability over time. Experienced runners commonly rotate shoes or vary training surfaces in order to prevent chronic running injuries. In situations involving repetitive loading, movement variability could provide protection from the accumulation of trauma to the involved tissues. This postulated protective mechanism was characterized by McCaw (1989, p. 102) as "intra-individual within-activity cross-training."

These observations, along with the previously discussed associations between variability and biological health, permit the formulation of a hypothesis relating variability and musculoskeletal injury (James, 1996). The hypothesis predicts that musculoskeletal health is maintained in a repetitive submaximal loading environment by variation above some critical value of the characteristics of loading (e.g., stress magnitude, frequency, direction). Conversely, too little variability is predicted to contribute to the accumulation of trauma over time by not permitting adequate tissue adaptation time between loading events. This hypothesis is presented graphically in figure 2.2. In the figure, the abscissa represents time, and the ordinate represents a generalized characteristic of loading (stress magnitude in the current example). Two boundaries within the figure are indicated by the upper and lower dotted lines. The lower boundary denotes the minimum amount of stress required to elicit physiological adaptation within the tissue. Stress values below this line would be insufficient to provide the adaptation stimulus. The upper boundary indicates the physiological limit (threshold) of the tissue. Stress values above the threshold result in tissue yielding on a macroscopic level or complete failure. The region above the physiological limit boundary is designated the acute injury region, which represents traumatic single-loading-event injuries. In reality, the stress magnitude of this threshold fluctuates with the integrity of the tissue (i.e., lesser value for weakened tissue) and thus

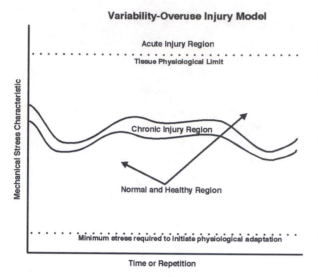

**Figure 2.2** Variability model for musculoskeletal health and injury. The model illustrates the hypothesis that variability in musculoskeletal loading can be preventative of certain types of overuse injuries.

From James, 1996, pp. 6, 105.

is dynamic across time. The area between the upper and lower boundaries is the normal and healthy region, as indicated in the figure. Performances across time that are of sufficient magnitude to elicit adaptation, but that do not result in acute injury, fall within this area. Variability of movement results in a broad distribution of stress values within this region, which is postulated to represent a normal, healthy amount of stress variability. The area or channel bounded by the undulating lines denotes the chronic injury region. Occasional performances within the channel are expected, and would also be considered normal stress magnitudes within the normal and healthy region. However, long-term repeated performances within the channel, with little variation, are hypothesized to contribute to overuse injury by the mechanisms that have been described. In the figure, the fluctuations in stress magnitude and varying widths of the channel are meant to characterize its fluidity (evolving, changing) across time. The critical amount of variation necessary to promote tissue health at any given instant in time is represented theoretically by the width of the chronic injury region at that time, but is not actually known. Factors believed to influence the width of the chronic injury region at a given time are the same as those that influence tissue health and adaptation in general. For example, factors that affect neuromuscular control (e.g., strength, coordination, fatigue), tissue integrity (e.g., tissue strength, previous injury), physiological adaptation (e.g., cellular health and adaptation, nutritional status), and general training variables (e.g., frequency, intensity, duration) determine the state of the tissue and its tolerance to stress. Healthy, well-adapted tissue would likely be able to tolerate more repetitive loading due to less variability than weakened or unhealthy tissue. Hence, the width of the chronic injury region could be narrower for healthy tissue in an optimal environment. Changing one or more of the involved factors would alter the tissue's sensitivity to invariance, and thus change the width of the chronic injury region across time.

The hypothesized variability-overuse injury relationship is depicted more directly in figure 2.3. In this figure, the abscissa represents the variability of the relevant stress characteristic, and the ordinate is the possibility of incurring overuse injury. At small magnitudes of variability, the likelihood of overuse injury is greatest. The potential for overuse injury decreases as the variability increases, but at a possible cost of increasing the likelihood of a single aberrant performance that could result in an acute injury. Hypothetically, the curve is asymptotic to both axes. The probability for overuse injury is never zero as long as stress is present. Similarly, variability is always present because of the biomechanical complexity of the involved systems.

Although the **variability and overuse injury hypothesis** is speculative, experimental data from James, Dufek, and Bates (2000) have supported its plausibility and have indicated the need for additional rigorous empirical data to test the hypothesis. In the study, healthy control subjects were compared to healthy overuse injury-prone individuals during a controlled vertical landing activity from three heights. Variability in ankle, knee, and hip joint moment of force time to peak, impulse, and peak values were quantified during the initial landing phase (0-100 ms postcontact) for each subject and landing height by computing for each subject the mean absolute difference of each landing trial within a condition from the condition mean. Results

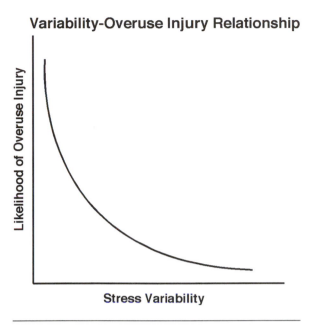

**Figure 2.3**  Hypothetical relationship between stress variability and likelihood of overuse injury.

of multivariate analysis of variance (MANOVA) and follow-up analyses indicated statistical differences between the two subject groups for some variability variables at some heights. However, the directions of the differences for the three loading parameters (time to peak, impulse, peak) were not always consistent with the prediction specified by the hypothesis. Variability in the time to peak ankle joint moment at the lowest landing height was greater for the healthy control group compared to the injury-prone group, thus providing direct support for the hypothesis. Variability in a linear combination of the impulse variables indicated group differences when all heights were considered collectively using the MANOVA procedure, but follow-up tests failed to elicit information about differences for individual joint moment variables or directions of those differences. This second result suggests that variability in the quantified variables differentiated the groups, but failed to provide enough information to support or refute the hypothesis. Conversely, variability in the peak ankle joint moment variable at the medi-um height indicated group differences; but greater variability was observed for the injury-prone group, thus contradicting the hypothesis. Collectively, these results suggest that variability may be able to distinguish the healthy control and injury-prone individuals; but the relevant variables, directions of the differences, and cause-effect relationships between variability and overuse injury have yet to be determined. Future descriptive studies can confirm the relationship between variability and overuse injury conditions and whether or not the hypothesis (healthy individuals exhibit greater variability) is plausible. Careful prospective studies are needed to establish the causal relationship, if one exists, between musculoskeletal loading variability and overuse injury.

# Methodological Considerations of Movement Variability

Variability has an unusual relationship with the methodology used to study movement. Variability can be both the subject of interest and a factor that constrains the effectiveness of the methodological process. This section briefly reviews the role of variability in traditional statistics. Then variability is considered relative to its role in some common biomechanics methodologies.

## Statistical Considerations of Variability and Error

Traditional parametric statistical tests of the type commonly used in biomechanics research (e.g., analysis of variance, *t*-test, correlation, and regression) are each specific formulations of a more general mathematical model called the General Linear Model (Aron and Aron, 1994). The General Linear Model is a linear mathematical relationship between a dependent variable and one or more independent (predictor) variables (Aron and Aron, 1994). The mathematics of the model permits quantification of differences between two or more mean values representing different treatment groups (e.g., *t*-test, analysis of variance), degree of relationship among two or more variables (e.g., correlation), and prediction of a dependent variable from one or more independent variables (e.g., regression). The equation representing the General Linear Model is of the form:

$$Y = a + b_1X_1 + b_2X_2 + b_3X_3 + \ldots + E \qquad (2.3)$$

where $Y$ is the actual value of some dependent variable; $a$ is the y-intercept or fixed influence of all independent variables on the dependent variable; $b_1$, $b_2$, and $b_3$ are coefficients of the influence of each independent variable upon the dependent variable; $X_1$, $X_2$, and $X_3$ are values of each independent variable; and E is the error term, which represents the sum of all other influences on the dependent variable not included in the model (Aron and Aron, 1994). In correlation and regression, the model is used to develop the equation of a line of best fit of the predicted values of the dependent variable ($Y'$) compared to the observed values of the same variable ($Y$) of the experimental data for each value of the independent variable ($X$), based on a least-squares error criterion. In this application, the lesser the error, the better the fit of the line and the better the prediction. In $t$-test and analysis of variance, the model permits estimation of the experimental error within a treatment level (i.e., subjects treated alike) and the variance between groups of subjects receiving different treatments. In hypothesis testing, treatment effects are determined by the ratio of the unsystematic component of variation (experimental error) compared to the systematic component (treatment effects). The greater the experimental error in relation to the treatment variance, the less likely the statistical test is to detect differences. When the null hypothesis is true, the ratio is

$$E_{unsys} / E_{unsys} \qquad (2.4)$$

where $E_{unsys}$ is the unsystematic experimental error, thus indicating that there is no variance between treatment groups that is indistinguishable from the error within groups (Keppel, 1982). When the null hypothesis is false, the ratio is

$$(VAR_{sys} + E_{unsys}) / E_{unsys} \qquad (2.5)$$

where $VAR_{sys}$ is the systematic variance due to treatment, thus indicating that the variance between treatment groups is greater than the variance within groups (Keppel, 1982). Statistical significance is dependent on both the size of the treatment variance and the magnitude of the experimental error, as shown in equation 2.5. Careful researchers design experiments to maximize the systematic treatment variance and minimize the error variance.

The unsystematic error (E) in the General Linear Model is influenced by the amount of **intra-** and **interindividual movement variability** and other sources of variance. The error within an experiment is the difference between the hypothetical true score performance on a dependent variable and the observed score on that same variable (Thomas and Nelson, 2001). Several sources of the unsystematic experimental error are recognized, including measurement error, variations within the testing environment, and individual differences (Keppel, 1982). Thomas and Nelson (2001) identified errors from these same categories that are associated with the testing, scoring, instrumentation, and participant. Testing errors are primarily associated with the giving and following of instructions. Scoring errors pertain to accuracy of the scoring itself and the competency of the scorers, including such factors as properly obtaining and recording values. Instrumentation errors are caused by improper calibration, failure of equipment, or inability of the instrument to discriminate among relevant performance behaviors (Thomas and Nelson, 2001); errors in data reduction, processing, and modeling procedures that are common in biomechanics research may also be included in this category. Participant errors are attributable to many factors, including the individual's motivation, health, knowledge, fluctuations in memory, fatigue, and performance (Thomas and Nelson, 2001).

Participant errors also include the movement variability discussed in this chapter. Hence, statistically, movement variability is commonly treated as a source of experimental error and is undesirable relative to the ability to predict future scores or detect differences among groups. Using traditional methodology, it is impossible to partition the sources of experimental error in order to obtain true inherent movement variation; thus measurements of movement variability also contain elements of the other sources of measurement and environmental variance. This error composition has the following consequences: (1) greater movement variability increases the magnitude of unsystematic experimental error within the General Linear Model; and (2) when movement variability is the focus of the study, this source of variance can never be completely discriminated from the other sources of experimental error. In either case, the best strategy for the researcher is to control the sources of variability or error that threaten the validity of the experiment. For example, when a large amount of movement variability threatens the detection of group differences, the researcher should consider alternative experimental designs or statistical approaches that can better account for the individual variability. Similarly, when movement variability is the subject of interest, the researcher should demonstrate greater care in order to minimize the non-movement sources of unsystematic experimental error, such as those due to testing, scoring, and instrumentation. The next section deals with some of these methodological issues. Additionally, chapter 1 provides an in-depth discussion of experimental and statistical design issues.

## Considerations of Variability and Error in Biomechanics Methodology

As a quantitative science, biomechanics research relies heavily on the ability to obtain, process, and analyze empirical data. Error is inherent in these processes. Common methodological practices such as digitizing and data filtering are particularly problematic and can alter the magnitude of the observed movement variability by changing the composition of the overall variability component (movement variance plus error) of the data. In the case of digitizing (converting the locations of anatomical or other landmarks obtained from film or video images to corresponding x-, y- or x-, y-, z-coordinates in an appropriate axis system), inconsistent and incorrect identification of targeted landmarks adds variability to the data in the form of error or noise that is assumed to be random. This error cannot be easily differentiated from the actual movement variability of interest; thus the magnitude of the experimentally obtained kinematic variability quantified from film or video contains a component of measurement error. Two common methods for managing digitizing error are to (1) ignore it, assuming that it is random and therefore equal across all levels of the experiment, and (2) attempt to remove or reduce it by digitally filtering the data in either the time or frequency domain (other smoothing methods are also appropriate, but are not discussed here).

The first of these options is the simplest and most direct to implement, as no post-digitizing treatment of the data is required before the desired analysis is conducted or the variables of interest are computed. Buzzi and colleagues (2001) used this procedure in their nonlinear analysis of knee and ankle angle data. In some cases this option is appropriate, but one major limitation is the difficulty in using the raw displacement data to compute numerical derivatives (e.g., velocity, acceleration) because of the higher-order increase in error from one derivative to the next. Additionally, since no attempt is made to remove or reduce the error, even raw linear and angular displacement data may be difficult to interpret relative to their functional significance when the error magnitude is great. In variability analyses using traditional variables, such as standard deviation or coefficient of variation, one possibility for accounting for the random digitizing error is to digitize a nonmoving reference marker from which one can make an estimate of digitizing error. One could then remove the digitizing error (arithmetically, statistically, etc.) from the variability observed in the movement of the other relevant landmarks. In nonlinear signal processing, the presence of measurement error affects the components of the phase portrait (plot of a variable vs. its derivative), thereby necessitating special techniques for reconstruction of the phase portrait in pseudo-phase space (Mullin, 1993). Mullin (1993) recommended two such techniques, the method of delays and singular value decomposition; the reader is directed to this and other works on nonlinear signal processing for additional discussion.

Attempting to remove or reduce the measurement error by digitally filtering the data in either the time or frequency domain is the second option commonly used in biomechanical analyses to manage digitizing error. In the time domain, a low pass digital filter is used to remove higher-frequency components from within the digital signal. The concept is that the components of the signal contributed by actual biological movement reside on the lower end of the frequency spectrum, while random error (noise) is composed of higher-frequency components. The filter effectively removes the frequency components above a critical cutoff value either selected by the researcher or based on an error optimization technique (e.g., optimal filter; see Winter, 1990, and chapter 9 for further discussions of digital filtering).

In practice, there is an overlap of the frequencies within the signals of the actual movement components and noise; hence the filter either allows some noise to pass through, attenuates some of the biological signal, or does both. Just as in non-variability kinematic studies, in variability experiments the selection of the cutoff value is critical for achieving the proper balance between minimizing the error and attenuating the true signal. The effects of permitting too much error to pass through are the same as those previously discussed when no filtering is performed at all, albeit with less severe consequences. When the filter cutoff value is set too low, too much of the true signal of a single movement trial is removed, thus reducing the observed variability across multiple trials. For example, in the case of over smoothing (i.e., cutoff value too low), the magnitude of a movement is reduced as a function of the amount of the signal removed. As the cutoff value is progressively lowered, the time series representation of the movement magnitude converges to an invariant straight line, thus reducing the observed variability of the characteristic of interest across the multiple trials. Use of an optimal filter would seem ideal except that the researcher would not be able to conclude with certainty that the variability observed experimentally is due to movement variability and not due to variance induced or reduced by the differential cutoff values selected for each movement trial by the automated algorithm. The relationship between the variance in the optimally selected cutoff values and the variability observed in the actual signal is not well known. Does the optimal filter induce greater, lesser, or no change in the variability of the filtered signal? One could argue that because the goal of the optimal filter procedure is to select the best cutoff value, the identified value provides the best compromise between maintaining the information in the true signal and removing the majority of the noise. In other words, the optimal filter could act to preserve the actual variability within the true signal; conversely, it might not. This relationship requires additional examination.

Similar filtering techniques can be implemented in the frequency domain (e.g., power spectrum), where the time series signal is decomposed into a series of sine and cosine waveforms and the power of each frequency band is computed over a selected frequency range (Mullin, 1993). A critical cutoff value can be identified from the power spectrum, and data below the value can be preserved while information above the critical value is discarded. The Fourier transform and its discrete implementation the Fast Fourier Transform (FFT) are widely used for this and a variety of other signal-processing purposes. However, FFT, time series digital filter, and other common signal-processing procedures are linear techniques, and some experts have suggested that they are not well suited to examining nonlinear (e.g., chaotic) phenomena (Mullin, 1993). Conversely, other authors have suggested that Fourier analysis and similar methods are appropriate regardless of whether or not the data are chaotic (Williams, 1997). Regardless of the view taken, when the variability associated with nonlinear processes is the focus of examination, a variety of linear and nonlinear signal-processing techniques are desirable for understanding the data and permitting the distinction between chaos and noise (Mullin, 1993; Williams, 1997). Several works, including those of Mullin (1993), Buzzi and colleagues (2001), and Williams (1997) and several chapters in the current volume, use or discuss such techniques.

Another issue relevant to variability, error, and common biomechanics methodology is the number of trials (observed performances) required of a subject in an experiment. The number of trials can vary from one to a seemingly unlimited number, depending on the research question and the constraints imposed by the task and experimental protocol. For example, in a continuous activity such as running, more than one trial is desirable because each stride is both error producing and error correcting, resulting in variability of movement (Bates, De Vita, and Kinoshita, 1983). One trial is not necessarily representative of the individual's performance over time. By chance, any one trial could be representative of an average performance but might also reflect an atypical pattern. The greater the movement variability exhibited by the individual, the greater the likelihood of sampling an atypical performance from the population of all possible performances.

Thus, except in special circumstances (e.g., only one trial is available, performance of the one trial is the subject of interest), it is generally good practice to obtain a number of trials of the performance in order to ensure that the population of trials is adequately represented. This situation is analogous to the practice of selecting the appropriate number of individuals (sample size) in an experiment to represent a larger population of individuals for the purpose of statistical inference. In most cases, random selection of the sample of individuals is desirable. Similarly, random selection of a sample of trials from within an individual is preferred. Bates (1996) hypothesized that individuals produce performances quantitatively as if they are random trial generators and has reported data to support this contention. Relative to the number of trials required of a subject, the critical question is how many trials are necessary to characterize a representative and stable performance. Although it may be that no two trials are identical on the magnitude of a selected variable of interest, the variability does not increase continuously but eventually stabilizes, defining a functional variability range (Bates, De Vita, and Kinoshita, 1983). In running and other repetitive activities, previous research has shown the number of trials per subject-condition used in various experiments to vary widely, including such values as 1 trial (running; Nigg et al., 1978), 3 trials (jumping; Jensen and Phillips, 1991), 5 trials (running; Cavanagh and Lafortune, 1980), 9 trials (walking; Winter, 1984), 10 trials (running, landing; Dufek, Bates, Stergiou, and James, 1995), 15 trials (running; McCaw, 1989), and 25 trials (running; Lees and Bouracier, 1994), to over 100 trials (landing; Sveistrup, Davis, Derrick, James, and Bates, 1989).

The number of trials selected should be based on sound methodological principles and not be simply a matter of convenience for the researcher. Belli and colleagues (1995) suggested that intra-individual step variability should be considered in studies examining mechanical efficiency in treadmill running, indicating that 32 to 64 consecutive steps are necessary if one is to obtain better than 1% accuracy on mean and variability values. Previously, Bates, Osternig, Sawhill, and James (1983) concluded that a minimum of 8 nonconsecutive steps (trials) was necessary to obtain stable data in 43 ground reaction force variables during overground running. For the ground reaction force data, the criterion used to evaluate stability for each variable was based on a variable's 10-trial mean value. Stability was defined as having occurred when successive mean deviations were within one-quarter standard deviation of the mean value for each variable (Bates, Osternig, Sawhill, and James, 1983). These experimental results were supported by a follow-up study that used a computer model to generate both error-free and error-included data for different numbers of trials ranging from 0 to 100. Results of the model suggested that 8 to 10 trials were necessary to achieve the one-quarter standard deviation criterion value (Bates, De Vita, and Kinoshita, 1983). However, the estimation of the critical number of trials necessary to achieve stability is dependent on both the criterion value selected (e.g., one-quarter standard deviation) and the number of trials used to determine the mean value. Hence, the number of trials necessary for performance stability as determined by this method slightly underestimates the actual number of trials required (Bates, De Vita, and Kinoshita, 1983). Hamill and McNiven (1990) used a similar technique to determine that 10 trials were necessary to achieve stability in selected ground reaction force variables during walking.

## Example 2.1

Although Bates, Osternig, Sawhill, and James (1983) and Hamill and McNiven (1990) reported results for running and walking ground reaction force variables, respectively, similar methods have not been applied to landing data. This example presents data from a controlled landing activity in order to (1) demonstrate the method for determining the number of trials necessary in an experiment and (2) determine whether the results from landing are similar to those reported for running and walking. The data presented are from five subjects who performed multiple bilateral drop-landing trials from a height of 0.60 m with the left foot landing on a force platform and the collection system set to sample at 1,000 Hz (refer to James, 1991, for a complete methodological description). The peak vertical ground reaction force magnitude associated with heel contact (i.e., second peak) was selected as the variable of interest for this example (figure 2.4). The resulting individual and mean values for 30 trials for the five subjects are shown in table 2.1. The number of trials representing performance stability was determined by the method described by Bates, Osternig, Sawhill, and James (1983), in which data from 10 trials were used to determine the stability criteria for running. In the current example, the data from 30 trials are presented in the order in which they were obtained experimentally, but the initial evaluation was limited to the first 10 trials of the data set in order to permit comparison with the study by Bates and colleagues.

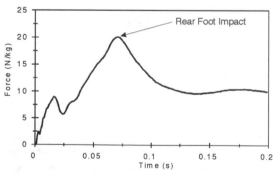

**Figure 2.4** Example vertical ground reaction force time history for drop landing from 0.60 m (left foot). The second peak (rearfoot impact) is labeled.

Data from James, 1991.

The first step was to compute the mean, standard deviation, and one-quarter standard deviation values for the first 10 trials for each subject (table 2.1). Second, the cumulative mean, standard deviation, and mean deviation values were computed for the 10 trials (table 2.2). A cumulative mean value was calculated as the average of each trial with all previous trials. This calculation was repeated in succession for all trials from 1 to 10 for each subject, and the final cumulative mean value is identical to the overall 10-trial mean. The cumulative standard deviation was computed similarly. A mean deviation value was calculated as the absolute difference between the cumulative mean of the corresponding trial and the mean of all 10 trials. Finally, stability was estimated as one greater than the smallest trial number for which all successive mean deviations were smaller than the criterion value of one-quarter of the 10-trial standard deviation for that particular subject (Bates, Osternig, Sawhill, and James, 1983).

As table 2.2 shows, the critical mean deviations that satisfied the established criteria were 1.3, 1.2, 1.4, 1.3, and 0.7 N/kg for subjects 1 through 5, respectively. The corresponding trial numbers for the critical mean deviation values were 9, 8, 4, 6, and 8 for subjects 1 through 5, respectively. Therefore, the minimum number of trials

**Table 2.1 Peak Vertical Ground Reaction Force Values Associated With Heel Contact During Drop Landing From 0.60 m**

| Trial | Subj 1 | Subj 2 | Subj 3 | Subj 4 | Subj 5 |
|---|---|---|---|---|---|
| 1 | 33.4 | 16.9 | 18.7 | 31.7 | 25.7 |
| 2 | 37.4 | 22.2 | 25.4 | 28.0 | 18.2 |
| 3 | 33.1 | 16.2 | 28.4 | 29.8 | 18.7 |
| 4 | 36.4 | 20.3 | 19.1 | 32.4 | 20.1 |
| 5 | 32.9 | 21.8 | 25.9 | 27.1 | 20.7 |
| 6 | 26.0 | 22.4 | 26.8 | 29.0 | 17.6 |
| 7 | 35.0 | 23.3 | 26.2 | 36.5 | 20.9 |
| 8 | 28.5 | 22.9 | 16.7 | 29.5 | 21.3 |
| 9 | 32.1 | 26.2 | 34.9 | 36.0 | 17.8 |
| 10 | 20.1 | 27.6 | 20.3 | 29.7 | 16.5 |
| $M_{10}$ | **31.5** | **22.0** | **24.2** | **31.0** | **19.8** |
| $SD_{10}$ | **5.3** | **3.6** | **5.5** | **3.2** | **2.6** |
| $1/4\ SD_{10}$ | **1.3** | **0.9** | **1.4** | **0.8** | **0.7** |
| 11 | 20.1 | 19.7 | 19.3 | 30.6 | 19.1 |
| 12 | 19.0 | 20.1 | 26.2 | 22.3 | 20.6 |
| 13 | 28.3 | 24.0 | 27.0 | 31.7 | 19.5 |
| 14 | 18.1 | 28.6 | 27.0 | 21.2 | 21.2 |
| 15 | 17.7 | 17.4 | 23.6 | 25.6 | 18.1 |
| 16 | 28.5 | 19.2 | 26.0 | 25.5 | 19.6 |
| 17 | 34.9 | 22.8 | 22.5 | 28.9 | 15.4 |
| 18 | 16.0 | 27.4 | 26.0 | 26.3 | 18.2 |
| 19 | 26.7 | 16.3 | 30.0 | 27.5 | 15.4 |
| 20 | 28.2 | 22.4 | 26.7 | 30.1 | 17.1 |
| $M_{20}$ | **27.6** | **21.9** | **24.8** | **29.0** | **19.1** |
| $SD_{20}$ | **6.9** | **3.7** | **4.4** | **3.8** | **2.4** |
| $1/4\ SD_{20}$ | **1.7** | **0.9** | **1.1** | **1.0** | **0.6** |
| 21 | 22.4 | 20.4 | 27.6 | 28.3 | 19.0 |
| 22 | 24.0 | 24.9 | 24.4 | 26.1 | 16.1 |
| 23 | 23.6 | 22.7 | 30.1 | 27.3 | 19.0 |
| 24 | 22.0 | 19.3 | 32.3 | 20.6 | 11.2 |
| 25 | 28.2 | 19.5 | 24.5 | 26.7 | 18.9 |
| 26 | 21.4 | 21.7 | 19.8 | 20.1 | 16.2 |
| 27 | 25.6 | 24.4 | 25.9 | 22.1 | 17.1 |
| 28 | 21.6 | 24.4 | 22.4 | 31.5 | 17.1 |
| 29 | 21.8 | 17.2 | 28.1 | 20.8 | 19.6 |
| 30 | 21.9 | 25.8 | 26.6 | 34.6 | 19.7 |
| $M_{30}$ | **26.2** | **21.9** | **25.3** | **27.9** | **18.5** |
| $SD_{30}$ | **6.1** | **3.4** | **4.1** | **4.4** | **2.5** |
| $1/4\ SD_{30}$ | **1.5** | **0.9** | **1.0** | **1.1** | **0.6** |

M and SD are mean and standard deviation, respectively. Subscripts 10, 20, and 30 represent the number of trials used to compute each variable. All values are reported in Newtons per kilogram of body mass.

Data from James, 1991.

*(continued)*

# Table 2.2 Cumulative Mean, Standard Deviation, and Mean Deviation Values for 10 Trials of the Landing Ground Reaction Force Data

| Trial | Subj 1 | | | Subj 2 | | | Subj 3 | | | Subj 4 | | | Subj 5 | | |
|---|---|---|---|---|---|---|---|---|---|---|---|---|---|---|---|
| | M | SD | D | M | SD | D | M | SD | D | M | SD | D | M | SD | D |
| 1 | 33.4 | 0.0 | 2.0 | 16.9 | 0.0 | 5.1 | 18.7 | 0.0 | 5.6 | 31.7 | 0.0 | 0.7 | 25.7 | 0.0 | 5.9 |
| 2 | 35.4 | 2.8 | 3.9 | 19.5 | 3.8 | 2.4 | 22.0 | 4.8 | 2.2 | 29.8 | 2.6 | 1.1 | 22.0 | 5.3 | 2.2 |
| 3 | 34.6 | 2.4 | 3.2 | 18.4 | 3.3 | 3.5 | 24.1 | 5.0 | 0.1 | 29.8 | 1.8 | 1.2 | 20.9 | 4.2 | 1.1 |
| 4 | 35.1 | 2.1 | 3.6 | 18.9 | 2.8 | 3.1 | 22.9 | 4.8 | 1.4 § | 30.5 | 2.0 | 0.5 | 20.7 | 3.4 | 0.9 |
| 5 | 34.6 | 2.1 | 3.2 | 19.5 | 2.8 | 2.5 | 23.5 | 4.4 | 0.8 | 29.8 | 2.3 | 1.2 | 20.7 | 3.0 | 0.9 |
| 6 | 33.2 | 4.0 | 1.7 | 20.0 | 2.7 | 2.0 | 24.0 | 4.1 | 0.2 | 29.7 | 2.0 | 1.3 § | 20.2 § | 2.9 | 0.4 |
| 7 | 33.5 | 3.7 | 2.0 | 20.4 | 2.8 | 1.5 | 24.3 | 3.9 | 0.1 | 30.6 | 3.2 | 0.3 | 20.3 | 2.7 | 0.5 |
| 8 | 32.8 | 3.9 | 1.4 | 20.8 | 2.7 | 1.2 § | 23.4 | 4.5 | 0.9 § | 30.5 | 3.0 | 0.5 | 20.4 | 2.5 | 0.7 § |
| 9 | 32.8 | 3.6 | 1.3 § | 21.4 | 3.1 | 0.6 | 24.7 | 5.7 | 0.4 | 31.1 | 3.3 | 0.1 | 20.1 | 2.5 | 0.4 |
| 10 | 31.5 | 5.3 | 0.0 | 22.0 | 3.6 | 0.0 | 24.2 | 5.5 | 0.0 | 31.0 | 3.2 | 0.0 | 19.8 | 2.6 | 0.0 |
| ¼ SD₁₀ | 1.3 | | | 0.9 | | | 1.4 | | | 0.8 | | | 0.7 | | |

M and SD are mean and standard deviation, respectively. D is the deviation (absolute difference) rounded to one decimal place of the individual trial mean from the 10-trial mean; apparent arithmetic discrepancies are due to rounding error. § = critical deviations.

Data from James, 1991.

required for stability was 10, 9, 5, 7, and 9 for each subject, respectively, with a five-subject average of 8.0 ± 2.0 trials (table 2.3). These results compare favorably to those reported for running by Bates, Osternig, Sawhill, and James (1983), who indicated that a minimum of 8 trials was necessary to reach performance stability with use of the 10-trial mean as the reference value.

**Table 2.3    Average Number of Trials for Stability of the Ground Reaction Force Variable**

| Number of trials | M | SD | 95% confidence lower limit | 95% confidence upper limit |
|---|---|---|---|---|
| 10 | 8.0 | 2.0 | 6.2 | 9.8 |
| 20 | 12.6 | 3.5 | 9.5 | 15.7 |
| 30 | 17.4 | 5.7 | 12.4 | 22.4 |
| *M* | **12.7** | | **9.4** | **16.0** |
| *SD* | **4.7** | | **3.1** | **6.3** |

Number of trials is the number of trials per subject used to quantify the individual subject mean and performance stability values. M (mean), SD (standard deviation), and 95% confidence limits shown in this table were computed for the group of five subjects.

Data from James, 1991.

Although specific values for the results of a 20-trial mean stability determination were not reported in the study by Bates and colleagues, the authors did indicate that such an evaluation was conducted for several variables and that the results were similar to those obtained using the 10-trial mean. To examine the effects of additional trials on the stability estimation procedure for the landing data, the ground reaction force values were reevaluated using both 20- and 30-trial data sets. These results (summarized in table 2.3) suggest that as the number of trials used for the reference value increases, so does the minimum number of trials necessary to reach performance stability. Table 2.3 shows that as the number of reference trials increased from 10 to 20, the mean of the minimum number of trials required for stability increased by 4.6 trials. Similarly, an increase in the number of reference trials from 20 to 30 resulted in an additional increase of 4.8 trials necessary for stability. Hypothetically, the principles of sampling would suggest that the greater number of trials used in the estimation, the greater the accuracy (i.e., the more representative the sample). Therefore, these results suggest either that it is less easy to obtain performance stability in landing than during running, that some other confounding factor affected the results (e.g., learning or fatigue), or that the values reported by Bates, Osternig, Sawhill, and James (1983) underestimated the actual number of trials required for performance stability. However, regardless of the actual number of trials necessary to represent a stable performance, it is clear from this discussion that numerous trials are necessary and that the use of one, two, three, five, or even up to eight trials might not provide a representative sampling of a subject's performance because of the inherent movement variability characteristics of the individual during the performance. These issues should be explored further for a wider variety of activities, movement parameters, and variables.

# Traditional Methods for Quantifying Variability

There are numerous methods and quantities for representing variability. The variability in kinematic, kinetic, and temporal variables can be computed using both traditional and nontraditional approaches. Traditional methods originating from descriptive statistics can be applied to both traditional and nontraditional variables, are considered most appropriate for quantifying the total variability within a system, and can be applied to both discrete and continuous data. Nontraditional methods are those that use techniques from the study of nonlinear dynamical systems to isolate chaotically deterministic variability from other variability components contained within the movement process. Some of these nontraditional methods are discussed in greater detail in chapters 3 and 6.

Several of the traditional methods for quantifying variability, along with numerical examples, are presented here for reference. Two standard data sets (appendix B) are used to demonstrate the application of these methods to biomechanical data. The first data set is from one male subject (age 24 years, height 1.83 m, mass 79.4 kg) performing 10 bilateral step-off landing trials from a height of 0.45 m (maximum vertical jump height) with the right foot landing on a force platform and the collection system set to sample at 1,000 Hz (for a complete methodological description see James, 1996). The resultant ground reaction force magnitude generated during the initial landing phase (0-100 ms postcontact) was selected to demonstrate both discrete and continuous (time series) applications of the variability quantification techniques. The second data set is from one male subject (age 25 years, height 1.89 m, mass 80.0 kg; midfoot striker) running on a treadmill (3.73 m/s) for several minutes. Ten consecutive running strides from the midportion (steady state) of the run were selected for analysis. Ankle and knee joint angle values were computed from the video records (right sagittal, 60 Hz) of the performance (for a complete methodological description see Poklikuha, 2001). The ankle and knee angle values were selected to demonstrate variability quantification techniques useful for relative motion (i.e., angle-angle) analyses.

## Discrete Methods

Discrete variables are commonly used in biomechanics research; examples include such measures as the value of a joint angle at a specific instant in time (e.g., knee angle at foot contact in landing), the timing of an event (e.g., time to maximum foot pronation following foot strike in running), and the peak magnitude of a variable of interest (e.g., peak vertical ground reaction force during the support phase of walking), to name just a few. The variability of discrete variables across trials can be quantified using at least five common descriptive statistics explained in most introductory statistical textbooks, including range, variance, standard deviation, coefficient of variation, and interquartile range. To illustrate each of these techniques, the magnitudes of a discrete variable (peak resultant ground reaction force magnitude associated with heel contact during landing) were extracted from the 10 landing curves shown in figure 2.5 and are presented in table 2.4. The complete force time records are provided in appendix B. Qualitatively, the variability is evident among the trials presented. In all cases, quantification of the variability variables was applied to the sample of trials provided, which was assumed to represent a larger population of trials possible by the individual. The significance of this assumption is that the formulas for sample variability were used rather than the formulas for population variability. While this approach was appropriate for the given data, in some applications, use of the corresponding population formulas may be more appropriate.

The first variability quantity used to describe the peak force data is the range. The range is simply the dif-

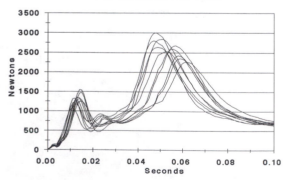

**Figure 2.5** Ten resultant ground reaction force time histories for step-off landing from 0.45 m (right foot).

Data from James, 1996.

**Table 2.4    Peak Resultant Ground Reaction Force Values Associated With Rearfoot Impact for One Subject During Step-Off Landing From 0.45 m**

| Trial | Force peak (N) |
|-------|----------------|
| 1 | 2351 |
| 2 | 3005 |
| 3 | 2801 |
| 4 | 2420 |
| 5 | 2841 |
| 6 | 2264 |
| 7 | 2588 |
| 8 | 2633 |
| 9 | 2674 |
| 10 | 2529 |
| M | **2610** |

M is the mean. Force values are in Newtons (N).

Data from James, 1996.

ference between the greatest and least values and is computed by subtraction of the least value from the greatest value.

For the landing data in table 2.4, the range is 741 N. Range is a good indicator of the spread of the data, with a large range implying that the data are spread over a large interval and a small range indicating that the values are more concentrated (Milton, 1999).

**Variance** is a measure of variability that uses the sum of the squared deviations between the individual values and the sample mean divided by the appropriate degrees of freedom for the sample as shown in equation 2.6:

$$s^2 = \frac{\sum_{i=1}^{n}(x_i - M)^2}{n-1} \tag{2.6}$$

where $s^2$ is the sample variance, $x_i$ is $i$th sample data value, $M$ is the sample mean, and $n$ is the sample size. The total variance for the landing data is 53,382 N$^2$ as shown in table 2.5.

From the variance, the standard deviation is easily computed:

$$SD = s = (s^2)^{1/2} \tag{2.7}$$

where $SD$ and $s$ represent the sample standard deviation and $s^2$ is the sample variance as calculated in equation 2.6. The resulting standard deviation for the landing data is 231.0 N as shown in table 2.5 and is customarily reported to one decimal place beyond that of the original data (Milton, 1999). Although variance and standard deviation are computed similarly, one important practical difference is that the units for variance are "squared." Thus, for the force data example, the variance units are N$^2$, while those for standard deviation are N. In many cases, this practical difference permits easier interpretation of standard deviation as a measure of variability because the units are consistent with those of the original data and the mean value.

Occasionally, it is useful to quantify the amount of variability relative to the magnitude of the mean. As an example, one might compare the variability between two performances with very different mean scores. A direct, nonadjusted comparison of variability values may not provide enough information to allow one to interpret whether any variability differences (or non-differences) are due to inherent properties of the movement or due to the magnitudes of movement within each performance (i.e., mean). Range, variance, and standard deviation are all absolute measures of variability. The most common quantity that represents a relative (normalized) variability measure is the coefficient of variation, which is the variability (standard deviation) converted to a percentage of the mean value, as shown in equation 2.8:

$$CV = (SD / M) \times 100 \tag{2.8}$$

where $CV$ is the sample coefficient of variation, $SD$ is the sample standard deviation from equation 2.7, and $M$ is the sample mean. The coefficient of variation for the data set in the example is 8.9% as shown in table 2.5.

One major limitation of the previously described variability measures is that they are all substantially affected by extreme performances (i.e., outliers). A single unrepresentative value can inflate the variability magnitude, thus misrepresenting the amount of variability in most of the data (Milton, 1999). The interquartile range

**Table 2.5   Examples of Discrete Variability Quantities Using the Ground Reaction Force Data From Landing**

| Trial | Force peak (N) | $x_i - M$ | $(x_i - M)^2$ | Data sorted (ascending) |
|---|---|---|---|---|
| 1 | 2,351 | – 259 | 67,081 | 2,264 |
| 2 | 3,005 | 395 | 156,025 | 2,351 |
| 3 | 2,801 | 191 | 36,481 | 2,420 |
| 4 | 2,420 | – 190 | 36,100 | 2,529 |
| 5 | 2,841 | 231 | 53,361 | 2,588 |
| 6 | 2,264 | – 346 | 119,716 | 2,633 |
| 7 | 2,588 | – 22 | 484 | 2,674 |
| 8 | 2,633 | 23 | 529 | 2,801 |
| 9 | 2,674 | 64 | 4,096 | 2,841 |
| 10 | 2,529 | – 81 | 6,561 | 3,005 |
| $M =$ | 2,610 | $\Sigma(x_i - M)^2 =$ | 480,434 | |
| Range = | 741 | $n - 1 =$ | 9 | |
| | | $s^2 =$ | 53,382 N$^2$ | |
| | | $SD =$ | 231.0 N | |
| | | $CV =$ | 8.9% | |
| | | $IQR =$ | 381 N | |

M is the mean, $x_i$ is an individual data sample, n is the number of samples, $s^2$ is the variance, SD is the standard deviation, CV is the coefficient of variation, and IQR is the interquartile range.

Data from James, 1996.

is a variability measure that is resistant to outliers and represents the length of the interval that contains approximately the middle 50% of the data (Milton, 1999). If the interquartile range is small, then the data are concentrated close to the center of the distribution. If the interquartile range is large, then the data are widely dispersed, indicating greater variability (Milton, 1999). The procedure for finding the interquartile range is to complete the following series of simple steps:

1. Compute the location of the median value within the sorted data (ascending order) using:

$$\text{median location} = (n + 1) / 2 \qquad (2.9)$$

where $n$ is the sample size. If necessary, truncate the result by dropping the decimal value (0.5) to form an integer approximation of the median location.

2. Find the quartile location ($Q$) by:

$$Q = (\text{median location} + 1) / 2 \qquad (2.10)$$

3. Find the data value occupying the upper end of the first quartile ($QI$) by taking the value in position $Q$. If $Q$ is not an integer, then average the two data values that $Q$ falls between.

4. Find the data value occupying the upper end of the third quartile (*Q3*) by taking the value in position $(n + 1) - Q$. Again, if *Q* is not an integer, then average the two data values on either side.

5. Calculate the interquartile range by:

$$IQR = Q^3 - Q1 \qquad (2.11)$$

where IQR is the interquartile range, *Q3* is the value of the data point at the upper end of the third quartile, and *Q1* is the value of the data point at the upper end of the first quartile. Approximately 25% and 75% of the data will fall below and above *Q1* and *Q3*, respectively (Milton, 1999).

---

**Example 2.2**

For the example ground reaction force data, the sorted (ascending) values are shown in the rightmost column of table 2.5. The following demonstrates each of the steps for calculating interquartile range:

1. median location = (10 + 1) / 2 = 5.5 = 5
2. Q = (5 + 1) / 2 = 3
3. Q1 = value of data point in position 3 = 2,420 N
4. Q3 = value of data point in position (10 + 1) – 3 = 2,801 N
5. IQR = 2,801 – 2,420 = 381 N

---

Of the five discrete methods described for quantifying variability, the standard deviation and coefficient of variation variables are probably the most widely used in biomechanics and other movement science research. For example, numerous studies in the literature have used standard deviation (e.g., Bates, Osternig, Sawhill, and James, 1983; Ivry and Corcos, 1993; Schmidt et al., 1979) and coefficient of variation (e.g., Belli, Lacour, Komi, and Candau, 1995; Heiderscheit, 2000; Winter, 1992) for both discrete and continuous data. Movement science researchers have not used range, variance, and interquartile range frequently for quantifying variability for nonstatistical purposes.

## *Continuous Methods*

Collectively, the range, variance, standard deviation, coefficient of variation, and interquartile range quantities provide a thorough description of the variability of discrete variables across multiple performance trials. However, discrete variables do not capture all of the information relevant for understanding the variability of a movement pattern performed across time. Continuous methods (i.e., time series graphs, angle-angle plots, and others) provide an alternative for representing movement and movement variability as a function of time or other movement parameter and can represent both spatial and temporal characteristics (Hamill, Haddad, and McDermott, 2000). Many of the variability statistics used for discrete variables can be applied to the analysis of continuous variables. For example, multiple time series curves plotted on the same graph (e.g., figure 2.5, p. 46) can also be presented as a single ensemble or average curve plotted with the multiple-trial variability represented as a variability "band" (figure 2.6). The ensemble average is computed as the mean across trials for each data point (sample) in the curve, as shown in equation 2.12:

$$M_i = \frac{\sum\limits_{j=1}^{n} x_{ij}}{n} \tag{2.12}$$

where $M_i$ is the mean for the $i$th sample, $x_{ij}$ is the data value for the $i$th sample and $j$th trial, and $n$ is the number of trials. The variability band can be determined in several ways including point-by-point and curve-average methods. Point-by-point methods consist of computing the variability across all trials for each data sample (e.g., 10-trial standard deviation for data point 1, 10-trial standard deviation for data point 2), resulting in a unique variability value for each group of data points representing an instant in time. Curve-average methods compute the average variability across all samples, thus representing the total variability of the continuous curve as a single value. Both standard deviation and coefficient of variation quantities are commonly applied to both the point-by-point and curve-average methods. The standard deviation and coefficient of variation point-by-point formulas are shown in equations 2.13 and 2.14:

$$SD_i = \left( \frac{\sum\limits_{j=1}^{n} (x_{ij} - M_i)^2}{n-1} \right)^{1/2} \tag{2.13}$$

where $SD_i$ is the standard deviation for the $i$th sample, $x_{ij}$ is the data value for the $i$th sample and $j$th trial, $M_i$ is the 10-trial mean for the $i$th sample, and $n$ is the number of trials.

$$CV_i = (SD_i / M_i) \times 100 \tag{2.14}$$

where $CV_i$ is the coefficient of variation for the $i$th sample, and $SD_i$ and $M_i$ are the corresponding standard deviation and mean values for the $i$th sample.

The formulas for the standard deviation and coefficient of variation curve-average methods are shown in equations 2.15 and 2.16; the resulting quantities are equivalent

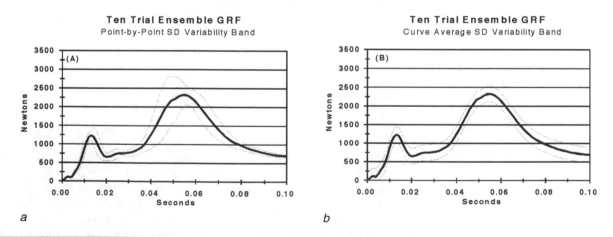

**Figure 2.6** Ten trial ensemble resultant ground reaction force time histories for landing. (a) Standard deviation variability band calculated using a point-by-point method. (b) Standard deviation variability band calculated using a curve average method.

Data from James, 1996.

to non-normalized and normalized root mean square standard deviation values, respectively, across the interval of interest (Winter, 1984).

$$SD_{avg} = \left( \frac{\sum_{i=1}^{k} SD_i^2}{k} \right)^{1/2}$$

(2.15)

where $SD_{avg}$ is the average of the individual point-by-point standard deviation values across all $k$ samples composing the continuous curve, and $SD_i$ is the standard deviation value for the $i$th sample.

$$CV_{avg} = \frac{SD_{avg}}{\frac{\sum_{i=1}^{k} |M_i|}{k}} \cdot 100$$

(2.16)

where $CV_{avg}$ is the average coefficient of variation for all $k$ samples composing the continuous curve, $SD_{avg}$ is the average standard deviation across all samples, and $M_i$ is the mean value for the $i$th sample (Winter, 1984).

One obtains the graphical representation of these variability values (i.e., variability band) by plotting the multiple-trial ensemble average curve along with two additional curves representing the ensemble average ($M_i$) plus and minus the standard deviation value for each data point ($SD_i$), that is, $M_i \pm SD_i$. Note that while the coefficient of variation values can be computed, the standard deviation values are best used for graphical representation of the variability band due to the congruence of units. Range, interquartile range, and other variability measures using the same units as the ensemble data also can be used graphically. Ensemble graphs for the ground reaction force data in the example presented earlier (figure 2.5, p. 46) are shown in figure 2.6. Figure 2.6$a$ is the ensemble graph with a point-by-point standard deviation variability band; figure 2.6$b$ is the same ensemble graph with the computed curve-average standard deviation band. The variability band computed using the point-by-point method varies in width across time, widening when the values for corresponding data points diverge and narrowing when the values converge. The width of the variability band at any location along the curve is evaluated only along the vertical (y) axis direction, but is influenced by both data magnitude and temporal deviations. The curve-average method maintains a constant width of the variability band at each data point in the y-axis direction. This band represents the average variability across all data points and thus is representative of the entire curve, but it fails to elicit detailed information about the variability occurring at specific locations along the curve.

Application of continuous methods is advisable for data that have been either normalized to a set number of data points between two critical events (e.g., 100% stride in running) or temporally aligned to a single critical event (e.g., ground contact in landing). The data used to create figure 2.6 (appendix B, table B.1) were temporally aligned at the initial ground contact event at the beginning of the landing phase.

While time series graphs (i.e., plot of a variable vs. time) are among the most commonly used methods for representing continuous biomechanical data, relative motion plots or angle-angle diagrams (i.e., plot of one angular variable vs. another) are also invaluable. Angle-angle diagrams graph the angular movement of one body segment or joint against another segment or joint (intra- or interlimb) and provide an excellent representation of the coordination between the involved variables (Enoka, 1994; Hamill et al., 2000; Sparrow, Donovan, van Emmerik, and Barry, 1987).

An angle-angle graphical depiction is particularly useful for evaluating continuous cyclical movement patterns in which multiple cycles (trials) are obtained and plotted on the same figure. An example of an angle-angle diagram for 10 strides of running is shown in figure 2.7, which plots the knee angle (proximal joint) on the x-axis and the ankle angle (distal joint) along the y-axis. The data used to create figure 2.7 are provided in appendix B, table B.2. Qualitatively, the variability across strides is evident as no one ankle-knee sequence corresponds perfectly to another.

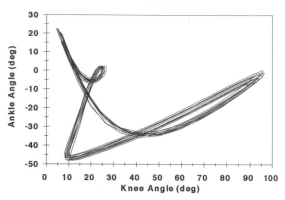

**Figure 2.7**   Ten-stride (cycle) ankle angle diagram for running. Zero degrees represents anatomic position. Positive values indicate dorsiflexion and flexion for the ankle and knee, respectively.

Data from Poklikuha, 2001.

Grieve (1968) is credited with devising the angle-angle diagram to evaluate walking patterns, and several others have since suggested methods for quantifying the coordination that is qualitatively observable in the relative patterns (Sidaway, Heise, and Schoenfelder-Zohdi, 1995; Sparrow et al., 1987; Whiting and Zernicke, 1982). Whiting and Zernicke (1982) used a pattern recognition chain-encoding technique adapted from Freeman (1961) to compute the degree of similarity among cycles. The chain-encoding technique was used to transform the continuous cyclical curves into a series of discrete digital elements. The measure of similarity between two patterns was determined by computation of a recognition coefficient, defined as the peak value of a cross-correlation function ($R_{xy}$) with values closer to 1.0 indicating greater similarity between the curves (Whiting and Zernicke, 1982). Hershler and Milner (1980), Barry (1982), and Sparrow and colleagues (1987) examined additional variables obtained from the chain-encoding technique, including pattern centroids, areas, heights, widths, and perimeter lengths as measures of similarity (Sparrow et al., 1987).

Variability in any one of these measures across movement cycles can be computed using discrete methods to represent the consistency of the performance. However, limitations in these techniques prevent their applicability in all instances of angle-angle analyses. For example, the cross-correlation technique computes the degree of linear relationship between the two angle dependent variables and as such assumes that the data fit a linear model. In reality, most angle-angle depictions of movement are more complex and are nonlinear. Variables such as angle-angle centroids, areas, heights, widths, and perimeter lengths permit objective comparison among individual movement cycles; but collapsing a complex angle-angle relationship into one or more single-quantity descriptors obscures information about the nature of the coordinative pattern as it evolves over time.

Hamill and colleagues (2000) explained a measure of relative motion between the joint variables on an angle-angle plot called the coupling angle. The coupling angle is the orientation from the right horizontal axis of the vector between two adjacent points on the angle-angle diagram. Point-by-point calculation of the coupling angle results in multiple values to describe a single angle-angle movement cycle. Additionally, the length of the vector between the two adjacent points on the graph reflects the directional concentration of the data (Hamill et al., 2000). Both quantities reflect the relative amount of movement between the involved joints, and quantification of differences across cycles indicates the variability in the movement patterns across time. However, quantification of the mean and standard deviation of the coupling angle requires special consideration, as this variable is considered a circular variable (see chapter 6 on circular statistics).

More conventionally, one can compute the variability for each joint angle variable on the angle-angle plot independently using a discrete method applied to the continuous time series curve as described previously (i.e., ensemble curve with a variability band). The individual time series curves along with the ensemble graphs for the ankle and knee are shown in figure 2.8, and the data are provided in appendix B, table B.2. However, the variability in each of these graphs is one-dimensional, as there is deviation only in the y-axis direction. Hence, the variability as depicted in the time series graph does not fully represent the combined variability when the ankle and knee are plotted in two dimensions on the angle-angle diagram. One solution is to compute and graph an angle-angle ensemble curve, complete with a variability band that is free to vary in two dimensions. One obtains the angle-angle ensemble curve by first computing a multiple-trial time series ensemble for each variable, then plotting the data from the two individual ensemble curves on the angle-angle diagram (i.e., the angle-angle ensemble diagram is actually the ensemble angle-ensemble angle diagram). Examples of this ensemble angle-angle diagram are shown in figure 2.9, along with graphical depictions of two different variability bands. The data used to create these graphs are provided in appendix B, tables B.2, B.3, B.4, and B.5.

Sidaway and colleagues (1995) presented a method for quantifying the variability of the angle-angle diagram that is appropriate for both linear and nonlinear data and is analogous to the curve-average variability method described previously for single-variable time series data. The method involves computing the root mean square error of each individual cycle from the angle-angle ensemble mean, averaging the error across cycles, then normalizing the root mean square average value to the resultant excursion of the mean angle-angle curve over the entire cycle. The result of this

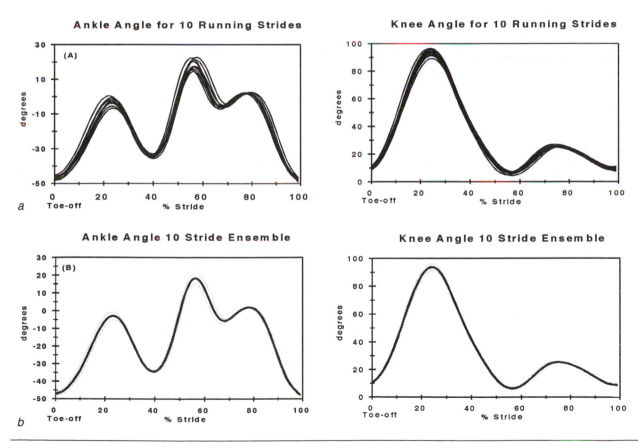

**Figure 2.8**   (a) Ankle and knee angle time histories for 10 complete and consecutive running strides. (b) Ensemble average values for the corresponding ankle and knee data. The standard deviation variability band was calculated using a point-by-point method.

Data from Poklikuha, 2001.

**Ensemble Angle-Angle Diagram**
Curve-Average Variability Band

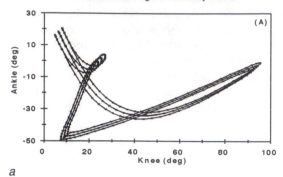

a

**Ensemble Angle-Angle Diagram**
Point-by-Point Variability Band

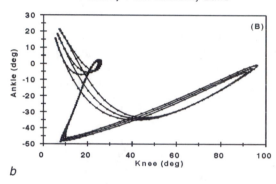

b

**Figure 2.9** Ensemble ankle-knee angle-angle diagrams. *(a)* Variability band calculated using a curve-average method representing the root mean square error across the entire curve. *(b)* Variability band calculated using a point-by-point method representing the root mean square error at each data point.

Data from Poklikuha, 2001.

analysis is shown graphically in figure 2.9a (appendix B, tables B.3 and B.4), and the steps are summarized as follows:

1. Compute the time-varying ensemble average of each variable as shown in equation 2.12 (p. 50), and use the two resulting data series to create a mean angle-angle plot.

2. For each instant in time (data sample) for each movement cycle, calculate the two-dimensional resultant deviation between the data value of the individual cycle and the value of the multiple-cycle mean as shown in equation 2.17:

$$R_{ij} = ((x_{ij} - Mx_i)^2 + (y_{ij} - My_i)^2)^{1/2} \quad (2.17)$$

where $R_{ij}$ is the resultant deviation for the $i$th sample and $j$th cycle; $x_{ij}$ and $y_{ij}$ are the data values for the $i$th sample and $j$th cycle for the x- and y-axis variables, respectively; and $Mx_i$ and $My_i$ are the corresponding $x$ and $y$ multiple-cycle mean values for the $i$th sample.

3. Compute the root mean square deviation for each cycle as shown in equation 2.18:

$$RMS_j = \left( \frac{\sum_{i=1}^{k} R_{ij}^2}{k} \right)^{1/2} \quad (2.18)$$

where $RMS_j$ is the root mean square deviation for the $j$th cycle; $R_{ij}$ is the resultant deviation for the $i$th sample and $j$th cycle; and $k$ is the number of samples.

4. Calculate the average root mean square value across all cycles as shown in equation 2.19:

$$RMS_{avg} = \frac{\sum_{j=1}^{n} RMS_j}{n} \quad (2.19)$$

where $RMS_{avg}$ is the multiple-cycle average root mean square; $RMS_j$ is the root mean square value for the $j$th cycle; and $n$ is the number of cycles.

5. Normalize the $RMS_{avg}$ value to the resultant excursion of the mean angle-angle plot over the entire cycle as shown in equation 2.20:

$$RMS_N = RMS_{avg} / L \quad (2.20)$$

where $RMS_N$ is the normalized root mean square value; $RMS_{avg}$ is the multiple-cycle average root mean square; and $L$ is the resultant excursion of the mean angle-angle curve (Sidaway et al., 1995).

This final normalization step was suggested to account for possible differences in joint ranges of motion that might occur under different conditions or between individuals (Sidaway et al., 1995). This step is conceptually similar to dividing the standard deviation by the mean value when computing a coefficient of variation. The appropriateness of the normalization procedure is a decision that the researcher should make based on the nature of the data and the characteristics of interest. Hence, the normalization step may not be necessary in all instances. Therefore, for the current use the $RMS_{avg}$ (equation 2.19) is considered the fundamental quantity that represents the composite absolute variability across all cycles.

To graph the variability band along with the ensemble $RMS_{avg}$ curve, the orthogonal $x$ and $y$ vector components of the $RMS_{avg}$ value are needed. The components ($RMSx_{avg}$ and $RMSy_{avg}$) can be added to the corresponding mean values to compute the coordinates necessary to plot the mean "plus" and "minus" curves. For example, each coordinate $(x_i, y_i)$ for the "plus" curve is calculated by $Mx_i + RMSx_{avg}$ and $My_i + RMSy_{avg}$ for the $x$ and $y$ components, respectively. Coordinates for the "minus" curve are determined similarly, except that the appropriate root mean square value is subtracted from each $x$ and $y$ mean. The inconvenience in this process is the calculation of $RMSx_{avg}$ and $RMSy_{avg}$. Conceptually, $RMSx_{avg}$ and $RMSy_{avg}$ can be computed using the cosine and sine, respectively, of the average angle of $RMS_{avg}$ from the right horizontal axis. This angle can be determined using the following steps:

1. For each data sample for each movement cycle, calculate the one-dimensional (x or y) squared-deviation between the data value of the individual cycle and the value of the multiple-cycle mean as shown in equations 2.21 and 2.22:

$$Rx_{ij}^2 = (x_{ij} - Mx_i)^2 \tag{2.21}$$

$$Ry_{ij}^2 = (y_{ij} - My_i)^2 \tag{2.22}$$

where $Rx_{ij}$ and $Ry_{ij}$ are the deviations in the x and y directions, respectively, for the $i$th sample and $j$th cycle, $x_{ij}$ and $y_{ij}$ are the data values for the $i$th sample and $j$th cycle for the x- and y-axis variables, and $Mx_i$ and $My_i$ are the corresponding multiple-cycle mean values for the $i$th sample.

2. Compute the root mean square deviation for each cycle and component:

$$\text{RMS}x_j = \left( \frac{\sum_{i=1}^{k} Rx_{ij}^2}{k} \right)^{1/2} \tag{2.23}$$

$$\text{RMS}y_j = \left( \frac{\sum_{i=1}^{k} Ry_{ij}^2}{k} \right)^{1/2} \tag{2.24}$$

where $RMSx_j$ and $RMSy_j$ are the component root mean square deviations for the $j$th cycle, $Rx_{ij}$ and $Ry_{ij}$ are the component deviations for the $i$th sample and $j$th cycle, and $k$ is the number of samples.

3. Calculate the resultant root mean square value for each cycle:

$$RMS_j = (RMSx_j^2 + RMSy_j^2)^{1/2} \tag{2.25}$$

where $RMS_j$ is the resultant root mean square value for the $j$th cycle, and $RMSx_j$ and $RMSy_j$ are the individual component root mean square values for the same cycle. Notice that the values for $RMS_j$ using this approach are equal to those as computed in equation 2.18.

4. Using $RMS_j$, compute the average resultant root mean square (i.e., $RMS_{avg}$) across all cycles as shown in equation 2.19.

5. For each cycle, calculate the angle ($\Phi$) of the resultant $RMS_j$:

$$\Phi_j = \tan^{-1} (RMSy_j / RMSx_j) \qquad (2.26)$$

where $\Phi_j$ is the angle from the right horizontal of $RMS_j$ and $RMSx_j$ and $RMSy_j$ are the orthogonal $x$ and $y$ components of $RMS_j$. The value of the angle will be in the range $0 \leq \Phi \leq 90$, indicating that $RMS_j$ falls within the first quadrant. This angle does not represent the average angle of the individual deviations ($R_{ij}$), but serves to weight the magnitude of the $x$ and $y$ components in order to determine the location of the coordinates of the curves comprising the variability band.

6. Compute the average of $\Phi_j$ across all $n$ cycles, as shown:

$$\phi_{avg} = \frac{\sum_{j=1}^{n} \phi_j}{n} \qquad (2.27)$$

Because $\Phi$ is constrained to the range $0°$ to $90°$, it is not considered a circular variable; thus a simple average is sufficient to represent all cycles collectively.

7. Use $\Phi_{avg}$ to compute the $x$ and $y$ components of $RMS_{avg}$:

$$RMSx_{avg} = RMS_{avg} \cdot \cos (\Phi_{avg}) \qquad (2.28)$$

$$RMSy_{avg} = RMS_{avg} \cdot \sin (\Phi_{avg}) \qquad (2.29)$$

These values are added to and subtracted from the x- and y-coordinates of the ensemble curve to produce the "plus" and "minus" coordinates of variability band. The width of the band ($RMS_{avg}$) remains constant in the direction of $\Phi_{avg}$ across the ensemble curve.

One disadvantage of the average root mean square quantity is that a single value is used to represent the variability among a group of cycles across time. For movements in which the amount of variability may vary with time, the average root mean square value can mask important information about changes in the amount of variability as the movement is executed. An alternative is to use a point-by-point method that computes a root mean square value for each data sample, thus permitting the characterization of changes in variability as the movement evolves over time. The advantage to this technique is that amounts of variability during different parts of the movement cycle (or at discrete instants in time of the cycle) can be evaluated. The result of such an analysis is shown graphically in figure 2.9*b* (appendix B, table B.5). The steps are summarized here. The first two steps are identical to those used to calculate the average root mean square value as described by Sidaway and colleagues (1995). The remainder of the steps used established single-variable statistical methods applied to two-dimensional variables and are quantitatively similar to the methods for analyzing movement accuracy, bias, and consistency in two dimensions described by Hancock, Butler, and Fischman (1994).

1. Compute the time-varying ensemble average of each variable as shown in equation 2.12 (p. 50) and use the two resulting data series to create a mean angle-angle plot.

2. For each instant in time (data sample) for each movement cycle, calculate the two-dimensional resultant deviation between the data value of the individual cycle and the value of the multiple-cycle mean (equation 2.17, p. 54).

3. Compute the root mean square deviation value across cycles for each data sample:

$$\text{RMS}_i = \left( \frac{\sum_{j=1}^{n} R_{ij}^2}{n} \right)^{1/2} \tag{2.30}$$

where $RMS_i$ is the root mean square deviation for the $i$th sample, $R_{ij}$ is the resultant deviation for the $i$th sample and $j$th cycle, and $n$ is the number of cycles. This $RMS_i$ value is equivalent to computing the magnitude of the resultant standard deviation of each sample from the individual joint angle component standard deviation values. In other words,

$$RMS_i = SD_R = (SD_{xi}^2 + SD_{yi}^2)^{1/2} \tag{2.31}$$

where $RMS_i$ is the root mean square across trials for each $i$th sample (as described previously), $SD_R$ is the resultant standard deviation for the two angle variables, and $SD_{xi}$ and $SD_{yi}$ are the population standard deviation values (i.e., population formula; $n$ degrees of freedom, instead of $n - 1$) for the $x$ and $y$ variables, respectively, at the $i$th sample. The sample standard deviation formula can also be used to compute $SD_{xi}$ and $SD_{yi}$, and is appropriate when the cycles obtained experimentally are considered to represent a sample of the larger population of cycles possible by the individual. If the sample formula for standard deviation is used, $SD_R$ no longer equals $RMS_i$, but instead is a sample approximation.

4. Graph the values by adding and subtracting $SD_R$ from the ensemble angle-angle curve data values to compute values for two additional lines, one consisting of the ensemble data plus one standard deviation and the other consisting of the ensemble data minus one standard deviation. This addition or subtraction can be done using vector algebra, with each axis direction treated independently. For example, the x-coordinates of the line in the "plus" direction can be determined by

$$x_{pi} = M_{xi} + SD_{xi} \tag{2.32}$$

while the x-coordinate data in the "minus" direction can be determined by

$$x_{mi} = M_{xi} - SD_{xi} \tag{2.33}$$

where $x_{pi}$ and $x_{mi}$ are the x-coordinates for points on the "plus" and "minus" variability lines, respectively, for each $i$th data sample, and $M_{xi}$ and $SD_{xi}$ are the multiple-cycle mean and standard deviation values for the $i$th data point in the x-axis direction. The corresponding formulas for the y direction are shown in equations 2.34 and 2.35:

$$y_{pi} = M_{yi} + SD_{yi} \tag{2.34}$$

$$y_{mi} = M_{yi} - SD_{yi} \tag{2.35}$$

where $y_{pi}$ and $y_{mi}$ are the y-coordinates for points on the "plus" and "minus" variability lines, respectively, for each $i$th data point, and $M_{yi}$ and $SD_{yi}$ are the multiple-cycle mean and standard deviation values for the $i$th data point in the y-axis direction.

One can evaluate the magnitudes of the resulting $RMS_i$ or $SD_R$ values directly for the variability of movement or normalize them to some relevant quantity. For example, the normalizing value could account for the movement range of the individual joint variables (e.g., resultant joint excursion) as suggested by Sidaway and colleagues (1995). Similarly, one might also use an appropriate mean value (e.g., resultant mean) of the coupled variables. The choice of the appropriate normalizing coefficient is a function of the research question and whether the absolute or relative variability is the quantity of interest.

Another type of relative motion plot especially common in analysis of nonlinear data is the phase portrait, which plots a variable of interest (e.g., knee angle) versus its derivative (e.g., knee angular velocity). Phase portraits are used as indicators of coordination both within and between joints or body segments. A common variable used to quantify the phase portrait is the phase angle, which is the angle from the right horizontal axis of the vector connecting the origin of the plot to each x, y data point. Phase angles can be used to quantify the phasic coordination of interjoint or intersegment motions by calculation of the continuous relative phase. Briefly, continuous relative phase is calculated as the point-by-point difference in the phase angle of one joint or segment compared to the phase angle of another joint or segment (Byrne, Stergiou, Blanke, Houser, and Kurz, 2002; Hamill et al., 2000; Stergiou, Jensen, Bates, Scholten, and Tzetzis, 2001). Although one must consider several issues (e.g., normalization methods, sinusoidal characteristics of the curve, circular data) relative to computation of continuous relative phase, both mean (e.g., mean absolute relative phase) and variability (e.g., standard deviation) of the continuous relative phase can be computed on a point-by-point basis for the coupled joints or segments (Byrne et al., 2002; Hamill et al., 2000; Stergiou et al., 2001). The variability computed from continuous relative phase is another reflection of the total variability of the involved systems as the movement pattern evolves over time.

# Summary

The biological variability present within the neuromotor system is believed to be a function of both the deterministic evolutionary processes of the movement and error. Additionally, variations in the performance environment and task, which both function externally to the individual, influence the amount and characteristics of the movement variability observed. Measurement of movement introduces an additional source of variability (measurement error), which can be difficult to distinguish from the underlying biological processes. Investigators in a variety of human movement science disciplines have examined the sources and nature of variability and have learned much in the past three decades, although much work still lies ahead. The specific research questions examined determine which variables are most appropriate for quantifying variability. In some cases, examination of the total variability (excluding, controlling, or minimizing measurement error) is appropriate, and in other situations the individual deterministic and stochastic components of variability are of greater interest. From an applied perspective, the consequence of variability can be examined relative to both biomechanical function (e.g., performance, injury) and methodology. Several techniques (parameters) can be used to quantify total variability (e.g., range, standard deviation, coefficient of variation, interquartile range), and these techniques can be applied to both discrete and continuous data in both traditional and nontraditional ways. Investigators are encouraged to seek and use the appropriate methodology for examining movement and movement variability relative to the underlying theoretical construct utilized and the specific research questions asked.

# Work Problems

**1**

Using the knee angle data given (first 10% of stride for three running cycles), calculate the following: (a) point-by-point standard deviation and coefficient of variation values and (b) curve-average standard deviation and coefficient of variation values.

| Knee Angle (in Degrees) for Running | | | | |
|---|---|---|---|---|
| | **% Stride** | **Cycle 1** | **Cycle 2** | **Cycle 3** |
| Toe-off | 0 | 11.3 | 11.2 | 10.6 |
| | 1 | 12.5 | 12.9 | 12.1 |
| | 2 | 14.0 | 14.9 | 13.8 |
| | 3 | 15.8 | 17.4 | 16.0 |
| | 4 | 18.3 | 20.5 | 18.8 |
| | 5 | 21.1 | 24.0 | 22.0 |
| | 6 | 24.4 | 28.0 | 35.6 |
| | 7 | 28.0 | 32.4 | 29.6 |
| | 8 | 32.1 | 37.1 | 33.9 |
| | 9 | 36.5 | 42.3 | 38.5 |

**2**

Using the knee angle ($x$) data given in work problem 1 and the ankle angle ($y$) data given next, calculate the following: (a) point-by-point variability ($RMS_i$) and (b) curve-average variability ($RMS_{avg}$) values.

| Ankle Angle (in Degrees) for Running | | | | |
|---|---|---|---|---|
| | **% Stride** | **Cycle 1** ($y$) | **Cycle 2** ($y$) | **Cycle 3** ($y$) |
| Toe-off | 0 | −46.1 | −45.4 | −47.3 |
| | 1 | −46.0 | −44.9 | −46.8 |
| | 2 | −45.8 | −44.2 | −46.1 |
| | 3 | −45.2 | −43.3 | −45.1 |
| | 4 | −44.4 | −42.0 | −43.7 |
| | 5 | −43.3 | −40.4 | −42.1 |
| | 6 | −42.0 | −38.5 | −40.3 |
| | 7 | −40.4 | −36.4 | −38.4 |
| | 8 | −38.7 | −33.9 | −36.2 |
| | 9 | −36.7 | −31.2 | −33.8 |

# Suggested Readings and Other Resources

Glass, L., and Mackey, M.C. (1988). *From Clocks to Chaos*. Princeton, NJ: Princeton University Press. Discusses the cyclical nature of many biological systems and the rhythms of life. Provides a basis for understanding physiological variability and its relationship to health and disease.

Mullin, T. (1993). *The Nature of Chaos*. Oxford: Clarendon Press. Applications and examples of nonlinear mathematics to a variety of scientific problems in biology, chemistry, engineering, and physics.

Newell, K.M., and Corcos, D.M. (1993). *Variability and Motor Control*. Champaign, IL: Human Kinetics. A comprehensive examination of intra-individual movement variability from a motor control perspective. Considers both variability as noise and variability as a reflection of system dynamics.

Williams, G.P. (1997). *Chaos Theory Tamed*. Washington, DC: Joseph Henry Press. A primer for scientists and students working outside of the field of mathematics who desire an understanding of the concepts and vocabulary of chaos.

Winter, D.A. (1990). *Biomechanics and Motor Control of Human Movement*. New York: Wiley InterScience. A graduate-level biomechanics text that integrates methodological approaches with biological function. It contains a concise explanation of digital filtering.

# References

Aron, A., and Aron, E.N. (1994). *Statistics for Psychology*. Englewood Cliffs, NJ: Prentice Hall.

Babloyantz, A., and Destexhe, A. (1986). Strange attractors in the human cortex. In L. Rensing, U. Heiden, and M. Mackey (eds.), *Temporal Disorder in Human Oscillatory Systems: Proceedings of an International Symposium* (pp. 48-56). New York: Springer Verlag.

Barry, E.B. (1982). *Characterisations of Gait*. Unpublished master's thesis, University of New South Wales.

Bates, B.T. (1996). Single-subject methodology: An alternative approach. *Medicine and Science in Sports and Exercise, 28,* 631-638.

Bates, B.T., De Vita, P., and Kinoshita, H. (1983). The effect of intra-individual variability on sample size. In B. Nigg and B. Kerr (eds.), *Biomechanical Aspects of Sport Shoes and Playing Surfaces* (pp. 191-198). Calgary, Alberta: University Press.

Bates, B.T., Osternig, L.O., Sawhill, J.A., and James, S.L. (1983). An assessment of subject variability, subject-shoe interaction and evaluation of running shoes using ground reaction force data. *Journal of Biomechanics, 16,* 181-191.

Belli, A., Lacour, J.R., Komi, P.V., and Candau, R. (1995). Mechanical step variability during treadmill running. *European Journal of Applied Physiology, 70,* 510-517.

Bernstein, N. (1967). *Coordination and Regulation of Movement*. New York: Pergamon Press.

Bingham, G.P., Schmidt, R.C., Turvey, M.T., and Rosenblum, L.D. (1991). Task dynamics and resource dynamics in the assembly of a coordinated rhythmic activity. *Journal of Experimental Psychology, 17,* 359-381.

Buzzi, U.H., Stergiou, N., Giakas, G., Dierks, T.A., and Georgoulis, A.D. (2001). The effect of ACL reconstruction on locomotor variability. *Proceedings of the 25th Annual Meeting of the American Society of Biomechanics* (pp. 295-296). San Diego: University of California San Diego.

Byrne, J.E., Stergiou, N., Blanke, D., Houser, J.J., and Kurz, M.J. (2002). Comparison of gait patterns between young and elderly women: An examination of coordination. *Perceptual and Motor Skills, 94,* 265-280.

Cavanagh, P.R., and Lafortune, M.A. (1980). Ground reaction forces in distance running. *Journal of Biomechanics, 13,* 397-406.

Collins, J.J., Imhoff, T.T., and Grigg, P. (1996). Noise-enhanced information transmission in rat Sa1 cutaneous mechanoreceptors via aperiodic stochastic resonance. *Journal of Neurophysiology, 76,* 642-645.

Dufek, J.S., Bates, B.T., Stergiou, N., and James, C.R. (1995). Interactive effects between group and single-subject response patterns. *Human Movement Science, 14,* 301-323.

Enoka, R.M. (1994). *Neuromechanical Basis of Kinesiology*. Champaign, IL: Human Kinetics.

Freeman, H. (1961). A technique for the classification and recognition of geometric patterns. *Proceedings of the 3rd International Congress on Cybernetics* (pp. 348-368). Namur, Belgium: International Association of Cybernetics.

Gallez, D., and Babloyantz, A. (1991). Predictability of human EEG: A dynamical approach. *Biological Cybernetics, 64,* 381-391.

Gallistel, C.R. (1981). *The Organization of Action: A New Synthesis*. Hillsdale, NJ: Erlbaum.

Glass, L., and Mackey, M.C. (1988). *From Clocks to Chaos*. Princeton, NJ: Princeton University Press.

Goldberger, A., Bhargava, V., West, B., and Mandell, A.J. (1986). Some observations on the question: Is ventricular fibrillation "chaos"? *Physica, 19(D),* 282-289.

Goldberger, A.L., Rigney, D.R., and West, B.J. (1990). Chaos and fractals in human physiology. *Scientific American, 262,* 42-49.

Goldberger, A.L., and West, B.J. (1987). Chaos in physiology: Health or disease? In H. Degn, A.V. Holden, and L.F. Olsen (eds.), *Chaos in Biological Systems* (pp. 1-4). New York: Plenum Press.

Grieve, D.W. (1968). Gait patterns and the speed of walking. *Biomedical Engineering, 3,* 119-122.

Hamill, J., and McNiven, S.L. (1990). Reliability of selected ground reaction force parameters during walking. *Human Movement Science, 9,* 117-131.

Hamill, J., Haddad, J.M., and McDermott, W.J. (2000). Issues in quantifying variability from a dynamical systems perspective. *Journal of Applied Biomechanics, 16,* 407-418.

Hancock, G.R., Butler, M.S., and Fischman, M.G. (1994). The ABC's of two-dimensional error scores: Measures and analyses of accuracy, bias, and consistency. Symposium paper presented at the meeting of the American Alliance for Health, Physical Education, Recreation and Dance, Denver.

Heiderscheit, B.C. (2000). Movement variability as a clinical measure for locomotion. *Journal of Applied Biomechanics, 16,* 419-427.

Hershler, C., and Milner, M. (1980). Angle-angle diagrams in the assessment of locomotion. *American Journal of Physical Medicine, 59,* 109-125.

Higgins, J.R. (1977). *Human Movement: An Integrated Approach.* St. Louis: Mosby.

Ivry, R., and Corcos, D.M. (1993). Slicing the variability pie: Component analysis of coordination and motor dysfunction. In K.M. Newell and D.M. Corcos (eds.), *Variability and Motor Control* (pp. 415-447). Champaign, IL: Human Kinetics.

James, C.R. (1991). *Effects of Fatigue on Mechanical and Muscular Components of Performance During Drop Landings.* Unpublished master's thesis, University of Oregon, Eugene.

James, C.R. (1996). *Effects of Overuse Injury Proneness and Task Difficulty on Joint Kinetic Variability During Landing.* Unpublished doctoral dissertation, University of Oregon, Eugene.

James, C.R., and Bates, B.T. (1997). Experimental and statistical design issues in human movement research. *Measurement in Physical Education and Exercise Science, 1*(1), 55-69.

James, C.R., Dufek, J.S., and Bates, B.T. (2000). Effects of injury proneness and task difficulty on joint kinetic variability. *Medicine and Science in Sports and Exercise, 32*(11), 1833-1844.

Jensen, J.L., and Phillips, S.J. (1991). Variations on the vertical jump: Individual adaptations to changing task demands. *Journal of Motor Behavior, 23,* 63-74.

Keele, S.W., Davidson, M., and Hayes, A. (1998). Sequential representation and the neural basis of motor skills. In J.P. Piek (ed.), *Motor Behavior and Human Skill* (pp. 3-28). Champaign, IL: Human Kinetics.

Kelso, J.A.S., and Ding, M. (1993). Fluctuations, intermittency, and controllable chaos in biological coordination. In K.M. Newell and D.M. Corcos (eds.), *Variability and Motor Control* (pp. 291-316). Champaign, IL: Human Kinetics.

Kelso, J.A.S., Scholz, J.P., and Schoner, G. (1986). Nonequilibrium phase transitions in coordinated biological motion: Critical fluctuations. *Physics Letters A, 118,* 279-284.

Kelso, J.A.S., and Tuller, B. (1984). A dynamical basis for action systems. In M. Gazziniga (ed.), *Handbook of Cognitive Neuroscience* (pp. 321-356). New York: Plenum Press.

Keppel, G. (1982). *Design and Analysis: A Researcher's Handbook.* Englewood Cliffs, NJ: Prentice Hall.

Lees, A., and Bouracier, J. (1994). The longitudinal variability of ground reaction forces in experienced and inexperienced runners. *Ergonomics, 37,* 197-206.

Lundy-Ekman, L., Ivry, R., Keele, S., and Woollacott, M. (1991). Timing and force control deficits in clumsy children. *Journal of Cognitive Neuroscience, 3,* 370-377.

McCaw, S.T. (1989). *Bilateral Lower Extremity Function During the Support Phase of Running.* Unpublished doctoral dissertation, University of Oregon, Eugene.

Milton, J.S. (1999). *Statistical Methods in the Biological and Health Sciences.* Boston: McGraw-Hill.

Mullin, T. (1993). A dynamical systems approach to time series analysis. In T. Mullin (ed.), *The Nature of Chaos* (pp. 23-50). Oxford: Clarendon Press.

Newell, K.M., and Corcos, D.M. (1993). Issues in variability and motor control. In K.M. Newell and D.M. Corcos (eds.), *Variability and Motor Control* (pp. 1-12). Champaign, IL: Human Kinetics.

Newell, K.M., and Slifkin, A.B. (1998). The nature of movement variability. In J.P. Piek (ed.), *Motor Behavior and Human Skill* (pp. 143-160). Champaign, IL: Human Kinetics.

Newell, K.M., van Emmerik, R.E.A., and Sprague, R.L. (1993). Stereotypy and variability. In K.M. Newell and D.M. Corcos (eds.), *Variability and Motor Control* (pp. 37-52). Champaign, IL: Human Kinetics.

Nigg, B.M., Eberle, G., Freg, D., Luethi, S., Segesser, B., and Weber, B. (1978). Gait analysis and sport shoe construction. In E. Asmussen and K. Jorgensen (eds.), *Biomechanics VI-A* (pp. 303-309). Baltimore: University Park Press.

Nordin, M., and Frankel, V.H. (2001). *Basic Biomechanics of the Musculoskeletal System* (3rd ed.). Philadelphia: Lippincott Williams & Wilkins.

Poklikuha, G. (2001). *Effects of Cycling Pedal Rate on Running Kinematics and Neuromuscular Timing in Triathletes.* Unpublished master's thesis, Texas Tech University, Lubbock.

Schmidt, R.A., Zelaznik, H., Hawkins, B., Frank, J.S., and Quinn, J.T. (1979). Motor-output variability: A theory for the accuracy of rapid motor acts. *Psychological Review, 86,* 415-451.

Scholz, J.P. (1990). Dynamic pattern theory—some implications for therapeutics. *Physical Therapy, 70,* 827-843.

Schoner, G., and Kelso, J.A.S. (1988). Dynamic pattern generation in behavioral and neural systems. *Science, 239,* 1513-1520.

Schroeder, M. (1991). *Fractals, Chaos, Power Laws: Minutes from an Infinite Paradise.* New York: Freeman.

Sidaway, B., Heise, G., and Schoenfelder-Zohdi, B. (1995). Quantifying the variability of angle-angle plots. *Journal of Human Movement Studies, 29,* 181-197.

Sparrow, W.A., Donovan, E., van Emmerik, R., and Barry, E.B. (1987). Using relative motion plots to measure changes in intra-limb and inter-limb coordination. *Journal of Motor Behavior, 19,* 115-129.

Stam, K.J., Tavy, D.L.J., Jelles, B., Achterbeeke, H.A.M., Slaets, J.P.J., and Keunen, R.W.M. (1994). Non-linear dynamical analysis of multichannel EEG: Clinical applications in dementia and Parkinson's disease. *Brain Topography, 7,* 141-150.

Stergiou, N., Buzzi, U.H., Hageman, P.A., and Heidel, J. (2000). A chaotic analysis of gait parameters in different age groups. *Proceedings of the 24th Annual Meeting of the American Society of Biomechanics* (pp. 75-76). Chicago: University of Illinois at Chicago.

Stergiou, N., Jensen, J.L., Bates, B.T., Scholten, S.D., and Tzetzis, G. (2001). A dynamical systems investigation of lower extremity coordination during running over obstacles. *Clinical Biomechanics, 16,* 213-221.

Sveistrup, H., Davis, H., Derrick, T., James, R., and Bates, B.T. (1989). Effects of repeated trials in vertical ground reaction force measurements and lower limb IEMG profiles of a drop landing. Presented to the Canadian Association of Sports Sciences. Montreal, Quebec, November.

Thomas, J.R., and Nelson, J.K. (2001). *Research Methods in Physical Activity* (4th ed.). Champaign, IL: Human Kinetics.

Tsuji, H., Venditti, F.J., Manders, E.S., Evans, J.C., Larson, M.G., Feldman, C.L., and Levy, D. (1994). Reduced heart rate variability and mortality risk in an elderly cohort: The Framingham heart study. *Circulation, 90,* 878-883.

Turvey, M.T. (1990, August). Coordination. *American Psychologist,* pp. 938-953.

van Emmerik, R.E.A., Sprague, R.L., and Newell, K.M. (1993). Quantification of postural sway profiles in tardive dyskinesia. *Movement Disorders, 8,* 305-314.

van Emmerik, R.E.A., and van Wegen, E.E.H. (2000). On variability and stability in human movement. *Journal of Applied Biomechanics, 16,* 394-406.

Whiting, H.T.A. (1984). *Human Motor Actions: Bernstein Reassessed.* New York: North-Holland.

Whiting, W.C., and Zernicke, R.F. (1982). Correlation of movement patterns via pattern recognition. *Journal of Motor Behavior, 14,* 135-142.

Whiting, W.C., and Zernicke, R.F. (1998). *Biomechanics of Musculoskeletal Injury.* Champaign, IL: Human Kinetics.

Williams, K.R. (1993). Biomechanics of distance running. In M.D. Grabiner (ed.), *Current Issues in Biomechanics* (pp. 3-31). Champaign, IL: Human Kinetics.

Williams, G.P. (1997). *Chaos Theory Tamed.* Washington, DC: Joseph Henry Press.

Winter, D.A. (1984). Kinematic and kinetic patterns in human gait: Variability and compensating effects. *Human Movement Science, 3,* 51-76.

Winter, D.A. (1990). *Biomechanics and Motor Control of Human Movement.* New York: Wiley InterScience.

Winter, D.A. (1992). Foot trajectory in human gait: A precise and multifactorial motor control task. *Physical Therapy, 72,* 45-56.

Zanone, P.G., and Kelso, J.A.S. (1992). Learning and transfer as dynamical paradigms for behavioral change. In G.E. Stelmach and J. Requin (eds.), *Tutorials in Motor Behavior, II* (pp. 563-582). Amsterdam: North-Holland.

Zernicke, R.F., and Loitz, B.J. (1992). Exercise-related adaptations in connective tissue. In P. Komi (ed.), *Strength and Power in Sport* (pp. 77-95). Champaign, IL: Human Kinetics.

# CHAPTER 3

# Nonlinear Tools in Human Movement

Nicholas Stergiou, PhD, Ugo H. Buzzi, MS, Max J. Kurz, MS, and Jack Heidel, PhD

Variability is inherent within all biological systems and can be characterized as the normal changes that occur in motor performance across multiple repetitions of a task. Until recently, variability was interpreted as the result of random processes (**noise**; Glass and Mackey, 1988). However, recent literature from a number of scientific domains has shown that many phenomena previously described as noisy are actually the result of nonlinear interactions and have a **deterministic origin** (Gleick, 1987; Glass and Mackey, 1988; Amato, 1992). Thus, one can gain important information regarding the system's behavior by examining the "noisy" component of the measured signal.

Variability has also been linked to the health of a biological system. Specifically, Goldberger and colleagues (1984, 1988) observed different variability patterns in heart rhythms among healthy and diseased patients and suggested that a healthy system has a certain amount of inherent variability. This healthy variability is not random but contains order, and can be characterized via nonlinear mathematical descriptors. Therefore, nonlinear methods have been used increasingly to describe complex conditions that cannot be well characterized by linear techniques. For example, nonlinear analysis has recently appeared in research on heart rate irregularities, sudden cardiac death syndrome, blood pressure control, brain ischemia, epileptic seizures, and several other phenomena (Babloyantz and Destexhe, 1986; Goldberger and West, 1987; Skarda and Freeman, 1987; Goldberger et al., 1988; Wagner et al., 1996; Goldstein et al., 1998; Lanza et al., 1998; Toweill and Goldstein, 1998; Buchman et al., 2001; Slutzky et al., 2001) for the purposes of understanding the biological complexity of these conditions and eventually developing prognostic and diagnostic tools.

Nonlinear methods that are being used to examine variability in biological rhythms such as heart rate or blood pressure also may be useful in examining human movement and its complexity. The study of variability in human movement can lead to a better understanding of the holistic behavior of the motor system. Davis (1993) identified the following as questions that need to be answered regarding variability in human movement.

1. What is the difference, if any, between the variability associated with learning a new skill and the variability associated with disease/health?

2. What are the sources of variability, and how do they interact in the production of the observed variation in movement?

3. Does variability reflect the stability of the motor system?

4. How noisy can a motor system be and still produce stable movement patterns?

5. What are the stability-efficiency compromises in the system?

6. Are there characteristics of within-subject variability that can be used to build better statistical decision tools?

To answer these questions, human movement researchers have recently used techniques from nonlinear dynamics as the methodological framework. This has been most evident in research on gait and postural sway. The use of nonlinear dynamics has the potential to provide new insights into the ways in which the nervous system controls the complexities of human movement.

A frequently asked question is why we should use techniques from nonlinear dynamics instead of the common traditional linear tools (e.g., standard deviation [SD], coefficient of variation) to examine variability in human movement. There are several reasons for the selection of nonlinear tools as an alternative methodology for examining variability. To provide a rationale, we will use stride-to-stride variability during gait as an example. However, one can apply the same logic to any cyclic human movement that produces continuous data.

1. Traditional linear tools can mask the true structure of variability in the gait pattern. Typically, in gait analysis laboratories, kinematic (and/or kinetic) data from a few strides are averaged to generate a "mean" picture of the subject's gait. In this averaging procedure, which is frequently accompanied by normalization, the temporal variations of the gait pattern are lost. In contrast, nonlinear techniques focus on understanding how gait changes over time (Hausdorff et al., 1996; Dingwell et al., 2000; Buzzi, 2001; Buzzi et al., in press).

2. From a statistical standpoint, the valid use of traditional linear tools to study variability assumes that variations between strides are random and independent (of past and future strides). However, recent studies (Hausdorff et al., 1996; Dingwell and Cusumano, 2000; Buzzi, 2001; Stergiou et al., in review-a) have shown that such variations are distinguishable from noise and may have a deterministic origin. Thus, they are neither random nor independent.

3. Traditional linear tools provide different answers than do nonlinear tools regarding stability and complexity (Dingwell et al., 2000; Slifkin and Newell, 2000; Stergiou et al., in review-b; see also the section "Approximate Entropy" later in this chapter).

4. The complexity of human gait and the multitude of feedback control systems involved in gait production make gait similar in many respects to other physiological life rhythms (e.g., heartbeat) in which variability has been described as exhibiting deterministic dynamics (Goldberger et al., 2002). The underlying fractal-like morphology of many structures of human physiology (lungs, neurons, etc.) increases the likelihood that gait is controlled by such dynamics (Goldberger et al., 1990). Deterministic dynamics offer to the moving system the ability to be adaptable and flexible in an unpredictable and ever changing environment. Nonlinear tools have the ability to describe these dynamics.

In the following sections of this chapter we present the most commonly used nonlinear tools and illustrate their application to several practical problems. The use of these tools is based on examination of the structural characteristics of a *time series* that is embedded in an appropriately constructed *state space*.

# Time Series

A **time series** is simply a list of numbers assumed to measure some process sequentially in time. Thus, a time series can be represented by a function $x(t)$ where $t$ has the discrete values $t = t_0, t_1, t_2, \ldots$ where the time interval $\Delta t = t_{i+1} - t_i$ between adjacent events is assumed to be constant. In figures 3.1, 3.2, and 3.3 we present three known time series (signals). Figure 3.1 shows a simple periodic function from $\sin(t/5)$ at integer values of $t$ from 0 to 999. Figure 3.2 shows a prime example of a chaotic system, the Lorenz attractor. Figure 3.3 shows a time series from random numbers with a Gaussian distribution centered on zero and a SD of 1.0.

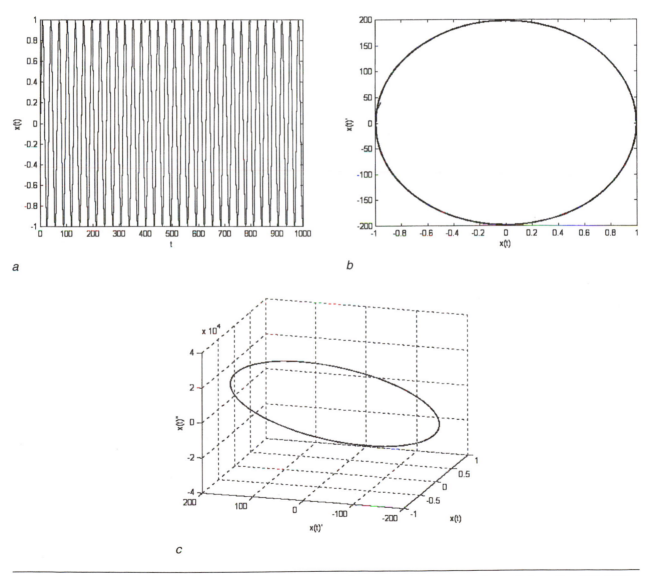

**Figure 3.1**    *(a)* The time series from a simple periodic function from $\sin(t/5)$ at integer values of $t$ from 0 to 999. *(b)* The corresponding phase plane plot. *(c)* A three-dimensional plot of the same time series where the second derivative is plotted versus the first derivative and the position time series data.

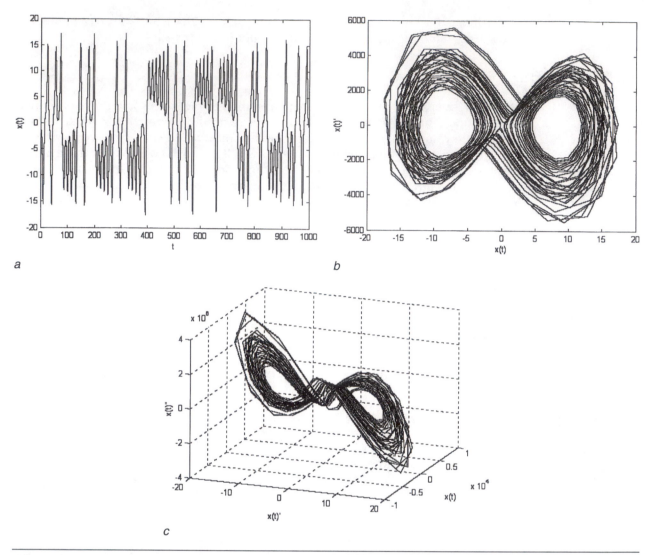

**Figure 3.2** *(a)* A time series from a chaotic system, the Lorenz attractor. *(b)* The corresponding phase plane plot. *(c)* A three-dimensional plot of the same time series where the second derivative is plotted versus the first derivative and the position time series data.

# State Space

To examine the dynamic behavior of a system from time series data, we have to investigate the structural characteristics of the time series. The first step in accomplishing this task is to appropriately reconstruct the state space where the behavior of the system is embedded. **State space** is a vector space where the dynamical system (e.g., a swaying body during posture, a moving body during locomotion) can be defined at any point (Abarbanel, 1996). For example, the natural sway of the body, which is reflected in the center-of-pressure time series, is a rhythmic activity and can be modeled as an inverted pendulum (Yamada, 1995; Harbourne and Stergiou, in press). Such a model can produce a limit cycle motion (e.g., closed periodic orbits) in state space. We can then examine the characteristics of that state space in order to gain insight into the motor control of posture.

The **phase space** (or phase plane) plot is a representation of the behavior of the dynamic system in state space. Typically, it takes the form of a two-dimensional plot of

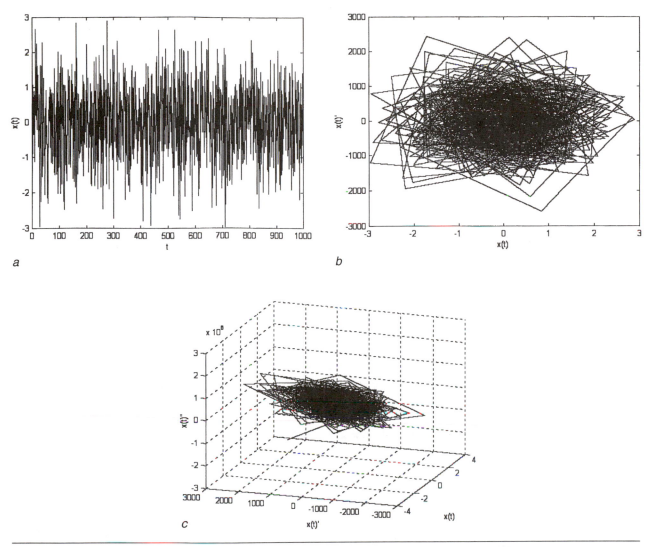

**Figure 3.3** *(a)* A time series from random numbers with a Gaussian distribution centered on zero and a SD of 1.0. *(b)* The corresponding phase plane plot. *(c)* A three-dimensional plot of the same time series where the second derivative is plotted versus the first derivative and the position time series data.

the position *X* of the time series (on the horizontal axis) versus the first derivative *X'* (on the vertical axis). In figures 3.1*b*, 3.2*b*, and 3.3*b*, we can see the phase space plots from periodic, chaotic, and random time series and observe their trajectories over a given time period. It is interesting that the periodic data give a closed orbit with complete overlapping of the trajectories and absolutely no divergence. The phase plot of the chaotic data provides an elegant picture that is not apparent when just the time series is observed (figure 3.2). In contrast, the phase space plot of the random data shows no clear pattern. In figures 3.1*c*, 3.2*c*, and 3.3*c*, we see a three-dimensional plot where the second derivative is plotted versus the first derivative and the position time series data. The second derivative is calculated as the difference between the slopes of the lines connecting each data point within its two nearest neighbors. These types of plots can be useful in certain instances in which the structure of the time series is not obvious in two dimensions.

A cross section of the phase plane has also been used to gain additional insight into the structure of the time series. Such cross sections are called return maps (Denton et al., 1990; Sprott and Rowlands, 1992). Denton and associates (1990) claimed that return maps are useful when the true structure of the data is masked by noise. Another form of representation is that of Poincare sections, in which the phase plane plot is cut by a line perpendicular to the trajectories. The $X$ data points of that line are plotted as $X_n$ on the horizontal axis and $X_{n+1}$ on the vertical axis (Denton et al., 1990; Sprott and Rowlands, 1992).

As mentioned previously, the phase plane plot is just a representation of the state space. To classify time series with nonlinear tools (e.g., Lyapunov exponent), we must first appropriately reconstruct the state space where the behavior of the system is embedded (Abarbanel, 1996). In other words, we must describe the geometric structure of the dynamical system by calculating the $d$ dimensional state space. The embedding dimension $d$ is the number of successive points $X_i = (x(t_i), x(t_i + T\Delta t), x(t_i + 2T\Delta t), \ldots, x(t_i + dT\Delta t))$, where $T$ is the time delay, a positive integer, that gives the multiple $T\Delta t$, which determines how many of the data points are used in the analysis. If $T$ is too small, $T = 1, 2, 3, \ldots$, then successive points in the state space may be too close together to be sufficiently independent. If $T$ is too large, the successive points may be so independent as to be essentially random. Thus, $T$ is selected so as to minimize the average mutual information. The calculation of an appropriate $T$ is usually determined through estimation of the first minimum of the average mutual information as described by Abarbanel (1996). The Tools for Dynamics software and the cspW (Contemporary Signal Processing for Windows; www.zweb.com/apnonlin) algorithms, described by Abarbanel (1996), were used to calculate the time delay $T$ from the time series data presented in figures 3.1, 3.2, and 3.3. The results were values of 12, 3, and 1 for the periodic, chaotic, and random time series data, respectively.

Subsequently, knowing the proper time delay $T$, one can calculate the embedding dimension $d$ for the dynamical system. The **embedding dimension** describes the minimum number of variables that is required to form a valid state space from a given time series. In other words, it is a description of the number of dimensions needed to unfold the structure of a given dynamical system. The embedding dimension $d$ was calculated using the "global false nearest neighbors" algorithm (Abarbanel, 1996). If $d$ is too small, then too many false nearest neighbors will arise when the point $X_i$ is considered as a point in the $d$ dimensional state space. If $d$ is too large, the points become so distant in the $d$ dimensional space that again they are essentially random. Thus, $d$ is chosen as the smallest integer such that the percentage of global false nearest neighbors is essentially zero. The same software was used to calculate the embedding dimension $d$ for the known time series presented in figures 3.1, 3.2, and 3.3. The results are presented in figure 3.4. The embedding dimension $d$ (plotted on the X-axis) is that dimension where the percentage of "false nearest neighbors" (plotted on the Y-axis) drops to zero, or the closest to zero. As we can see in figure 3.4, for all the time series, an embedding dimension $d$ equal to 3 will be sufficient to describe the geometric structure of the dynamical system. However, an interesting observation is that for the random time series we do not observe a zero percentage of "false nearest neighbors." Instead, we see an increase at the higher dimensions. Abarbanel (1996) explained that such a pattern is typical with data that have increased noise. In this case, the random time series data are in fact noise.

In various human movement experiments in which the embedding dimension $d$ was estimated, the following results were obtained. In rhythmic upper extremity movements, Mitra and colleagues (1997) found $d$ to be around 4. In lower extremity movements during locomotion, Dingwell and Cusumano (2000) found $d$ to be around 5. This latter value was similar to the results from our laboratory (Buzzi et al., 2001; Buzzi, 2001; Dierks et al., 2001; Buzzi et al., in press; Stergiou et al., in review-a). Lastly, in body sway during sitting postural control, Harbourne and Stergiou (in press) found $d$ to be around 5.

Once the time delay and the embedding dimension are determined, the $d$ dimensional system formed by taking successive points $X_0 = (x(t_0), \ldots, x(t_0 + dT\Delta t))$, $X_1 = (x(t_1),$

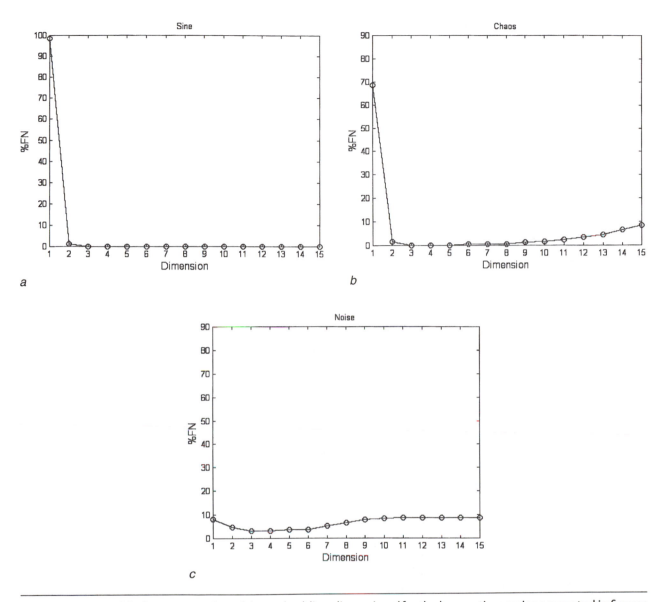

**Figure 3.4**   A graphical representation of the embedding dimension *d* for the known time series presented in figures 3.1, 3.2, and 3.3: *(a)* periodic, *(b)* chaotic, and *(c)* random time series data. In all three graphs, the embedding dimension *d* (plotted on the X-axis) is that dimension where the percentage of "false nearest neighbors" (plotted on the Y-axis) drops to zero, or the closest to zero.

. . . , $x(t_1 + dT\Delta t))$, . . . from the time series data $x(t_0)$, $x(t_1)$, . . . is used to compute the largest Lyapunov exponent and the correlation dimension. Incorporating an accurate number of embedded dimensions when calculating measures such as the Lyapunov exponent and correlation dimension ensures that the dynamical properties of the system remain unchanged when one is deriving multidimensional dynamic information from an unidimensional time series (Rapp, 1994; Abarbanel, 1996; Dingwell et al., 2000).

# Lyapunov Exponent

The **Lyapunov exponent** (LyE) is a measure of the rate at which nearby trajectories in state space diverge (Wolf et al., 1985; Rosenstein et al., 1993; Abarbanel, 1996). In other words, the LyE quantifies the exponential separation of trajectories with time in state space. As nearby points separate, they diverge rapidly and produce instability. The LyE estimates this instability, which is largely affected by the initial conditions of the system (Dingwell et al., 2000). Specifically, the LyE is calculated as the slope of the average logarithmic divergence of the neighboring trajectories in the state space (Wolf et al., 1985; Rosenstein et al., 1993; Dingwell and Cusumano, 2000). We can illustrate this with an example. Examining the periodic time series presented in figure 3.1, we observe no divergence. All trajectories perfectly overlap each other, and thus we should expect the LyE to have a zero value. However, this should not be the case with the chaotic data (figure 3.2) and the random data (figure 3.3). Indeed, this intuition is correct. Using the Chaos Data Analyzer software (www.webassign.net/pasnew/cdapro/cdapro.html; sprott.physics.wisc.edu/cda.htm) developed by Sprott and Rowlands (1992), one can calculate the LyE for these time series for the embedding dimensions from figure 3.4. The resulting LyE values are –0.001, 0.100, and 0.469 for the periodic, the chaotic, and the random time series, respectively (table 3.1). It is evident from these values that the higher the instability and the divergence, the larger the value of the LyE.

| Table 3.1    Values From the Known Time Series Presented in Figures 3.1, 3.2, and 3.3 for the Nonlinear Tools Described in This Chapter | | | |
|---|---|---|---|
| **Variables** | **Sine wave** | **Lorenz attractor** | **Random data** |
| Lyapunov exponent | –0.001 | 0.100 | 0.469 |
| Surrogate Lyapunov exponent | 0.113 | 0.341 | 0.470 |
| Correlation dimension | 1.000 | 1.941 | 4.723 |
| Approximate entropy | 0.169 | 0.449 | 1.663 |

The number of input data points for LyE computations varies from 1,000 to 10,000 (Grassberger and Procaccia, 1983a; Wolf et al., 1985; Sugihara and May, 1990), and some researchers have used even larger data sets. For example, whereas Timmer and associates (2000) used 30,000 data points to examine pathological tremors, Tsonis (1992) suggested using as few as 1,584 data points when the embedding dimension $d$ of the system is equal to 3 ($N = 10^{2+0.4d}$; formula adapted from Tsonis, 1992). Regarding research on the dynamics of gait, Dingwell and colleagues (2000, 2001) used 8,000 data points corresponding to 100 gait cycles. To clarify this topic, we conducted a study in our laboratory and found that the value of the LyE stabilizes after 35 gait cycles. This value corresponds to 7,000 data points at a sampling frequency of 180 Hz (Keenan and Stergiou, 2002b).

To further clarify the use of the LyE, let us consider its application in postural control. Variability in posture, as reflected in high excursions in postural sway, has been associated with stability (Odenrick and Sandstedt, 1984; Riach and Hayes, 1987). However, traditional measures of postural sway quantify only magnitude of sway and not the temporally evolving dynamics of postural control. A dynamical system (e.g., a swaying body during posture) is highly dependent on initial conditions, which are the constraints (e.g., strength limits, joint flexibility, perceptual abilities) that underly its function (Abarbanel, 1996; Dingwell et al., 2000; Stergiou et al., in review-a). Stability can be defined as the sensitivity of a dynamical system to perturbations, and local stability is the sensitivity of the system to internal perturbations (e.g., the natural fluctuations that occur during posture) (Dingwell et al., 2000). The effects of these natural fluctuations are what researchers are trying to evaluate with different measures of postural sway (Horak and Diener, 1994; Buchanan and Horak, 2001). However, one can estimate local stability directly using the LyE (Dingwell et al., 2000; Dingwell and Cusumano, 2000; Harbourne and Stergiou, in press).

On this basis, Yamada (1995) examined postural stability from center of pressure time series data during standing in normal adults using the LyE. The author reported that determinism exists in the body sway. Recently, Harbourne and Stergiou (in press) evaluated local stability during sitting postural development. They examined five normal infants longitudinally at three stages of sitting: stage 1, when infants could hold up their head and upper trunk but could not sit independently; stage 2, when infants began to sit independently for a brief period of time; and stage 3, when infants could sit independently. In this study, the infant was held at the trunk in the sitting position in the middle of a force platform. A small, flat contact switch was held between the infant and the investigator's hand. Once the infant was stable and calm, the investigator released the trunk support as much as possible (thus releasing the contact switch, which was used as an event marker for the exact time of release), and data were collected while the infant attempted to maintain sitting postural control. The parent either talked to the infant or displayed toys to hold the infant's attention to maintain a forward orientation. When the child was too young to sit independently, the investigator maintained contact at the chest area as lightly as possible to allow the child to attempt the skill but not be in danger of falling. At the third stage, when infants could easily sit independently and reach for toys, the parents either held the toys further away, or simply talked to the infant to keep the infant from leaning and reaching for the toy. The authors used the LyE to analyze center of pressure time series data during sitting in terms of stability of the neuromuscular system. They found significant differences ($p = 0.004$) in the LyE values between the three stages of development. Specifically, LyE values decreased from 0.0693 at stage 1 to 0.0263 at stage 3. These results indicated a more stable strategy of maintaining postural control through development.

The LyE has also been used with locomotor dynamics. Several researchers evaluated time series data of gait kinematics to examine local dynamic stability in individuals of different age groups, individuals with peripheral neuropathy, and individuals with anterior cruciate ligament (ACL) reconstruction (Dingwell et al., 2000; Buzzi et al., 2001; Stergiou et al., in review-a). **Local dynamic stability** is defined as the sensitivity of the system to small perturbations, such as the natural stride-to-stride variations present during locomotion (Dingwell et al., 2000). Thus, researchers seek to understand how subjects compensate to these variations. Local dynamic stability can be quantified using the LyE. Dingwell and associates (2000) investigated the effect of diabetic neuropathy on local dynamic stability in continuous overground walking. They analyzed time series from sagittal plane joint angles of the lower extremity and triaxial accelerations of the trunk collected during a 10-min walk at self-selected speeds. The results showed that neuropathic patients exhibited smaller LyE values and increased local dynamic stability in comparison with matched healthy controls. These patients also exhibited slower walking velocities. This latter result was explained as a compensatory strategy to maintain stability. Dingwell and associates (2001) used the same experimental paradigm to examine the differences in local dynamic stability between continuous overground and treadmill walking. They found that treadmill walking resulted in smaller LyE values and increased local dynamic stability. The authors suggested caution with the use of a motorized treadmill to explore stability and variability of locomotion.

Gait dynamics of pathological groups have also been examined in terms of local dynamic stability. Specifically, Stergiou and colleagues (2000) evaluated the effect of aging on gait dynamics. Twenty females (10 women aged 20-37, 10 women aged 71-79) walked on a treadmill at a self-selected pace while continuous kinematic data were collected from the right lower extremity. The time series from 30 gait cycles for the hip, knee, and ankle Y-coordinates (vertical displacement) and the relative knee angles for each subject were analyzed. The results showed significantly larger LyE values for the elderly group, and the authors concluded that aging might affect stability during gait. In other studies from the same laboratory (Stergiou et al., in review-a; Buzzi et al., 2001), the effect of an ACL reconstruction on local dynamic stability was evaluated. Ten subjects who had undergone ACL reconstruction using an autogenous patellar tendon graft and 10 healthy gender-matched controls walked on a motorized treadmill at a self-selected pace while continuous kinematic data (the same as in the earlier study) were collected from both of their lower extremities. The ACL group had significantly smaller LyE values for both legs. It was concluded that the ACL reconstruction affected the gait dynamics by increasing local dynamic stability.

Lastly, the LyE has been used to examine the dynamics of other movement patterns such as rhythmic upper extremity movements (Mitra et al., 1997), handwriting (Longstaff and Heath, 1999), and pathological tremors (Timmer et al., 2000).

# Surrogation

As mentioned earlier, periodic signals will result in LyE values that are negative or zero. It has been suggested that a positive LyE may indicate the presence of chaos within a time series (Kaplan and Glass, 1995; Mitra et al., 1997). However, completely random data also produce a positive outcome (Theiler et al., 1992). Consequently, several groups who reported chaos in their time series had to reconsider their findings (Palus, 1996; Theiler and Rapp, 1996). To avoid this problem, it is important to validate results against surrogate data to distinguish a true deterministic origin from randomness (Rapp, 1994).

Surrogation is a technique that can accurately determine if the source of a given time series is actually deterministic in nature (Theiler et al., 1992; Dingwell and Cusumano, 2000). The technique compares the actual data and a random data set that has a structure similar to that of the original data set. Surrogation removes the deterministic structure from the original data set, generating a random equivalent with the same mean, variance, and power spectra as the original. Subsequently, LyE values for all surrogate time series are computed and compared to the LyE values of their original counterparts. Significant differences in the LyE values between the original and surrogate counterparts indicate that the original data are not randomly derived and therefore may be deterministic in nature.

We performed the surrogation technique on the known time series data shown in figures 3.1, 3.2, and 3.3 using the algorithms described by Theiler and colleagues (1992), implemented in MATLAB (MathWorks). An outline of the procedure used is presented in figure 3.5. The interested reader can find MATLAB code for the surrogation technique in Schiff and colleagues (1994). The LyE values from both sets of time series (original and surrogate) were calculated using the Chaos Data Analyzer software mentioned earlier. As expected for the random data, we found no differences between the original data and their surrogate counterparts (0.469 vs. 0.470; table 3.1). However, the periodic and the chaotic data revealed large differences (−0.001 vs. 0.113; 0.100 vs. 0.341), clearly showing what we already knew, that these data have a deterministic origin. Figure 3.6 illustrates the effect of surrogation on the three known time series. It can be seen that the periodic and the chaotic data were largely distorted by surrogation. Specifically, the deterministic structure of the chaotic data was gone. However, the random data remain very stochastic, as expected.

Several researchers have utilized the surrogation technique to validate LyE results and identify the true nature of their time series. Dingwell and Cusumano (2000) indicated that stride-to-stride fluctuations were clearly distinguishable from linearly autocorrelated Gaussian noise. They suggested that locomotor variability might reflect deterministic central nervous system processes. Similar results were found

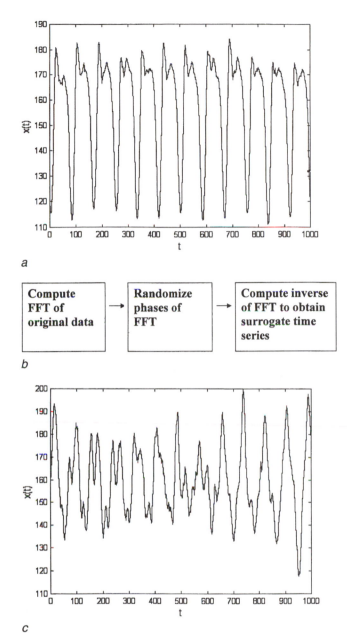

*a*

| Compute FFT of original data | → | Randomize phases of FFT | → | Compute inverse of FFT to obtain surrogate time series |
|---|---|---|---|---|

*b*

*c*

**Figure 3.5** An outline of the surrogation procedure. *(a)* Start with the original time series data. *(b)* Generate phase-randomized surrogate fluctuations from original data. *(c)* Surrogate time series has the same mean, variance, and power spectra as the original time series.

for fluctuations measured from kinematic time series data during walking and running on a motorized treadmill (Keenan and Stergiou, 2002a; Buzzi et al., in press; Stergiou et al., in review-b). Buzzi and colleagues (2001, 2002) found that locomotor variability of the kinematic patterns in Parkinson's patients during treadmill walking in the off cycle of their dopamine treatment may be deterministic in nature. This result suggested that the inherent variability generated by the neuromuscular system might still be intact in this pathological group. The investigators further concluded that if the basal ganglia were the nervous system source of this disorder, one could speculate that the mechanisms responsible for the deterministic nature of variability in gait are not housed there. Lastly, Harbourne and Stergiou (in press) showed that variability in the center of pressure time series data from infants during the development of independent sitting is not random noise but has a deterministic origin.

## Correlation Dimension

Another nonlinear tool that has been used extensively is the **correlation dimension** (CoD). The CoD approximates the fractal dimension of the region in state space occupied by the dynamical system (Grassberger and Procaccia, 1983b; Theiler, 1986). In other words, it is a measure of the dimensionality of a dynamical system. It can be used to evaluate how the data points in a time series from a dynamical system (e.g., center of pressure time series from a swaying body during posture) are organized within a state space (Sprott and Rowlands, 1992).

The CoD is usually calculated using the Grassberger and Procaccia (1983b) algorithm. Based on this algorithm, CoD is defined as

$$\mathrm{CoD} = \lim_{r \to 0} \frac{d \log C(r)}{d \log r} \qquad (3.1)$$

where $C(r)$ is the correlation integral that is calculated as

$$C(r) = \mathrm{const} \sum_{i=1}^{N} \sum_{j=i+\mu}^{N} \Theta(r - |x(i) - x(j)|) \qquad (3.2)$$

where $x(i)$ indicates the states embedded in the reconstructed state space (Abarbanel, 1996), $\Theta$ (. . .) is the function applied to count the number of pairs of points with the radius $r$, and $\mu$ is the Theiler (1986) correction employed to exclude temporally correlated points. In practicality, the second equation counts the number

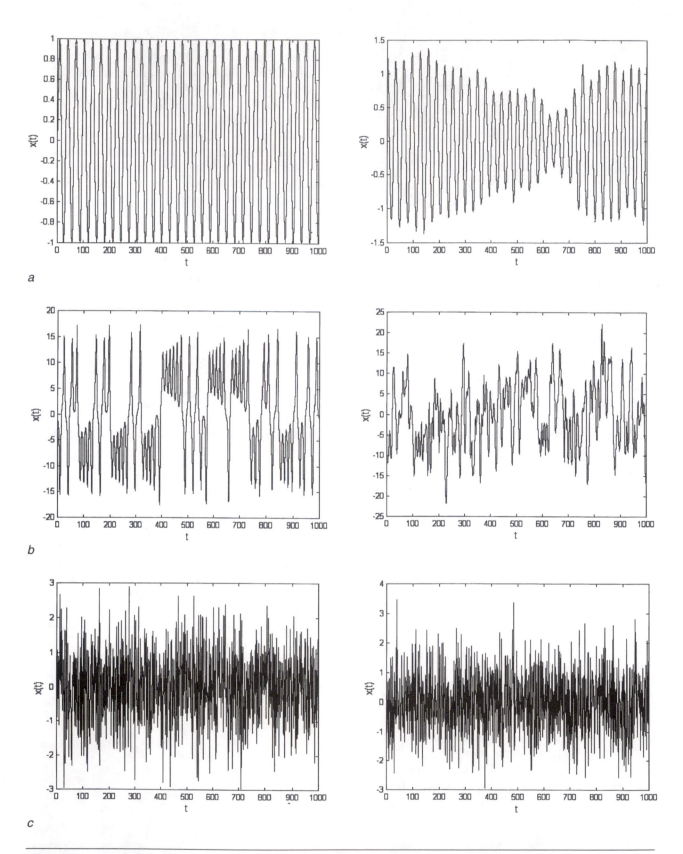

**Figure 3.6** A graphical representation of the effect of application of the surrogation technique to the known time series presented in figures 3.1, 3.2, and 3.3: *(a)* periodic, *(b)* chaotic, and *(c)* random time series data. In all three graphs, the original time series are in the right column; their surrogate counterparts are in the left column.

of pairs of points that are no greater than $r$ apart as a proportion of the total number of pairs of points in the time series. Therefore, CoD is an estimate of the average slope of a plot where $\log C(r)$ is graphed versus $\log(r)$ for a fixed value of the embedding dimension $d$.

It has been suggested that periodic data will result in a small integer value, while a non-integer value will indicate chaotic processes in a time series (Mandelbrot, 1982; Grassberger and Procaccia, 1983b). Using the known time series data given in figures 3.1, 3.2, and 3.3, we tested this hypothesis by calculating the CoD using the Chaos Data Analyzer software mentioned previously. We calculated the CoD values to be 1.000, 1.941, and 4.723 for the periodic, the chaotic, and the random time series, respectively (table 3.1). It is evident from these values that the suggestion is not true. It has been suggested that this CoD algorithm is very sensitive to changes in the time series length, experimental noise, and several other parameters (Abarbanel, 1996). Thus, the stability of the CoD results is questionable. However, the finding that dimensionality increased as the data became more complex is important. Thus, it may be better not to focus on the "nature" of the CoD value (integer or not) but to focus on its magnitude and how it changes over experimental conditions (Harbourne and Stergiou, in press).

Regarding the number of input data points for CoD computations, Grassberger and Procaccia (1983b) suggested a wide range of 1,000 to 10,000 data points, while Abraham and associates (1986) recommended a minimum of 20 cycles of motion. Regarding research on the dynamics of gait, Dingwell and colleagues (2000, 2001) used 8,000 data points that corresponded with 100 gait cycles. In studies conducted in our laboratory, we have typically used more than 2,000 data points or at least 30 gait cycles (Buzzi, 2001; Harbourne and Stergiou, in press; Stergiou et al., in review-a).

Newell and colleagues (1993) suggested that CoD can be used as a method to evaluate the number of degrees of freedom during posture. Specifically, they examined the motor control of patients with tardive dyskinesia by measuring standing postural sway and utilized the CoD to assess the dimensionality of the center of pressure time series data. The authors found that dimensionality was systematically lower in the pathological group (CoD values around 1.3) in comparison with healthy controls (CoD values around 2.2). Using similar methods, Harbourne and Stergiou (in press) examined longitudinally the development of sitting postural control in five infants with typical development. Results showed that dimensionality significantly decreased early in development (CoD values dropped from 4.1 to 3.8), indicating a constraint of the degrees of freedom. Subsequently, dimensionality significantly increased (CoD values increased from 3.8 to 4.1), indicating a possible release of the degrees of freedom as sitting independence emerged. Yamada (1995) also examined the dimensionality of postural control during quiet standing and during standing while swinging the arms. The similar values (range 2.1 to 2.5) obtained for CoD between the two conditions indicated that the structural properties (dimension) of the dynamical system remained the same.

Several researchers have also examined the dimensionality of locomotor patterns. Dingwell and Cusumano (2000) found differences between healthy controls and patients with peripheral neuropathy during overground walking. They concluded that sensory feedback plays an important role in the dimensional structure of walking kinematics. Stergiou and associates (2000) found significantly increased dimensionality in elderly (CoD values above 3) when compared with young adults (CoD values below 2) during treadmill walking. The authors attributed this result to the increased complexity and noise in the kinematic patterns of the elderly subjects.

Lastly, the CoD has been used to examine the dimensionality of other movement patterns such as handwriting (Longstaff and Heath, 1999). Specifically, Longstaff and Heath examined the time series generated from eight participants who wrote the pseudo-word "madronal" on a computer graphics table. The CoD values ranged between 2.7 and 3.7. The authors suggested that for handwriting, CoD values are small and fractional but with substantial individual variability. They concluded that handwriting is a low-dimensional movement pattern.

# Approximate Entropy

Entropy is a basic statistical concept that was first presented in information theory (Shannon and Weaver, 1963) as a measure of uncertainty or variability. In a similar sense, **approximate entropy** (ApEn) is a specific method to determine complexity that can quantify the regularity or predictability of a time series (Pincus, 1991; Pincus et al., 1991). A more predictable and regular time series is less complex than a less predictable and regular one. Approximate entropy measures the logarithmic probability that a series of data points a certain distance apart will exhibit similar relative characteristics on the next incremental comparison within the state space (Pincus, 1991, 1995). Time series with a greater likelihood of remaining the same distance apart upon comparison will result in lower ApEn values, while data points that exhibit large differences in distances between data points will result in higher values.

The mathematical definition of ApEn is discussed in far greater detail in studies by Pincus (1995), Pincus and Kalman (1997), and Pincus (2000), as well as in several other publications. Here, we present a compound version of this definition as adapted from these publications. Thus, to define ApEn [better identified as ApEn($m$, $r$, $N$)], we start with our $N$ input data points $u(1)$, $u(2)$, . . . , $u(N)$ and also incorporate two input parameters, $m$ and $r$. The input parameter $m$ is the length of compared runs, and $r$ is a tolerance. The first step is to form vector sequences $x(1)$ through $x(N - m - 1)$ from the $\{u(i)\}$, defined by $x(i) = [u(i), . . . , u(i + m - 1)]$. These vectors are basically $m$ consecutive $u$ values, beginning with the $i$th point. The second step is to define the distance $d[x(i),x(j)]$ between vectors $x(i)$ and $x(j)$ as the largest difference in their respective scalar components. The third step is to use the vector sequences $x(1)$ through $x(N - m - 1)$ to create (for each $i \# N - m + 1$).

$$C_i^m(r) = \left(\text{number of } x(j) \text{ such that } d\big[x(i),x(j)\big] \# r\right) / \left(N - m + 1\right) \qquad (3.3)$$

The $C_i^m(r)$ values measure (within the tolerance $r$) the regularity of patterns similar to a given pattern of window length $m$. The fourth step is to define $\Phi^m(r)$ as the average value of ln $C_i^m(r)$, where ln is the natural logarithm. Lastly, we define approximate entropy as

$$\text{ApEn}(m,r,N) = \Phi^m(r) - \Phi^{m+1}(r) \qquad (3.4)$$

Using ApEn, we basically calculate the logarithmic probability that runs of patterns that are close (e.g., within tolerance $r$) for $m$ observations remain close (with the same tolerance) on the next incremental comparisons. Small ApEn values indicate a higher probability of remaining close, or high regularity. For a more physiologically based explanation, readers may refer to a figure-based description of the algorithm presented by Pincus and Goldberger (1994).

Computer code to calculate the ApEn of a time series is available in FORTRAN (Pincus et al., 1991). The ApEn values typically range from 0 to 2. Values closer to 0 are consistent with greater periodicity (less complexity). Conversely, values nearing 2 represent greater irregularity (higher complexity) (Pincus, 1995, 2000). Using the known signals presented in figures 3.1, 3.2, and 3.3, we estimated the ApEn values to be 0.1689 for the periodic data, 0.4496 for the chaotic data, and 1.6625 for the random data (table 3.1). It is evident from these values that the higher the complexity, the larger the value of the ApEn.

The number of input data points for ApEn computations is typically between 50 and 5,000 (Pincus, 1995, 2000). The fact that such small data sets can be used for the calculation of ApEn is actually one of the advantages of ApEn in comparison with other nonlinear tools (e.g., Lyapunov exponent). For noisy and medium-sized data sets, ApEn has been shown to produce stable values (Pincus, 1991; Pincus and Keefe, 1992). In studies in our laboratory, we have typically used more than 2,000 data points (Buzzi, 2001; Harbourne and Stergiou, in press; Stergiou et al., in review-a).

As mentioned in the definition presented earlier, calculation of the ApEn requires the selection of two input parameters: $m$, the number of observation windows to be compared, and $r$, the tolerance factor. These two parameters, as well as the data length, must remain the same in all calculations to ensure meaningful comparisons (Pincus and Goldberger, 1994). It has been shown (Pincus and Goldberger, 1994; Pincus, 1995, 2000) that for $m$ values of one and two and $50 \# N \# 5{,}000$, values of $r$ between 0.1 and 0.25 SD of the $u(i)$ data (see mathematical definition presented earlier) produce good statistical validity for ApEn in many models. In all the human movement studies presented later in this chapter, a value of two was used for $m$, while values of $r$ were between 0.20 and 0.25.

The ApEn measure has been used in several medical settings during the last decade. It has been utilized to examine the effect of gender in growth hormone secretion (Pincus et al., 1996); the differences in heart rate control in normal and aborted-sudden infant death syndrome infants (Pincus et al., 1993); the effect of age on luteinizing hormone and follicular stimulating hormone serum concentrations in women and men (Pincus et al., 1997); the effect of aging on cardiovascular dynamics (Kaplan et al., 1991); and several other questions in enzymology, cardiology, and endocrinology (Pincus, 2000; Fleisher et al., 1993; Veldhuis and Pincus, 1998). These studies generally showed that sickness and aging correlate well with decreased ApEn values. These findings are in accordance with the general hypothesis proposed in the medical sciences that abnormal physiology is associated with more regularity (increased periodicity), while normal physiology is linked with greater complexity (more irregularity) (Pincus and Goldberger, 1994).

Some researchers have also used the ApEn measure to study questions related to human movement. In such studies, a change in complexity may be indicative of learning and a reorganization of the available degrees of freedom (Newell, 1997; Vaillancourt and Newell, 2000). Morrison and Newell (1996) utilized an ApEn analysis to determine the degree of regularity in acceleration signals. These signals were sampled from accelerometers positioned on young adults at the right and left upper limbs during a two-arm postural pointing task. The authors used the ApEn measure to infer the relative degree of active control during limb motion. In particular, the lower the ApEn value, the more active the control that was being exerted at that particular effector unit. The results revealed ApEn values between 1.75 and 1.5 and indicated more regularity at the upper arm and hand than at the forearm and finger, which suggested a compensatory synergy organized about the actions of the wrist and shoulder joints for this specific task. Newell and associates (2000a) used the ApEn measure to examine goniometer-generated signals from the upper limb of young adults. The authors investigated dimensional constraints of limb movements by comparing preferred and random oscillations of specific segments of the upper limb. The results revealed more irregularity (higher ApEn values) for the random trials and for the more distal limb segments. In another study using the same experimental apparatus, Newell and associates (2000b) investigated complexity in young adults who were taught to move randomly. The results showed small ApEn values (range 0.4 to 0.2), indicating that practice did not enhance the irregularity (randomness) of their movement.

Several human movement studies have also used the ApEn measure with pathological populations. Vaillancourt and Newell (2000) examined the complexity of resting and postural tremor in Parkinson's patients using finger accelerometer signals. The results showed more regularity (lower ApEn values and increased periodicity) for the Parkinson's disease group in comparison with age-matched controls. The authors suggested that a possible explanation for these results is a loss of independent sources of control due to Parkinson's. The same research group (Vaillancourt et al., 2001) explored the effect of Parkinson's disease on the regularity of force tremor. This tremor was produced using load cells that the subjects squeezed with their index finger and thumb at different effort levels. The results were similar to those in the study by Vaillancourt and Newell (2000). Parkinson's patients had significantly lower ApEn values than age-matched healthy controls. Buzzi (2001) explored complexity for the same pathology using gait-related variables. Specifically, Buzzi examined the angular displacements of the lower extremity joints for regularity changes in Parkinson's patients during treadmill walking at the off cycle of their dopamine treatment. The results revealed no significant differences in ApEn values (range 0.4 to 0.2) between the pathological group and an age-matched control. In addition, Buzzi (2001) found decreasing regularity from distal to proximal joints. This latter result agrees with findings from the study by Newell and colleagues (2000b) already cited. Using techniques similar to those in the study by Buzzi (2001), Stergiou and colleagues (in review-a) examined the effects of ACL reconstruction on gait dynamics. The ApEn values (range 0.4 to 0.2) for the ankle and knee joints showed a significant linear trend with greater periodicity from the reconstructed leg, to the healthy leg of the ACL group, to a healthy leg from a control group. The authors concluded that an ACL reconstruction using an autogenous patellar tendon graft results in a more periodic gait that may lead to loss of adaptability in the lower extremity.

In human movement research related to comparisons between populations of different ages, Newell (1997) used center of pressure data to examine the motor control of children, adults, and elderly persons by measuring standing postural sway. Using ApEn to assess complexity, the author found that 3-year-old children showed decreased complexity (values around 0.4) in comparison to that for 5 year olds (values around 0.5); the complexity was approximately the same for 5 year olds and adult subjects and then decreased again in elderly subjects (values around 0.35). The elderly subjects showed the smaller ApEn values. The results indicated that young children control the body's degrees of freedom by decreasing complexity and thus increasing stability. As children grow, they are able to be more flexible and have better control over the body's degrees of freedom, resulting in greater complexity. Harbourne and Stergiou (in press) used similar methodology to examine the development of independent sitting. The results showed increased regularity as a more stable sitting posture evolves, indicating a more periodic strategy of maintaining sitting postural control. The ApEn values ranged between 0.6 and 0.2.

Several investigations have addressed the relationship between ApEn and SD. These studies have shown that the two measures give different results. Specifically, in the study by Vaillancourt and associates (2001), the SD values were very similar across conditions (force tremor recordings) and subject groups. However, the ApEn values differed considerably, and the authors were able to distinguish between elderly healthy controls and Parkinson's patients through the use of regularity. Slifkin and Newell (2000), examining the effect of increases in continuous isometric force production on the variability of motor responses, found that SD values increased exponentially as a function of force level production. However, ApEn values gave very different results and revealed an inverted-U-shaped function. In a study in our laboratory (Stergiou et al., in review-b) we obtained similar results (figure 3.7). In this study we used tibial accelerometer data from 17 healthy adults who walked on a motorized treadmill at five different speeds (self-selected pace, 10% and 20% slower, and 10% and 20% faster). An inverted-U-shaped function was also revealed for the ApEn values as a function of speed level. However, the SD values increased with speed increases. On the basis of these studies, one can conclude that SD and ApEn do not correlate well and that extreme care should be exercised in their interpretation. If the focus of a research question is on predictability and regularity within a time series, then the ApEn should be used.

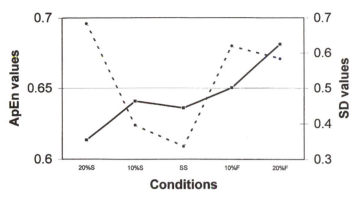

**Figure 3.7** Approximate entropy (primary Y-axis) and SD (secondary Y-axis) values from a study by Stergiou and associates (in review-b) in which healthy adults walked at different speeds on a motorized treadmill. These values are plotted versus the experimental conditions (speeds) used in the study (self-selected pace, 10% and 20% slower, and 10% and 20% faster). An inverted-U-shaped function was revealed for the ApEn values as a function of speed level production. However, the SD values increased with speed increases.

From Stergiou, Keenan, Dierks, and Kurz, in review-b.

In the spectral domain, small ApEn values often correspond to a narrow frequency range, while large ApEn values correlate with broadband spectra and a great frequency range (Pincus, 1997). However, the power spectrum analysis and the ApEn measure have produced conflicting results in several human movement studies. In the study by Vaillancourt and associates (2001) cited earlier, the frequency (modal and amplitude) analysis was unable to distinguish between Parkinson's patients and healthy elderly subjects regarding force tremor. This was not the case with ApEn, which was more sensitive to changes between groups. The authors concluded that tremor analyses should not be limited to just a frequency domain examination, but should incorporate measures of regularity. In a study by Slifkin and Newell (2000), ApEn and power spectrum analyses gave similar results (an inverted-U-shaped function) regarding variability and continuous force production. However, as mentioned previously, these results were completely different from those acquired with traditional measures of variability (such as SD). Thus, it seems that ApEn can complement a frequency analysis and provide useful information about the time domain. Furthermore, ApEn seems to be a more effective statistic to discern subtle differences between levels of the independent variable in question.

The ApEn measure has also been compared with other nonlinear measures such as the CoD (Grassberger and Procaccia, 1983b; Pincus, 2000) and the LyE (Pincus, 1997). It appears that these measures are more powerful than ApEn (and when $2 \geq m$) when purely **deterministic systems** are evaluated. However, in biological settings where we usually encounter mixed (stochastic and deterministic) systems, ApEn seems to provide higher statistical accuracy. In general, the use of ApEn has several advantages over use of other nonlinear methods. Pincus (1997) summarized these as follows:

1. Noise of magnitude below the specified $r$ tolerance does not affect ApEn

2. ApEn is not affected by outliers

3. Even small data sets (100 points) can be used for the estimation of ApEn with good statistical validity

4. ApEn is finite for random, noisy deterministic, and mixed data sets, which are usually the processes encountered in complex biological systems

5. Increased complexity within the data corresponds well with increases in ApEn. These advantages are the reason for the large number of studies in biological settings that have used ApEn

Lastly, we should mention an extension of the ApEn that gives an additional measure of regularity, the cross-approximate entropy (Cross-ApEn). The Cross-ApEn can actually measure the degree of asynchrony between two time series (Pincus and Kalman, 1997; Pincus, 2000). This measure is not discussed in this chapter. However, the reader is directed to some excellent sources (Roelfsema et al., 1988; Pincus and Singer, 1996; Vaillancourt and Newell, 2000) for discussions and applications of this method in medical and biological settings.

**Example 3.1**

The example presented next is from our laboratory research with Parkinson's disease (Buzzi, 2001). In this experiment we attempted to understand differences in locomotor variability using nonlinear tools.

Parkinson's disease has been described as a shift away from "chaotic activity" to a "pathological homeostasis" at the neuronal level (Toro et al., 1999). Parkinson's disease could then be caused by a loss of variability in some systems (Pool, 1989). There have also been attempts to identify the underlying dynamics of Parkinson's tremor (Timmer et al., 2000). However, attempts at examining Parkinson's gait have been limited to traditional measures of variability (e.g., SD, coefficient of variation). These measures, while providing accurate descriptions of variability, are not effective in explaining what is actually occurring or changing within a system. Perturbations to a dynamic system may lead to different patterns of macroscopic order that are not predictable by traditional methods (Goldberger et al., 1984). These macroscopic changes may arise directly from changes in the microscopic components of the system (Goldberger et al., 1984). The pathological microscopic changes that occur in the brain as a result of Parkinson's disease may reveal themselves in macroscopic changes during movement, particularly during gait. Analyzing these changes may provide a window for understanding the effects of dopamine deficiency on movement control. It is this knowledge that may prove useful in the development of a noninvasive diagnostic procedure for the early detection of Parkinson's disease or of a predisposition to the disease. We used nonlinear methods to assess changes in human behavior due to the onset of Parkinson's disease in order to understand how motor control strategies are affected.

Six subjects who had idiopathic Parkinson's disease (age = 66.0 ± 7.7 years) and six healthy controls (age = 63.5 ± 7.0 years) participated in the study. Parkinson's subjects undergoing dopamine therapy were asked to refrain from taking their medication the morning of data collection. Kinematic data from 30 continuous footfalls (average = 2,378 data points/trial) of the right lower extremity were collected using two cameras (60 Hz) while the subjects walked on a treadmill at their self-selected pace. Sagittal three-dimensional angles of the hip, knee, and ankle were calculated. Here we present data from just two representative subjects, one from each of the two groups, to illustrate the application of the methods described earlier.

Our first step is to plot and inspect the time series and their phase plane plots (figures 3.8 and 3.9). Although the time series do look different between the two subjects, they cannot assist us in distinguishing differences in variability. However, the phase plane plots clearly show that the Parkinson's subject has much more variable joint movement patterns. This is especially evident at the ankle (figure 3.9$a$) and the knee (figure 3.9$b$).

Subsequently, to classify variability in the time series using nonlinear tools (e.g., Lyapunov exponent), we had to appropriately reconstruct the state space where the behavior of the system is embedded. Thus, we calculated the number $d$ of embedded dimensions that are needed to unfold the structure of the dynamical system. The results are shown in figure 3.10. For the control subject, the number $d$ of embedded dimensions is 4, 3, and 3, for the ankle, the knee, and the hip, respectively. For the Parkinson's subject, the corresponding values are 5, 4, and 3. We can see that although the pathological subject had higher values, all values were still 5 at most. This value was used for the calculation of the LyE and CoD. However, a very interesting result is that for the ankle and the knee of the Parkinson's subject, there is an increase in the percentage of the false nearest neighbors at the higher dimensions (figure 3.10, $a$ and $b$). Abarbanel (1996) explained that such a pattern is typical with data that have increased noise. In this case, we have a quantifiable way to distinguish increased noise in the two joint movement patterns for the Parkinson's patient.

Our next step was to calculate the LyE and the CoD. For the LyE, the results indicated that the control subject was more locally stable at the knee and the ankle compared to the Parkinson's subject (table 3.2). That was not the case for the hip, where the Parkinson's

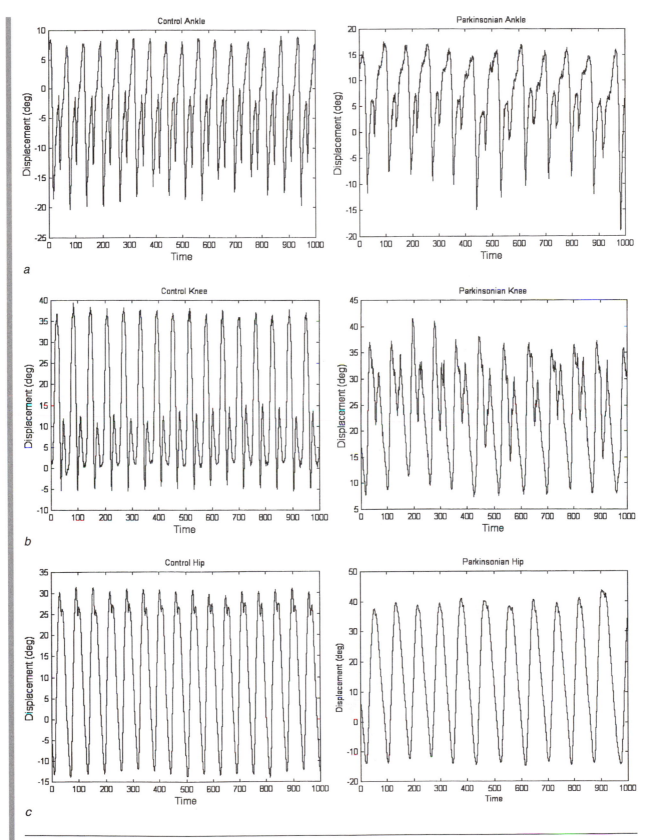

**Figure 3.8** Time series data from sagittal three-dimensional angles of the *(a)* ankle, *(b)* knee, and *(c)* hip from one control (left column) and one Parkinson's subject (right column).

From Buzzi, 2001, p. 39-43.

*(continued)*

**Control**

**Parkinson**

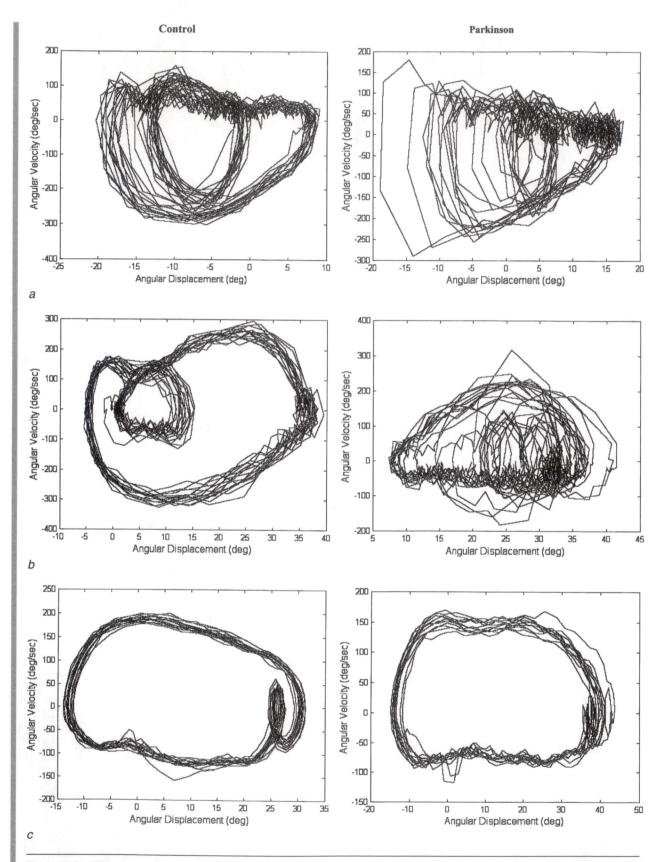

**Figure 3.9** The phase plane plots of the time series presented in figure 3.8. These time series data are from sagittal three-dimensional angles of the *(a)* ankle, *(b)* knee, and *(c)* hip from one control and one Parkinson's subject.

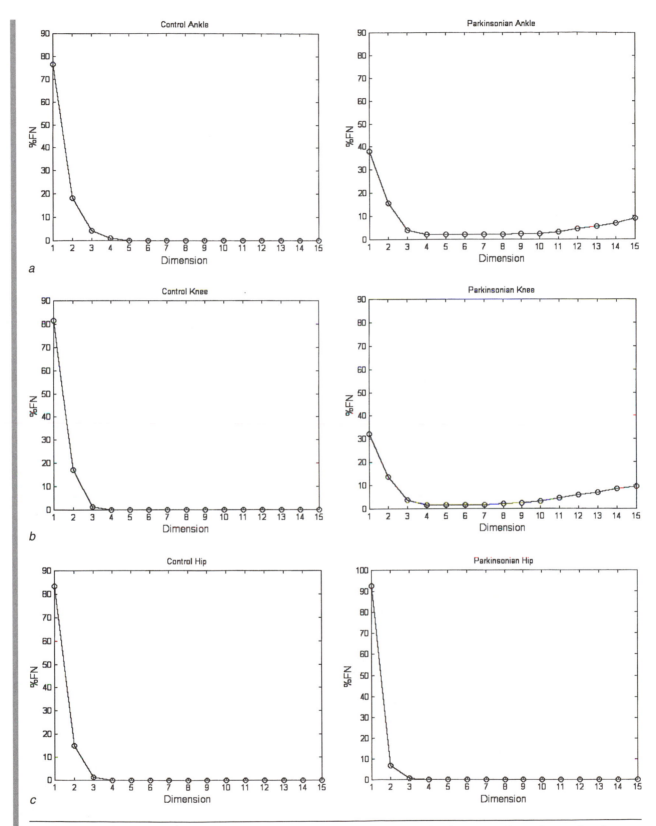

**Figure 3.10** The embedding dimensions *d* from the time series presented in figure 3.8. These time series data are from sagittal three-dimensional angles of the *(a)* ankle, *(b)* knee, and *(c)* hip from one control and one Parkinson's subject. The embedding dimension *d* (plotted on the X-axis) is that dimension where the percentage of "false nearest neighbors" (plotted on the Y-axis) drops to zero, or the closest to zero.

subject had a lower LyE value. In addition, the LyE values for both subjects increased from proximal to distal, indicating lower local stability toward the periphery (table 3.2). That was also the case with the CoD results (table 3.2). The increased dimensionality observed toward the ankle might indicate less motor control and more flexibility in the periphery. In addition, the Parkinson's subject had higher values than the control for the knee and the ankle, indicating less control in these two joint movement patterns.

To validate these results, the procedure of surrogation was used. The results showed clearly higher values for all surrogated time series (table 3.2), possibly indicating that the original time series were not randomly derived but that the fluctuations observed were the result of deterministic processes. This is an important finding for the Parkinson's subject, because it shows that the deterministic structure of the gait variability remains intact even without medication.

The last step in the analysis was to calculate the ApEn. These results showed that the knee and the ankle of both subjects had more complexity than the hip (table 3.2). However, the Parkinson's subject had higher values for these two joints, indicating even more complexity in the time series than the control subject. This result agreed with the finding from the LyE. An interesting observation was that the Parkinson's patient had a smaller value for the hip that also agreed with the LyE results. This finding possibly demonstrates an adaptation at the hip for the Parkinson's patient to compensate for the increased complexity and local stability at the more distal joints.

The preceding discussion clearly shows how a nonlinear evaluation can assist in identifying differences between a pathological subject and a healthy one. We used several nonlinear tools because we believe that a plurality of measuring techniques will enhance our ability to explain complex phenomena such as human gait. We further believe that these methods can provide a window into the control processes of the neuromuscular system and offer a quantitative assessment of neurologic status as well as the efficacy of therapy.

**Table 3.2   Values for the Nonlinear Tools Used for the Analysis of the Joint Angular Time Series Data From One Parkinson's Subject and One Healthy Age-Matched Control Subject During Treadmill Walking**

| Variables | Control | Parkinson's |
|---|---|---|
| **LYAPUNOV EXPONENT** | | |
| Hip | 0.117 | 0.102 |
| Knee | 0.166 | 0.191 |
| Ankle | 0.188 | 0.195 |
| **SURROGATE LYAPUNOV EXPONENT** | | |
| Hip | 0.213 | 0.169 |
| Knee | 0.289 | 0.315 |
| Ankle | 0.346 | 0.296 |
| **CORRELATION DIMENSION** | | |
| Hip | 2.012 | 2.015 |
| Knee | 3.151 | 3.441 |
| Ankle | 3.246 | 3.673 |
| **APPROXIMATE ENTROPY** | | |
| Hip | 0.314 | 0.278 |
| Knee | 0.353 | 0.505 |
| Ankle | 0.337 | 0.446 |

From Buzzi, 2001, p. 39-43.

## Other Tools

There are several other nonlinear tools that this chapter has not covered. Here we briefly outline those that are the most well known and widely used.

A method that Hausdorff and colleagues (1995, 1996, 1997) have used is **detrended fluctuation analysis** (DFA). This type of analysis is used to evaluate the presence of long-range, power-law correlations as part of multifractal cascades that exist over a wide range of time scales. Researchers have used DFA to examine human gait dynamics in patients with Parkinson's and Huntington's disease and in populations of different ages (Hausdorff et al., 1997, 1999; Goldberger et al., 2002).

A very interesting application of DFA is to distinguish between the noise induced by slow pace walking compared to normal walking (Hausdorff et al., 2001). These two types of walking also differ from random shuffling of the data, with normal walking having an intermediate value between the other two (slow pace walking and random shuffling data). One also observes a similar dichotomy with heartbeat data for the three classes of healthy young persons, healthy elderly persons, and heart failure patients, with the healthy young in the middle (Goldberger et al., 2002). We intend to investigate this unusual phenomenon in the near future.

Another method is the stationarity of time series (Rapp, 1994; Newell, 1997; Dingwell and Cusumano, 2000). Stationarity refers to the statistical similarity of successive parts of a time series. It has been suggested that all original time series should be checked for stationarity, since almost all nonlinear dynamical tools include an assumption of stationarity in their definitions (Rapp, 1994). Stationarity can be evaluated both in the time domain and in the embedded space, with the latter being the most appropriate. Since all nonlinear tools are calculated in the embedded space, it is logical that stationarity should be examined at this level (Rapp, 1994). The technique most frequently used to evaluate stationarity is a graphical method called recurrence diagrams (Eckmann et al., 1987). These diagrams are then examined for homogeneity and specific patterns to assess stationarity.

Dingwell and Cusumano (2000) applied this technique to evaluate sagittal kinematic time series during overground walking. They also examined differences between a group of subjects with peripheral neuropathy versus healthy controls. The authors found no evidence for lack of stationarity in the time series. However, they did notice that two pathological subjects exhibited nonstationarity at their ankles where the neuropathy was most severe. Newell (1997) also used stationarity measures to examine center of pressure time series. However, Newell did not use recurrence plots and evaluated stationarity using a statistical test proposed by Priestley and Subba-Rao (1969). Newell (1997) evaluated the time and the frequency domain of a number of age groups during standing and found a lack of stationarity in the time domain. However, the results for the frequency domain were inconclusive. The author also suggested that stationarity should not be treated as a validity tool, but that it has merit as a motor control technique to examine the organization of the dynamics in a system.

One can also evaluate complexity with several other methods besides the ApEn. A very interesting method was proposed by Tononi and colleagues (1994) and is based on statistical measures that capture regularities in the deviation from independence (mutual information) among subsets of a system. Mutual information is defined as a measure of statistical dependence that describes the amount of information available about the state of a subset through knowledge of the state of the rest of the system (and vice versa) (Tononi et al., 1998). Mutual information is zero in statistical independence. The uniqueness of this measure is that it can produce low values for both completely random and completely periodic systems. However, systems with high complexity but increased order (e.g., chaotic) can produce large values.

Lastly, an important issue with the use of nonlinear tools is that of filtering. Rapp (1994) characterized filtering as "a potentially dangerous activity." Rapp (1994) also explained that filtering of the time series can affect dimension estimates and other calculations. It was suggested that filtering in the time domain should be avoided and that the application of a geometric filter (a filter that operates in the embedding space) could be considered. In almost all the studies reviewed in this chapter, the time series data were analyzed unfiltered.

# Other Available Software and Algorithms

For all the nonlinear tools presented in this chapter, we have provided the algorithms and the related software for usage and application. However, there are several more pieces of software that the interested reader can acquire and explore. A publicly available software package is TISEAN (Hegger et al., 1999). The interested reader is referred to Timmer and colleagues (2000) for an application of this software. TISEAN has a variety of algorithms for data representation, prediction, noise reduction, and dimension and exponent estimation. This software can be downloaded at www.mpipks-dresden.mpg.de/~tisean/TISEAN_2.1/index.html.

In addition, we recommend www.physionet.org. This is a very well-constructed resource Web site that was developed as a result of an NIH grant by Dr. Ary L. Goldberger (Goldberger et al., 2002). In this location, the reader can find sample time series, a variety of publicly available software, and numerous publications on nonlinear dynamics.

# Summary

Inherent neuromuscular variability can be observed at many levels of movement organization. It is evident in the outcome of a motor performance and in the spatial orientation of body segments during a movement pattern. Parameters of human movement variability during repetitive continuous actions have recently been studied using nonlinear tools, and these studies have proved to be extremely insightful. The purpose of this chapter was to inform the reader about the most commonly used nonlinear tools for the examination of variability in human movement. Several examples illustrated the application of these tools. We believe this type of analysis can offer a different approach to research questions and help advance understanding of the control process of human movement and movement pathology.

We would like to thank Dr. Steven Pincus and Dr. Henry Abarbanel for kindly reviewing early drafts of this chapter.

# Suggested Readings and Other Resources

## Web Sites and URLs

http://amath.colorado.edu/faculty/jdm/faq.html. A Web site on frequently asked questions in nonlinear dynamics.

http://sprott.physics.wisc.edu/cda.htm. The Web site where the Chaos Data Analyzer software can be obtained.

www.physionet.org. A public service of the National Institutes of Health's National Center for Research Resources that contains tutorials, software, and publications about nonlinear dynamics.

www.zweb.com/apnonlin. The Web site of Applied Nonlinear Sciences where the Tools for Dynamics software and the Contemporary Signal Processing for Windows algorithms can be obtained.

## Books and Articles

Abarbanel, H.D.I. 1996. *Analysis of observed chaotic data.* New York: Springer-Verlag. An excellent book on the nonlinear analysis of time series.

Goldberger, A.L., L.A. Amaral, J.M. Hausdorff, P.Ch. Ivanov, C.K. Peng, and H.E. Stanley. 2002. Fractal dynamics in physiology: Alterations with disease and aging. *Proceedings of the National Academy of Science USA* 99(Suppl. 1):2466-2472. A comprehensive article on the application of the DFA technique.

Pincus, S.M., I.M. Gladstone, and R.A. Ehrenkranz. 1991. A regularity statistic for medical data analysis. *Journal of Clinical Monitoring* 7:335-345. An article that contains FORTRAN code for the approximate entropy.

Rapp, P.E. 1994. A guide to dynamical analysis. *Integrative Physiological and Behavioral Science* 29:311-327. A comprehensive article on the proper application of nonlinear techniques.

Schiff, S.J., T. Sauer, and T. Chang. 1994. Discriminating deterministic versus stochastic dynamics in neuronal activity. *Integrative Physiological and Behavioral Science* 29:246-261. An article that contains MATLAB code for the surrogation technique.

# References

Abarbanel, H.D.I. 1996. *Analysis of observed chaotic data.* New York: Springer-Verlag.

Abraham, N.B., A.M. Albano, B. Das, G. De Guzman, S. Yong, R.S. Gioggia, G.P. Puccioni, and J.R. Tredicce. 1986. Calculating the dimension of attractors from small data sets. *Physics Letters A* 144:217-221.

Amato, I. 1992. Chaos breaks out at NIH, but order may come of it. *Science* 257:747.

Babloyantz, A., and A. Destexhe. 1986. Low-dimensional chaos in an instance of epilepsy. *Proceedings of the National Academy of Science USA* 83:3513-3517.

Buchanan, J.J., and F.B. Horak. 2001. Transitions in a postural task: Do the recruitment and suppression of degrees of freedom stabilize posture? *Experimental Brain Research* 139:482-494.

Buchman, T.G., J.P. Cobb, A.S. Lapedes, and T.B. Kepler. 2001. Complex systems analysis: A tool for shock research. *Shock* 16:248-251.

Buzzi, U.H. 2001. *An investigation into the dynamics of Parkinsonian gait.* MS thesis, University of Nebraska at Omaha.

Buzzi, U.H., N. Stergiou, G. Giakas, and T.A. Dierks. 2001. The effect of ACL reconstruction on locomotor variability. *Proceedings of the 25th Annual Meeting of the American Society of Biomechanics,* 295-296. San Diego: University of California San Diego.

Buzzi, U.H., N. Stergiou, and E. Markopoulou. 2002. A dynamical analysis of parkinsonian gait. *Proceedings of the IV World Congress of Biomechanics.* Calgary: The University of Calgary.

Buzzi, U.H., N. Stergiou, M.J. Kurz, P.A. Hageman, and J. Heidel. In press. Nonlinear dynamics indicates aging affects variability during gait. *Clinical Biomechanics.*

Davis, H.P. 1993. *Simulation of performer variability.* PhD dissertation, University of Oregon.

Denton, T.A., G.A. Diamond, R.H. Helfant, S. Khan, and H. Karagueuzian. 1990. Fascinating rhythm: A primer on chaos theory and its application to cardiology. *American Heart Journal* 120:1419-1440.

Dierks, T.A., N. Stergiou, U.H. Buzzi, S.M. Keenan, and J. Heidel. 2001. The effect of speed on performer variability during locomotion. *Proceedings of the 25th Annual Meeting of the American Society of Biomechanics,* 293-294. San Diego: University of California San Diego.

Dingwell, J.B., and J.P. Cusumano. 2000. Nonlinear time series analysis of normal and pathological human walking. *Chaos* 10:848-863.

Dingwell, J.B., J.P. Cusumano, P.R. Cavanagh, and D. Sternad. 2001. Local dynamic stability versus kinematic variability of continuous overground and treadmill walking. *Journal of Biomechanical Engineering* 123:27-32.

Dingwell, J.B., J.P. Cusumano, D. Sternad, and P.R. Cavanagh. 2000. Slower speeds in patients with diabetic neuropathy lead to improved local dynamic stability of continuous overground walking. *Journal of Biomechanics* 33:1269-1277.

Eckman, J.P., S.O. Kamhorst, and D. Ruelle. 1987. Recurrence plots of dynamical systems. *Europhysics Letters* 1:973-977.

Fleisher, L.A., S.M. Pincus, and S.H. Rosenbaum. 1993. Approximate entropy of heart rate as a correlate of postoperative ventricular dysfunction. *Anesthesiology* 78:683-692.

Glass, L., and M.C. Mackey. 1988. *From clocks to chaos: The rhythms of life.* Princeton, NJ: Princeton University Press.

Gleick, J. 1987. *Chaos: Making a new science.* New York: Viking Penguin.

Goldberger, A.L., L.A. Amaral, J.M. Hausdorff, P.Ch. Ivanov, C.K. Peng, and H.E. Stanley. 2002. Fractal dynamics in physiology: Alterations with disease and aging. *Proceedings of the National Academy of Science USA* 99(Suppl. 1):2466-2472.

Goldberger, A.L., L.J. Findley, M.R. Blackburn, and A.J. Mandell. 1984. Nonlinear dynamics of heart failure: Implications of long-wavelength cardiopulmonary oscillations. *American Heart Journal* 107:612-615.

Goldberger, A.L., D.R. Rigney, J. Mietus, E.M. Antman, and S. Greenwald. 1988. Nonlinear dynamics in sudden cardiac death syndrome: Heart rate oscillations and bifurcations. *Experientia* 44(11-12):983-987.

Goldberger, A.L., D.R. Rigney, and B.J. West. 1990. Chaos and fractals in human physiology. *Scientific American* 262:42-49.

Goldberger, A.L., and B.J. West. 1987. Applications of nonlinear dynamics in clinical cardiology. *Annals of the New York Academy of Sciences* 504:195-213.

Goldstein, B., D. Toweill, S. Lai, K. Sonnenthal, and B. Kimberly. 1998. Uncoupling of the automatic and cardiovascular systems in acute brain injury. *American Journal of Physiology* 275 (*Regulatory Integrative Comparative Physiology* 44):R1287-R1292.

Grassberger, P., and I. Procaccia. 1983a. Characterization of strange attractors. *Physical Review Letters* 50:346-349.

Grassberger, P., and I. Procaccia. 1983b. Measuring the strangeness of strange attractors. *Physica D* 9:189-208.

Harbourne, R.T., and N. Stergiou. In press. Nonlinear analysis of the development of sitting postural control. *Developmental Psychobiology.*

Hausdorff, J.M., Y. Ashkenazy, C.K. Peng, P.Ch. Ivavov, H.E. Stanley, and A.L. Goldberger. 2001. When human walking becomes random walking: Fractal analysis and modeling of gait rhythm fluctuations. *Physica A* 302:138-147.

Hausdorff, J.M., S.L. Mitchell, R. Firtion, C.K. Peng, M.E. Cudkowicz, J.Y. Wei, and A.L. Goldberger. 1997. Altered fractal dynamics of gait: Reduced stride-interval correlations with aging and Huntington's disease. *Journal of Applied Physiology* 82:262-269.

Hausdorff, J.M., C.K. Peng, Z. Ladin, J.Y. Wei, and A.L. Goldberger. 1995. Is walking a random walk? Evidence for long-range correlations in stride interval of human gait. *Journal of Applied Physiology* 78: 349-358.

Hausdorff, J.M., P.L. Purdon, C.K. Peng, Z. Ladin, J.Y. Wei, and A.L. Goldberger. 1996. Fractal dynamics of human gait: Stability of long-range correlations in stride interval fluctuations. *Journal of Applied Physiology* 80:1448-1457.

Hausdorff, J.M., L. Zemany, C.K. Peng, and A.L. Goldberger. 1999. Maturation of gait dynamics: Stride-to-stride variability and its temporal organization in children. *Journal of Applied Physiology* 86:1040-1047.

Hegger, R., H. Kantz, and T. Schreiber. 1999. Practical implementation of nonlinear time series methods: The TISEAN package. *Chaos* 9:413-435.

Horak, F.B., and H.C. Diener. 1994. Cerebellar control of postural scaling and central set in stance. *Journal of Neurophysiology* 72:479-493.

Kaplan, D.T., M.I. Furman, S.M. Pincus, S.M. Ryan, L.A. Lipsitz, and A.L. Goldberger. 1991. Aging and the complexity of cardiovascular dynamics. *Biophysical Journal* 59:945-949.

Kaplan, D.T., and L. Glass. 1995. *Understanding nonlinear analysis.* New York: Springer-Verlag.

Keenan, S.M., and N. Stergiou. 2002a. The effect of speed on the deterministic origin of the variability present during human locomotion. *Proceedings of the IV World Congress of Biomechanics.* Calgary: The University of Calgary.

Keenan, S.M., and N. Stergiou. 2002b. The reliability of the Lyapunov Exponent during treadmill walking. *Proceedings of the IV World Congress of Biomechanics.* Calgary: The University of Calgary.

Lanza, G.A., V. Guido, M. Galeazzi, M. Mustilli, R. Natali, C. Ierardi, C. Milici, F. Burzotta, V. Pasceri, F. Tomassini, A. Lupi, and A. Meseri. 1998. Prognostic role of heart rate variability in patients with a recent acute myocardial infarction. *American Journal of Cardiology* 82:1323-1328.

Longstaff, M.G., and R.A. Heath. 1999. A nonlinear analysis of the temporal characteristics of handwriting. *Human Movement Science* 18:485-524.

Mandelbrot, B.B. 1982. *The fractal geometry of nature.* New York: Freeman Press.

Mitra, S., M.A. Riley, and M.T. Turvey. 1997. Chaos in human rhythmic movement. *Journal of Motor Behavior* 29:195-198.

Morrison, S., and K.M. Newell. 1996. Inter- and intra-limb coordination in arm tremor. *Experimental Brain Research* 110:455-464.

Newell, K.M. 1997. Degrees of freedom and the development of postural center of pressure profiles. In *Applications of nonlinear dynamics to developmental process modeling,* eds. K.M. Newell and P.C.M. Molenaar, 63-84. Mahwah, NJ: Erlbaum.

Newell, K.M., S. Challis, and S. Morrison. 2000a. Dimensional constraints on limb movements. *Human Movement Science* 19:175-201.

Newell, K.M., K.M. Deutsch, and S. Morrison. 2000b. On learning to move randomly. *Journal of Motor Behavior* 32:314-320.

Newell, K.M., R.E.A. van Emmerik, D. Lee, and R.L. Sprague. 1993. On postural stability and variability. *Gait and Posture* 1:225-230.

Odenrick, P., and P. Sandstedt. 1984. Development of postural sway in the normal child. *Human Neurobiology* 3:241-244.

Palus, M. 1996. Nonlinearity in normal human EEG: Cycles, temporal asymmetry, nonstationarity and randomness, not chaos. *Biological Cybernetics* 75:389-396.

Pincus, S.M. 1991. Approximate entropy as a measure of system complexity. *Proceedings of the National Academy of Science USA* 88:2297-2301.

Pincus, S.M. 1995. Approximate Entropy (ApEn) as a complexity measure. *Chaos* 5:110-117.

Pincus, S.M. 1997. Approximate entropy (ApEn) as a regularity measure. In *Applications of nonlinear dynamics to developmental process modeling,* eds. K.M. Newell and P.C.M. Molenaar, 243-268. Mahwah, NJ: Erlbaum.

Pincus, S.M. 2000. Irregularity and asynchrony in biologic network signals. *Methods of Enzymology* 321: 149-182.

Pincus, S.M., T.R. Cummins, and G.G. Haddad. 1993. Heart rate control in normal and aborted-SIDS infants. *American Journal of Physiology* 264 (*Regulatory Integrative Physiology* 33):R638-R646.

Pincus, S.M., E.F. Gevers, I.C.A.F. Robinson, G. Van der Berg, F. Roelfsema, M.L. Hartman, and J.D. Veldhuis. 1996. Females secrete growth hormone with more process irregularity than males in both humans and rats. *American Journal of Physiology* 270 (*Endocrinology Metabolism* 33):E107-E115.

Pincus, S.M., I.M. Gladstone, and R.A. Ehrenkranz. 1991. A regularity statistic for medical data analysis. *Journal of Clinical Monitoring* 7:335-345.

Pincus, S.M., and A.L. Goldberger. 1994. Physiological time-series analysis: What does regularity quantify? *American Journal of Physiology* 266 (*Heart Circulatory Physiology* 35):H1643-H1656.

Pincus, S.M., and R.E. Kalman. 1997. Not all (possibly) "random" sequences are created equal. *Proceedings of the National Academy of Science USA* 94:3513-3518.

Pincus, S.M., and D.L. Keefe. 1992. Quantification of hormone pulsatility via an approximate entropy algorithm. *American Journal of Physiology* 262 (*Endocrinology Metabolism* 25):E741-E754.

Pincus, S.M., and B.H. Singer. 1996. Randomness and degrees of irregularity. *Proceedings of the National Academy of Science USA* 93:2083-2088.

Pincus, S.M., J.D. Veldhuis, T. Mulligan, A. Iranmanesh, and W.S. Evans. 1997. Effects of age on the irregularity of LH and FSH serum concentrations in women and men. *American Journal of Physiology* 273 (*Endocrinology Metabolism* 36):E989-E995.

Pool, R. 1989. Is it healthy to be chaotic? *Science* 243:604-607.

Priestley, M.B., and T. Subba-Rao. 1969. A test for non-stationarity of time series. *Journal of the Royal Statistical Society Series B* 31:140-149.

Rapp, P.E. 1994. A guide to dynamical analysis. *Integrative Physiological and Behavioral Science* 29:311-327.

Riach, C.L., and K.C. Hayes. 1987. Maturation of postural sway in young children. *Developmental Medicine and Child Neurology* 29:650-658.

Roelfsema, F., S.M. Pincus, and J.D. Veldhuis. 1988. Patients with Cushing's disease secrete ACTH and cortisol jointly more asynchronously than healthy subjects. *Journal of Clinical Endocrinology and Metabolism* 83:688-692.

Rosenstein, M.T., J.J. Collins, and C.J. DeLuca. 1993. A practical method for calculating largest Lyapunov exponents from small data sets. *Physica D* 65:117-134.

Schiff, S.J., T. Sauer, and T. Chang. 1994. Discriminating deterministic versus stochastic dynamics in neuronal activity. *Integrative Physiological and Behavioral Science* 29:246-261.

Shannon, C.E., and W. Weaver. 1963. *The mathematical theory of communication.* Urbana, IL: University of Illinois Press.

Skarda, C.A., and W.J. Freeman. 1987. How brains make chaos in order to make sense of the world. *Behavioral and Brain Sciences* 10:161-195.

Slifkin, A.B., and K.M. Newell. 2000. Variability and noise in continuous force production. *Journal of Motor Behavior* 32:141-150.

Slutzky, M.W., P. Cvitanovic, and D.J. Mogul. 2001. Deterministic chaos and noise in three in vitro hippocampal models of epilepsy. *Annals of Biomedical Engineering* 29:607-618.

Sprott, J.C., and G. Rowlands. 1992. *Chaos data analyzer: User's manual.* New York: American Institute of Physics.

Stergiou, N., U.H. Buzzi, P.A. Hageman, and J. Heidel. 2000. A chaotic analysis of gait parameters in different age groups. *Proceedings of the 24th Annual Meeting of the American Society of Biomechanics,* 75-76. Chicago: University of Illinois at Chicago.

Stergiou, N., U.H. Buzzi, M.J. Kurz, and G. Giakas. In review-a. The effect of anterior cruciate ligament reconstruction on gait dynamics. *Journal of Biomechanical Engineering.*

Stergiou, N., S.M. Keenan, T.A. Dierks, and M.J. Kurz. In review-b. The effect of speed on performer variability during locomotion. *Journal of Motor Behavior.*

Sugihara, G., and R.M. May. 1990. Nonlinear forecasting as a way of distinguishing chaos from measurement error in time series. *Nature* 344:734-741.

Theiler, J. 1986. Spurious dimensions from correlation algorithms applied to limited time series data. *Physical Review* 34:2427-2433.

Theiler, J., S. Eubank, A. Longtin, B. Galdrikian, and J.D. Farmer. 1992. Testing for nonlinearity in time series: The method of surrogate data. *Physica D* 58:77-94.

Theiler, J., and P.E. Rapp. 1996. Re-examination of the evidence for low-dimensional, nonlinear structure in the human electroencephalogram. *Electroencephalography and Clinical Neurophysiology* 98:213-222.

Timmer, J., S. Haussler, M. Lauk, and C.H. Lucking. 2000. Pathological tremors: Deterministic chaos or nonlinear stochastic oscillators? *Chaos* 10:278-288.

Tononi, G., G.M. Edelman, and O. Sporns. 1998. Complexity and coherency: Integrating information in the brain. *Trends in Cognitive Sciences* 2:474-484.

Tononi, G., O. Sporns, and G.M. Edelman. 1994. A measure of brain complexity: Relating functional segregation and integration in the nervous system. *Proceedings of the National Academy of Science USA* 91: 5033-5037.

Toro, M.G., J.S. Ruiz, J.A. Talavera, and C. Blanco. 1999. Chaos theories and therapeutic commonalities among depression, Parkinson's disease, and cardiac arrhythmias. *Comprehensive Psychiatry* 40:238-244.

Toweill, D.L., and B. Goldstein. 1998. Linear and non-linear dynamics and the pathophysiology of shock. *New Horizons* 6:155-168.

Tsonis, A.A. 1992. *Chaos: From theory to applications.* New York: Plenum Press.

Vaillancourt, D.E., and K.M. Newell. 2000. The dynamics of resting and postural tremor in Parkinson's disease. *Clinical Neurophysiology* 111:2046-2056.

Vaillancourt, D.E., A.B. Slifkin, and K.M. Newell. 2001. Regularity of force tremor in Parkinson's disease. *Clinical Neurophysiology* 112:1594-1603.

Veldhuis, J.D., and S.M. Pincus. 1998. Orderliness of hormone release patterns: A complementary measure to conventional pulsatile and circadian analyses. *European Journal of Endocrinology* 138:358-362.

Wagner, C.D., B. Nafz, and P.B. Persson. 1996. Chaos in blood pressure control. *Cardiovascular Research* 31:380-387.

Wolf, A., J.B. Swift, H.L. Swinney, and J.A. Vastano. 1985. Determining Lyapunov exponents from a time series. *Physica D* 16:285-317.

Yamada, N. 1995. Chaotic swaying of the upright posture. *Human Movement Science* 14:711-726.

# PART II

## Methods to Examine Coordination and Stability in Human Movement

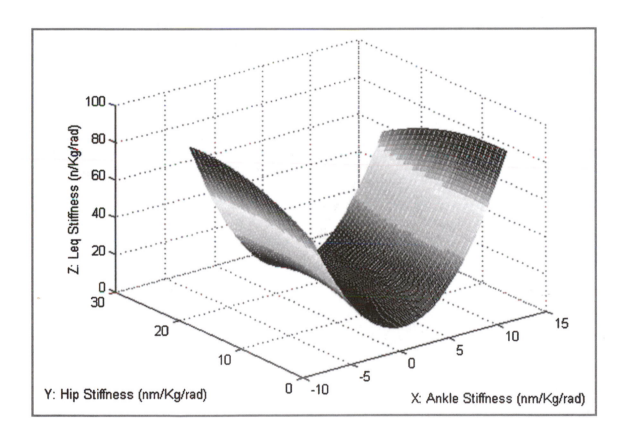

**T**his second section presents statistical and mathematical techniques used to evaluate coordination and variability as described by the Dynamic Systems Theory. This theory has provided a critical foundation for many recent developments in human movement science.

The authors of chapter 4 present an overview of the tenets of the Dynamic Systems Theory, discussing the specific tools and techniques of this theoretical perspective in detail. The chapter uses gait as the primary experimental paradigm to venture into the study of stability and change in movement coordination. Specifically, it provides methods to measure and quantify segmental and joint angle coordination in human movement. Through well-defined research examples, the authors illustrate how Dynamic Systems Theory tools can be used to explore the organization of the neuromuscular system during movement. They also provide directions showing how these tools can offer further insights into movement abnormalities and the effectiveness of rehabilitative and surgical protocols.

Dynamic Systems Theory tools often require the application of directional statistics. Directional statistics is a statistical domain that can provide techniques for making statistical inference about data that occur cyclically or take on values around a circle. Chapter 5 presents this domain in a clear tutorial fashion. The authors discuss all the directional statistics methods that are analogous to the standard tests used for linear variables (i.e., $t$-test, ANOVA). Examples of applications of these methods illustrate how they can be used in many types of human movement analysis. The authors conclude that the use of directional statistics will increase as the types of analyses included in the rest of the book become more popular.

Chapter 6 presents two novel mathematical methods to examine coordination and variability in human movement. These methods are the Response Surface Methodology and the spanning set model. The first method details a surface that can describe nonlinear interactions between functional components during movement. Additional features of the surface, such as the gradient vector and grid line curvature, can also provide important quantitative information. The authors explain in detail how the Response Surface Methodology can be used to examine coordination and stability under a wide range of experimental designs. The second method presented in the chapter is an alternative to the traditional tools (i.e., coefficient of variation) for the description of variability around the mean ensemble curve. The generation of a mean ensemble curve is a standard procedure in human movement sciences used to describe average behavior both within and between subjects during an experiment. The authors, in a clear tutorial fashion, explain how the spanning set method can assess variability around the mean ensemble curve. They employ examples from locomotion to illustrate the use of the techniques, also providing recommendations for the future application of these techniques to investigate gait pathologies.

# CHAPTER 4

# Applied Dynamic Systems Theory for the Analysis of Movement

Max J. Kurz, MS, and Nicholas Stergiou, PhD

Effective organization of the multiple degrees of freedom present in the neuromuscular system has been theoretically proposed as a necessity for healthy functional movement patterns (Bernstein, 1967; Turvey, 1990). On the basis of this notion, we could also theoretically suggest that the inability to synergistically orchestrate the many degrees of freedom present in the neuromuscular system is a hallmark of pathological movement patterns. Traditionally, biomechanical tools have been utilized to define the dynamic organization of the neuromuscular system. However, it becomes an overwhelming task to determine which biomechanical variables actually capture the state of the neuromuscular system. Hence scientists and clinicians have used a wide variety of measures such as joint kinematics, joint moments, and electromyography to define the organization of the neuromuscular system. These approaches have provided useful scientific information that has advanced our understanding of the organization of the system for healthy and pathological movement patterns. However, recent developments from the Dynamic Systems Theory (DST) have provided useful tools that allow for the theoretical expression of the behavior of the neuromuscular system in a low-dimensional term (i.e., one variable). Such an approach may offer a better way to gain scientific information on the organization of the system for completing functional movement patterns. DST tools may provide a means to select the proper biomechanical variables that capture the organization of the neuromuscular system.

According to the principles of DST, movement patterns arise from the synergistic organization of the neuromuscular system based on morphological factors (i.e., biological constructs), biomechanical factors (i.e., Newtonian laws), environmental factors (i.e., spatial and temporal configuration of events), and task constraints (e.g., walking at slow or fast speeds) (Higgins, 1985; Kugler and Turvey, 1987; Thelen et al., 1987; Beek and Beek, 1988; Clark et al., 1989; Thelen, 1989; Thelen and Ulrich, 1991; Lockman and Thelen, 1993). Therefore, DST embraces the idea that the generation of movement patterns is multifactorial and that movement involves the coupling of the multiple degrees of freedom present in the human body. Movement patterns are then the result of the individual muscles and neuropathways collectively working together to achieve a functional outcome that meets the constraints of the system. Such coordinative structures in the extremities often span more than one joint (Kelso, 1995).

Slight variations in the way the degrees of freedom are coupled together in the coordinative structure provide a rationale as to why no two steps are exactly alike during gait and why intersubject differences exist for completing the same movement pattern. Dynamic Systems Theory suggests that variations in movement patterns are attributable to the neuromuscular system's response to global (changes in the environment or task) and local perturbations (e.g., joint flexibility and proprioception) (Higgins, 1985; Kugler and Turvey, 1987; Thelen et al., 1987; Beek and Beek, 1988; Clark et al., 1989; Thelen, 1989; Thelen and Ulrich, 1991; Lockman and Thelen, 1993). This would suggest that variations in the way the neuromuscular system is organized may be related to health. For example, it is possible that abnormal movement patterns may be due to an inability of the coordinative structures to organize the degrees of freedom in an effective way to adapt to perturbations experienced.

To explain the principles of DST in this chapter, we will use gait as the movement pattern in question. With use of these principles, the collective state of the neuromuscular system has traditionally been modeled as a pendulum, with the legs or the segments of the leg oscillating in a quasi-sinusoidal fashion (Kelso and Tuller, 1984; Whitall, 1989; Holt and Jeng, 1992; Clark and Phillips, 1993; Thelen and Ulrich, 1991; Barela et al., 2000; Stergiou et al., 2001a, 2001b; Byrne et al., 2002). The use of the pendular model is based on the idea that the state of the system is dependent on energy exchange that occurs with each oscillation. Therefore, the principles of thermodynamics govern the behavior of the pendulum as time progresses. Such exchange of energy during the oscillatory motion indicates that the pendulum is a limit cycle system that is attracted to a closed, periodic orbit. Global and local perturbations affect the exchange of energy and determine the path of the pendulum (Schmidt et al., 1992). Additionally, an increased amount of variability in the periodic orbit is accompanied by an energy imbalance in the system. Through such non-equilibrium states, new behaviors arise in the pendular motion (Schmidt et al., 1992).

Modeling the lower extremity as a pendulum that has a constant energy flux is feasible because there is a constant exchange of energy during the gait cycle (Clark and Phillips, 1993). Energy in the pendular system (lower extremity) is dissipated during ground contact and put back into the system during push-off and the swing phase. A non-equilibrium state is evident if one observes the variations in the oscillatory behavior of the lower extremity segments during gait. Such non-equilibrium energy states are also evident in the biomechanical literature where there is an increased amount of variability prior to a bifurcation or hysteresis in the behavior of the neuromuscular system. For example, an increase in the variability of the lower extremity joint kinematics has been observed prior to the transition from walking to running (Diedrich and Warren, 1995). After this point of transition, the amount of variability present in the lower extremity joint kinematics decreases.

Although there are slight variations in the ways the degrees of freedom can be coupled, the neuromuscular system is attracted to a state of equilibrium (Schmidt et al., 1992). Dynamic Systems Theory proposes that when the neuromuscular system is globally or locally perturbed, it will spontaneously return to a stable state or point of equilibrium after the perturbation subsides (Scholz and Kelso, 1989; Schmidt et al., 1992). This stable state is referred to as an attractor. Attractors are the preferred movement patterns of the lower extremity during gait. Using the principles of DST, investigators have determined that the oscillations of the lower extremity can be mapped onto an attractor that has a periodic (limit cycle) shape (Clark and Phillips, 1993; figure 4.1). Modifications in the shape of the attractor trajectory indicate new behaviors (Winstein and Garfinkel, 1989; Clark and Phillips, 1993; Barela et al., 2000; Stergiou et al., 2001a; Byrne et al., 2002).

Figure 4.1 presents a phase plot for a lower extremity segment. The phase plot consists of the segment's displacement on the x-axis and the segment's velocity on the y-axis. Inspection of figure 4.1 indicates that the behavior of the lower extremity segments

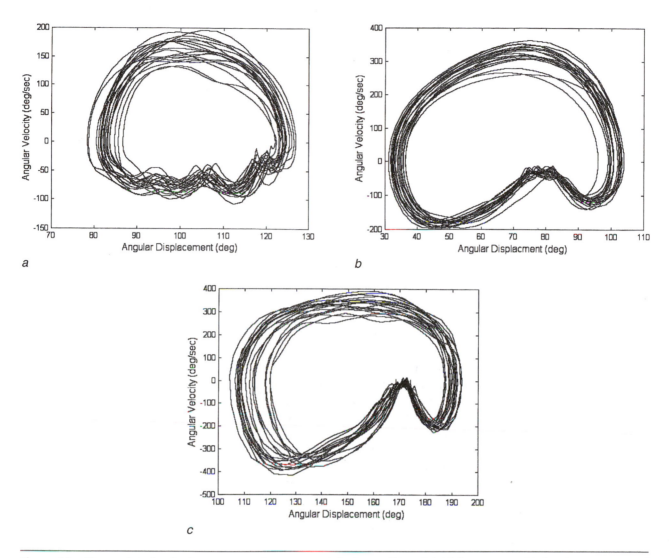

**Figure 4.1**   Phase plots for the lower extremity segments during walking: *(a)* thigh segment, *(b)* shank segment, and *(c)* foot segment.

during the gait cycle conforms to the shape of a limit cycle system that has a closed periodic orbit. The shape of the trajectory provides evidence that the behavior of the lower extremity segments can be described as a limit cycle oscillator. Additionally, it is evident that there are slight variations in the path of the trajectory for each gait cycle. Traditionally, such variations have been suggested to be "biological noise" in the neuromuscular system (Glass and Mackey, 1988). However, DST embraces these variations by suggesting that they are necessary for the neuromuscular system to adapt (neuromuscular flexibility) to global and local perturbations in the gait pattern, and for the exploration of new neuromuscular solutions for constraining the degrees of freedom (Kelso et al., 1991; Schmidt et al., 1992; Kelso and Ding, 1993; Kelso, 1995, 1997).

In movement science, it is most interesting to evaluate how two limit cycle oscillators are coupled together during the gait. Coupled limit cycles have three properties known as intrinsic dynamics: phase locking, entrainment, and structural stability (Kelso et al., 1981; Clark and Whitall, 1989). Phase locking is the fixed phase relationship between the coupled limit cycles that is stable. Entrainment is the interaction between the limit cycles, in which possible phase shift will occur in association with discontinuity and instability (high variability will be associated with discontinuities). Structural stability is the ability of the limit cycle or attractor to stay within the preferred phasing relationship despite perturbations or changes in initial conditions.

The coupling of two limit cycle oscillators during gait can be expressed through **relative phase** measures (Scholz and Kelso, 1989; Clark and Phillips, 1993; Barela et al., 2000; Stergiou et al., 2001a, 2001b; Byrne et al., 2002; Kurz and Stergiou, 2002). Observations of the relative phase relationship can provide quantitative information on how the segments of a joint (i.e., foot and shank of the ankle joint) or legs are coordinated during gait. We can explore stability and the organization of the relative phasing relationship by introducing perturbations in the gait cycle (Thelen et al., 1984; Clark and Whitall, 1989; Clark and Phillips, 1993). Adaptations in the relative phasing relationship to perturbations provide insight into the organization of the neuromuscular system during gait. Different relative phasing relationships that emerge provide insights into neuromuscular deficiencies (Barela et al., 2000). Additionally, variability in the phasing relationship provides scientific information about the stability of the selected gait pattern (Kelso et al., 1991; Kelso, 1995). An increased amount of variability in the relative phasing relationship is associated with an unstable movement pattern (Kelso et al., 1991; Kelso, 1995). Studies have indicated that variability in the relative phasing relationship increases prior to a bifurcation to a new behavior (Kelso et al., 1991; van Emmerik and Wagenaar, 1996; van Emmerik et al., 1999; Selles et al., 2001). Once the new behavioral state is reached, the amount of variability in the relative phasing relationship decreases (Haken et al., 1985; Kelso et al., 1986, 1991; Diedrich and Warren, 1995; van Emmerik and Wagenaar, 1996; van Emmerik et al., 1999; Selles et al., 2001). A loss of variability during transition states has been associated with pathological conditions. Such rigidity in the movement pattern has been found in subjects with Parkinson's disease (van Emmerik et al., 1999) and low back pain (Selles et al., 2001).

Dynamic Systems Theory provides additional measures of the stability of the relative phase relationship based on Haken's phase transition theory (Scholz and Kelso, 1989). According to this theory, the stability of the movement pattern is defined by the time taken for the relative phase relationship to return to its original state after the perturbation is removed from the system. This is referred to as the relaxation time. The longer the time interval (to return to the original relative phase relationship), the less stable the organization of the neuromuscular system (Scholz and Kelso, 1989).

Dynamic Systems Theory emphasizes the identification of a low-dimensional parameter (i.e., a single variable) that defines the dynamic state of the neuromuscular system (Barela et al., 2000). This variable is referred to as an order parameter. The **order parameter** compresses the multiple degrees of freedom contained in the movement pattern into one value. Previous work has demonstrated that the relative phase relationship between the lower extremity segments (e.g., shank-thigh) is an order parameter that defines the collective state of the neuromuscular system during gait (Clark et al., 1990; Clark and Phillips, 1993; Diedrich and Warren, 1995; Barela et al., 2000; Stergiou et al., 2001a, 2001b; Byrne et al., 2002). Selection of relative phase as an order parameter is based on the facts that the segments of the lower extremity conform to a limit cycle attractor, that relative phase variability increases prior to behavioral transitions, and that relative phase variability decreases once a new behavior is selected. Therefore, on the basis of the principles of DST, relative phase captures the dynamic organization of the neuromuscular system during gait in a low-dimensional term. Since relative phase encompasses angular displacement and velocity within one variable, some have argued that relative phase provides a better measure of the organization of the neuromuscular system than other biomechanical measures (Kelso, 1995; Barela et al., 2000). This rationale is supported by biological evidence of receptors in the joint that are responsive to changes in both displacement and velocity (McCloskey, 1978). Since relative phase accounts for such biological properties as one variable, it has a distinct advantage for determining the organization of the neuromuscular system during gait. Additionally, Barela and colleagues (2000) reported scientific evidence that relative phase provided a better measure of changes in the organization of the neuromuscular system than traditional biomechanical measures (i.e., joint angular displacement).

Changes from one relative phasing relationship to another are discontinuous, and they occur when a variable to which the neuromuscular system is sensitive and is scaled up or down through a critical threshold (Byrne et al., 2002). This variable is referred to as the **control parameter.** Changes in the control parameter's value cause the neuromuscular system to move between different phasing relationships (Clark and Phillips, 1993; Kelso, 1995). A variable is a control parameter if variability in the relative phase increases prior to a bifurcation in the behavior and the variable promotes new relative phasing relationships (behaviors). Several control parameters for the neuromuscular system during gait have been presented in the literature. Researchers have effectively used gait velocity (Diedrich and Warren, 1995; van Emmerik et al., 1999; van Emmerik and Wagenaar, 1996; Selles et al., 2001), obstacle height (Stergiou et al., 2001a, 2001b), and additional weight on the lower extremity (Thelen et al., 1984; Clark and Whitall, 1989; Clark and Phillips, 1993; Byrne et al., 2002) as control parameters for exploration of the control mechanisms of gait. Changes in the physiological state of the neuromuscular system also serve as control parameters. For example, neuromuscular development (Thelen et al., 1984; Thelen, 1986; Thelen et al., 1987; Clark and Whitall, 1989; Whitall, 1989; Clark and Phillips, 1993) and neuromuscular dysfunction (van Emmerik et al., 1999; Heiderscheit et al., 1998; Hamill et al., 1999; Barela et al., 2000; Byrne et al., 2002) have been used as control parameters for explaining how the neuromuscular system is organized during gait.

The purpose of this chapter is not to test the theoretical aspects of DST but rather to provide insight into application of DST tools for investigating normal and abnormal movement patterns. The chapter first details applications of the various tools of DST. Following this discussion we present as examples two investigations in which the gait of elderly persons and individuals with postsurgical anterior cruciate ligament reconstruction was evaluated with DST tools.

# Phase Portraits and Phase Angles

One can view the behavior of a dynamic system as a differential equation in which the changing state of the system is a function of a state vector (Arbarbanel, 1996). Although the differential equation for the system is typically unknown, plotting the current state of the system versus its rate of change helps us understand the behavior of the dynamic system (Rosen, 1970; Winstein and Garfinkel, 1989). This type of plot is referred to as a phase portrait (also called a phase plane). A **phase portrait** provides a qualitative picture of the organization of the neuromuscular system during gait. Changes in the configuration of the phase portrait provide initial insight into the control mechanisms of gait (Winstein and Garfinkel, 1989).

We can capture the dynamic state of the neuromuscular system by modeling the lower extremity segments as a series of coupled pendulums in which each segment oscillates about its respective joint (figure 4.2). For example, we can model the thigh segment as an inverted pendulum that oscillates about the knee joint.

Figure 4.2 details a two-dimensional lower extremity marker set for the analysis of gait in the sagittal plane. As indicated in the figure, the angular displacement of the segment is measured relative to a fixed horizontal reference. Plotting the segment's angular position versus its angular velocity

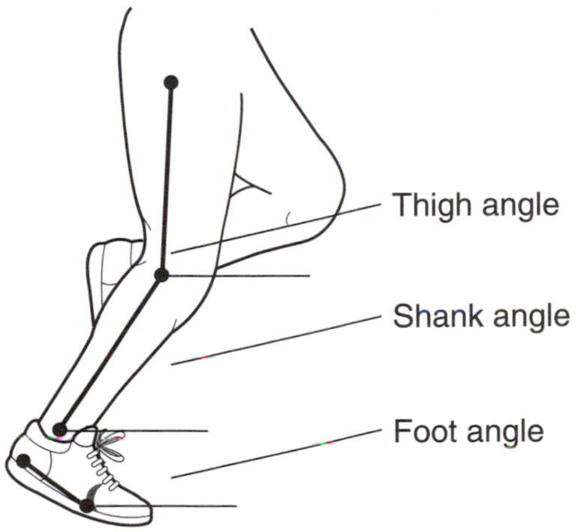

**Figure 4.2** Lower extremity segments modeled as inverted pendulums.

Reprinted from *Journal of Biomechanics*, 35(3), Max J. Kurz and Nicholas Stergiou, Effect of normalization and phase angle calculations on continuous relative phase, pp. 369-374, Copyright 2002, with permission from Elsevier.

during the gait cycle reveals that the segments conform to an oscillating limit cycle system (figure 4.3). In figure 4.3*a* we have created a phase portrait for a lower extremity segment over several gait cycles. The behavior of the segment in figure 4.3*a* evolves over time in a clockwise direction. The points where the trajectory crosses the x-axis (i.e., *y* values = 0) are related to transitions in the segment's movement pattern during the gait cycle. At these points, the angular velocity of the segment is actually zero. This indicates that there is a local minimum or maximum in the angular displacement of the segment. A higher number of zero crosses in the trajectory path within a gait cycle would suggest a greater number of changes in the dynamics of the segment (Winstein and Garfinkel, 1989). A zero cross occurs when the trajectory crosses the x-axis of the phase portrait. Additionally, cusps in the trajectory path indicate sudden interruption in the movement pattern (Winstein and Garfinkel, 1989). When one compares normal and abnormal gait patterns, the number of zero crosses and cusps in the phase portrait can provide initial evidence about neuromuscular control features of gait (Winstein and Garfinkel, 1989).

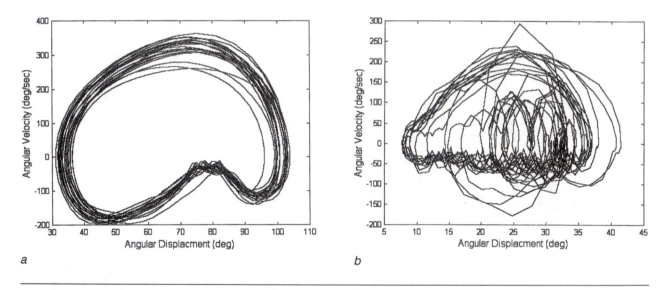

*a*                                                                     *b*

**Figure 4.3**   Shank-thigh phase portrait during gait: *(a)* normal healthy gait, *(b)* Parkinsonian gait.

When multiple gait cycles are plotted on the same phase portrait (as in figure 4.3), the amount of variability in the path of the trajectory can be used to qualitatively assess the stability of the neuromuscular system (Clark and Phillips, 1993). Slight variations in the trajectory are due to the neuromuscular system's response to global and local perturbations experienced during the gait cycle (Clark and Phillips, 1993). Such flexibility allows the neuromuscular system to maintain a stable and proficient movement pattern. However, excessive variability has been associated with instabilities in the behavior of the neuromuscular system (Clark and Phillips, 1993). Such instabilities are evident in the Parkinsonian gait portrayed in figure 4.3*b*. Compared to the healthy control, the Parkinsonian phase portrait has excessive variability that appears more random. This excessive variability may suggest a lack of control of the multiple degrees of freedom and may indicate disorder in the organization of the neuromuscular system. Although recent DST investigations have noted differences in variability between healthy and pathological gait patterns, further research is necessary to define how much variability is to be considered pathological.

As the neuromuscular system approaches a point of transition to a new behavior, there is an increased amount of variability in the phase portrait (Clark and Phillips,

1993; Kelso, 1995). Dynamic Systems Theory suggests that such variations in the behavior of the neuromuscular system are necessary for the exploration of new ways to orchestrate the numerous degrees of freedom for a stable movement pattern (Higgins, 1985; Thelen et al., 1987; Kugler and Turvey, 1987; Beek and Beek, 1988; Clark et al., 1989; Thelen, 1989; Thelen and Ulrich, 1991; Kelso et al., 1991; Lockman and Thelen, 1993; Kelso, 1997). Once the neuromuscular system finds the new stable state, the configuration of the phase portrait changes, and the amount of variability decreases (Clark and Phillips, 1993).

The **phase angle** of the phase portrait trajectory quantifies the behavior of the lower extremity segment and is used to calculate relative phase. To calculate the phase angle, the phase portrait trajectories are transformed from Cartesian (x, y) to polar coordinates, with a radius r and phase angle θ (Kelso et al., 1986; Scholz and Kelso, 1989; Scholz, 1990; Clark and Phillips, 1993). The angle formed by the radius and the horizontal axis is the phase angle of the trajectory (figure 4.4; equation 4.1):

$$\Theta_i = \tan^{-1}\left[\frac{y_i}{x_i}\right] \tag{4.1}$$

where $y$ is the angular velocity and $x$ is the angular displacement at the $i$th point of the trajectory. Figure 4.4 displays a phase portrait for a segment during gait. The angle formed between the x-axis and the vector r is called the phase angle. This angle quantifies where the trajectory is located in the phase portrait as time progresses. As indicated in figure 4.4, positive phase angles are calculated if the trajectory is within quadrant 1, and negative phase angles are calculated if the trajectory is within quadrant 4.

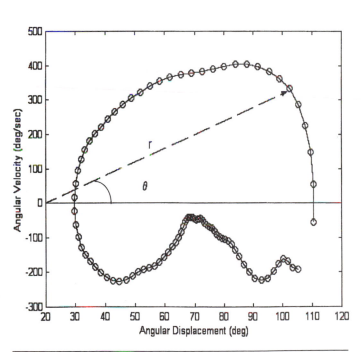

**Figure 4.4**   Shank phase angle.

## Relative Phase

Relative phase provides a measure of the interaction or coordination of two segments during the gait cycle (Haken et al., 1985; Kelso et al., 1986; Scholz and Kelso, 1989; Diedrich and Warren, 1995). To calculate relative phase, we subtract the phase angle of the proximal segment from that of the distal segment for each $i$th data point of the time-normalized gait cycle as indicated in equation 4.2:

$$\theta_{relative\ phase} = \Phi_{distal\ segment} - \Phi_{proximal\ segment} \tag{4.2}$$

where $\theta_{relative\ phase}$ is the relative phase angle between the distal and proximal segment, $\Phi_{distal\ segment}$ is the phase angle of the distal segment, and $\Phi_{proximal\ segment}$ is the phase angle of the proximal segment. The uniqueness of the relative phase measure is that it compresses four variables (i.e., proximal and distal segments' displacement and velocities) into one measure. Relative phase values that are zero degrees suggest that the two oscillating segments are in phase, while relative phase values that approach 180° are considered out of phase (Haken et al., 1985; Kelso et al., 1986; Scholz and Kelso, 1989; Diedrich and Warren, 1995). Positive relative phase values indicate that the distal segment is ahead of the proximal segment in phase space, and negative relative phase values indicate that the proximal segment is ahead in phase space (Clark and Phillips, 1993; Barela et al., 2000). The slope of the relative phase curve configuration indicates which segment is moving faster during periods of the gait cycle (Barela et al., 2000). A positive slope indicates that the distal segment is moving faster in phase space, while a negative slope indicates that the proximal segment is moving faster in phase space. The local minimum and maximum of the relative phase curve provide insight into changes in coordination between the two segments, since they represent reversals in the coordination dynamics (Barela et al., 2000). Changes in the timing of the reversals and the number of reversals help advance understanding of normal and pathological gait patterns.

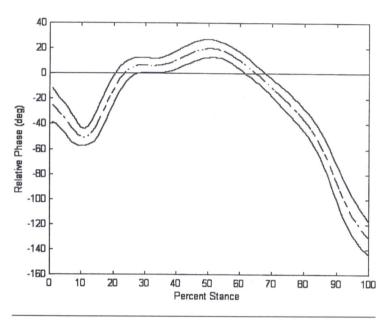

**Figure 4.5** Relative phase mean ensemble curve for the relative phase between the shank and thigh segments during the stance period of walking for a healthy exemplar subject. The dashed line represents the mean relative phase, and the bold lines represent the standard deviation about the mean ensemble.

Figure 4.5 displays the relative phase angle between two lower extremity segments during gait. Inspection of figure 4.5 indicates that the thigh was leading the shank segment during the initial portions of stance. At approximately 15% of the stance, the relationship between the shank and thigh reversed. Reversal in the relationship between the two segments is evident by the local minimum in the relative phase graph. The positive slope after the local minimum indicates that the shank was leading the thigh segment. The shank-thigh relationship became more in phase during midstance. Another segment reversal occurred at approximately 50% of the stance. At this local maximum, there was a reversal in which the thigh segment led the shank segment. The fact that the thigh was leading the shank is also evident by the negative slope of the relative phase curve following the local maximum. During the late portions of the stance, the relationship between the shank and thigh became progressively out of phase.

## Phase Portrait Normalization

Comparing the phase angles of two segments using relative phase provides a measure of coordination between the two segments during the gait cycle. There has been debate in the literature as to whether relative phase captures the collective state of the neuromuscular system (Schmidt et al., 1991; Fuchs et al., 1996) and whether current phase portrait normalization techniques for biomechanical data alleviate artifacts that may appear in the relative phase measure (Burgess-Limerick et al., 1993; Fuchs et al., 1996; Hamill et al., 2000; Kurz and Stergiou, 2002).

Initial criticism with respect to relative phase was based on integer multiple peaks that appear in the power spectrum when two segments are not oscillating with a 1:1 frequency (Schmidt et al., 1991). Schmidt and colleagues determined that the larger the difference in frequency between two oscillating segments, the greater the likelihood of additional peaks in the power spectrum at integer multiples of the dominant frequency. The authors suggested that such peaks indicate intracyclic organization of dynamic substructures that cannot be summarized with mean relative phase measures. Thus they suggested that relative phase by itself does not capture the coordination between the two segments, but rather that coordination is a more complicated behavior that encompasses the various spectral components (Schmidt et al., 1991).

Addressing these issues, Fuchs and colleagues (1996) provided a mathematical proof (using known nonlinear differential equations) that relative phase does provide a quantitative measure of coordination between two oscillators. Integer multiple artifacts present in the relative phase power spectrum of two segments oscillating at different frequencies are related to the phase portrait coordinate system used for the two oscillators. Fuchs and colleagues detailed the use of a nonlinear transformation of the phase portrait coordinates (i.e., phase portrait normalization) to remove the integer multiple artifacts in the relative phase power spectrum. The difficulty in using a nonlinear transformation as proposed by Fuchs and colleagues (1996) is that the nonlinear differential equation of the two oscillating lower extremity segments is relatively unknown with human experimental data. However, the authors determined that the dominant frequencies of the original signal were at least an order of magnitude larger than the integer multiple artifacts present in the power spectrum. Therefore, the dominant frequency components capture the nature of the relative phase relationship regardless of integer multiple artifacts present in the power spectrum. Fuchs and colleagues (1996) indicated that removing the integer multiple artifacts with a normalization technique may not be necessary for making conclusions regarding an in-phase or an out-of-phase segmental relationship.

Another criticism that has been prevalent in the literature is of investigations that do not normalize the amplitudes of the phase portrait coordinates prior to calculation of relative phase. The rationale for normalizing the phase plot coordinates is based on the assumption that the segment with the largest amplitude will dominate the relative phase measure, providing inadequate results for coordination (Burgess-Limerick et al., 1993; van Emmerik and Wagenaar, 1996; Hamill et al., 1999; Li et al., 1999; van Emmerik et al., 1999; Hamill et al., 2000). Normalization of the phase portrait should produce a scalar multiple of the original phase portrait trajectory and maintain the dynamic qualities of the segment. However, Kurz and Stergiou (2002) provided evidence that amplitude normalization techniques currently used in the literature may modify the dynamic qualities of the oscillating lower extremity segment. The reason is that these techniques normalize the phase portrait coordinates (velocity and displacement) with different scale factors. Since the aspect ratio of the phase portrait contains the dynamic qualities of the segment, normalizing the data with different scale factors may cause the dynamic qualities of the phase portrait to be lost. Figure 4.6 presents the effect of the various amplitude normalization techniques on the configuration of a partial limit cycle phase plot trajectory. Figure 4.6a is the original phase plot trajectory (non-normalized), while figure 4.6, b and c, represents two amplitude normalization techniques from the literature (van Emmerick and Wagenaar, 1996; Hamill et al., 1999; Li et al., 1999).

Evaluation of figure 4.6 indicates that the amplitude-normalized phase portrait trajectories (figure 4.6, b and c) are not scalar multiples of the original trajectory (figure 4.6a). Rather the amplitude normalization methods tend to distort the original shape of the phase plot trajectory. This is especially evident in figure 4.6c. According to Kurz and Stergiou (2002), the loss of the aspect ratio in the phase plot results in different calculated relative phase relationships. Changing the dynamic qualities of the oscillating segment is not the purpose of normalizing the phase portrait. Rather, the purpose of phase portrait normalization is to remove possible artifacts that can arise in the relative phase calculation. Based on the data presented by Kurz and Stergiou (2002), it can be suggested that current amplitude normalization techniques may modify the original dynamics of the segment. Thus it is important to take care when one is using the current normalization techniques presented in the literature.

Kurz and Stergiou (2002) provided additional evidence that amplitude differences between lower extremity segments may not actually be a problem in the calculation of relative phase as previously suspected. Since the arc tangent function is based on

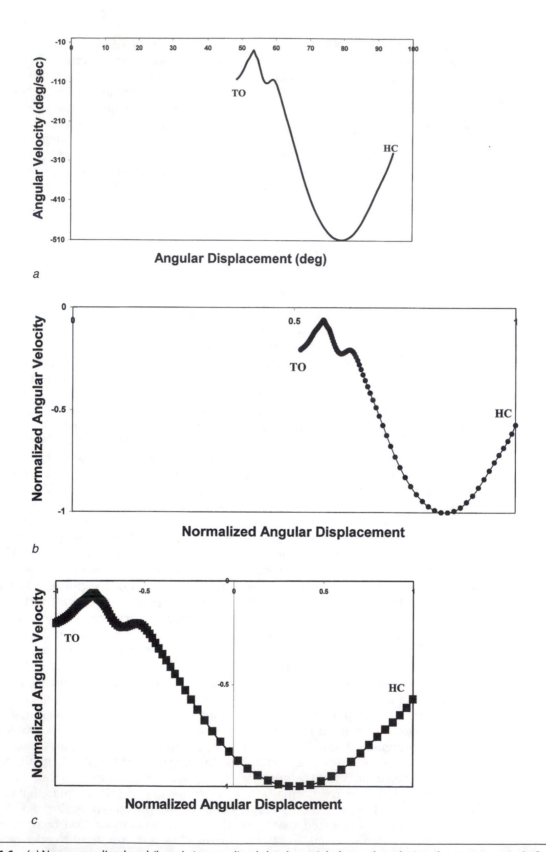

**Figure 4.6** *(a)* Non-normalized and *(b* and *c)* normalized shank partial phase plots during the stance period of running.

Reprinted from *Journal of Biomechanics,* 35(3), Max J. Kurz and Nicholas Stergiou, Effect of normalization and phase angle calculations on continuous relative phase, pp. 369-374, Copyright 2002, with permission from Elsevier.

a ratio (velocity/displacement), differences in amplitude are removed with the phase angle calculation. The arc tangent function "normalizes" differences in amplitude between the segments based on this ratio. Kurz and Stergiou (2002) suggested that because of the properties of the arc tangent function, amplitude normalization of the phase portrait is not necessary.

It is a misconception in the literature that artifacts in the relative phase measure arise from different amplitudes of the two oscillating segments. Artifacts in the relative phase measures arise when the two segments have different frequencies (Fuchs et al., 1996). However, no normalization procedure has been developed that removes these artifacts in human experimental data without modifying the aspect ratio of the original data. Further research is necessary to provide a normalization routine that maintains the aspect ratio of the oscillator regardless of the given frequency. However, the reader should note that normalization is not necessary to quantify whether two oscillating segments are in phase or out of phase based on the relative phase measure (Fuchs et al., 1996). Therefore, when one is using measures such as mean absolute relative phase and deviation phase, normalization may not be necessary.

## Mean Absolute Relative Phase

To quantify and statistically test differences between relative phase curves, it is necessary to characterize the curves by single numbers. **Mean absolute relative phase** (MARP) can be used to quantify whether the interacting segments display an in-phase or out-of-phase pattern during gait (Stergiou et al., 2001a, 2001b; Byrne et al., 2002). Mean absolute relative phase is calculated from the mean ensemble curve by averaging the absolute values of the ensemble curve points (equation 4.3).

$$\text{MARP} = \sum_{i=1}^{N} \frac{|\phi \, \text{relative phase}|}{N} \tag{4.3}$$

where $N$ is the number of points in the relative phase mean ensemble and $\phi$relative phase is the relative phasing relationship between two segments. A low MARP value indicates that the oscillating segments have a more in-phase relationship; a high MARP value indicates that the oscillating segments have a more out-of-phase relationship.

## Deviation Phase

One can determine variation in the organization of the neuromuscular system by calculating the **deviation phase** (DP) of the relative phase for the two interacting segments (van Emmerik and Wagenaar, 1996; Heiderscheit et al., 1998; Hamill et al., 1999; Stergiou et al., 2001a, 2001b; Byrne et al., 2002). Deviation phase provides a measure of stability of the organization of the neuromuscular system. Deviation phase is calculated by averaging the standard deviations of the ensemble relative phase curve points (equation 4.4).

$$\text{DP} = \frac{\sum_{i=1}^{N} |\text{SD}_i|}{N} \tag{4.4}$$

where $N$ is the number of points in the relative phase mean ensemble and SD is the standard deviation of the mean ensemble at the $i$th point. A low DP value indicates a more stable (less variable) organization of the neuromuscular system; a high DP value indicates less stability in the organization of the neuromuscular system.

# Point Estimate Relative Phase

**Point estimate relative phase** provides a measure of the relationship between two oscillating segments during the gait cycle based on their relative times to reach a local maximum or minimum. Local minimums and maximums typically represent key events in the gait cycle such as maximum knee flexion or initiation of ground contact. For example, the timing difference between the right and left legs making initial ground contact has been used in the exploration of gait patterns as the neuromuscular system matures (Thelen and Ulrich, 1991). Additionally, investigators have used point estimate measures to explore the relative timing of the lower extremity as gait velocity (Diedrich and Warren, 1995; van Emmerik and Wagenaar, 1996) is scaled as a control parameter. Point estimate provides a means to calculate the phase lag of two signals during the gait cycle. The point estimate relative phase (PRP) angle is calculated as follows (Kelso, 1995):

$$PRP = (t_1 - t_2 / T) \times 180° \tag{4.5}$$

where $t_1$ is the time to local minimum/maximum angle of segment one, $t_2$ is the time to local minimum/maximum angle of segment two, and $T$ is the period of segment one. Point estimate relative phase values that are close to zero indicate that the segment oscillations are in phase, while a value close to 180° indicates that the segment oscillations are out of phase. Several authors have argued that compared to relative phase, point estimate provides a better measure of coordination between two oscillating segments (Diedrich and Warren, 1995; Kelso, 1995). The authors' rationale is that point estimates are not dependent on the biological signal's having a stationary quasi-sinusoidal structure (Diedrich and Warren, 1995). However, we believe that this argument needs further investigation.

A graphical description of the PRP for two running conditions (barefoot and footwear) is detailed in figure 4.7. In this figure the times for the knee and ankle joints to reach a local maximum during the stance were used to calculate the PRP. Figure 4.7 and the data presented in table 4.1 indicate that in both conditions there was a phase lag in oscillatory behaviors of the knee and ankle joint. However, it is evident that the latency was smaller for the barefoot condition. The PRP values confirm this observation (table 4.1). Therefore, we could suggest that for these trials, the knee and ankle joints were more in-phase during barefoot running than during running with footwear.

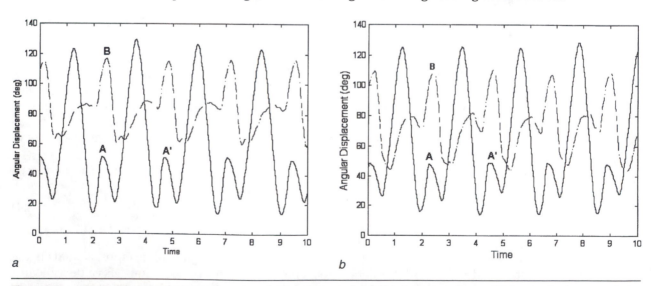

**Figure 4.7** Joint angle curves during running (a) barefoot and (b) with footwear. The solid line is the knee, and the dashed line is the ankle. A on the respective graphs represents maximum knee flexion in the stance; A' is the second occurrence of maximum knee flexion in the stance; B is maximum ankle dorsiflexion in the stance that occurs between A and A'. The difference in time between A' and A is the period of the knee signal.

**Table 4.1    Point Estimate Relative Phase and the Occurrences of Local Maximums for the Respective Joints During Running Barefoot and With Footwear**

| | Point estimate | | | |
|---|---|---|---|---|
| Condition | Ankle joint time (s) | Knee joint time (s) | Period(s) | Relative phase (degrees) |
| Barefoot | 2.383 | 2.267 | 2.403 | 8.69 |
| Footwear | 2.50 | 2.333 | 2.367 | 12.70 |

Knee and ankle joint times correspond to points A and B on the respective graphs. The period is the time interval from A to A′ in figure 4.7.

## Discrete Relative Phase

The local minimum and maximum of the relative phase curve help to elucidate the changes in coordination between the two segments since they represent reversals in the coordination dynamics (Barela et al., 2000). Discrete relative phase evaluates the local minimum and maximum of the relative phase curve configuration. This methodology has been used extensively in charting changes in the development of the neuromuscular system (Thelen and Ulrich, 1991; Clark and Phillips, 1993) and for distinguishing differences in the organization of the neuromuscular system for abnormal and normal gait patterns (Barela et al., 2000).

Figure 4.8 displays the relative phase for the shank and thigh during the stance period of walking for an elderly and a young subject. Using **discrete relative phase** measures one can conclude that there are differences in coordination between the two subjects at the local minimum and maximum identified. Thus, modifications in the behavior at these discrete points can reveal information about the dynamics of the neuromuscular system during gait (Thelen and Ulrich, 1991; Clark and Phillips, 1993; Barela et al., 2000).

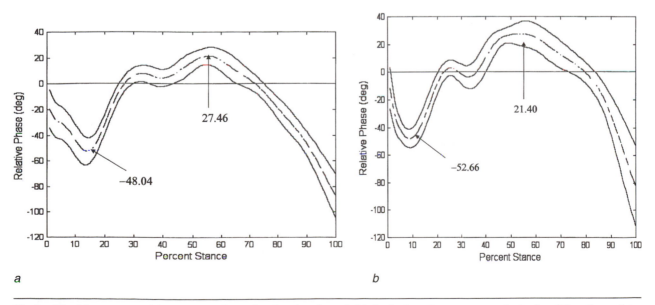

**Figure 4.8**   Mean ensemble relative phase for the shank and thigh during the stance period of walking from (a) a young subject and (b) an elderly subject. The dashed line represents the mean ensemble, and the solid lines are the standard deviations.

However, discrete measures do not reveal the organization of the neuromuscular system evolving over time since they detail only one point in the gait cycle.

# Complete Examples for the Application of Dynamic Systems Theory Tools

The following sections provide detailed examples of the use of DST tools to investigate the organization of the neuromuscular system for normal and abnormal gait. The investigation in the first example addressed issues related to the stability of elderly gait by evaluating relative phase variability. In this example, it was hypothesized that the elderly subjects would have greater variability in the lower extremity segmental couplings during the stance period of walking, suggesting an unstable gait pattern. The second investigation utilized relative phase to examine the effect of anterior cruciate ligament reconstruction on the organization of the neuromuscular system. Since anterior cruciate ligament reconstruction alters the biological properties of the lower extremity system, the hypothesis was that individuals with anterior cruciate ligament reconstruction would display neuromuscular adaptations while walking and running. Such information can aid in diagnosing the health of the reconstructed leg and the state of the neuromuscular system at various stages of rehabilitation.

## Example 4.1

Human locomotion emerges as a result of organizing the multiple degrees of freedom present in the human body (Bernstein, 1967; Turvey, 1990). Many of the research efforts in movement science are concerned with quantifying how the neuromuscular system is organized in individuals who display normal and abnormal gait. DST has provided tools for evaluating the organization of the neuromuscular system in low-dimensional terms (Kelso et al., 1991). DST proposes that change in the organization of the neuromuscular system occurs when a variable to which the neuromuscular system is sensitive is scaled up or down through a critical threshold (Byrne et al., 2002). This variable is referred to as a control parameter. In several investigations, neuromuscular age has been used as a control parameter for exploring the maturation of gait in children (Clark and Whitall, 1989; Whitall, 1989; Clark and Phillips, 1993 ). On the basis of this notion, we suggest that the aging process of the neuromuscular system continues to be a control parameter throughout one's life span. Previous investigations have noted changes in the behavior of the neuromuscular system as biosensors and muscular strength deteriorate in persons who are elderly (Whanger and Wang, 1974; Woollacott et al., 1986; Whipple et al., 1987; Bendall et al., 1989; Winter, 1991; Judge et al., 1996). Such modifications may alter the organization of the neuromuscular system and diminish the ability of the neuromuscular system to sense global and local perturbations experienced during the gait cycle. Potentially, these alterations to the biological system may promote a less stable gait pattern in persons who are elderly.

Researchers have suggested that the kinematic changes in elderly gait are related to attempts to increase stability in the gait cycle (Gabel and Nayak, 1984; Winter et al., 1990; Hausdorff et al., 1997; Maki, 1997; Kerrigan et al., 1998). From a DST standpoint, variability of segmental couplings can be used to determine the stability of gait (Diedrich and Warren, 1995; van Emmerik and Wagenaar, 1996). A low amount of variability in the segmental couplings indicates a stable movement pattern, while a high amount of variability indicates an unstable movement pattern (Diedrich and Warren, 1995; van Emmerik and Wagenaar, 1996). Evaluation of the variability present in the lower extremity segmental couplings can provide scientific evidence of the stability of the organization of the neuromuscular system during elderly gait.

The goal of the investigation in this example was to evaluate the stability of the organization of the neuromuscular system during elderly gait. With the use of DST tools, we hypothesized that elderly persons would have greater variability in the lower extremity segmental couplings during the stance period. Such increased variability in the segmental coupling pattern would suggest an unstable gait in elderly persons. We speculated that instabilities in the organization of the neuromuscular system may be due to aging.

## METHODS

Participants in this investigation included 10 young (age = 25.1 ± 5.3 years; height = 1.7 ± .05 m; weight = 63.93 ± 6.5 kg) and 10 elderly (age = 74.6 ± 5.3 years; height = 1.59 ± .05 m; weight = 64.07 ± 9.7 kg) subjects who had prior treadmill walking experience. The subjects walked on a motorized treadmill while kinematic data of the right lower extremity were collected using a 60-Hz high-speed video camera for 10 footfalls. Prior to videotaping, reflective markers were positioned on the subject's right lower extremity (figure 4.2, p. 97).

From the plane coordinates obtained, the sagittal foot, shank, and thigh angular displacements were calculated relative to a fixed horizontal axis (figure 4.2). All kinematic angular displacements and velocities during stance were normalized to 100 points using a cubic spline routine to enable mean ensemble curves to be derived for each subject-condition.

To examine the variability in the organization of the neuromuscular system for the groups, DST tools were employed (Scholz and Kelso, 1989; Kelso et al., 1991; Scholz, 1990). From the angular displacements and velocities, phase portraits were generated for the respective segments. Additionally, phase angles (equation 4.1) for each phase portrait were utilized to calculate the relative phase (RP). Relative phase detailed the organization of the neuromuscular system during the stance period (Scholz and Kelso, 1989; Kelso et al., 1991). Relative phase was calculated by subtracting the phase angles of the corresponding segments throughout stance:

$$\Phi_{\text{sagittal foot-shank rel. phase}} = \Phi_{\text{foot}} - \Phi_{\text{shank}} \tag{4.6}$$

$$\Phi_{\text{sagittal shank-thigh rel. phase}} = \Phi_{\text{shank}} - \Phi_{\text{thigh}} \tag{4.7}$$

where $\Phi_{\text{sagittal foot-shank rel. phase}}$ is the relative phase angle between the foot and shank segments, $\Phi_{\text{sagittal shank-thigh rel. phase}}$ is the relative phase angle between the shank and thigh segments, $\Phi_{\text{foot}}$ is the phase angle of the foot segment, $\Phi_{\text{shank}}$ is the phase angle of the shank segment, and $\Phi_{\text{thigh}}$ is the phase angle of the thigh segment. The RP curves for each segmental relationship were averaged across stance periods (trials), and mean ensemble curves were generated for each subject.

Stability in the organization of the neuromuscular system was determined by calculating the deviation phase (DP) of the RP curve for the two interacting segments. DP provided a measure of stability of the organization of the neuromuscular system during gait (Diedrich and Warren, 1995; van Emmerik and Wagenaar, 1996). DP was calculated by averaging the standard deviations of the ensemble RP curves (equation 4.4). A low DP value indicated a more stable (less variable) organization of the neuromuscular system, while a high DP value indicated a less stable (more variable) organization. Group means were calculated for the DP for each respective RP relationship. Statistical differences between the two groups (elderly vs. young subjects) for DP were noted with independent $t$-test ($p < 0.05$).

## RESULTS

Significant differences between the two groups were found for the foot-shank ($p < 0.01$) and shank-thigh ($p < 0.01$) RP relationships (figure 4.9). The results of this investigation indicate that the elderly subjects had greater variability in both lower extremity segmental relationships. An increase in variability of the couplings in the elderly subjects indicated a less stable organization of the neuromuscular system.

In general, variability was higher for the elderly subjects throughout the stance period for both joints. However, differences in variability of the segmental couplings were most pronounced during the early and late portions of the stance period for both groups (figure 4.10).

*(continued)*

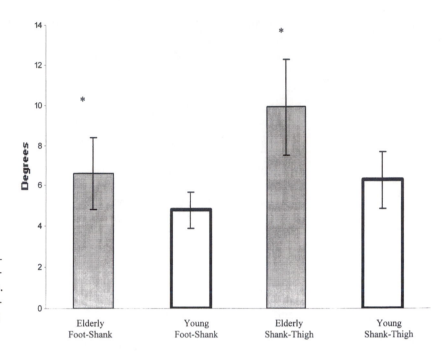

**Figure 4.9** Relative phase variability for young and elderly groups. Asterisk indicates a significant difference between the groups at the .01 alpha level.

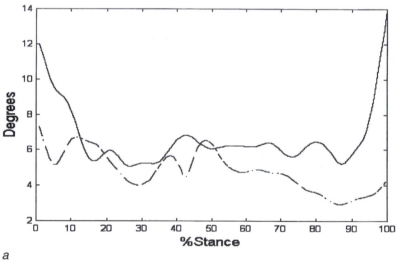

*a*

*b*

**Figure 4.10** Mean ensemble deviation phase curves during the stance period: *(a)* foot-shank deviation phase, *(b)* shank-thigh deviation phase. The dashed line represents the young subjects, and the bolded line represents the elderly subjects.

## DISCUSSION

Our hypothesis that the elderly subjects would have greater variability in the lower extremity segmental couplings during stance period was supported by our results. From a DST standpoint, variability in the segmental coupling pattern is associated with a search for a new stable attractor and with flexibility to overcome local and global perturbations experienced during gait (Clark and Phillips, 1993; Kelso, 1997). Several investigations have indicated that the lower extremity biosensors' ability to monitor feedback is diminished in persons who are elderly (Whanger and Wang, 1974; Woollacott et al., 1986). An inability to perceive local and global perturbations experienced during gait may promote a tendency in elderly persons to increase variability in order to remain flexible when faced with unforeseen perturbations. Additionally, since the biosensors cannot provide adequate information about the state of the selected gait pattern, the neuromuscular system may be in a constant flux searching for a gait pattern that is perceived as a stable attractor. Thus, it is possible that the increased flexibility and variability associated with searching for a stable attractor may place the neuromuscular system in a constant state of instability that may result in a fall.

Decreased stability during the early and late portions of the stance period was noted in both the elderly and control groups (figure 4.10). This change in variability may be due to the loose-packed position of the joints in these portions of the stance (Norkin and Levangie, 1992). For example, the knee is in a closed-packed position at full extension that is achieved during midstance in walking, and is in a loose-packed position during heel contact and toe-off. A loose-packed joint position would have a greater dependence on the surrounding joint musculature. Changing from a loose-packed to a closed-packed position may change the dynamics of the task by altering the forces acting at the joint. Furthermore, it has been noted in the literature that elderly persons have diminished muscular strength (Whipple et al., 1987; Bendall et al., 1989; Winter, 1991; Judge et al., 1996). Thus, changes in the system's constraints may influence the stability of the attractor for those who are elderly. This could possibly explain the additional increased variability present in the elderly RP during these portions of the stance period.

In summary, this investigation indicated that the elderly lower extremity displays a more variable neuromuscular organization than that of young persons. From a DST standpoint, the increased variability indicated that elderly persons have a less stable gait pattern during the stance period. Such increased variability may be due to muscular weakness and a decreased ability of the biosensors to sense global and local perturbations experienced during gait. The use of DST tools to examine the organization of the neuromuscular system offers new insights into the stability of elderly gait patterns that may not be evident otherwise.

## Example 4.2

Although differences in gait biomechanics following an anterior cruciate ligament (ACL) reconstruction have been well documented (De Vita et al., 1992; Timoney et al., 1993; De Vita et al., 1997, 1998), the underlying mechanisms responsible for these differences are not well understood. Previous investigations have suggested that differences in gait biomechanics for the ACL-reconstructed limb may be due to neuromuscular adaptations (Berchuck et al., 1990; Andriacchi and Birac, 1993; De Vita et al., 1997). However, limited progress has been made in exploring the nature of these adaptations (De Vita et al., 1997). The lack of progress in this area may be attributable to the fact that it is difficult to determine which biomechanical variables capture the adaptations of the neuromuscular system.

DST suggests that gait patterns arise from the synergistic organization of the neuromuscular system based on morphological factors (i.e., biological constructs), biomechanical factors (i.e., Newtonian laws), environmental factors (i.e., spatial and temporal configuration of events), and task constraints (e.g., walking at slow or fast speeds) (Higgins, 1985; Thelen et al.,

*(continued)*

1987; Beek and Beek, 1988; Clark et al., 1989; Thelen, 1989; Thelen and Ulrich, 1991; Lockman and Thelen, 1993). From a DST perspective, ACL reconstruction may change the biological constraints of the system. ACL reconstruction involves surgical alterations of the biological structure at both the donor site (i.e., patellar tendon) and the reconstructed ACL. Additionally, several investigators have noted that ACL reconstruction can alter joint mechanoreceptors in the ligaments and meniscus of the knee (Friden et al., 2001). Such modifications may result in a different organization of the neuromuscular system. Potentially, DST tools such as RP may be able to provide further insight into neuromuscular adaptations after ACL reconstruction.

The purpose of the investigation in this example was to evaluate the organization of the neuromuscular system for individuals with ACL reconstruction as they walked and ran. Since ACL reconstruction alters the biological constraints of the lower extremity, we hypothesized that persons with ACL reconstruction would display different neuromuscular adaptations during stance period while walking and running.

## METHODS

Ten subjects who had undergone ACL reconstruction using the central third of their patellar tendon participated in this investigation (seven females, three males; mean age, 23.9 years; mean mass, 81.1 kg; mean height, 177.3 cm). Ten healthy gender-matched subjects who had never had any kind of orthopedic or neurological condition volunteered for the control group (mean age, 21.7 years; mean mass, 67.2 kg; mean height, 171.9 cm).

Subjects walked and ran on a motorized treadmill while kinematic data of the lower extremity were collected using a high-speed (60-Hz) camera. Before videotaping, reflective markers were positioned on the subject's right lower extremity (figure 4.2, p. 97). Fifteen consecutive footfalls were collected for each walking and running condition.

From the plane coordinates obtained, the sagittal foot, shank, and thigh angular displacements were calculated relative to a fixed horizontal axis (figure 4.2). All kinematic angular displacements and velocities during the stance period were normalized to 100 points using a cubic spline routine to enable mean ensemble curves to be derived for each subject-condition.

To examine adaptations in the organization of the neuromuscular system for the respective groups, DST tools were employed (Scholz and Kelso, 1989; Scholz, 1990; Kelso et al., 1991). The phase portraits for the sagittal foot, shank, and thigh segments were generated from the respective angular displacements and velocities. Additionally, phase angles (equation 4.1) for each of the respective segment phase plots were utilized to calculate the RP. RP detailed the organization of the neuromuscular system during the stance period (Scholz and Kelso, 1989; Kelso et al., 1991). RP was calculated by subtracting the phase angles of the corresponding segments throughout the stance:

$$\Phi_{\text{sagittal foot-shank rel. phase}} = \Phi_{\text{foot}} - \Phi_{\text{shank}} \qquad (4.8)$$

$$\Phi_{\text{sagittal shank-thigh rel. phase}} = \Phi_{\text{shank}} - \Phi_{\text{thigh}} \qquad (4.9)$$

where $\Phi_{\text{sagittal foot-shank rel. phase}}$ is the relative phase angle between the foot and shank segments, $\Phi_{\text{sagittal shank-thigh rel. phase}}$ is the relative phase angle between the shank and thigh segments, $\Phi_{\text{foot}}$ is the phase angle of the foot segment, $\Phi_{\text{shank}}$ is the phase angle of the shank segment, and $\Phi_{\text{thigh}}$ is the phase angle of the thigh segment. RP represented the phasing relationships between the actions of the two interacting segments at every point during the stance. Values close to zero degrees indicated that the two segments moved in the same fashion, or were in-phase. Values close to 180° indicated that the two segments moved in opposite directions or out-of-phase. The RP curves for each segmental relationship were averaged across stance periods (trials), and mean ensemble curves were generated for each subject-condition.

To statistically test differences between the RP curves, it was necessary to characterize the curves by single numbers. Therefore, the mean absolute values of the ensemble RP curve values (MARP) were calculated by averaging the absolute values of the ensemble curve points for the stance period (equation 4.3). A low MARP value indicated a more in-phase relationship between the interacting segments; a high MARP value indicated that the neuromuscular organization was more out-of-phase. Group means were calculated for the MARP of each segmental relationship. Statistical differences between the two groups (ACL reconstruction vs. control) were noted with an independent $t$-test ($p < 0.05$).

## RESULTS

Graphical evaluation of the phase plots for the respective segments during walking indicated that the healthy control subjects and the subjects with reconstructed ACLs tended to conform to different attractors (figure 4.11). This observation was based on the fact that the shapes of the trajectories were quite different for the two groups. Differences in the configuration of the phase plot were most evident at the shank and thigh segments (figure 4.11). The ACL-reconstruction shank phase plot had a configuration that lacked the distinct cusp during midstance (figure 4.11b). Additionally, the shank segment (figure 4.11b) had a concave-down configuration in the late portion of the stance, while the healthy control shank segment had a concave-up configuration. The thigh segment for subjects with ACL reconstruction had more local minimums and maximums throughout the phase plot trajectory compared to the healthy control (figure 4.11c).

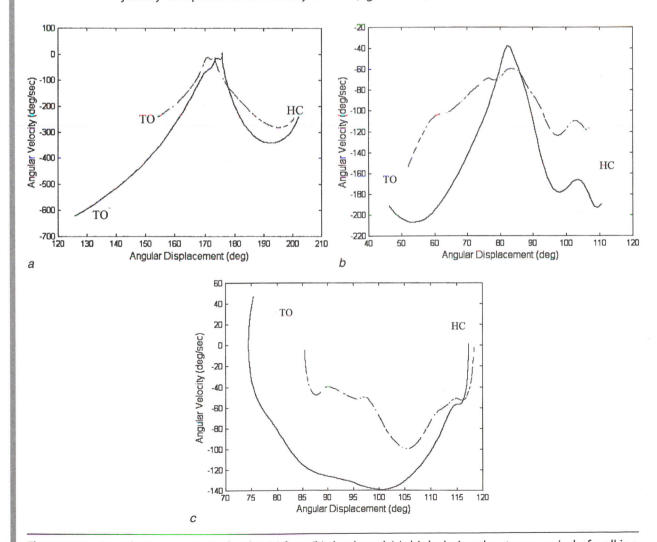

**Figure 4.11**  Partial phase portraits for the (a) foot, (b) shank, and (c) thigh during the stance period of walking. The solid line represents an exemplar healthy control subject, and the dashed line represents an exemplar subject with ACL reconstruction. HC represents heel-contact and TO represents toe-off.

*(continued)*

Further graphical differences in the shapes of the phase plot configuration were evident during running (figure 4.12). The ACL-reconstruction foot phase plot was grossly different late in stance, where a local minimum was present and there was a change in direction (zero cross) in the trajectory (figure 4.12*a*). The shank segment was also quite different (figure 4.12*b*); the healthy control had a zero cross late in the stance period, and the two subjects had different local minimums and maximums throughout the phase plot trajectory. The thigh phase plots had a similar general configuration (figure 4.12*c*), but the ACL-reconstruction phase plot was missing a local maximum early in the stance period. Based on qualitative evaluation of the configuration of the phase plots for the respective segment-conditions (walk and run), it was apparent that the organization of the neuromuscular system was different for the ACL-reconstruction and healthy control subjects.

Evaluation of the graphical configuration of the RP curves for walking and running provided further evidence that the organization of the neuromuscular system for the ACL reconstruction subjects and healthy controls was different (figures 4.13 and 4.14). During

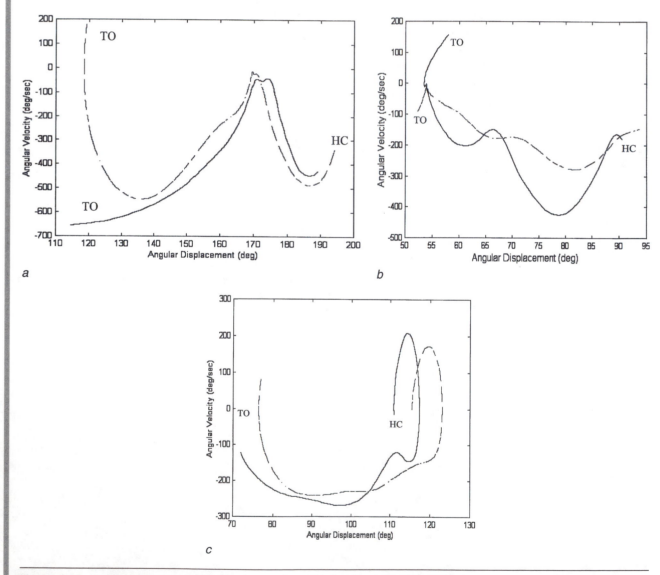

*a*

*b*

*c*

**Figure 4.12**   Partial phase portraits for the *(a)* foot, *(b)* shank, and *(c)* thigh during the stance period of running. The solid line represents an exemplar healthy control subject, and the dashed line represents an exemplar ACL-reconstruction subject. HC represents heel-contact and TO represents toe-off.

midstance there were noticeable differences in the relative phasing between the foot and the shank (figure 4.13a). The RP for the ACL-reconstruction exemplar subject had a diminished local minimum and a more out-of-phase pattern during midstance. Additionally, there were differences in the timings and values at the local minimum and maximums.

Further qualitative differences were noted for the shank-thigh RP relationship (figure 4.13b). During the midstance there was not a distinct local maximum in the RP relationship for the individual with ACL reconstruction. In fact, inspection of the RP curve for the ACL reconstruction subject indicated that there were multiple fluctuations between local minimum and maximum during midstance. Local minimums and maximums suggest a change in direction of the relationship between the two segments.

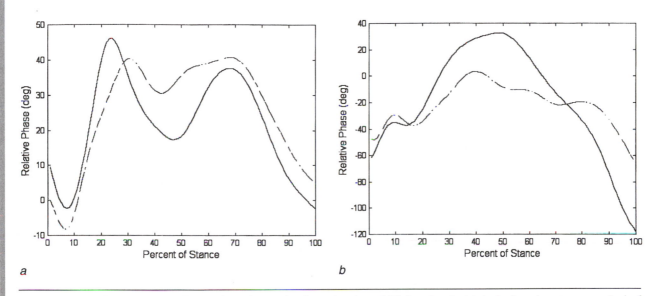

**Figure 4.13**    Relative phase relationships for *(a)* the foot-shank and *(b)* the shank-thigh during the stance period of walking. The solid line represents an exemplar healthy control subject, and the dashed line represents an exemplar ACL-reconstruction subject.

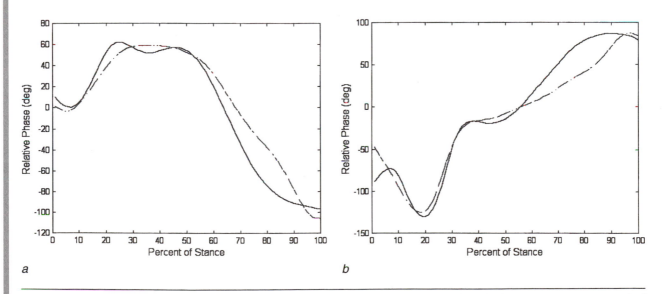

**Figure 4.14**    Relative phase relationships for *(a)* the foot-shank and *(b)* the shank-thigh during the stance period of running. The solid line represents an exemplar healthy control subject, and the dashed line represents an exemplar ACL-reconstruction subject.

*(continued)*

During running, there were further notable qualitative differences between the ACL-reconstruction and healthy control RP for the foot-shank and shank-thigh (figure 4.14). The ACL-reconstruction foot-shank RP lacked a local minimum during midstance (figure 4.14a). Additionally, the ACL-reconstruction foot-shank had a slight concave-up configuration during the late portion of the stance while the healthy control foot-shank had a concave-down configuration (figure 4.14a).

Further differences were noted in the shank-thigh RP (figure 4.14b). Early in the stance there was a noticeable difference in the RP configuration where the ACL-reconstruction RP lacked a local maximum. Additionally, the shank-thigh RP from midstance to the local maximum during late stance was quite different for the ACL-reconstructed subject. The qualitative changes in the RP configurations made it apparent that the ACL reconstruction and healthy control subjects had different neuromuscular system organizations for completing walking and running.

Statistical MARP differences between the groups were noted for both the foot-shank ($p < 0.05$) and the shank-thigh ($p < 0.05$; figure 4.15a) for walking. During running, statistical differences between the groups were noted only for the foot-shank ($p < 0.05$; figure 4.15b).

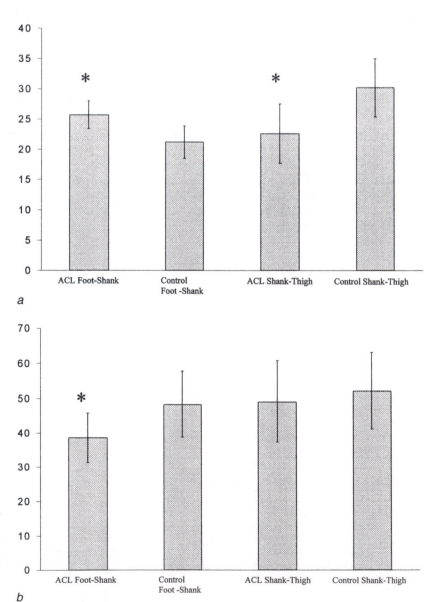

**Figure 4.15** Mean absolute relative phase (MARP) for healthy controls and ACL-reconstructed groups: (a) walking and (b) running.

Asterisk indicates a significant difference between the groups at the .05 alpha level.

## DISCUSSION

Our hypothesis that individuals with ACL reconstruction would have different neuromuscular adaptations during stance period while walking and running was supported in this investigation. From a DST standpoint, these neuromuscular adaptations may be related to changes in the biological constraints during surgical reconstruction of the joint. Previous studies have indicated that ACL surgery alters the structural features of the knee joint and sensory information (Friden et al., 2001). Essentially, changes in the biological constraints of the joint may force the neuromuscular system to seek a new adaptation for completing a functional gait pattern. Since the subjects in this investigation were completely rehabilitated from their surgery (average 3.4 years postsurgery), this would suggest that these adaptations in the organization of the neuromuscular system may be a learned response from multiple previous gait cycles.

The results of this investigation also indicated that neuromuscular adaptations are not localized at the knee joint. During walking, the subjects with ACL reconstruction displayed significantly different adaptations that encompassed both the ankle and the knee joint. This would suggest that identification of neuromuscular adaptations should include an assessment of the entire lower extremity. Interestingly, no significant differences were evident for the shank-thigh RP between the ACL-reconstruction subjects and healthy controls during running. This lack of difference may possibly be attributable to the increased inertial contributions of the segments during running. An alternative explanation is that adaptive changes in the other joints, such as the ankle and possibly the hip, helped to maintain control at the knee. Overall our results emphasize the importance of evaluating adaptations that occur at all the lower extremity joints. Focusing the evaluation primarily on the knee joint may not reveal the true neuromuscular adaptations from ACL reconstruction. A clinical evaluation that focuses only on the dynamics of the knee may miss neuromuscular adaptations that occur elsewhere as a result of ACL reconstruction.

Traditional clinical measures of joint function in persons with ACL reconstruction have been based on isokinetic testing of muscular strength of the rehabilitated knee joint (Snyder-Mackler et al., 1993, 1995). Isokinetic testing provides a functional assessment of the knee and of the surgical and rehabilitative protocols applied. However, isokinetic testing tells nothing about the actual adaptations that may occur during daily dynamic activities (e.g., locomotion). Although the rehabilitated knee may have acceptable strength, the state of the neuromuscular system after rehabilitation may still may be unknown with isokinetic testing. The use of DST tools enables exploration of neuromuscular adaptations encompassing all the lower extremity joints. Potentially, DST may provide a better way to classify the various stages of rehabilitation and various ACL surgical interventions.

In summary, the investigation in this example indicated that individuals with an ACL reconstruction display neuromuscular adaptations during the stance period while walking and running. These neuromuscular adaptations may be related to the surgical reconstruction of the joint. RP measures are novel compared to traditional biomechanical measures (e.g., joint displacement and joint moments) in that they compress displacement and velocity of the respective segments into a low-dimensional term. The results of this investigation indicate that such an approach will help to classify neuromuscular adaptations at various stages of rehabilitation and with various surgical interventions. Further research using DST tools is necessary to advance understanding of the organization of the neuromuscular system after an ACL reconstruction.

# Summary

As mentioned at the beginning of this chapter, a functional gait pattern involves the integration of multiple degrees of freedom that arise from internal factors, environmental factors, and the task constraints. It can be suggested that organizing all these degrees of freedom is a major problem especially in abnormal gait patterns. Such difficulties were evident in the two investigations presented in this chapter as examples. This chapter presented DST tools that can be used to explore how the neuromuscular system is organized during gait. The uniqueness of the RP measure presented in this chapter is that it compresses four variables (i.e., proximal and distal segments' displacement and velocities) into one measure. Incorporating velocity and displacement into one measure has an advantage over other biomechanical measures as suggested by evidence in the literature that there are both displacement and velocity receptors in the joint structures (McCloskey, 1978). Additionally, by examining the phase angles of two interacting segments we can better understand coordination. Effectively one can view coordination as the "mastery of redundant degrees of freedom" in the human body (Bernstein, 1967; Turvey, 1990). With the DST tools presented in this chapter, we hope that we have inspired the reader to further investigate the organization of the neuromuscular system during gait. Perhaps the use of DST tools will provide further insights into movement abnormalities and the effectiveness of rehabilitative and surgical protocols.

# Work Problems

Included in appendix C are two sample kinematic data sets for a subject walking and running. The kinematic data were collected at 60 Hz. The respective data sets contain sagittal angles (relative to the right horizontal) and angular velocities for the thigh and shank segments during the stance period. Using the data sets provided, calculate the following DST measures for walking and running.

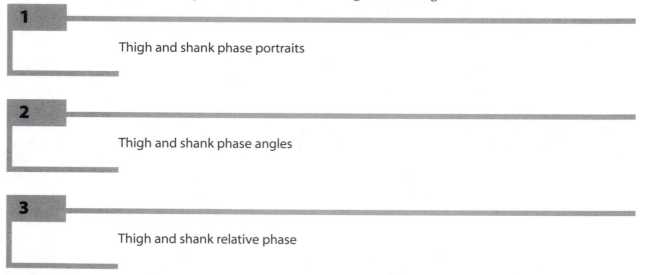

**1** Thigh and shank phase portraits

**2** Thigh and shank phase angles

**3** Thigh and shank relative phase

# Suggested Readings and Other Resources

Fuchs, A., V.K. Jirsa, H. Haken, and J.A.S. Kelso. 1996. Extending the HKB model of coordination movement to oscillators with different eigenfreqencies. *Biological Cybernetics* 74:21-30. This investigation provides the reader with further insight on relative phase artifacts and issues related to normalization.

Kelso, J.A.S. 1995. *Dynamic Patterns.* Boston: MIT Press. This book provides the reader with additional insight on the theoretical issues of DST.

Kurz, M.J., and N. Stergiou. 2002. Effect of normalization and phase angle calculations on continuous relative phase. *Journal of Biomechanics* 35(3):369-374. This investigation offers further information on the calculation of relative phase and issues related to amplitude normalization.

Stergiou, N., J.L. Jensen, B.T. Bates, S.D. Scholten, and G. Tzetzis. 2001. A dynamical systems investigation of lower extremity coordination. *Clinical Biomechanics* 16:213-221. This study provides the reader with further insight on the application of DST tools for the evaluation of running mechanics. The investigators propose a viable alternative to examine questions in running injuries.

Stergiou, N., S.D. Scholten, J.L. Jensen, and D. Blanke. 2001. Intralimb coordination following obstacle clearance during running: The effect of obstacle height. *Gait and Posture* 13:210-220. This investigation offers readers further insight for the application of DST tools to evaluate behavioral transitions during locomotion.

# References

Andriacchi, T., and D. Birac. 1993. Functional testing in the anterior cruciate ligament-deficient knee. *Clinical Orthopaedics and Related Research* 228:40-47.

Arbarbanel, H.D.I. 1996. *Analysis of Observed Chaotic Data.* New York: Springer-Verlag.

Barela, J.A., J. Whitall, P. Black, and J.E. Clark. 2000. An examination of constraints affecting the intralimb coordination of hemiparetic gait. *Human Movement Science* 19:251-273.

Beek, P.J., and W.J. Beek. 1988. Tools for constructing dynamical models of rhythmic movement. *Human Movement Science* 7:301-342.

Bendall, M.J., E.J. Bassey, and M.B. Pearson. 1989. Factors affecting walking speed of elderly people. *Age and Ageing* 18:327-332.

Berchuck, M., T. Andriacchi, B. Bach, and B. Reider. 1990. Gait adaptations by patients who have a deficient anterior cruciate ligament. *Journal of Bone and Joint Surgery* 72A:871-887.

Bernstein, N. 1967. *Coordination and Regulation of Movement.* New York: Pergamon Press.

Byrne, J.E., N. Stergiou, D. Blanke, J.J. Houser, and M. Kurz. 2002. Comparison of gait patterns between young and elderly women: An examination of coordination. *Perceptual and Motor Skills* 94:265-280.

Burgess-Limerick, R., B. Abernethy, and R.J. Neal. 1993. Relative phase quantifies interjoint coordination. *Journal of Biomechanics* 26(1):91-94.

Clark, J.E., and S.J. Phillips. 1993. A longitudinal study of intralimb coordination in the first year of independent walking: A dynamical systems analysis. *Child Development* 64:1143-1157.

Clark, J.E., S.J. Phillips, and R. Petersen. 1989. Developmental stability in jumping. *Developmental Psychology* 25:929-935.

Clark, J.E., T.L. Truly, and S.J. Phillips. 1990. On the development of walking as a limit cycle system. In: E. Thelen and L. Smith (Eds.), *Dynamical Systems in Development: Application* (pp. 71-93). Cambridge: MIT Press.

Clark, J.E., and J. Whitall. 1989. Changing patterns of locomotion: From walking to skipping. In: M.H. Woollacott and A. Shumway-Cook (Eds.), *Development of Posture and Gait Across the Life Span* (pp. 25-47). Columbia, SC: University of South Carolina Press.

De Vita, P., P. Blankenship-Hunter, and W. Skelly. 1992. Effects of a functional knee brace on the biomechanics of running. *Medicine and Science in Sports and Exercise* 24:797-806.

De Vita, P., T. Hortobagyi, and J. Barrier. 1998. Gait biomechanics are not normal after anterior cruciate ligament reconstruction and accelerated rehabilitation. *Medicine and Science in Sports and Exercise* 30(10):1481-1488.

De Vita, P., T. Hortobagyi, J. Barrier, M. Torry, K.L. Glover, D.L. Speroni, J. Money, and M.T. Mahar. 1997. Gait adaptations before and after anterior cruciate ligament reconstruction surgery. *Medicine and Science in Sports and Exercise* 29(7):853-859.

Diedrich, F.J., and W.H. Warren. 1995. Why change gaits? Dynamics of the walk-run transition. *Journal of Experimental Psychology: Human Perception and Performance* 21(1):183-202.

Friden, T., D. Roberts, E. Ageberg, M. Walden, and R. Zatterstrom. 2001. Review of knee proprioception and the relation to extremity function after an anterior cruciate ligament rupture. *Journal of Orthopaedic and Sports Physical Therapy* 31(10):567-576.

Fuchs, A., V.K. Jirsa, H. Haken, and J.A.S. Kelso. 1996. Extending the HKB model of coordination movement to oscillators with different eigenfreqencies. *Biological Cybernetics* 74:21-30.

Gabel, A., and U.S.L. Nayak. 1984. The effect of age on variability in gait. *Journal of Gerontology* 39(6):662-666.

Glass, L., and M.C. Mackey. 1988. *From Clocks to Chaos: The Rhythms of Life.* Princeton, NJ: Princeton University Press.

Haken, H., J.A.S. Kelso, and H. Bunz. 1985. A theoretical model of phase transitions in human hand movements. *Biological Cybernetics* 51:347-356.

Hamill, J., J.M. Haddad, and W.J. McDermott. 2000. Issues in quantifying variability from a dynamical systems perspective. *Journal of Applied Biomechanics* 16:407-418.

Hamill, J., R.E.A. van Emmerik, B.C. Heiderscheit, and L. Li. 1999. A dynamical systems approach to lower extremity running injuries. *Clinical Biomechanics* 14:297-308.

Hausdorff, J.M., H.K. Edelberg, S.L. Mitchell, A.L. Goldberger, and J.Y. Wei. 1997. Increased gait unsteadiness in community-dwelling elderly fallers. *Archives of Physical Medicine and Rehabilitation* 78:278-283.

Heiderscheit, B.C., J. Hamill, and R.E.A. van Emmerik. 1998. Q-angle influence on the variability of lower extremity coordination during running. *Medicine and Science in Sports and Exercise* 31(9):1313-1319.

Higgins, S. 1985. Movement as an emergent form: Its structural limits. *Human Movement Science* 4:119-148.

Holt, K.G., and S.F. Jeng. 1992. Advances in biomechanical analysis of the physically challenged child: Cerebral palsy. *Pediatric Exercise Science* 4:213-235.

Judge, J.O., R.B. Davis, and S. Ounpuu. 1996. Step length reductions in advanced age: The role of ankle and hip kinetics. *Journal of Gerontology* 51A(6):303-312.

Kelso, J.A.S. 1995. *Dynamic Patterns.* Boston: MIT Press.

Kelso, J.A.S. 1997. Relative timing in brain and behavior: Some observations about the generalized motor program and self-organized coordination dynamics. *Human Movement Science* 16:453-460.

Kelso, J.A.S., J.J. Buchanan, and S.S. Wallace. 1991. Order parameters for the neural organization of single, multijoint limb movement patterns. *Experimental Brain Research* 85:432-444.

Kelso, J.A.S., and M. Ding. 1993. Fluctuations, intermittency, and controllable chaos in biological coordination. In: K.M. Newell and D.M. Corcos (Eds.), *Variability and Motor Control* (pp. 291-316). Champaign, IL: Human Kinetics.

Kelso, J.A.S., K.G. Holt, P. Rubin, and P.N. Kugler. 1981. Patterns of human interlimb coordination emerge from the properties of non-linear, limit cycle oscillatory processes: Theory and data. *Journal of Motor Behavior* 13:226-261.

Kelso, J.A.S., E.L. Saltzman, and B. Tuller. 1986. The dynamical perspective on speech production: Data and theory. *Journal of Phonetics* 14:29-59.

Kelso, J.A.S., and B. Tuller. 1984. A dynamical basis for action systems. In: M. Gazzaniga (Ed.), *Handbook of Cognitive Neuroscience* (pp. 321-356). New York: Plenum Press.

Kerrigan, D.C., M.K. Todd, U.D. Croce, L.A. Lipisitz, and J.A. Collins. 1998. Biomechanical gait alterations independent of speed in healthy elderly: Evidence for specific limiting impairments. *Archives of Physical Medicine and Rehabilitation* 79:317-322.

Kugler, P.N., and M.T. Turvey. 1987. *Information, Natural Law, and the Self-assembly of Rhythmic Movement.* Hillsdale, NJ: Erlbaum.

Kurz, M.J., and N. Stergiou. 2002. Effect of normalization and phase angle calculations on continuous relative phase. *Journal of Biomechanics* 35(3):369-374.

Li, L., E.C.H. van den Bogert, G.E. Caldwell, R.E.A. van Emmerik, and J. Hamill. 1999. Coordination patterns of walking and running at similar speed and stride frequency. *Human Movement Science* 18:67-85.

Lockman, J.L., and E. Thelen. 1993. Developmental biodynamics: Brain, body, behavior connections. *Child Development* 64:953-959.

Maki, B.E. 1997. Gait changes in older adults: Predictors of falls or indicators of fear? *Journal of the Geriatric Society* 45(4):313-320.

McCloskey, D.I. 1978. Kinesthetic sensibility. *Physiological Reviews* 58:763-820.

Norkin, C.C., and P.K. Levangie. 1992. *Joint Structure and Function: A Comprehensive Analysis.* Philadelphia: Davis.

Rosen, R. 1970. *Dynamical Systems Theory in Biology: Vol. 1. Stability and Its Application.* New York: Wiley.

Schmidt, R.C., P.J. Beek, P.J. Treffner, and M.T. Turvey. 1991. Dynamic substructure of coordinated rhythmic movements. *Journal of Experimental Psychology: Human Perception and Performance* 17(3):635-651.

Schmidt, R.C., P.J. Treffner, B.K. Shaw, and M.T. Turvey. 1992. Dynamical aspects of learning an interlimb rhythmic movement pattern. *Journal of Motor Behavior* 24:67-83.

Scholz, J.P. 1990. Dynamic pattern theory—some implications for therapeutics. *Physical Therapy* 70(12): 827-842.

Scholz, J.P., and J.A.S. Kelso. 1989. A quantitative approach to understanding the formation and change of coordinated movement patterns. *Journal of Motor Behavior* 21(2):122-144.

Selles, R.W., R.C. Wagenaar, T.H. Smit, and P.I.J.M. Wuisman. 2001. Disorders in trunk rotation during walking in patients with low back pain: A dynamical systems approach. *Clinical Biomechanics* 16:175-181.

Snyder-Mackler, L.S., S. Binder-McLeod, and P. Williams. 1993. Fatigability of human quadriceps femoris muscle following anterior cruciate ligament reconstruction. *Medicine and Science in Sports and Exercise* 25:783-789.

Snyder-Mackler, L., A. Delitto, S. Bailey, and S. Stralka. 1995. Strength of the quadriceps femoris muscle and functional recovery after reconstruction of the anterior cruciate ligament. *Journal of Bone and Joint Surgery* 77-A:1166-1173.

Stergiou, N., J.L. Jensen, B.T. Bates, S.D. Scholten, and G. Tzetzis. 2001a. A dynamical systems investigation of lower extremity coordination during running over obstacles. *Clinical Biomechanics* 16:213-221.

Stergiou, N., S.D. Scholten, J.L. Jensen, and D. Blanke. 2001b. Intralimb coordination following obstacle clearance during running: The effect of obstacle height. *Gait and Posture* 13:210-220.

Thelen, E. 1986. Treadmill-elicited stepping in seven-month-old infants. *Child Development* 57:1498-1506.

Thelen, E. 1989. The (re)discovery of motor development: Learning new things from an old field. *Developmental Psychology* 25:946-949.

Thelen, E., D.M. Fisher, and R. Ridley-Johnson. 1984. The relationship between physical growth and new-born reflex. *Infant Behavior and Development* 7:479-493.

Thelen, E., J.A.S. Kelso, and A. Fogel. 1987. Self-organizing systems and infant motor development. *Developmental Review* 7:479-493.

Thelen, E., and B.D. Ulrich. 1991. Hidden skills: A dynamic analysis of treadmill stepping during the first year. *Monographs of the Society for Research in Child Development* 56(223).

Timoney, J.M., W.S. Inman, P.M. Quesada, P.F. Sharkey, R.L. Barrack, H.B. Skinner, and A.H. Alexander. 1993. Return of normal gait patterns after anterior cruciate ligament reconstruction. *American Journal of Sports Medicine* 21:887-889.

Turvey, M.T. 1990. Coordination. *American Psychologist* 45(8):938-953.

van Emmerik, R.E.A., and R.C. Wagenaar. 1996. Effects of walking velocity on relative phase dynamics in the trunk in human walking. *Journal of Biomechanics* 29(9):1175-1184.

van Emmerik, R.E.A., R.C. Wagenaar, A. Winogrodzka, and E.C. Wolters. 1999. Identification of axial rigidity during locomotion in Parkinson disease. *Archives of Physical Medicine and Rehabilitation* 80:186-191.

Whanger, A., and H.S. Wang. 1974. Clinical correlates of vibratory sense in elderly psychiatric patients. *Journal of Gerontology* 29:39-45.

Whipple, R.H., L.I. Wolfson, and P.M. Amerman. 1987. The relationship of knee and ankle weakness to falls in the nursing home residents: An isokinetic study. *Journal of the American Geriatrics Society* 35:13-20.

Whitall, J. 1989. A developmental study of interlimb coordination in running and galloping. *Journal of Motor Behavior* 21(4):409-428.

Winstein, C.J., and A. Garfinkel. 1989. Qualitative dynamics of disordered human locomotion: A preliminary investigation. *Journal of Motor Behavior* 21(4):373-391.

Winter, D.A. 1991. *The Biomechanics and Motor Control of Human Gait: Normal, Elderly, and Pathological.* Waterloo, Ontario: University of Waterloo Press.

Winter, D.A., A.E. Patla, J.S. Frank, and S.E. Walt. 1990. Biomechanical walking pattern changes in the fit and healthy elderly. *Physical Therapy* 70:340-347.

Woollacott, M.H., A. Shumway-Cook, and L.M. Nasher. 1986. Aging and postural control: Changes in sensory organization and muscular coordination. *International Journal of Aging and Human Development* 23:97-114.

CHAPTER **5**

# Directional Statistics

Patrick J. Sparto, PhD, PT, and Robert H. Schor, PhD

We use statistics to make statements about the probability associated with observations and data. Many types of data can be associated with some subset of the real numbers; examples include counts of numbers of events or measurements of time, length, weight, or similar physical quantities. Such data are characterized by being ordered, with a minimum and maximum value, and by a measure of "central tendency," whether expressed as an **arithmetic mean** or as a **nonparametric** median. Other types of data, however, are cyclical, or repetitive, data in which the concept of "maximum value" may not be defined. Examples include direction of heading, daily (or weekly or annual) time-of-occurrence, or the phase associated with the response to a periodic stimulus. Such data are more naturally represented when they are plotted as points around a circle rather than as points along a line segment. The treatment of the probabilities of data that show such repetitive qualities is the basis of directional (or "circular") statistics.

## Why Are Directional Statistics Needed?

For a quick example of the difference between "linear" and "circular" statistics, consider the following problem: What is the mean time of occurrence of births in Chicago? If we ask this question focusing on a specific day, say April 13, 1987, we can guess that the mean time will be somewhere around noon, assuming that there were approximately as many births before noon as after noon on that particular day. On the other hand, if we consider the question in general, for no specific day, the concept of the "mean time of birth" becomes much more difficult to define; indeed, if we assume that, on average, the same number of births occurs during each hour of the day, it is not clear that we can even define such a mean.

In the first part of this example, we were able to convert the circular problem of time of day to a linear problem by considering that the day begins and ends at midnight—we effectively "cut" the circular plot at midnight and unrolled the time data into a straight line. For such "linearized" data, one can discuss familiar statistical quantities such as mean and standard deviation. In many situations, one can also show that the data samples follow a normal distribution. Suppose we ask "When do people in Chicago eat lunch?" If we "unroll" the day by cutting the circle at midnight, we might find that the distribution of lunchtimes is fairly well approximated by a normal distribution with a mean near noon and a standard deviation of perhaps an hour. Such a distribution will be insignificant by six standard deviations away from the mean (i.e., at 6 A.M. and 6 P.M.). On the other hand, if we ask "When do people in the United States eat a meal?" the distribution of times will clearly be much wider. In particular, if we attempted to approximate this distribution by a normal distribution with a standard deviation of, say, 5 hours, then the plot of such a normal distribution would extend more than 24 hours before effectively vanishing. The tails of the distribution, when it is wrapped around the 24-hour circle, are going to alter the probabilities in ways we will explore in this chapter.

There is a well-developed theory for "linear" statistics, including fairly simple derivations of probability densities for standard models. The probability density function for the normal (Gaussian) distribution, $f(x)$, is characterized by two parameters—the arithmetic mean $\mu$ and the standard deviation $\sigma$.

$$f(x) = \frac{1}{\sigma\sqrt{2\pi}}\exp\left(-\frac{(x-\mu)^2}{2\sigma^2}\right) \qquad (5.1)$$

Recall that this function gives the probability of observing an event in a small region near "$x$." Examination of the function shows that it reaches its maximum at $x = \mu$, that it is symmetric about this mean (the squaring operation treats positive and negative deviations from the mean as the same), and that $\sigma$ governs how quickly the function gets small (to look at this another way, $\sigma$ acts like a scale factor, asking "How many standard deviations are we from the mean?"). The factor in front of the exponential arises from the requirement that the integral of this distribution function over all possible values of $x$ (e.g., from $-\infty$ to $+\infty$) is unity. Other distributions, such as the Student's-t distribution and the $\chi^2$ distribution, arise in the development of statistical testing and can be derived in a straightforward manner.

In contrast, models for circular distributions are not as intuitive, nor as well studied. One that is often encountered is the von Mises distribution, which can be considered a "circular analog" of the normal distribution. As we will see later, many tests for circular parameters are based on computations derived from the von Mises distribution. One will also encounter "wrapped" distributions, such as the wrapped normal distribution already mentioned.

The von Mises distribution, $f(\phi)$, like the normal distribution, is characterized by two parameters, a mean angle $\bar{\theta}$ and a "concentration parameter" $\kappa$ (which is analogous to the reciprocal of the variance).

$$f(\phi) = \frac{1}{2\pi I_0(\kappa)}\exp\left(\kappa\cos\left(\phi - \bar{\theta}\right)\right) \qquad (5.2)$$

For comparison, a plot of both a normal and a von Mises distribution function is shown in figure 5.1. Both a conventional "linear" plot (with angle varying between $-180°$ and $+180°$) and a polar "circular" plot are presented. The parameters of the functions are a mean angle ($\mu$ or $\bar{\theta}$) of 30°, a standard deviation ($\sigma$) of 50°, and a concentration parameter ($\kappa$) of 1.3. Like the normal distribution, the von Mises distribution attains its maximum at the mean and is symmetric about the mean (the symmetry here comes from the cosine function, which is symmetric about 0). The leading coefficient that involves a modified Bessel function of order zero again arises from the requirement that the integral of this function, here from 0 to $2\pi$, must be unity.

The lack of simple mathematical models for circular statistical situations means that the rich variety of tests for linear models is often not available when one is dealing with circular data. Indeed, as we will see later, one technique that is quite effective when the data can be "bunched" in a sector of the circle is to cut the circle, straighten it out, and thereby transform the problem into an equivalent "linear statistics" case.

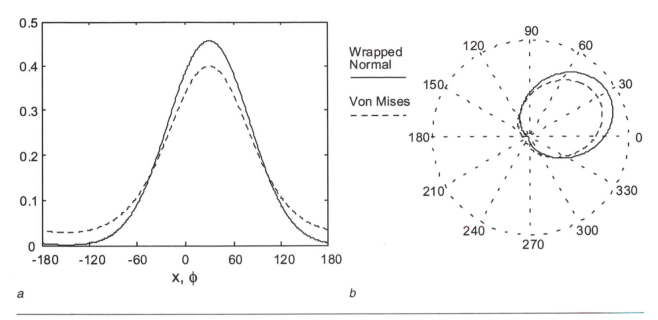

**Figure 5.1** Probability density functions of circular distributions. The probability density functions associated with two common circular distributions, the wrapped normal distribution (a normal distribution that has been "wrapped" around a circle) and the Von Mises distribution, have been plotted in both (a) rectangular and (b) polar coordinates. For both distributions, the "mean" parameter has been set to 30°. The wrapped normal distribution has a standard deviation of 50°, while the von Mises distribution has a concentration parameter κ equal to 1.3 (the reciprocal of the variance of the normal distribution).

# Examples of Directional Statistics

Directional statistics rather naturally arise when one is studying directional phenomena—many of the techniques that have been developed for using circular statistics come from studies of animal orienting behavior. Questions such as "Does the dance pattern of honey bees convey information about the location of the nectar?" or "Do homing pigeons use the magnetic field of the earth to orient their flight?" can lead to data amenable to analysis using directional statistics.

In addition to such "spatial" questions, directional statistics can play a role in two types of temporal data. The first is in analyses of circadian (or other time-periodic) behavior. A key question is often the simple one of "Is there a periodicity to the behavior?" If we predetermine the period of interest (e.g., if we rephrase the question and ask "Does this behavior occur at the same time every day?"), we are now asking if the data depart from a **uniform distribution** on the circle. A related temporal question arises in the study of the response to a periodic—typically a sinusoidal—stimulus, in which the parameters of the response are typically reported as a gain and phase (with respect to the stimulus waveform). Phase is a circular quantity that, if analyzed in isolation from the response gain, needs to be handled using the techniques of directional statistics. In particular, the computation of a "mean phase" through simply forming the arithmetic mean of a set of phase data has little statistical validity. In the field of human movement analysis, the technique of dynamical systems analysis as described in chapter 4 is an ideal application for directional statistics (Kurz and Stergiou, 2002; van Emmerik and Wagenaar, 1996; Kelso, 1984; Schoner and Kelso, 1988; Scholz et al., 1987).

# Representation of Circular and Axial Data

All directional (or circular) data can be represented in polar coordinates by a unit vector at some angle θ with respect to the +x-axis. This form of representation captures and maintains the "circular" characteristic of these data, indicating, for example, that points at 359° and 1° are very close to each other. As these data are characterized by a single parameter (here called θ), one could also represent the data by "unrolling" the circle and letting θ vary over its range (e.g., 0°-360° or 0-24 hours); in this latter case, the reader needs to realize that the ends of the line are really "next to" each other.

If the number of data points is fairly small, each point can be plotted as the end of a unit radius oriented at the characteristic polar angle θ (figure 5.2a). If necessary, the (unit) radius for each point can be varied slightly (say from 0.9 to 1.1) to avoid plotting points on top of each other (figure 5.2b). Alternatively, one can construct a circular histogram by dividing the circle into equal-sized bins, counting the number of data points in each bin, then drawing a circular wedge whose width spans the bin and whose height is proportional to the number of points in that bin (figure 5.2c).

For some kinds of circular data, the "sign" of the direction is not important. For example, we might be concerned with the axis of sway for a person standing with one foot in front of the other. When we characterize circular data by describing its plane, we lose the distinction, for example, between right and left or between forward and

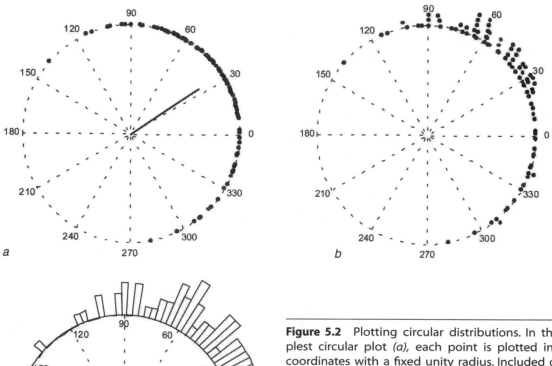

**Figure 5.2** Plotting circular distributions. In the simplest circular plot *(a)*, each point is plotted in polar coordinates with a fixed unity radius. Included on this plot is a line representing the mean vector (with each of the data points considered as a vector of unit length); the direction of this vector is a measure of central tendency for the data, and its length is inversely related to the dispersion of the data. In *(b)*, overlapping points have been plotted with differing radii to show all of the points more clearly. One can also construct a "circular histogram" *(c)* by counting the number of data points that fall in each circular bin.

backward. With data of this type we are more concerned with the *axis* (or plane) of the movement than with the true direction; data like these are called *axial data*. If we parameterize the orientation of these axes, instead of having angular values ranging over 360° we have axial values ranging over 180° (since both 0° and 180° lie on the same axis).

There are three "natural" ways to plot axial data. If we again parameterize the axis by an angle θ, we can plot the data on a circle using diameters at this orientation (figure 5.3*a*). Note that with axial data, we plot diameters instead of the radii we used in figure 5.2. A second way is to plot the points as radii on a half-circle, drawing the reader's attention to the fact that the ends of the semicircle are really adjacent to each other (figure 5.3*b*). We can graphically enforce this restriction by plotting the points using 2θ as the angular parameter, so that our points will, in fact, go around the circle (figure 5.3*c*).

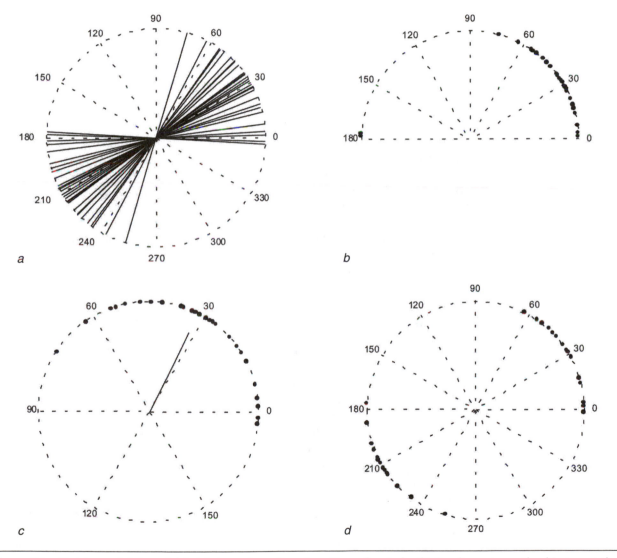

**Figure 5.3**  Plotting axial data. *(a)* Axial data plotted as diameters of a circle. *(b)* An alternate method is to cut the circle in half and plot the ends of the axes that fall in this semicircle. *(c)* This type of plot can also be "stretched" to include the full circle by doubling the polar angles before plotting the points and relabeling the angle scale from 0 to 180. Also shown on this figure is the mean vector calculated for this plot (see example 5.1). *(d)* Data that are inherently axial in nature often show a bimodal distribution. Such data can be converted into unimodal data by the "doubling the angle" procedure used to create figure 5.3*c*.

A closely related phenomenon is data that show a strong **bimodal** distribution. For example, some animals orient with respect to polarized light (the blue sky is naturally polarized). Here, the data will exhibit the full range of 360°, but they will cluster into two groups 180° from each other. Indeed, each observation can be considered a random sample from an axial distribution, where instead of plotting the axis, we randomly choose one or the other end of the diameter. An example is shown in figure 5.3*d*, where the points represent the ends of the axes illustrated in figure 5.3*a*. For such bimodal data, replotting using twice the angle converts the data into the **unimodal** representation of figure 5.3*c*.

# Descriptive Statistics

Data are "interesting" (or, alternatively, "significant") when they are bunched up in some way, that is, when they are not uniformly distributed. For linear data, the two most common statistics used to describe the general location of the data are the (arithmetic) mean (or "average") and the median. Given a data sample of *n* points, we can compute the sample mean as follows:

$$\mu = \frac{\sum_{i=1}^{n} x_i}{n} \tag{5.3}$$

We determine the median by sorting the samples and looking at the middle; if *n* is odd, we choose this middle value, whereas if *n* is even, common practice is to use the mean of the two middle values. Both of these measures are guaranteed to lie somewhere "in the middle" of the data; indeed, if the data are distributed in a symmetric manner, the two measures will largely coincide. The choice of which measure to use is based, in part, on the nature of the data themselves (many subjective scales—for example "bad, OK, good, excellent"—don't lend themselves to computation of a mean), as well as the apparent distribution of the data (a "very wild" outlier point can cause the mean to move outside the cluster formed by the rest of the data).

## *Central Tendency Measures*

How do we compute a measure of **central tendency** for circular data? One trick is to cut the circle somewhere, transforming the circular problem into a linear one. For example, we could cut the circle of figure 5.2*a* at –180°, and would then have points that appear to follow a normal distribution with mean near 30°. Note, however, that this method depends critically on where the circle is cut: If it is opened at 0°, so that the angles go from 0° to 360°, this will result in a bimodal distribution, and the arithmetic mean will yield a value near 120°. For an even better example, consider what the circular mean should be if the data are uniformly distributed, that is, if no particular angle is favored. One would argue that, unlike the situation with the linear case (where there is always a central value, even when the data are uniformly distributed), it does not make sense to define a "mean" when circular data are uniformly distributed.

If we consider our data points (e.g., those in figure 5.2) as unit vectors arrayed around the circle, we can define, and compute, a "mean vector" by simply adding together the vectors and dividing by *n*. Letting $\theta_i$ be the circular data value, we can write down the x-y coordinates of the vector, and from there can calculate the mean vector ($\bar{X}$, $\bar{Y}$), which we then express in polar coordinates ($\bar{\rho}$, $\bar{\theta}$) to get the mean vector length and mean angle, respectively.

$$x_i = \cos\theta_i \qquad \overline{X} = \frac{1}{n}\sum x_i$$

$$y_i = \sin\theta_i \qquad \overline{Y} = \frac{1}{n}\sum y_i \tag{5.4}$$

$$\overline{\rho} = \sqrt{\overline{X}^2 + \overline{Y}^2}$$

$$\overline{\theta} = \tan^{-1}\frac{\overline{Y}}{\overline{X}}$$

In the case of uniformly distributed data, the mean $\overline{X}$ and $\overline{Y}$ vector components will be zero, giving us a resultant vector of zero length $\overline{\rho}$ and an indeterminate angle $\overline{\theta}$. However, for all other cases, we can compute a mean angle and use it as a measure of central tendency. Such a mean was computed for the data shown in figure 5.2 and is indicated by a plot of the mean vector.

The median circular value $\hat{\theta}$ also has some interesting peculiarities. Consider a cut along a diameter that divides the circle in half; if the number of points is odd, make one end of the diameter go through one of the points; if the number is even, make one end of the diameter go halfway between a pair of adjacent points. If the number of points in each half-circle is the same, then the end of the diameter that is closer to the majority of the points is a median. A surprising property of circular distributions is that such a median is not unique, even with an odd number of points. The simplest example is three points around a circle such that the resulting triangle connecting the points has angles less than 90°—each of the points will fit the definition of a median in this case.

When the data, by inspection, clearly have a median value, the following algorithm will allow practical computation. The first step is to find the biggest "space" between adjacent points (one can do this by computing a series that is the difference between adjacent values, taking care, if there are values at, for example, 179° and –179°, to compute the difference as –2°, not 358°). The next step is to "cut the circle" in this space and readjust the angular scale so that it monotonically increases (thus if the cut was at –90°, the scale would run from –90° to 270°). The data are now arranged on a linear scale, and one can compute the (linear) median in the usual manner.

## Dispersion Measures

For linear data, common measures of **dispersion** include the variance, standard deviation, and range (or percentiles). Analogous measures can be defined for circular data. Although the linear standard deviation and variance are used to construct confidence intervals and perform hypothesis testing, the circular standard deviation and variance play a less direct role and are used mainly for descriptive purposes.

Recall that the mean angle of a sample of circular data was computed as the angular (polar) coordinate of the resulting mean data vector; we also obtained $\overline{\rho}$ as the length of this mean vector. This parameter will vary from 0 to 1; it will be 0 when the points are uniformly distributed (that is, maximally dispersed around the circle), and it will be 1 when they are all the same (a case of "minimum dispersion"). (Note that the inverse is not true—a mean length of 0 does not necessarily mean a uniform distribution.)

Since $\overline{\rho}$ is maximal when the data are most concentrated, $1 - \overline{\rho}$ will grow as the data are more dispersed. One definition for circular variance and circular standard deviation (Batschelet, 1981, p. 34) is as follows:

$$s_b^2 = 2(1 - \overline{\rho})$$

$$s_b = \sqrt{2(1 - \overline{\rho})} \tag{5.5}$$

The factor of 2 in the equation makes the circular variance conform to the linear variance when the data are clustered (i.e., when deviations of all the angles from the mean are small). As $\overline{\rho}$ goes from 1 (most concentrated, least "variance") to 0, the variance $s^2$ goes from 0 to 2.

An alternate definition for the circular standard deviation (Mardia and Jupp, 2000, p.18) is the following:

$$s_m = \sqrt{-2 \ln \overline{\rho}} \tag{5.6}$$

This definition has the property that the standard deviation becomes infinite when the vector length vanishes.

Circular variance and standard deviations are basically descriptive measures that can be applied to circular data—which definition one chooses is largely a matter of taste. As we will see later in the chapter, the mean vector length $\overline{\rho}$ is used in tests of significance and the calculation of confidence intervals.

A more straightforward measure of dispersion is the range of the circular data. This is the smallest interval that contains all of the data points. A practical way to compute the range is to look for adjacent points that have the largest space between them. These same points are the endpoints of the range. In particular, if $\phi$ is the largest space between adjacent points, the range will be 360°-$\phi$.

## Analyzing Axial or Strongly Bimodal Data

Axial or bimodal data, such as illustrated in figure 5.3, present a problem when one considers a mean angle or a measure of dispersion. If the data are truly axial and are plotted as shown in figure 5.3a, one can certainly naturally define the nonparametric measures for the median and range of the axial data. Alternatively, for the purposes of computation, one converts the bimodal (or axial) data set into a unimodal (or radial) data set by doubling the angular parameter.

The equations for the computation of the mean and standard deviation for axial data, corresponding to equations 5.4 and 5.5, are as follows:

$$x_{axial,i} = \cos 2\theta_i \quad \overline{X}_{axial} = \frac{1}{n}\sum x_{axial,i}$$

$$y_{axial,i} = \sin 2\theta_i \quad \overline{Y}_{axial} = \frac{1}{n}\sum y_{axial,i}$$

$$\overline{\rho}_{axial} = \sqrt{\overline{X}_{axial}^2 + \overline{Y}_{axial}^2} \tag{5.7}$$

$$\overline{\theta}_{axial} = \frac{1}{2}\tan^{-1}\frac{\overline{Y}_{axial}}{\overline{X}_{axial}}$$

$$s_{axial,b} = \frac{1}{2}\sqrt{2(1 - \overline{\rho}_{axial})} \tag{5.8}$$

$$s_{axial,m} = \frac{1}{2}\sqrt{-2 \ln \overline{\rho}_{axial}} \tag{5.9}$$

Table 5.1 gives the measures of central tendency and dispersion for the unimodal and bimodal data illustrated in figures 5.2 and 5.3.

**Table 5.1  Descriptive Statistics**

| Parameter | Unimodal data N($\mu = 30°$; $\sigma = 50°$) | Bimodal, axial data N($\mu = 30°$; $\sigma = 20°$) |
|---|---|---|
| Mean angle, $\bar{\theta}$ | 33.5° | 31.8° |
| Median angle, $\hat{\theta}$ | 35.8° | 32.4° |
| Standard deviation (Batschelet, 1981), $S_b$ | 40.5° | 18.3° |
| Standard deviation (Mardia and Jupp, 2000), $S_m$ | 43.5° | 19.3° |
| Range | 218° | 77° |
| Mean resultant length, $\bar{\rho}$ | .75 | .80 |

The mean and median angles, standard deviation, range, and mean resultant length are common descriptive statistics used to describe directional statistics. The values are calculated from the data displayed in figures 5.2 (unimodal) and 5.3 (bimodal or axial).

# Second-Order Analysis

In the area of human movement analysis, it is very common for subjects to perform multiple trials of a distinct experimental condition so that a better estimate of the particular measure for each subject can be obtained. Batschelet (1981) refers to this mean value of multiple trials obtained from one subject as first-order analysis. The first-order mean direction and vector length can be determined using equation 5.4. On the other hand, the grand mean direction from a set of first-order mean directions obtained from $n$ subjects is called second-order analysis.

Consider the following hypothetical example. Researcher A wants to know the maximum knee extension torque produced by subjects with advanced knee osteoarthritis. The researcher recruits 10 subjects and measures the isometric knee extension strength five times. Assuming that the knee strength is a linear variable, the computation of the mean knee strength is straightforward. The arithmetic mean strength during the five repetitions is calculated for each subject (first-order analysis), and then the grand mean is computed from these 10 individual subject mean values (second order). However, this same procedure cannot be applied in the circular case, because we would need to assume that each individual mean direction had a mean resultant length of 1.

The computation of the grand mean direction for second-order analysis is straightforward. Continuing with the previous example, suppose that the researcher also measured the knee flexion/extension excursion over five consecutive gait cycles and assumed that this measure was a circular variable. For each subject, the first-order analysis would compute the mean direction and mean resultant length of knee excursion from the five repetitions, using equation 5.4. In order to find the grand mean across all subjects, one would construct a Cartesian vector from each of the 10 polar pairs of mean resultant length and mean direction ($\bar{\rho}_i$, $\bar{\theta}_i$). After representing each vector in Cartesian coordinates, it is easy to obtain the grand mean vector ($\bar{X}_{so}$, $\bar{Y}_{so}$) and thus grand length and grand direction ($\bar{\rho}_{so}$, $\bar{\theta}_{so}$). Although ideally the number of trials for each subject should be equal, Zar (1999, p. 610) states that unequal trial numbers do not substantially alter the outcome.

$$\overline{X}_{so} = \frac{\sum\limits_{i=1}^{n} \overline{\rho}_i \cos\overline{\theta}_i}{n}$$

$$\overline{Y}_{so} = \frac{\sum\limits_{i=1}^{n} \overline{\rho}_i \sin\overline{\theta}_i}{n}$$

$$\overline{\rho}_{so} = \sqrt{\overline{X}_{so}^{\,2} + \overline{Y}_{so}^{\,2}}$$

$$\overline{\theta}_{so} = \tan^{-1}\frac{\overline{Y}_{so}}{\overline{X}_{so}}$$

(5.10)

# Tests of Uniformity

Given a sample of directional (or cyclical, that is "time-directional") data, we want to be able to draw some conclusions about these data. For example, we might want to ask "Are the animals going north?" or "Are rats more active at certain times of the day?" or "Is the phase of postural sway to a sinusoidally moving visual stimulus the same in children as in adults?"

These questions all involve making inferences on measures of central tendency of our data. However, the first question that needs to be addressed is "Is there any central tendency of the data?" The **null hypothesis** against which we need to test is that the data are randomly distributed, so that the underlying distribution function is a uniform distribution (that is, all directions are equally probable).

As is true with any data sample, whether from a linear or a circular distribution, a helpful first step is to plot the data and see if any patterns, or "clumps," emerge. Recall that in discussing measures of central tendency earlier, we defined the mean angle $\overline{\theta}$ and calculated the mean vector length $\overline{\rho}$, which we related to the circular standard deviation. In particular, we noted that if the data are uniformly distributed, the mean vector length $\overline{\rho}$ is zero. We also noted, however, that the converse is not necessarily true; that is, a zero vector length does not necessarily imply that the underlying distribution is uniform. An intuitive example of this is the data set illustrated in figure 5.3c, a symmetric bimodal distribution.

## Tests for Unimodal Data

A number of tests can be employed to test the hypothesis that the samples are drawn from a uniform distribution (this, of course, is the null hypothesis we hope to refute). Note that if the null hypothesis is rejected, then the alternate hypothesis is that the data possess a "net direction"—a concept that makes sense for unimodal, but not multimodal, data. We will therefore assume that an examination of the data does not reveal multiple clumps in the sample distribution. Before conducting such a test, one must set the significance level, $\alpha$, equal to the probability that finding a true difference is due to chance (usually set at .05).

## Rayleigh Test

A particularly simple test is the Rayleigh (1880) test, which uses the mean vector length statistic, $\overline{\rho}$ (Batschelet, 1981, p. 54; Mardia and Jupp, 2000, p. 94). As the data vary from uniformly distributed to tightly clumped (and thus highly nonuniform), $\overline{\rho}$ varies from 0 to 1. Assuming that the underlying data are drawn from a von Mises distribution, we can compute a "Rayleigh's $R$," or "Rayleigh's $z$," statistic, whose distributions are known:

$$R = n\overline{\rho}$$

$$z = n\overline{\rho}^2 \qquad (5.11)$$

Batschelet (1981, Table H) gives a table of the probabilities associated with various values of $\overline{\rho}$ and $n$, and Zar (1999, Table B.34) has a table with critical values of $z_{\alpha,n}$. An approximation of the probability of Rayleigh's $R$ (for $n > 5$) is as follows:

$$p = \exp\left(\sqrt{1 + 4n + 4(n^2 - R^2)} - (1 + 2n)\right) \qquad (5.12)$$

Furthermore, Mardia and Jupp (2000) note that for large $n$, $z$ is asymptotically distributed as $\dfrac{\chi_2^2}{2}$.

## Rayleigh Test for Uniformity Against a Specified Mean Angle, the V Test

The Rayleigh test determines whether the data samples are consistent with the assumption of a uniform distribution of angles; the alternative hypothesis is that some unspecified angle is preferred. In some cases there is reason to assume, a priori, that the data should have a specific mean direction. For example, if homing pigeons are released south of their home loft, one might assume not only that they would not fly "randomly," but also that they would tend to fly north to the loft.

In the $V$ test (Greenwood and Durand, 1955; Batschelet, 1981, p. 58; Mardia and Jupp, 2000, p. 98), one computes the projection of the mean vector (whose polar components are given by $\overline{\rho}$ and $\overline{\theta}$) onto a unit vector in the direction of the chosen mean, $\overline{\theta}_0$.

$$V = \overline{\rho}\cos(\overline{\theta} - \theta_0) \qquad (5.13)$$

If the data cluster around $\theta_0$, then $V$ will be near 1; deviations of the data from the expected mean angle, as well as dispersion of the data, will lead to a decrease in V. One then computes a test statistic:

$$v = V\sqrt{2n} \qquad (5.14)$$

Tables of critical values of $v_{\alpha,n}$ may be found in Batschelet (1981, Table I) and Zar (1999, Table B.35). Furthermore, for large $n$, $v$ approaches a one-tailed normal deviate:

$$v_{\alpha,n} \approx Z_\alpha \qquad (5.15)$$

It is important when performing this test to note that the angle $\theta_0$ must be chosen in advance. Furthermore, the V test is only a test of randomness; it is *not* a test of whether or not the sample mean deviates from an expected mean $\theta_0$. Once we have established (by means of the $V$ test or other tests) that the data samples are oriented, we should use the **parametric** one-sample test to ascertain whether or not the sample mean deviates from an expected value.

## Tests for Bimodal and Multimodal Data

If the data represent axial quantities (e.g., those shown in figure 5.3d) or are otherwise obviously bimodal, the Rayleigh and $V$ tests, as presented, may not be able to demonstrate a deviation from the assumption of a uniform distribution. What one does in this case is to transform the data by doubling all of the angles, thereby converting the bimodal data into a unimodal set (by mapping points at opposite ends of a diameter to the same place on the circle). One now performs a Rayleigh (or $V$) test and tries to reject the null hypothesis that the transformed data are uniformly distributed.

Should one encounter trimodal data, or data of higher modality, one can try to transform them into a unimodal sample by tripling (or otherwise suitably multiplying) the data values before performing the Rayleigh test. However, these higher-order procedures themselves serve to spread the data around the circle, making it more difficult to determine departures from uniformity for multimodal data.

## Distribution-Free (Nonparametric) Tests

There are a number of tests that do not assume any particular distribution of the underlying data (as seen earlier, the Rayleigh test assumed that if the data were not uniform, then they were unimodal and had a von Mises distribution). These tests can therefore also be directly employed with bimodal and multimodal data as well.

## The Hodges-Ajne Test

If the data are distributed uniformly, then any diameter of the circle will divide the sample of points into equal halves. Conversely, if a diameter can be found such that many more points are on one side of the diameter than on the other, the distribution of points is unlikely to be uniform.

To perform the test (Hodges, 1955; Ajne, 1968; Batschelet, 1981, p. 61; Mardia and Jupp, 2000, p. 105), we plot the data on a circle and construct a diameter that divides the points into the two most unequal halves (e.g., the data shown in figure 5.2 can be split by a diameter such that two points are on one side and 98 points are on the other). If $n$ is the total number of points and $m$ is the observed minimum number of points in a semicircular arc, the probability of seeing an $m$ this small is given by

$$p = 2^{1-n}(n-2m)\binom{n}{m} \tag{5.16}$$

where $\binom{n}{m}$ is the binomial coefficient.

$$\binom{n}{m} = \frac{n!}{m!(n-m)!} \tag{5.17}$$

Tables of the critical values for $m$ can also be found in Batschelet, 1981 (Table J); Mardia and Jupp, 2000 (Appendix 2.7); and Zar, 1999 (Table B.36). If $m$ is found to be greater than $n/3$, then equation 5.16 cannot be used.

A minor variation of this test, proposed by Batschelet (1981), can be used to ask the alternative question "Do the observed data cluster around some previously given angle?" One performs the Hodges-Ajne test, but fixes the diameter to be at right angles to the chosen direction.

## Rao Spacing Test

If the data are uniformly distributed around the circle, then the arcs between successive data points should be one-$n^{th}$ of a circle. Rao's test (Rao, 1969; Batschelet, 1981, p. 66; Mardia and Jupp, 2000, p. 108) is based on the total deviation of the arcs from their expected value. If we let $T_i$ be the arc lengths between successive data points, then

$$U = \tfrac{1}{2}\sum\left|T_i - \frac{360}{n}\right| \tag{5.18}$$

where the vertical bars indicate absolute value (n.b.: for this test, the angles must be in degrees). Tables of the critical values of this statistic can be found in Batschelet, 1981 (Table L) and Mardia and Jupp, 2000 (Appendix 2.9).

For data that are multimodal but do not possess a strong symmetry (as do axial data, for example), the Rayleigh test and the Hodges-Ajne test may fail to reject the null hypothesis. The Rao test, with its minimal assumptions, may be the only test that can demonstrate a deviation from uniformity under these circumstances. As with linear statistics, however, if the data are consistent with a parametric model (e.g., if the data are unimodal and von Mises-like in the circular case, or unimodal and bell shaped in the linear case), the parametric tests will be more powerful, able to show a significant deviation with fewer data points.

## Example 5.1

Suppose we mark out a circle 20 cm in diameter on the sidewalk and observe the heading of the next 30 ants that enter this circle. The headings of the 30 ants are given in table 5.2; the data are plotted in figure 5.3*d* (p. 125). Because the data are strongly bimodal, we also plot the data using twice the angle as the plotting parameter, giving figure 5.3*c*.

For both sets of data, we compute the mean heading and the associated circular standard deviation. If we let $\overline{\cos\theta}$ represent the mean of the $\cos\theta$ column, we can compute the mean vector length and mean direction:

$$\overline{\rho}_\theta = \sqrt{\overline{\cos\theta}^2 + \overline{\sin\theta}^2} = .01$$

$$\overline{\rho}_{2\theta} = \sqrt{\overline{\cos(2\theta)}^2 + \overline{\sin(2\theta)}^2} = .80$$

$$\overline{\theta}_\theta = \tan^{-1}\frac{\overline{\sin\theta}}{\overline{\cos\theta}} = -73$$

$$\overline{\theta}_{2\theta} = \frac{1}{2}\tan^{-1}\frac{\overline{\sin 2\theta}}{\overline{\cos 2\theta}} = 32$$

$$s_\theta = \sqrt{2(1-\overline{\rho}_\theta)} = 80$$

$$s_{2\theta} = \frac{1}{2}\sqrt{2(1-\overline{\rho}_{2\theta})} = 18$$

When the entire data set is considered (radial case), the length of the mean vector is essentially zero, with a correspondingly large circular standard deviation. When we work with doubled angles in the axial assumption case, the resultant mean vector length is considerably larger. Furthermore, the calculated circular mean goes through the intuitive "middle" of the data points (see figure 5.3*c*). We now want to do a Rayleigh test on these two data sets to see if the ants are moving in random directions or are perhaps following a path (table 5.3).

We can conclude that these data, when analyzed as axial data (doubling the angles so that the two modes fall on top of each other), show a strong directionality. Note that even though the radial case clearly shows evidence of directionality, the Rayleigh test fails to detect it because the assumption that the underlying population is unimodal has been violated.

If we attempted to perform the Hodges-Ajne test on these data (finding a diameter that divides the data into the two most unequal halves), the radial data set would not be judged different from a uniform distribution due to the symmetry of the data set. However, the Rao test, which directly tests the assumption that the data are evenly spaced around the circle,

*(continued)*

| Table 5.2 | Representation of Radial and Axial Data | | | | | | |
|---|---|---|---|---|---|---|---|
|  | θ | cos θ | sin θ |  | 2θ | cos2θ | sin2θ |
|  | 201.3 | −.93 | −.36 |  | 42.6 | .74 | .68 |
|  | 212.5 | −.84 | −.54 |  | 65.0 | .42 | .91 |
|  | 51.3 | .63 | .78 |  | 102.6 | −.22 | .98 |
|  | 187.1 | −.99 | −.12 |  | 14.2 | .97 | .25 |
|  | 216.5 | −.80 | −.59 |  | 73.0 | .29 | .96 |
|  | 47.2 | .68 | .73 |  | 94.4 | −.08 | 1.00 |
|  | 213.5 | −.83 | −.55 |  | 67.0 | .39 | .92 |
|  | 206.3 | −.90 | −.44 |  | 52.6 | .61 | .79 |
|  | 55.1 | .57 | .82 |  | 110.2 | −.35 | .94 |
|  | 224.5 | −.71 | −.70 |  | 89.0 | .02 | 1.00 |
|  | −1.9 | 1.00 | −.03 |  | −3.8 | 1.00 | −.07 |
|  | 198.2 | −.95 | −.31 |  | 36.4 | .80 | .59 |
|  | 1.2 | 1.00 | .02 |  | 2.4 | 1.00 | .04 |
|  | 207.3 | −.89 | −.46 |  | 54.6 | .58 | .82 |
|  | 32.3 | .85 | .53 |  | 64.6 | .43 | .90 |
|  | 31.2 | .86 | .52 |  | 62.4 | .46 | .89 |
|  | 176.7 | −1.00 | .06 |  | −6.6 | .99 | −.11 |
|  | 28.1 | .88 | .47 |  | 56.2 | .56 | .83 |
|  | 209.2 | −.87 | −.49 |  | 58.4 | .52 | .85 |
|  | 13.4 | .97 | .23 |  | 26.8 | .89 | .45 |
|  | 215.8 | −.81 | −.58 |  | 71.6 | .32 | .95 |
|  | 35.9 | .81 | .59 |  | 71.8 | .31 | .95 |
|  | 233.8 | −.59 | −.81 |  | 107.6 | −.30 | .95 |
|  | 3.3 | 1.00 | .06 |  | 6.6 | .99 | .11 |
|  | 233.8 | −.59 | −.81 |  | 107.6 | −.30 | .95 |
|  | 44.3 | .72 | .70 |  | 88.6 | .02 | 1.00 |
|  | 62.5 | .46 | .89 |  | 125.0 | −.57 | .82 |
|  | 253.7 | −.28 | −.96 |  | 147.4 | −.84 | .54 |
|  | 16.2 | .96 | .28 |  | 32.4 | .84 | .54 |
|  | 41.4 | .75 | .66 |  | 82.8 | .13 | .99 |
| $\overline{\cos\theta}, \overline{\sin\theta}$ |  | .00 | −.01 | $\overline{\cos 2\theta}, \overline{\sin 2\theta}$ |  | .35 | .71 |

The raw data are displayed in figure 5.3. The original (radial) data are in the column marked θ, with the doubled values (axial) in the 2θ column. θ and 2θ are in degrees. The horizontal (cos θ) and vertical (sin θ) components are computed, as well as the mean values.

should work for both the original data set and its axial (2θ) representation. To do this test, we sort the data, then replace the data points with the intervals between adjacent points. If the null hypothesis is correct and the data are distributed uniformly, then there will be 360° / 30 = 3° between adjacent points. We then compute the Rao statistic U according to equation 5.18. By simply looking at figure 5.3c and 5.3d, we can see that the spacing between adjacent points is far from uniform; indeed, the value of the U statistic is 194° for the "axial" data (figure 5.3c) and 290° for the "radial" data (figure 5.3c). According to table L in

**Table 5.3    Rayleigh's *R***

|  | Radial assumption | Axial assumption |
|---|---|---|
| $R = n\bar{p}$ | 0.45 | 23.9 |
| *p* (from Table H, Batschelet, 1981) | >.99 | <.01 |
| *p* (from equation 5.12) | >.99 | <.01 |

The *R* test statistic and *p*-values for the radial and axial data plotted in figure 5.3 and listed in table 5.2.

Batschelet, 1981, the critical value for U at the *p* < .01 criterion for *n* = 30 is 168.8°. Since the statistic for both sets of data, including the symmetric bimodal distribution of figure 5.3*d*, surpasses this critical value, we can conclude that these data are definitely *not* uniformly distributed.

# One-Sample Comparisons

A previous section delineated measures of central tendency and dispersion as ways of describing key features of the data: "Where are they, and how are they concentrated?" While these measures are useful, the researcher usually has a specific question about the data that he or she wants to answer. To return to an example mentioned earlier, one may want to ask, "Is the mean time of lunch different than 12:00?" As in linear statistics, one asks this question in order to try to make some inference about the data. This section begins to describe methods for making inferences about directional data, that is, hypothesis testing.

The most basic question one can ask is "Is the mean direction, $\bar{\theta}$, of a sample of directional data different from some hypothetical angle ($\theta_0$)?" The analogous test in linear statistics is the one-sample *t*-test. To answer this question, we assume that the data are drawn from a sample that is relatively unimodal and symmetric (i.e., one that can be approximated by a von Mises distribution). The most straightforward method for doing this is to construct a **confidence interval** about the mean $[\bar{\theta} - \delta, \bar{\theta} + \delta]$ and then determine if the confidence interval contains the hypothetical direction. Charts for determining $\delta$ based on $\bar{p}$ and *n* are found in Batschelet, 1981 (p. 86) for 95% and 98% confidence intervals. A drawback to using this method is that the concentration parameter of the von Mises distribution, $\kappa$, is assumed to be known. In most cases, $\kappa$ will not be known explicitly. In this situation, one can compute an approximate confidence interval by using values from the $\chi^2$ distribution (Mardia and Jupp, 2000, p. 124):

$$\delta = \cos^{-1}\left(\left[\frac{2n(2n^2\bar{p}^2 - n\chi^2_{1,1-\alpha})}{n^2\bar{p}^2(4n - \chi^2_{1,1-\alpha})}\right]^{0.5}\right), \quad \bar{p} \leq 0.667$$

$$\delta = \cos^{-1}\left(\frac{[n^2 - n^2(1-\bar{p}^2)\exp(\chi^2_{1,1-\alpha}/n)]}{n\bar{p}}\right), \quad \bar{p} > 0.667$$

(5.19)

If one wishes to perform a test rather than construct a confidence interval to determine if the mean direction from a sample is different from a given direction, one can use the following equations proposed by Upton (1973). First define the quantity $\bar{C}$:

$$\overline{C} = \frac{1}{n}\sum_{i=1}^{n}\cos(\theta_i - \theta_0) \tag{5.20}$$

The test statistic $P_1$ has a different formula depending on the value of $\overline{C}$. This test is appropriate if $n \geq 5$.

$$P_1 = \frac{4n(\overline{\rho}^2 - \overline{C}^2)}{2 - \overline{C}^2}, \quad \overline{C} \leq .667$$

$$P_1 = \frac{2n^3}{n^2 + n^2\overline{C}^2 + 3n}\log\frac{1-\overline{C}^2}{1-\overline{\rho}^2}, \quad \overline{C} > .667 \tag{5.21}$$

$P_1$ is approximately distributed as $\chi_1^2$; therefore the **p-value** can be explicitly determined.

## Example 5.2

Application of a small current (<1 mA) to the skin overlying the mastoid processes using surface electrodes has been shown to induce stereotypical postural deviations that are perpendicular to the naso-occipital axis if the polarities over each mastoid are opposite (Coats, 1972; Day et al., 1997). In a hypothetical experiment, Researcher A wanted to examine whether the direction of postural sway that was elicited when a group of subjects turned their heads 90° to the right (i.e., nose in line with right shoulder, defined as 0°) was different from the anterior-posterior body axis (i.e., in the direction of the long axis of the feet, $\theta_0 = 90°$). Twenty subjects participated in the experiment, and the direction of sway obtained consisted of the following values (figure 5.4):

$\theta$ = 107, 93, 101, 85, 94, 73, 107, 125, 103, 113, 71, 89, 128, 116, 97, 106, 129, 43, 109, 39

After verifying that the distribution was not uniform and observing that the sample was relatively symmetric and unimodal, Researcher A found that the mean direction $\overline{\theta} = 97$ and mean vector length $\overline{\rho} = .92$. She then inserted the appropriate values into equation 5.19:

$$\delta = \cos^{-1}\left(\frac{[20^2 - 20^2(1-.92^2)\exp(3.84/20)]}{20 \times .92}\right)$$

$$\delta = 11.6°$$

Hence the confidence interval for the mean direction is (85°, 109°). Since this interval (indicated by the shaded region in figure 5.4) contained the hypothetical angle of 90°, Researcher A concluded that the direction of sway was not different from the anterior-posterior body axis.

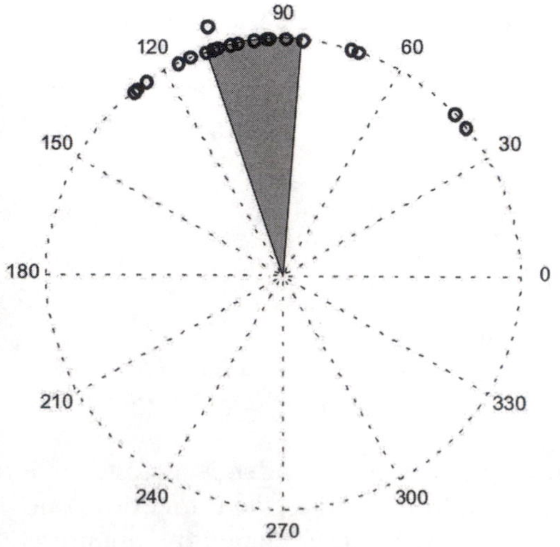

**Figure 5.4** Confidence interval for $\overline{\theta}$. This is one sample of data representing the direction of sway obtained from subjects receiving electrical stimulation to their vestibular nerves while their heads were rotated over their shoulder. Subjects typically sway in the direction of their interaural axis; 90° and 270° represent the anterior-posterior axis. The shaded area is the confidence interval, which contains the hypothetical direction of the interaural axis (90°). Thus the direction of sway was not different from the interaural axis.

# Comparisons of Two or More Samples

The next logical step in the development of directional statistics is testing to see if the mean angles of two or more samples of data are different. For example, one may ask "Is the mean time of arrival at work for junior faculty the same as for full professors?" These tests are analogous to the parametric $t$-test and nonparametric Mann-Whitney test for two samples found in linear statistics. For three or more samples, the linear statistic counterparts are the parametric one-way analysis of variance and nonparametric Kruskal-Wallis analysis of variance. In this section we demonstrate the method for two samples and then extend the models for three or more samples. There are a few restrictions on the use of these tests (Batschelet, 1981, p. 93). For one, each data point must be drawn at random and be independent of any other data point. In addition, the two samples that are being compared must be independent. The latter condition warrants against the use of these measures in typical within-subjects pre-post analyses. These so-called **paired** sample tests are covered in a later section of this chapter.

## Parametric Two-Sample Test

The parametric two-sample test (i.e., Watson-Williams test) examines the null hypothesis that the mean directions $\bar{\theta}_1$ and $\bar{\theta}_2$ are equal to one another (Watson and Williams, 1956; Batschelet, 1981, p. 95; Mardia and Jupp, 2000, p. 129; Zar, 1999, p. 625). In this test, it is assumed that the mean resultant lengths $\bar{\rho}_1$ and $\bar{\rho}_2$ are relatively equal and greater than .7. First, we must define the following quantities for the $n_1$ data points of sample 1 and $n_2$ points of sample 2:

$$C_1 = \sum_{i=1}^{n1} \cos \theta_{1,i} \quad C_2 = \sum_{j=1}^{n2} \cos \theta_{2,j} \quad C_T = C_1 + C_2$$

$$S_1 = \sum_{i=1}^{n1} \sin \theta_{1,i} \quad S_2 = \sum_{j=1}^{n2} \sin \theta_{2,j} \quad S_T = S_1 + S_2 \tag{5.22}$$

$$n_T = n_1 + n_2$$

Then the resultant lengths $\rho_1$, $\rho_2$, and $\rho_T$ (subscript $T$ for the total combined sample) are computed (n.b.: these are not the mean resultant length $\bar{\rho}$):

$$\rho_1 = \sqrt{\left(C_1^2 + S_1^2\right)} \quad \rho_2 = \sqrt{\left(C_2^2 + S_2^2\right)} \quad \rho_T = \sqrt{\left(C_T^2 + S_T^2\right)} \tag{5.23}$$

In order to compute the test statistic, we must estimate the concentration parameter $\tilde{\kappa}_T$ for the total sample. To do this, we assume $\bar{\rho} = (\rho_1 + \rho_2)/n_T$. If one knows the $\bar{\rho}$, then one can estimate $\tilde{\kappa}_T$ with sufficient accuracy using table 5.4. To make the test more conservative, one can select a greater $\bar{\rho}$ and $n_T$.

| Table 5.4 | Concentration Parameter, $\tilde{\kappa}$ | | | | | | | | | | | |
|---|---|---|---|---|---|---|---|---|---|---|---|---|
| | | | | | | $\bar{\rho}$ | | | | | | |
| $n_T$ | .45 | .50 | .55 | .60 | .65 | .70 | .75 | .80 | .85 | .90 | .95 | .99 |
| 10 | 0.83 | 1.00 | 1.18 | 1.37 | 1.59 | 1.86 | 2.19 | 2.66 | 3.39 | 4.84 | 9.30 | 45.3 |
| 20 | 0.93 | 1.09 | 1.26 | 1.45 | 1.67 | 1.94 | 2.28 | 2.76 | 3.53 | 5.07 | 9.79 | 47.8 |
| ∞ | 1.01 | 1.16 | 1.33 | 1.52 | 1.74 | 2.01 | 2.37 | 2.87 | 3.68 | 5.31 | 10.3 | 50.3 |

The value of the estimated concentration parameter $\tilde{\kappa}$, as a function of $\bar{\rho}$ and $n_T$. If values for $\bar{\rho}$ and $n_T$ are not listed, select the value according to the next highest $\bar{\rho}$ and $n_T$ to arrive at a more conservative test.

Finally, the test statistic $P_2$ is approximately distributed as the $F_{1,n_T-2}$:

$$P_2 = \left(1 + \frac{3}{8\tilde{\kappa}_T}\right)\frac{(n_T - 2)(\rho_1 + \rho_2 - \rho_T)}{n_T - \rho_1 - \rho_2}$$

(5.24)

If $P_2$ is greater than the critical value $F_{1,n_T-2}$, then the null hypothesis that the two mean directions are equal can be rejected.

## Example 5.3

The vestibulo-ocular reflex maintains images focused on the retina by moving the eyes at a velocity equal to and opposite in magnitude to the head velocity. Compared with control subjects, people with a vestibular system dysfunction commonly are found to have an abnormal phase in the vestibulo-ocular reflex compared with controls (Furman and Cass, 1996). In order to determine whether this phase was significantly different from that in controls, Researcher A collected vestibulo-ocular reflex data from 15 patients ($\theta_1$) and 15 controls ($\theta_2$) (figure 5.5).

$$\theta_1 = 152, 162, 128, 142, 145, 127, 125, 167, 150, 138, 165, 154, 132, 176, 155$$

$$\overline{\theta}_1 = 148, \overline{\rho}_1 = .96$$

$$\theta_2 = 182, 179, 163, 157, 168, 166, 162, 173, 182, 164, 162, 167, 175, 161, 158$$

$$\overline{\theta}_2 = 168, \overline{\rho}_2 = .99$$

$$C_1 = -12.25 \quad C_2 = -14.52 \quad C_T = -26.78$$

$$S_1 = 7.70 \quad S_2 = 3.11 \quad S_T = 10.81$$

$$\rho_1 = 14.47 \quad \rho_2 = 14.85 \quad \rho_T = 28.88$$

$$\overline{\rho} = .98 \quad \tilde{\kappa}_T = 21$$

$$P_2 = \left(1 + \frac{3}{8 \times 21}\right)\frac{(30 - 2)(14.47 + 14.85 - 28.88)}{30 - 14.47 - 14.85}$$

$$P_2 = 18.9, \quad p = 1.6 \times 10^{-4}$$

Since Researcher A found that $P_2$ was greater than the critical $F_{1,28} = 4.2$, she concluded that the phase of the vestibulo-ocular reflex in patients was different from that in controls.

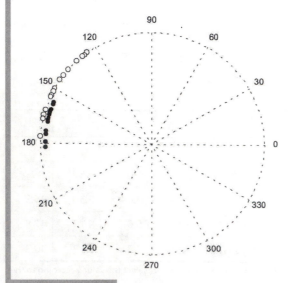

**Figure 5.5** Parametric two-sample test. This plot shows two samples of data representing the calculated phase of the horizontal vestibulo-ocular reflex (VOR) in patients with a vestibular disorder (larger radius) and healthy controls (smaller radius) while subjects rotated at 0.05 Hz. A two-sample test indicated that the two samples of data were not from the same distribution. Therefore, this sample of patients with a vestibular disorder displayed a phase lead of the horizontal VOR with respect to healthy controls.

## Parametric Multisample Tests

It is easy to extend this test for the analysis of three or more samples (Watson and Williams, 1956; Batschelet, 1981, p. 98; Mardia and Jupp, 2000, p. 134; Zar, 1999, p. 627). The null hypothesis is that the mean directions are all equal, versus the alternative that at least one is different. However, this test does not tell which mean direction is different. We recommend performing multiple pairwise comparisons (i.e., two-sample tests) and using a Bonferroni correction on the $\alpha$ to determine which mean directions may be different. The same assumptions regarding the dispersion apply, except that Stephens (1972) found that the test works well for $\bar{\rho} > 0.45$. Assuming there are $k$ samples, equations 5.22 and 5.23 become:

$$C_1 = \sum_{i=1}^{n1} \cos\theta_{1,i} \ldots C_k = \sum_{j=1}^{nk} \cos\theta_{k,j} \quad C_T = C_1 + \ldots + C_k$$

$$S_1 = \sum_{i=1}^{n1} \sin\theta_{1,i} \ldots S_k = \sum_{j=1}^{nk} \sin\theta_{k,j} \quad S_T = S_1 + \ldots + S_k \tag{5.25}$$

$$n_T = n_1 + \ldots + n_k$$

$$\rho_1 = \sqrt{\left(C_1^2 + S_1^2\right)} \ldots \rho_k = \sqrt{\left(C_k^2 + S_k^2\right)} \quad \rho_T = \sqrt{\left(C_T^2 + S_T^2\right)} \tag{5.26}$$

$\tilde{\kappa}_T$ is determined as before by using $\bar{\rho} = (\rho_1 + \ldots + \rho_k)/n_T$. The test statistic $P_k$ is approximately distributed as $F_{k-1,n_T-k}$:

$$P_k = \left(1 + \frac{3}{8\tilde{\kappa}_T}\right) \frac{(n_T - k)\left(\sum\limits_{i=1}^{k}\rho_i - \rho_T\right)}{(k-1)\left(n_T - \sum\limits_{i=1}^{k}\rho_i\right)} \tag{5.27}$$

## Nonparametric Tests

The preceding sections describe tests used to compare mean directions for samples that appear to be distributed normally on the circle (i.e., von Mises distribution). For example, the parametric two-sample test requires that (1) $\bar{\rho}_1 = \bar{\rho}_2 > .7$ and (2) the data be from unimodal and symmetric samples. If any of the assumptions regarding the parametric tests cannot be justified, then it is appropriate to use the following nonparametric tests. It is important to note that nonparametric tests do not examine whether there is a difference in mean direction. Rather, they test if the two samples come from the same population. If the samples are found to be different, one cannot exclude the possibilities that the samples differ in the mean direction, dispersion, or both. It is necessary to examine the data graphically in order to determine the cause for the difference in the distribution of the samples.

## Nonparametric Two-Sample Test

The nonparametric two-sample (uniform scores) test (Batschelet, 1981, p. 101; Mardia and Jupp, 2000, p. 147; Zar, 1999, p. 633) was proposed by Wheeler and Watson (1964) and separately by Mardia (1967). The null hypothesis is that there is no difference in the populations from which the two samples were drawn. Consider two independent random samples: $\theta_{1,1}, \ldots, \theta_{1,n_1}$ and $\theta_{2,1}, \ldots, \theta_{2,n_2}$. In order to perform the test, we must combine and rank order the angles from both samples: $r_1, \ldots, r_{(n_1+n_2)}$. In the case of a tie between angles from different samples, we randomly choose one of the angles. Next, we transform the ranks from the first sample $r_{1,1}, \ldots, r_{1,n_1}$ to:

$$\beta_{1,i} = \frac{2\pi r_{1,i}}{n_1 + n_2} \tag{5.28}$$

If the $\beta_{1,i}$ are uniformly distributed, we cannot reject the null hypothesis. In order to examine this, we use the test statistic $NP_2$, which is simply the square of the resultant length of the $\beta_{1,i}$:

$$NP_2 = \left(\sum_{i=1}^{n_1} \cos \beta_{1,i}\right)^2 + \left(\sum_{i=1}^{n_1} \sin \beta_{1,i}\right)^2 \tag{5.29}$$

For combined samples with $(n_1 + n_2) < 20$, Batschelet (1981, Table Q) and Mardia and Jupp (2000, Appendix 2.13) provide the critical values for $NP_2$. If the total $(n_1 + n_2) \geq 20$, then a transformed statistic $NP_2^*$ can be computed and compared against the $\chi_2^2$ distribution:

$$NP_2^* = \frac{2(n_1 + n_2 - 1)NP_2}{n_1 n_2} \tag{5.30}$$

## Example 5.4

Using the data from example 5.3 (p. 138), Researcher A performed the uniform scores test. The transformed uniform score angles for the patient ($\beta_1$) and control ($\beta_2$) groups are shown in table 5.5. From there, equations 5.29 and 5.30 are used to compute the statistics.

$$NP_2 = (1.91)^2 + (5.87)^2$$
$$NP_2 = 38.0$$

$$NP_2^* = \frac{2 \times (15 + 15 - 1) \times 38.0}{15 \times 15} = 9.8$$

Since Researcher A found that $NP_2^*$ was greater than the critical $\chi_2^2 = 6.0$, she concluded that the phase of the vestibulo-ocular reflex in patients was different than in controls.

### Table 5.5 Nonparametric Two-Sample Test

| $\theta_1$ | $r_1$ | $\beta_1$ | $\theta_2$ | $r_2$ | $\beta_2$ |
|---|---|---|---|---|---|
| 125 | 1 | 12 | 157 | 12 | 144 |
| 127 | 2 | 24 | 158 | 13 | 156 |
| 128 | 3 | 36 | 161 | 14 | 168 |
| 132 | 4 | 48 | 162 | 16 | 192 |
| 138 | 5 | 60 | 162 | 17 | 204 |
| 142 | 6 | 72 | 163 | 18 | 216 |
| 145 | 7 | 84 | 164 | 19 | 228 |
| 150 | 8 | 96 | 166 | 21 | 252 |
| 152 | 9 | 108 | 167 | 23 | 276 |
| 154 | 10 | 120 | 168 | 24 | 288 |
| 155 | 11 | 132 | 173 | 25 | 300 |
| 162 | 15 | 180 | 175 | 26 | 312 |
| 165 | 20 | 240 | 179 | 28 | 336 |
| 167 | 22 | 264 | 182 | 29 | 348 |
| 176 | 27 | 324 | 182 | 30 | 360 |

Two samples of data representing the calculated phase lag of the horizontal vestibulo-ocular reflex in patients with a vestibular disorder and healthy controls. Data are combined, ranked $r_i$ from 1 to 30, and then transformed into uniform scores $\beta_i$.

## Nonparametric Mulitsample Test

The nonparametric two-sample test can easily be generalized to the multisample case (Batschelet, 1981, p. 104; Mardia and Jupp, 2000, p. 156). We combine and rank order the $k$ samples as before. In the case of ties, we randomly choose angles from the samples. The ranks of the angles from the $j^{th}$ sample are defined as:

$$\beta_{j,i} = \frac{2\pi r_{j,i}}{\sum\limits_{j=1}^{k} n_j} \qquad (5.31)$$

The squares of the individual vector lengths $\rho_j^2$ are calculated according to:

$$\rho_j^2 = \left( \sum_{i=1}^{n_j} \cos \beta_{j,i} \right)^2 + \left( \sum_{i=1}^{n_j} \sin \beta_{j,i} \right)^2 \qquad (5.32)$$

Finally, the test statistic $NP_k$ is determined by:

$$NP_k = 2 \sum_{j=1}^{k} \frac{\rho_j^2}{n_j} \qquad (5.33)$$

If $k = 3$ and the largest $n \leq 5$, then we use the critical values provided by Batschelet (1981, Table R) or Mardia and Jupp (2000, Appendix 2.16). Otherwise, we compare $NP_k$ to the critical value for $\chi^2_{2k-2}$.

# Hypothesis Testing for Second-Order Analysis

Methods have also been worked out for performing hypothesis tests on data obtained from a second-order sample—that is, multiple trials obtained from several subjects. As before, the one-sample test involves finding whether or not the data are directed; if so, a confidence interval for the grand mean direction can be computed and compared to a hypothetical direction. Given that the data are relatively clustered (e.g., mean directions within 90°), the basic premise of this method is that the Cartesian representation of the individual mean vectors forms a bivariate normal distribution. The confidence ellipse centered at the grand mean vector $(\bar{X}_{so}, \bar{Y}_{so})$ is created using a series of algebraic equations that are presented in the next section (Zar, 1999, p. 611; Batschelet, 1981, p. 262). If the confidence ellipse does not contain the origin, then it is concluded that the sample is oriented about a mean direction. Confidence limits for the grand mean direction are determined by drawing tangents from the confidence ellipse to the origin.

## Parametric One-Sample Test for Second-Order Analysis

In order to ascertain whether a set of $n$ mean angles has any direction at all, Hotelling (1931) constructed a test statistic based on the $F$-distribution (Zar, 1999, p. 638). First, compute the individual mean vectors of each sample, $(\bar{X}_i, \bar{Y}_i) = (\bar{\rho}_i \cos \bar{\theta}_i, \bar{\rho}_i \sin \bar{\theta}_i)$. Then we define the following quantities, which are the familiar formulas for computing the sum of squares:

$$SSx = \sum_{i=1}^{n} \bar{X}_i^2 - \frac{\left( \sum\limits_{i=1}^{n} \bar{X}_i \right)^2}{n}$$

$$SSy = \sum_{i=1}^{n} \bar{Y}_i^2 - \frac{\left( \sum\limits_{i=1}^{n} \bar{Y}_i \right)^2}{n} \qquad (5.34)$$

$$SSxy = \sum_{i=1}^{n} \bar{X}_i \bar{Y}_i - \frac{\left( \sum\limits_{i=1}^{n} \bar{X}_i \right)\left( \sum\limits_{i=1}^{n} \bar{Y}_i \right)}{n}$$

After also calculating the grand means (equation 5.10), the test statistic $PSO_1$ is found from the following:

$$PSO_1 = \frac{n^2 - 2n}{2}\left[\frac{\overline{X}_{so}^2 SSy - 2\overline{X}_{so}\overline{Y}_{so}SSxy + \overline{Y}_{so}^2 SSx}{SSxSSy - SSxy^2}\right] \qquad (5.35)$$

If $PSO_1$ is greater than the critical value of the $F_{2,n-2}$, the null hypothesis that there is no mean direction can be rejected.

## Confidence Interval for Second-Order Analysis

If there is a mean direction, the following set of algebraic equations can be used to compute the confidence interval. If the interval includes a hypothetical direction, then we can say that the sample of data was not different from the hypothetical direction.

$$A = \frac{n-1}{SSx}$$

$$B = \frac{(n-1)SSxy}{SSxSSy}$$

$$C = \frac{n-1}{SSy}$$

$$D = \frac{2(n-1)\left[1 - \dfrac{SSxy^2}{SSxSSy}\right]F_{2,n-2,1-\alpha}}{n(n-2)} \qquad (5.36)$$

$$H = AC - B^2$$

$$G = A\overline{X}_{so}^2 + 2B\overline{X}_{so}\overline{Y}_{so} + C\overline{Y}_{so}^2 - D$$

$$J = H\overline{X}_{so}^2 - CD$$

$$K = \sqrt{DGH}$$

$$L = H\overline{X}_{so}\overline{Y}_{so} + BD$$

$$\delta_1 = \tan^{-1}\left(\frac{L+K}{J}\right), \delta_2 = \tan^{-1}\left(\frac{L-K}{J}\right)$$

Onc $\delta_1$ and $\delta_2$ are determined, the angles in each set $[\delta_1, \delta_1 + \pi]$, $[\delta_2, \delta_2 + \pi]$ that are closest to the mean angle $\theta_{so}$ form the confidence interval.

## Example 5.5

In the next hypothetical example, Researcher A wanted to replicate the experiment of Sparto et al. (1997) and determine, using dynamical systems analysis, the coordination of the lower extremities during lifting and lowering tasks. She measured the hip and knee joint angular excursion as 10 subjects repetitively lifted and lowered a box from the floor to waist height. She then computed the average relative phase angle between the hip and knee over each cycle. She wanted to determine whether or not there was a direction associated with these relative phase angles. The values for the subjects are shown in table 5.6.

| Subject | Cycle1 | Cycle2 | Cycle3 | Cycle4 | Cycle5 | $\bar{\theta}$ | $\bar{\rho}$ | $\bar{X}$ | $\bar{Y}$ |
|---|---|---|---|---|---|---|---|---|---|
| 1 | 7 | −19 | 17 | 20 | −4 | 4.3 | .97 | .97 | .07 |
| 2 | 37 | 37 | 14 | 21 | 18 | 25.4 | .99 | .89 | .42 |
| 3 | 12 | 28 | 5 | 71 | 13 | 25.0 | .92 | .83 | .39 |
| 4 | 17 | 35 | 16 | 13 | 0 | 16.2 | .98 | .94 | .27 |
| 5 | 20 | −9 | 28 | 48 | 3 | 17.9 | .94 | .90 | .29 |
| 6 | 31 | 39 | −17 | −13 | 25 | 13.2 | .92 | .89 | .21 |
| 7 | 8 | 27 | 30 | 28 | 40 | 26.6 | .98 | .88 | .44 |
| 8 | 27 | 37 | −8 | 15 | 12 | 16.7 | .97 | .92 | .28 |
| 9 | −18 | 20 | −5 | 43 | 0 | 7.7 | .93 | .92 | .13 |
| 10 | 24 | 19 | −2 | −40 | 14 | 3.7 | .92 | .92 | .06 |

**Table 5.6 Second-Order Analysis**

The relative phase angle data obtained from 10 subjects performing five consecutive cycles of repetitive lifting are shown. Mean angle $\bar{\theta}$ and mean resultant length $\bar{\rho}$ computed from the mean vector of five angles, as well as the Cartesian representation $(\bar{X}, \bar{Y})$.

The grand mean relative phase angle was 15.8°, indicating that knee extension was leading hip extension. Next, the sum of squares and location of the center of the confidence ellipse were computed.

$$SSx = 0.013$$
$$SSy = 0.173$$
$$SSxy = -0.032$$
$$(\bar{X}_{so}, \bar{Y}_{so}) = (0.91, 0.26)$$

Using these values, the test statistic was determined.

$$PSO_1 = \frac{10^2 - 2 \times 10}{2}\left[\frac{0.91^2 \times 0.173 - 2 \times 0.91 \times 0.26 \times (-0.032) + 0.26^2 \times 0.013}{0.013 \times 0.173 - (-0.032)^2}\right]$$

$$PSO_1 = 5539, \quad p = 2.7 \times 10^{-13}$$

Since Researcher A found that $PSO_1$ was greater than the critical $F_{2,8} = 4.5$, she concluded that the relative phase angle was directed. Although space limits prevent us from reproducing all the algebraic equations, the calculations showed that the confidence interval was from 7.2 to 24.2.

## Parametric Two-Sample Test for Second-Order Analysis

We can extend the previous parametric one-sample test to two samples (Zar, 1999, p. 641). For instance, suppose the researcher were to examine the knee excursion in two groups of subjects: one group with knee osteoarthritis ($n_1$ subjects) and a control group without knee osteoarthritis ($n_2$ subjects). We could ask the question: Is the mean knee excursion in the first group significantly different from that in the second group? It is assumed that both groups have equal variances and covariance. First the grand means $(\bar{X}_{so,1}, \bar{Y}_{so,1})$ and $(\bar{X}_{so,2}, \bar{Y}_{so,2})$ and sum of squares of the groups are computed as before, and then three new quantities are tabulated:

$$
\begin{aligned}
SSx_c &= SSx_1 + SSx_2 \\
SSy_c &= SSy_1 + SSy_2 \\
SSxy_c &= SSxy_1 + SSxy_2
\end{aligned}
\tag{5.37}
$$

The test statistic $PSO_2$ is:

$$
PSO_2 = \frac{n_1 + n_2 - 3}{2\left(\dfrac{1}{n_1} + \dfrac{1}{n_2}\right)} \left[ \frac{\left(\bar{X}_{so,1} - \bar{X}_{so,2}\right)^2 SSy_c - 2\left(\bar{X}_{so,1} - \bar{X}_{so,2}\right)\left(\bar{Y}_{so,1} - \bar{Y}_{so,2}\right)SSxy_c + \left(\bar{Y}_{so,1} - \bar{Y}_{so,2}\right)^2 SSx_c}{SSx_c SSy_c - SSxy_c^2} \right]
\tag{5.38}
$$

If $PSO_2$ is greater than the critical value of the $F_{2, n_1 + n_2 - 3}$, the null hypothesis that the mean directions are equal can be rejected.

## Nonparametric One-Sample Test for Second-Order Analysis

If the assumption that the data are normally distributed along both axes cannot be justified, one should use the method of Moore (1980; Batschelet, 1981, p. 212; Zar, 1999, p. 639). Instead of computing $(\bar{X}_i, \bar{Y}_i)$ as before, one rank orders the $n$ mean angles according to the order of the smallest to largest mean resultant length $(\bar{\rho}_i)$. The next step is to compute the following:

$$
\bar{X}_r = \frac{\sum_{i=1}^{n} i \cos \bar{\theta}_i}{n}
$$

$$
\bar{Y}_r = \frac{\sum_{i=1}^{n} i \sin \bar{\theta}_i}{n}
\tag{5.39}
$$

$$
NPSO_1 = \sqrt{\frac{\bar{X}_r^2 + \bar{Y}_r^2}{n}}
$$

where $i$ is the rank order. $NPSO_1$ is then compared to the critical values found in Batschelet (1981, Table P) and Zar (1999, Table B.39).

---

**Example 5.6**

Researcher A performed the nonparametric one-sample test on the same values she obtained from example 5.5. The rank order according to the mean resultant length can be seen in table 5.7, as well as the ranked pairs $(\bar{X}_i, \bar{Y}_i)$. The mean values and test statistic were found to be:

$$
\bar{X}_r = 5.21
$$

$$
\bar{Y}_r = 1.59
$$

$$
NPSO_1 = \sqrt{\frac{5.21^2 + 1.59^2}{10}} = 1.72
$$

Since Researcher A found that $NPSO_1 = 1.72$ was greater than the critical $NPSO_1 = 1.048$, she concluded that the relative phase angle was directed.

| Table 5.7 | Nonparametric One-Sample Test for Second-Order Analysis | | | | |
|---|---|---|---|---|---|
| Rank | Subject | $\bar{\rho}$ | $\bar{\theta}$ | $\bar{X}_r$ | $\bar{Y}_r$ |
| 1 | 3 | .92 | 25.0 | 0.91 | 0.42 |
| 2 | 6 | .92 | 13.2 | 1.95 | 0.46 |
| 3 | 10 | .92 | 3.7 | 2.99 | 0.19 |
| 4 | 9 | .93 | 7.7 | 3.96 | 0.54 |
| 5 | 5 | .94 | 17.9 | 4.76 | 1.54 |
| 6 | 8 | .97 | 16.7 | 5.75 | 1.72 |
| 7 | 1 | .97 | 4.3 | 6.98 | 0.52 |
| 8 | 4 | .98 | 16.2 | 7.68 | 2.23 |
| 9 | 7 | .98 | 26.6 | 8.04 | 4.04 |
| 10 | 2 | .99 | 25.4 | 9.03 | 4.29 |

The mean vector data from example 5.5 (table 5.6) are transformed into ranks.

## Nonparametric Two-Sample Test for Second-Order Analysis

If any of the assumptions for the parametric two-sample test in the preceding section are not met, then the following nonparametric procedure advocated by Mardia (1967) can be used (Zar, 1999, p. 643). First we find the mean center from the combined sample of mean angles from both groups. In reference to this location, we compute new headings for each of the mean angles. Finally, we apply the nonparametric uniform scores two-sample test to these new directions.

# Paired-Sample Tests

An important experimental design in human movement literature is the paired $t$-test, which examines whether or not measurements taken from the same individuals during two separate conditions are different. In this case, the two samples are not independent because it is assumed that two measurements taken from the same individual will be highly correlated. Thus the independence requirement for use of the two-sample tests as described earlier is not satisfied. The unit of analysis is the individual subject; one is more interested in within-subjects effects than between-subjects effects. In our ongoing example, Researcher A wants to know if the knee excursion in subjects with knee osteoarthritis is significantly greater six weeks after total knee replacement compared with before the surgery.

The parametric and nonparametric one-sample tests described in equations 5.35 and 5.39, respectively, can be used with the following modifications for paired data (Zar, 1999, pp. 645-647). The variables $(X_{p,i}, Y_{p,i})$ are defined simply as the differences between pairs:

$$X_{p,i} = \cos(\theta_{2,i}) - \cos(\theta_{1,i})$$
$$Y_{p,i} = \sin(\theta_{2,i}) - \sin(\theta_{1,i})$$

(5.40)

For the parametric test, replace $(\bar{X}_i, \bar{Y}_i)$ in equation 5.34 with $(X_{p,i}, Y_{p,i})$. Compute the sums of squares and replace $(\bar{X}_{so,i}, \bar{Y}_{so,i})$ with $(\bar{X}_p, \bar{Y}_p)$ in equation 5.35 to determine the test statistic.

In the nonparametric case, $(\rho_{p,i}, \theta_{p,i})$ are calculated according to the following equation:

$$\rho_{p,i} = \sqrt{X_{p,i}^2 + Y_{p,i}^2}$$

$$\theta_{p,i} = \tan^{-1} \frac{Y_{p,i}}{X_{p,i}}$$

(5.41)

then the $\rho_{p,i}$ are ranked from smallest to largest, and $\bar{\theta}_i$ is replaced by $\theta_{p,i}$ in equation 5.39.

## Example 5.7

To continue with the experiment in example 5.5, Researcher A wanted to examine the average phase angle of the hip during the lifting portion of the cycle at the beginning and end of a series of many repetitions. The data are shown in table 5.8 and in figure 5.6.

| Table 5.8 | Paired-Sample Test | | | | |
|---|---|---|---|---|---|
| Subject | $\theta_1$ | $\theta_2$ | $\theta_1 - \theta_2$ | $X_p$ | $Y_p$ |
| 1 | 86 | 89 | −3 | −.052 | .002 |
| 2 | 71 | 108 | −37 | −.635 | .006 |
| 3 | 89 | 111 | −22 | −.376 | −.066 |
| 4 | 98 | 109 | −11 | −.186 | −.045 |
| 5 | 77 | 121 | −44 | −.740 | −.117 |
| 6 | 91 | 108 | −17 | −.292 | −.049 |
| 7 | 94 | 118 | −24 | −.400 | −.115 |
| 8 | 69 | 73 | −4 | −.066 | .023 |
| 9 | 70 | 96 | −26 | −.447 | .055 |
| 10 | 88 | 93 | −5 | −.087 | −.001 |

Two samples of data representing the average phase angle of the hip obtained at the beginning and end of a cyclical lifting experiment. The horizontal and vertical components $(X_p, Y_p)$, computed from the difference between angles, are then used in Hotelling's test.

The mean phase angle was 83° at the start of the task and 103° at the end of the task, indicating that hip extension occurred later in the cycle at the end of the task. Next, the sum of squares and location of the center of the confidence ellipse were computed.

$$SSx = 0.509$$

$$SSy = 0.030$$

$$SSxy = 0.050$$

$$(\bar{X}_p, \bar{Y}_p) = (-0.33, -0.03)$$

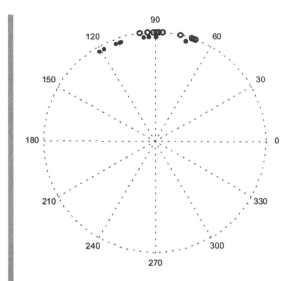

**Figure 5.6** Paired-sample test. This plot shows two samples of data representing the average phase angle of the hip relative to the vertical position of a box during a cyclical lifting and lowering experiment. At the beginning of the lifting trial (larger radius), there was less of a phase lag than at the end of the lifting trial (smaller radius).

Using these values, the next step was to employ the parametric one-sample test for second-order analysis.

$$PSO_1 = \frac{10^2 - 2 \times 10}{2}\left[\frac{(-0.33)^2 \times 0.030 - 2 \times (-0.33) \times (-0.03) \times (0.05) + (-0.03)^2 \times 0.509}{0.509 \times 0.030 - (0.050)^2}\right]$$

$$PSO_1 = 8.45, \quad p = .011$$

Since Researcher A found that $PSO_1$ was greater than the critical $F_{2,8} = 4.5$, she concluded that the average phase angle at the end of the task was significantly greater than at the beginning of the task.

Similar modifications can be made to handle paired data that result from second-order analyses by replacing $(X_{p,i}, Y_{p,i})$ with $(X_{p,so,i}, Y_{p,so,i})$ (Zar, 1999, pp. 646-647):

$$X_{p,so,i} = \bar{\rho}_{2,i}\cos(\bar{\theta}_{2,i}) - \bar{\rho}_{1,i}\cos(\bar{\theta}_{1,i})$$

$$Y_{p,so,i} = \bar{\rho}_{2,i}\sin(\bar{\theta}_{2,i}) - \bar{\rho}_{1,i}\sin(\bar{\theta}_{1,i})$$

(5.42)

# Correlations

In this last section, we examine how two variables relate to one another when at least one of the variables is distributed along the circle. For example, Researcher A may want to know if the knee excursion in a group of patients is related to their strength. Correlational analysis does not measure dependence of one variable on another, only how strongly two measures are associated. This section shows how to determine the association between two variables in cases in which the two variables consist of paired measurements from the same individuals. The two variables can both be distributed on the circle (circular-circular), or one can be on the circle and the other on the line (linear-circular). The treatment of this topic by Mardia and Jupp (2000) is excellent. We will begin with the parametric statistics.

## *Parametric Linear-Circular Correlation*

A linear-circular **correlation** examines the relationship between a linear random sample $x_1, \ldots, x_n$ and a circular random sample $\theta_1, \ldots, \theta_n$ (Batschelet, 1981, p. 191;

Mardia and Jupp, 2000, p. 245). In the parametric case, it is assumed that the linear sample is normally distributed. The null hypothesis is that there is no association between the two variables. First, several quantities need to be defined, where *corr* is the correlation coefficient:

$$r_{xc} = corr(x, \cos\theta), \quad r_{xs} = corr(x, \sin\theta), \quad r_{cs} = corr(\cos\theta, \sin\theta) \tag{5.43}$$

The linear-circular correlation coefficient $R_{x\theta}$ is then computed:

$$R_{x\theta} = \sqrt{\frac{r_{xc}^2 + r_{xs}^2 - 2r_{xc}r_{xs}r_{cs}}{1 - r_{cs}^2}} \tag{5.44}$$

A drawback to using this method is that only the strength of the association is examined—there is no indication whether the data are positively or negatively correlated. However, one can examine the polarity of the data simply by viewing a plot. In order to determine if the correlation is significantly different from 0, that is, if the null hypothesis is rejected, the test statistic $R_{LCP}$ for linear, circular parametric data is compared with the critical value obtained from the $F_{2,n-3}$ distribution (Mardia, 1976):

$$R_{LCP} = \frac{(n-3)R_{x\theta}^2}{1 - R_{x\theta}^2} \tag{5.45}$$

## Example 5.8

Researcher A wanted to know whether the method of lifting used by the subjects, as determined by the relative phase between the knee and the hip, was related to the amount of lifting power used by the subjects. She used the second-order sample of mean angles from the 10 subjects as in example 5.5 and also computed their average lifting power during the lift. The data are shown in table 5.9 and in figure 5.7. The computed quantities are:

$$r_{xc} = -.453, \quad r_{xs} = .493, \quad r_{cs} = -.974$$

$$R_{x\theta} = \sqrt{\frac{(-.453)^2 + (.493)^2 - 2(-.453)(.493)(-.974)}{1 - (-.974)^2}} = .508$$

$$R_{LCP} = \frac{(10-3)(.508)^2}{1 - (.508)^2}$$

$$R_{LCP} = 2.433, \quad p = .158$$

Since Researcher A found that $R_{LCP}$ was less than the critical $F_{2,7} = 4.7$, she concluded that the method of lifting was not significantly associated with the lifting power.

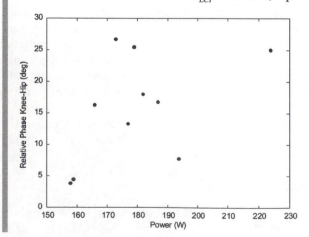

**Figure 5.7** Parametric linear-circular correlation. The relationship between the amount of power transferred to a box (abscissa) and the relative phase angle between the knee and hip (ordinate) during a repetitive lifting and lowering task. A test of parametric linear-circular correlation indicated that the relationship between the two measures was not significant.

| Table 5.9 | Parametric Linear-Circular Correlation | | | |
|---|---|---|---|---|
| Subject | *Power* | θ | cos(θ) | sin(θ) |
| 1 | 159 | 4.3 | .997 | .075 |
| 2 | 179 | 25.4 | .903 | .429 |
| 3 | 224 | 25.0 | .906 | .423 |
| 4 | 166 | 16.2 | .960 | .279 |
| 5 | 182 | 17.9 | .952 | .307 |
| 6 | 177 | 13.2 | .974 | .228 |
| 7 | 173 | 26.6 | .894 | .448 |
| 8 | 187 | 16.7 | .958 | .287 |
| 9 | 194 | 7.7 | .991 | .134 |
| 10 | 158 | 3.7 | .998 | .065 |

The association between lifting power and relative knee-hip phase angle (θ) is listed, as are the horizontal and vertical components.

## Parametric Circular-Circular Correlation

It is also possible to investigate the relationship between two circular variables. The circular-circular correlation measures the association between two independently drawn circular random samples $\theta_1, \ldots, \theta_n$ and $\phi_1, \ldots, \phi_n$ (Batschelet, 1981, p. 190; Mardia and Jupp, 2000, p. 248). The null hypothesis is that there is no association between the two variables. As in the linear-circular case, several quantities need to be defined:

$$r_{cc} = corr(\cos\theta, \cos\phi), \quad r_{sc} = corr(\sin\theta, \cos\phi), \quad r_\theta = corr(\cos\theta, \sin\theta)$$
$$r_{cs} = corr(\cos\theta, \sin\phi), \quad r_{ss} = corr(\sin\theta, \sin\phi), \quad r_\phi = corr(\cos\phi, \sin\phi)$$

(5.46)

The circular-circular correlation coefficient $R^2_{\theta\phi}$ is then computed. It is important to observe that this correlation coefficient is a square term while the linear-circular coefficient is to the power of 1.

$$R^2_{\theta\phi} = \frac{\left[(r^2_{cc} + r^2_{cs} + r^2_{sc} + r^2_{ss}) + 2r_\theta r_\phi(r_{cc}r_{ss} + r_{cs}r_{sc}) - 2r_\phi(r_{cc}r_{cs} + r_{sc}r_{ss}) - 2r_\theta(r_{cc}r_{sc} + r_{cs}r_{ss})\right]}{(1 - r^2_\theta)(1 - r^2_\phi)}$$

(5.47)

A significant association is shown if the test statistic $R_{CCP}$ is greater than the critical value determined by $\chi^2_4$ (Jupp and Mardia, 1980):

$$R_{CCP} = nR^2_{\theta\phi}$$

(5.48)

## Nonparametric Linear-Circular Correlation

If linear or circular samples such as those discussed in the previous section do not have unimodal and symmetric distributions, then it is wise to perform the nonparametric counterpart of linear-circular correlation (Mardia, 1976; Batschelet, 1981, p. 194; Mardia and Jupp, 2000, p. 246). The first step is to pairwise rank the *i* observations so that the linear variables go from smallest to largest: $x_{1,r} \leq x_{2,r} \cdots \leq x_{n,r}$. Then we take the corresponding ranks of the original angular data $r_1, \ldots, r_n$ and transform into uniform angles:

$$\beta_i = \frac{2\pi r_i}{n} \tag{5.49}$$

Define $T_c$ and $T_s$:

$$T_c = \sum_{i=1}^{n} i\cos\beta_i, \quad T_s = \sum_{i=1}^{n} i\sin\beta_i \tag{5.50}$$

We determine the nonparametric linear-circular correlation coefficient, $R_{x\theta,np}$, after first computing the factor $a_n$, which scales the correlation coefficient between 0 and 1.

$$a_n = \frac{1}{1 + 5\cot^2\left(\frac{\pi}{n}\right) + 4\cot^4\left(\frac{\pi}{n}\right)}, \quad n \ even$$

$$a_n = \frac{2\sin^4\left(\frac{\pi}{n}\right)}{\left(1 + \cos\left(\frac{\pi}{n}\right)\right)^3}, \quad n \ odd \tag{5.51}$$

$$R_{x\theta,np} = a_n\left(T_c^2 + T_s^2\right) \tag{5.52}$$

The test statistic $R_{LCNP}$ for linear-circular nonparametric data is compared with critical values tabulated by Batschelet (1981, Table X) and Mardia and Jupp (2000, Appendix 2.17).

$$R_{LCNP} = \frac{24\left(T_c^2 + T_s^2\right)}{n^2(n+1)} \tag{5.53}$$

### Example 5.9

Researcher A used the same values as in example 5.8 to compute the nonparametric linear-circular correlation. The data for the linear ranks $i$, angular ranks $r_i$, and other defined quantities are detailed in table 5.10.

$$T_c = -5.09, \quad T_s = -10.41$$

$$a_n = \frac{1}{1 + 5\cot^2\left(\frac{\pi}{10}\right) + 4\cot^4\left(\frac{\pi}{10}\right)} = .0025$$

$$R_{x\theta,np} = 0.0025 * \left((-5.09)^2 + (-10.41)^2\right) = 0.33$$

$$R_{LCNP} = \frac{24(25.91 + 108.37)}{10^2(10+1)}$$

$$R_{LCNP} = 2.93$$

Since Researcher A found that $R_{LCNP}$ was less than the critical value 5.48, she concluded that the method of lifting was not significantly associated with the lifting power.

| i | Subject | Power | θ | r | β | i × cos (β) | i × sin (β) |
|---|---------|-------|-----|----|-----|-------------|-------------|
| **Table 5.10** | **Nonparametric Linear-Circular Correlation** | | | | | | |
| 1 | 10 | 158 | 3.7 | 1 | 36 | 0.81 | 0.59 |
| 2 | 1 | 159 | 4.3 | 2 | 72 | 0.62 | 1.90 |
| 3 | 4 | 166 | 16.2 | 5 | 180 | −3.00 | 0.00 |
| 4 | 7 | 173 | 26.6 | 10 | 360 | 4.00 | 0.00 |
| 5 | 6 | 177 | 13.2 | 4 | 144 | −4.05 | 2.94 |
| 6 | 2 | 179 | 25.4 | 9 | 324 | 4.85 | −3.53 |
| 7 | 5 | 182 | 17.9 | 7 | 252 | −2.16 | −6.66 |
| 8 | 8 | 187 | 16.7 | 6 | 216 | −6.47 | −4.70 |
| 9 | 9 | 194 | 7.7 | 3 | 108 | −2.78 | 8.56 |
| 10 | 3 | 224 | 25.0 | 8 | 288 | 3.09 | −9.51 |

The data from example 5.8 were transformed into linear ranks ($i$), circular ranks ($r$), and uniform scores ($\beta$) for computation of nonparametric linear-circular correlation coefficient.

## Nonparametric Circular-Circular Correlation

The final statistical test examines the correlation between two circular samples when at least one of the samples does not come from unimodal and symmetric distributions. The construction of the nonparametric circular-circular correlation (Mardia, 1975; Batschelet, 1981, p. 184; Mardia and Jupp, 2000, p. 250) is similar to that of the nonparametric linear-circular test just presented. We indicate the ranks of both samples $\theta_1, \ldots, \theta_n$ and $\phi_1, \ldots, \phi_n$ by $r_1, \ldots, r_n$ and $t_1, \ldots, t_n$, respectively. Then we convert the circular rankings into uniform angles by:

$$\beta_i = \frac{2\pi r_i}{n}, \quad \gamma_i = \frac{2\pi t_i}{n} \tag{5.54}$$

Four intermediate calculations are necessary:

$$T_{cc} = \frac{1}{n}\sum_{i=1}^{n}\cos\beta_i\cos\gamma_i, \quad T_{cs} = \sum_{i=1}^{n}\cos\beta_i\sin\gamma_i$$

$$T_{sc} = \frac{1}{n}\sum_{i=1}^{n}\sin\beta_i\cos\gamma_i, \quad T_{ss} = \sum_{i=1}^{n}\sin\beta_i\sin\gamma_i \tag{5.55}$$

The positive rank circular-circular correlation is designated as $\bar{R}_+^2$ and the negative rank circular-circular correlation is $\bar{R}_-^2$:

$$\bar{R}_+^2 = \left(T_{cc} + T_{ss}\right)^2 + \left(T_{sc} - T_{cs}\right)^2$$

$$\bar{R}_-^2 = \left(T_{cc} - T_{ss}\right)^2 + \left(T_{sc} + T_{cs}\right)^2 \tag{5.56}$$

The nonparametric circular-circular correlation coefficient $R_{\theta\phi,np}$ is the larger of the two:

$$R_{\theta\phi,np}^2 = \max\left(\bar{R}_+^2, \bar{R}_-^2\right) \tag{5.57}$$

In order to determine if the association between the two circular samples is significantly different from 0, we compare $R^2_{\theta\phi,np}$ to critical values listed by Batschelet (1981, Table W) and Mardia and Jupp (2000, Appendix 2.18) for $n \leq 10$. Otherwise, we can compute the $p$-value from:

$$p = 1 - \left(1 - \exp\left((1-n)R^2_{\theta\phi,np}\right)\right)^2 \qquad (5.58)$$

# Summary

Directional statistics are a useful and necessary tool for many types of human movement analysis. Generally, there are analogs for many of the standard statistical tests used for linear variables. Most of these have been covered in this chapter. A quick-decision algorithm can be made to determine what type of directional statistical analysis applies for a given set of data. The first step for many people is to plot and compute the descriptive statistics for the data. By plotting the data, one can readily see if the data are unimodal, bimodal, or multimodal—which then leads to the use of the appropriate measures of central tendency and dispersion. Next, it is wise to perform a test of uniformity in order to see if the data are "directed" in some manner. If the data are not directed, then usually no further analysis is needed. On the other hand, if data are directed, one needs to determine whether one wants to test a hypothesis, or see if there is an association between two groups of data. The chapter covered several hypothesis tests. In order to perform these, first one needs to ask several questions.

1. Do the data consist of one, two, or many samples?
2. If two or more samples, are the groups independent or paired (dependent)?
3. Do the data come from a first- or second-order analysis?
4. Can the data be fit reasonably by a von Mises distribution (parametric), or are nonparametric tests needed?

A way to answer this last question is by considering whether or not the cumulative distribution function of the data has a nice peak and is symmetric. Finally, if the aim is to examine the association between two samples of data, the correlation functions in the last section of this chapter are appropriate.

This chapter should provide all the necessary ingredients for developing a toolbox of the basic directional statistical analyses. The authors have built their own using the basic MATLAB software with the inclusion of the Statistics Toolbox. Please refer to the following Web site (www.shrs.pitt.edu/physicaltherapy/about/faculty/sparto.html) to download functions that the authors have written for MATLAB. The Web site also has the scripts that were used to generate the examples and the work problems. As the types of analyses included in this book become more popular, the use of directional statistics needed to make conclusions about the data will become more prevalent as well.

# Work Problems

**1**

Given the following unimodal sample, plot and compute the mean and median angles, mean vector length, standard deviation, and range.

$$\theta = [-47, 67, 29, 44, 63, 86, 56, 1, 26, 90, 84, 26, 79, -24, 21, 66, -60, -5, 5, 4,$$
$$43, 13, 4, -51, 57, 29, 88, 31, 27, 70]$$

**2**

Given the following unimodal sample, plot and compute the mean and median angles, mean vector length, standard deviation, and range.

$$\theta = [199, 291, 120, 357, 148, 143, 289, 244, 106, 24, 297, 128, 154, 227, 151]$$

**3**

Given the following bimodal sample, plot and compute the mean and median angles, mean vector length, and standard deviation. Perform computations twice, assuming data are both radial and axial.

$$\theta = [68, 47, 95, 97, 99, 78, 84, 42, 76, 98, 121, 61, 59, 116, 56, 108, 124, 56, 57, 204, 212,$$
$$272, 231, 293, 229, 229, 355, 255, 247, 244, 249, 251, 268, 221, 199, 257, 297, 311]$$

**4**

Eight subjects performed a test five different times. From the resulting angles, compute the first-order mean angle and mean vector length. Then compute the second-order mean angle, mean vector length, and standard deviation.

$$S1 = [152, 94, 32, 132, 193]$$
$$S2 = [154, 299, 265, -11, 93]$$
$$S3 = [129, 70, 138, -18, 133]$$
$$S4 = [160, 266, 172, 169, 107]$$
$$S5 = [146, 187, 5, 253, 234]$$
$$S6 = [139, 68, 212, 122, 168]$$
$$S7 = [105, 265, -2, 130, 78]$$
$$S8 = [206, 87, 156, -15, 92]$$

**5**

Ten subjects performed a test three different times. From the resulting angles, compute the first-order mean angle and mean vector length. Then compute the second-order mean angle, mean vector length, and standard deviation.

$$S1 = [310, 319, 252]$$
$$S2 = [293, 289, 266]$$
$$S3 = [334, 335, 296]$$
$$S4 = [364, 263, 340]$$
$$S5 = [326, 295, 311]$$
$$S6 = [362, 313, 276]$$
$$S7 = [287, 269, 299]$$
$$S8 = [292, 297, 310]$$
$$S9 = [246, 283, 310]$$
$$S10 = [267, 311, 276]$$

**6**

For the sample of data provided in work problem 1, perform the Rayleigh test to determine if these data are directed. What do you conclude from the test?

**7**

For the sample of data provided in work problem 2, perform the Rayleigh test to determine if these data are directed. What do you conclude from the test?

**8**

For the sample of data provided in work problem 3, perform the Rayleigh test to determine if these data are directed, assuming that the data are (a) radial and (b) axial. What do you conclude from the test?

**9**

For the sample of data provided in work problem 1, perform the V test to determine if these data are directed, assuming that the hypothetical direction is 45°. What do you conclude from the test?

**10**

For the sample of data provided in work problem 1, perform the Hodges-Ajne nonparametric test to determine if these data are directed. What do you conclude from the test?

**11**

For the sample of data provided in work problem 2, perform the Hodges-Ajne nonparametric test to determine if these data are directed. What do you conclude from the test?

**12**

For the sample of data provided in work problem 3, perform the Hodges-Ajne nonparametric test to determine if these data are directed. What do you conclude from the test?

**13**

For the sample of data provided in work problem 1, perform the nonparametric Rao spacing test to determine if these data are directed. What do you conclude from the test?

**14**

For the sample of data provided in work problem 2, perform the nonparametric Rao spacing test to determine if these data are directed. What do you conclude from the test?

**15**

For the bimodal sample of data provided in work problem 3, perform the nonparametric Rao spacing test to determine if these data are directed. What do you conclude from the test? Compare results with those for work problem 8 (radial case) and work problem 12.

**16**

(a) For the sample of directed data provided in work problem 1, construct a 95% ($\alpha$ = .05) confidence interval. (b) Then, using two methods, determine if the sample of data is significantly different from the following hypothetical directions: 45°, 57.3°. (c) What do you conclude from the tests?

**17**

There is reason to believe that the data from work problem 3 are axial. In work problem 8, it was shown that the data were directed if the data were considered to be axial. Determine a confidence interval and see if the alignment of the data is different from a hypothetical direction of 75°. Verify your results by computing the $p$-value of the parametric one-sample test (equation 5.21).

**18**

Assuming that the following samples of data can be approximated by a von Mises distribution, determine if mean directions differ. What test would you use? State the null hypothesis, perform the appropriate test, and make conclusions.

$$\theta_1 = [-47, 67, 29, 44, 63, 86, 56, 1, 26, 90, 84, 26, 79, -24, 21, 66, -60, -5, 5,$$
$$4, 43, 13, 4, -51, 57, 29, 88, 31, 27, 70]$$

$$\theta_2 = [111, 38, 68, 59, 97, 81, 57, -6, 88, 56, 69, 84, 103, 79, 32, 53, 107,$$
$$102, 53, 98, 12, 49, 87, -16, 27]$$

**19**

Assuming that the following samples of data can be approximated by a von Mises distribution, determine if mean directions differ.

$$\theta_1 = [-47, 67, 29, 44, 63, 86, 56, 1, 26, 90, 84, 26, 79, -24, 21, 66, -60, -5,$$
$$5, 4, 43, 13, 4, -51, 57, 29, 88, 31, 27, 70]$$

$$\theta_3 = [15, 15, 48, 22, 15, -29, 60, 36, 86, 37, 34, 71, 42, 15, 53, 70, -25,$$
$$54, 31, 69, 40, 56, 35, 25, 14]$$

**20**

Consider all three samples $\theta_1$, $\theta_2$, and $\theta_3$ from work problems 18 and 19. How would you test to see if there was a difference in the mean directions of the samples? Interpret the findings of the test.

**21**

The data samples $\theta_1$ and $\theta_2$ from work problem 18 were multiplied by 1.5 and 1.8, respectively. Do the new samples still obey the assumptions for using the parametric two-sample test? Why or why not? Perform the nonparametric two-sample test on these new samples.

**22**

The data samples $\theta_1$, $\theta_2$, and $\theta_3$ from work problem 20 were multiplied by 1.5, 1.8, and 2.0, respectively. Perform the nonparametric multisample test on these new samples. Is there a difference in the directions of the samples?

**23**

What type of test would you perform to determine if the data from work problem 5 were directed? Perform the test and interpret the findings.

**24**

After the sample shown in work problem 5 was collected, a second sample of data was collected and a researcher wanted to determine if the mean directions were different. Perform the parametric two-sample test for second-order analysis on these samples of data. For sample 2, the following was calculated:

|     | $\bar{\theta}_2$ | $\bar{p}_2$ |
| --- | --- | --- |
| S1  | 291.1 | .77 |
| S2  | 300.7 | .94 |
| S3  | 259.7 | .61 |
| S4  | 287.6 | .97 |
| S5  | 291.7 | .78 |
| S6  | 313.4 | .84 |
| S7  | 266.8 | .69 |
| S8  | 251.6 | .52 |
| S9  | 238.7 | 1.00 |
| S10 | 259.1 | .95 |

$\bar{\theta}_{so,2} = 277.0°; \bar{p}_{so,2} = .74; s_{b,so,2} = 41.3°.$

## 25

Assume that the data from work problem 5 are nonparametric. Perform a nonparametric one-sample test for second-order analysis.

## 26

Assume that the data from work problem 24 are nonparametric. Perform a nonparametric two-sample test for second-order analysis. Why is the result different from that for work problem 24?

## 27

Assume that sample $\theta_2$ from work problem 18 and sample $\theta_3$ from work problem 19 were collected from the same subjects and thus are not independent. What type of analysis should you perform to see if there is a difference between the directions of the two samples? Perform both parametric and nonparametric tests and interpret the findings.

## 28

Assume that the samples from work problem 24 were collected from the same subjects and thus are not independent. Perform both parametric and nonparametric tests and interpret the findings.

**29**

You want to determine if there is a relationship between the linear and circular variables given next. Perform both parametric and nonparametric tests and interpret your findings.

$$X = [111, 38, 68, 59, 97, 81, 57, 6, 88, 56, 69, 84, 103, 79, 32, 53,$$
$$107, 102, 54, 98, 12, 49, 87, 16, 27]$$

$$\theta = [15, 16, 48, 22, 17, 29, 60, 36, 86, 37, 34, 71, 42, 18,$$
$$53, 70, 25, 54, 31, 69, 40, 56, 35, 26, 14]$$

**30**

The data in work problem 29 were rearranged slightly to give the following data. Determine if there is a relationship between the linear and circular variables given. Perform both parametric and nonparametric tests and interpret your findings.

$$X = [6, 12, 16, 27, 38, 32, 54, 68, 59, 49, 88, 56, 69, 57, 53,$$
$$84, 81, 98, 87, 103, 79, 107, 102, 111, 97]$$

$$\theta = [15, 16, 22, 17, 29, 18, 35, 26, 14, 48, 25, 36, 37, 34, 42,$$
$$53, 31, 60, 40, 56, 86, 71, 70, 54, 69]$$

**31**

The linear variable of work problem 29 was transformed into circular data. Determine if there is a relationship between the circular variables given. Perform both parametric and nonparametric tests and interpret your findings.

$$\theta_1 = [111, 38, 68, 59, 97, 81, 57, 6, 88, 56, 69, 84, 103, 79,$$
$$32, 53, 107, 102, 54, 98, 12, 49, 87, 16, 27]$$

$$\theta_2 = [15, 16, 48, 22, 17, 29, 60, 36, 86, 37, 34, 71, 42, 18,$$
$$53, 70, 25, 54, 31, 69, 40, 56, 35, 26, 14]$$

**32**

The linear variable of work problem 30 was transformed into circular data. Determine if there is a relationship between the circular variables given. Perform both parametric and nonparametric tests and interpret your findings.

$$\theta_1 = [6, 12, 16, 27, 38, 32, 54, 68, 59, 49, 88, 56, 69, 57, 53,$$
$$84, 81, 98, 87, 103, 79, 107, 102, 111, 97]$$

$$\theta_2 = [15, 16, 22, 17, 29, 18, 35, 26, 14, 48, 25, 36, 37, 34, 42,$$
$$53, 31, 60, 40, 56, 86, 71, 70, 54, 69]$$

# List of Symbols

$\alpha$—a priori significance level

$\beta,\gamma$—uniform scores of rank-ordered data

$\chi^2_{n,1-\alpha}$— value of chi-squared distribution with $n$ degrees of freedom, $\alpha$ significance level

$\delta$—delta angle used to construct confidence interval for first-order data

$\delta_{so}$—delta angle used to construct confidence interval for second-order data

$\kappa$—theoretical concentration parameter of a von Mises distribution

$\widetilde{\kappa}$—estimated concentration parameter of a von Mises distribution

$\mu$—mean of a normal distribution

$\theta,\phi$—samples of circular data

$\theta_0$—hypothetical angle or direction

$\theta_p$—angular difference between 2 paired data points

$\bar{\theta}$—mean direction of a von Mises distribution of unimodal data

$\bar{\theta}_{axial}$—mean direction of a von Mises distribution of axial data

$\bar{\theta}_{so}$—grand mean direction obtained from second-order analysis

$\hat{\theta}$—median direction

$\rho_p$—vector length of difference between 2 paired circular data points

$\bar{\rho}$—mean vector length of a von Mises distribution

$\bar{\rho}_{axial}$—mean vector length of a von Mises distribution of axial data

$\bar{\rho}_{so}$—grand mean vector length obtained from second-order analysis

$\sigma$—standard deviation of a normal distribution

$a_n$—scaling factor for nonparametric linear-circular correlation coefficient

$\exp(a)$—exponential of $a$

k—number of samples in a multisample comparison

$m$—minimum number of data points in any semicircular arc (Hodges-Ajne test)

$n$—number of data points in a sample

$p$-value—probability of erroneously rejecting the null hypothesis

$r,t$—rank order of a sample of linear or circular data

$s_b^2, s_b$—circular variance and standard deviation, as defined by Batschelet (1981)

$s_m^2, s_m$—circular variance and standard deviation, as defined by Mardia and Jupp (2000)

$v$—Rayleigh test of uniformity statistic when a hypothetical mean angle is given

$x$—sample of linear data

$z$—Rayleigh's $z$ statistic

$\bar{C}$—mean vector component of circular sample aligned with hypothetical direction

$F_{a,b,1-\alpha}$—value of F-distribution with $a,b$ degrees of freedom and $\alpha$ significance level

$I_0(\kappa)$—modified Bessel function of order 0

$NP_2$—nonparametric two-sample test statistic, $n < 20$

$NP_2^*$—nonparametric two-sample test statistic, $n > 19$

$NP_k$—nonparametric k-sample test statistic

$NPSO_1$—nonparametric second-order one-sample test statistic

$P_1$—parametric one-sample test statistic

$P_2$—parametric two-sample test statistic

$P_k$—parametric k-sample test statistic

$PSO_1$—parametric second-order one-sample test statistic

$PSO_2$—parametric second-order two-sample test statistic

$R$—Rayleigh test of uniformity statistic

$R_{CCNP}$—nonparametric circular-circular correlation test statistic

$R_{CCP}$—parametric circular-circular correlation test statistic

$R_{LCNP}$—nonparametric linear-circular correlation test statistic

$R_{LCP}$—parametric linear-circular correlation test statistic

$R^2_{\theta\phi,np}$—nonparametric circular-circular correlation coefficient

$R_{\theta\phi}^2$—parametric circular-circular correlation coefficient

$R_{x\theta,np}$—nonparametric linear-circular correlation coefficient

$R_{x\theta}$—parametric linear-circular correlation coefficient

$SSx$, $SSy$, $SSxy$— sums of squares for $\bar{X}$, $\bar{Y}$, $\bar{X}\bar{Y}$

$T$—arc length between successive data points in a circular sample

$U$—Rao spacing test of uniformity test statistic

$V$—component of a mean vector aligned with a hypothetical direction

$(X_p, Y_p)$—Cartesian coordinates of a sample of paired circular data

$(X_{p,so}, Y_{p,so})$—Cartesian coordinates of a second-order sample of paired circular data

$(\bar{X}, \bar{Y})$—mean vector of a sample of circular data

$(\bar{X}_p, \bar{Y}_p)$—mean vector of differences between paired circular data points

$(\bar{X}_r, \bar{Y}_r)$—mean vector of a ranked second-order sample of circular data

$(\bar{X}_{so}, \bar{Y}_{so})$—grand mean vector of a second-order sample of circular data

$Z_\alpha$—z-score of a normal distribution

# Suggested Readings and Other Resources

## Books

Batschelet, E. 1981. *Circular Statistics in Biology*. London: Academic Press. For a thorough treatment of the theory and application of circular statistics, this is a classic text. This book is somewhat more tutorial and descriptive than others on this list and provides many examples that are worked through.

Mardia, K.V., and P.E. Jupp. 2000. *Directional Statistics*. Chichester, England: Wiley. This text is another standard that provides a more detailed theoretical and mathematical background, including solutions for calculating the distributions to allow one to obtain probability ("*p*") values for the various tests.

Zar, J.H. 1999. *Biostatistical Analysis*. Upper Saddle River, NJ: Prentice Hall. This book is a classic reference for all types of statistical analysis. There are two excellent chapters on circular distributions (one on descriptive statistics, one on hypothesis testing), and the work of Batschelet and Mardia as well as that of Jupp is covered in Zar's chapters. This book has a decent number of work problems.

## URLs

To download directional statistics functions written by the authors for MATLAB (with Statistics Toolbox installed), access the following URL: www.shrs.pitt.edu/physicaltherapy/about/faculty/sparto.html

# References

Ajne, B. 1968. A simple test for uniformity of a circular distribution. *Biometrika* 55: 343-354.

Batschelet, E. 1981. *Circular Statistics in Biology*. London: Academic Press.

Coats, A.C. 1972. The sinusoidal galvanic body-sway response. *Acta Otolaryngologica* 74: 155-162.

Day, B.L., A. Severac-Cauquil, L. Bartolomei, M.A. Pastor, and I.N. Lyon. 1997. Human body-segment tilts induced by galvanic stimulation: A vestibularly driven balance protection mechanism. *Journal of Physiology* 500: 661-672.

Furman, J.M., and S.P. Cass. 1996. Laboratory evaluation. I. Electronystagmography and rotational testing. In *Disorders of the Vestibular System*, ed. R.W. Baloh and G.M. Halmagyi, 191-210. New York: Academic Press.

Greenwood, J.A., and D. Durand. 1955. The distribution of length and components of the sum of n random unit vectors. *Annals of Mathematical Statistics* 26: 233-246.

Hodges, J.L. 1955. A bivariate sign test. *Annals of Mathematical Statistics* 26: 523-527.

Hotelling, H. 1931. The generalization of the student's ratio. *Annals of Mathematical Statistics* 2: 360-378.

Jupp, P.E., and K.V. Mardia. 1980. A general correlation coefficient for directional data and related regression problems. *Biometrika* 67: 163-173.

Kelso, J.A.S. 1984. Phase transitions and critical behavior in bimanual coordination. *American Journal of Physiology: Regulatory Integrative and Comparative Physiology* 15: R1000-R1004.

Kurz, M.J., and N. Stergiou. 2002. Effect of normalization and phase angle calculations on continuous relative phase. *Journal of Biomechanics* 35: 369-374.

Mardia, K.V. 1967. A non-parametric test for the bivariate two-sample location problem. *Journal of the Royal Statistical Society Series B* 29: 320-342.

Mardia, K.V. 1975. Statistics of directional data. *Journal of the Royal Statistical Society Series B* 37: 349-393.

Mardia, K.V. 1976. Linear-circular correlation coefficients and rhythmometry. *Biometrika* 63: 403-405.

Mardia, K.V., and P.E. Jupp. 2000. *Directional Statistics*. Chichester: Wiley.

Moore, F.R. 1980. A modification of the Rayleigh test for vector data. *Biometrika* 67: 175-180.

Rao, J.S. 1969. *Some Contributions to the Analysis of Circular Data*. PhD dissertation, Indian Statistical Institute. Calcutta.

Rayleigh, Lord. 1880. On the resultant of a large number of vibrations of the same pitch and of arbitrary phase. *Philosophical Magazine* 10: 73-78.

Scholz, J.P., J.A.S. Kelso, and G. Schoner. 1987. Nonequilibrium phase transitions in coordinated biological motion: Critical slowing down and switching time. *Physiology Letters* A123: 390-394.

Schoner, G., and J.A.S. Kelso. 1988. Dynamic pattern generation in behavioral and neural systems. *Science* 239: 1513-1520.

Sparto, P.J., M. Parnianpour, T.E. Reinsel, and S. Simon. 1997. The effect of fatigue on multijoint kinematics, coordination, and postural stability during a repetitive lifting test. *Journal of Orthopaedic and Sports Physical Therapy* 25: 3-12.

Stephens, M.A. 1972. Multi-sample tests for the von Mises distribution. *Journal of the American Statistical Association* 67: 456-461.

Upton, G.J.G. 1973. Single-sample tests for the von Mises distribution. *Biometrika* 61: 87-99.

van Emmerik, R.E.A., and R.C. Wagenaar. 1996. Effects of walking velocity on relative phase dynamics in the trunk in human walking. *Journal of Biomechanics* 29: 1175-1184.

Watson, G.S., and E. Williams. 1956. On the construction of significance tests on the circle and sphere. *Biometrika* 43: 344-352.

Wheeler, S., and G.S. Watson. 1964. A distribution-free two-sample test on a circle. *Biometrika* 51: 256-257.

Zar, J.H. 1999. *Biostatistical Analysis*. Upper Saddle River, NJ: Prentice Hall.

# CHAPTER 6

# Mathematical Measures of Coordination and Variability in Gait Patterns

Max J. Kurz, MS, and Nicholas Stergiou, PhD

Gait patterns may be explained as simple, as in the case of the bouncing gait of the kangaroo, or more complex, as in the case of the cantering gait of a horse. In the bouncing gait, the legs display a symmetrical in-phase pattern. In the more complex cantering gait, there is an in-phase pattern between the right front and left legs, and a half-period out-of-phase movement between the left front and right back legs. Although it is apparent that a greater number of legs can result in greater diversity of locomotive patterns, even bipedal gait patterns can show a diversity of movement strategies. For example, various locomotive strategies are present as humans hop, walk, skip, and run.

Locomotion in general is easy to observe, but the organization of the neuromuscular system for selecting the various modes of locomotion is not easy to understand. Over the years, researchers have made considerable efforts to understand how quadrupeds and bipeds select various modes of locomotion. Such efforts have used kinematic (i.e., step interval, joint range of motion, etc.) and kinetic parameters (i.e., ground reaction forces, joint moments, etc.) to explain what drives the neuromuscular system to select a gait pattern (Taylor et al., 1970; McMahon, 1975; Kram and Taylor, 1990; Farley et al., 1993; Farley and Morgenroth, 1999). Additionally, recent research efforts have used measures of kinematic variability to explore the stability of various modes of locomotion (Winter, 1983; Gabel and Nayak, 1984; Diedrich and Warren, 1995; Maki, 1997; Heiderscheit, Hamill, and van Emmerik, 1999; van Emmerick et al., 1999; Hamill et al., 1999; Byrne et al., 2002; Kurz and Stergiou, 2003; Kurz, Stergiou, and Blanke, 2003).

Although there has been progress in determining the requisites for locomotion, many unanswered questions remain. For example, which kinematic and kinetic variables are important for the selection of a mode of locomotion and the transition from one mode to the next? How can we classify a stable gait pattern? What kinematic or kinetic variables determine speed of a mode of locomotion? How do these kinematic and kinetic variables interact for a selected gait pattern? What is the relationship between economy and kinematic and kinetic variables selected? This chapter presents novel mathematical measures for further exploration of such questions. Initial discussion focuses on how Response Surface Methodology can be used to model the interactive behaviors of the joints for the selection of various modes of locomotion. This is followed by discussion of how spanning set methodology can be used to explore the characteristics of variability in the gait pattern from the mean ensemble curve.

# Response Surface Methodology

Traditional analysis of the behavior of the neuromuscular system during locomotion has focused on mean kinematic and kinetic variables. These measures have provided useful scientific information about the changes in neuromuscular strategies for various modes of locomotion under a wide range of perturbations. However, on the basis of mean values alone, it is difficult to determine how these variables interact with each other for the selection of a gait pattern. It is well known that the behavior of one joint is influenced by the behavior of another joint in the lower extremity via the kinetic chain. Response Surface Methodology (RSM) provides a means to model the interactions of the joints and helps one determine which variables have the greatest influence on the selected mode of locomotion (Hill and Hunter, 1966; Box and Draper, 1987; Khuri and Cornell, 1987; McClave and Benson, 1988; Kurz, Porter and Steele, 2000; Kurz, Stergiou, and Scott, 2002; Kurz, Stergiou, and Millhollin, 2002; Kurz, Stergiou, and Porter, 2003). Response Surface Methodology models the behavior of the joints with a second-order nonlinear interactive regression model. Equation 6.1 details the RSM model for two joints and the criterion.

$$f(x,y) = b_0 + b_1x + b_2y + b_3xy + b_4x^2 + b_5y^2 \tag{6.1}$$

where $f(x,y)$ is the criterion, $x$ and $y$ are predictors, $xy$ is the interactive term, $x^2$ and $y^2$ are nonlinear terms, and $b_n$ are the coefficients for each respective term. The $x$ and $y$ terms in the model can represent kinematic variables (i.e., step length, joint range of motion, angular velocities, etc.) or kinetic variables (i.e., joint moments, joint powers, etc.). The criterion is typically an outcome variable of the mode of locomotion such as speed, economy, or ground reaction forces. The only stipulation is that there must be a relationship between the selected criterion and the respective predictors. Note that the RSM equation (equation 6.1) is based on parabolic terms (i.e., $x^2$ and $y^2$). Each parabolic term represents the behavior of one of the lower extremity joints investigated. Theoretically, the preferred behavior of the joint is located at the local minimum or maximum of the respective parabola. The remaining terms in the equation (i.e., $x$, $y$, and $xy$) influence the curvature and directionality (**concave up** or **concave down**) of the respective parabolic terms in the RSM model. Since the RSM model is based on a quadratic equation, the shape of the surface generally conforms to one of the following configurations: elliptic paraboloid (figure 6.1) or hyperbolic paraboloid (saddle configuration; see figure 6.5).

With use of the RSM model, a theoretical surface can be developed that graphically details the interactive behavior of the lower extremity joints during locomotion (figure 6.1).

In figure 6.1, the x- and y-axes represent the behavior of two joints (kinematic or kinetic) during locomotion, and the z-axis represents the criterion (outcome variable). Response Surface Methodol-

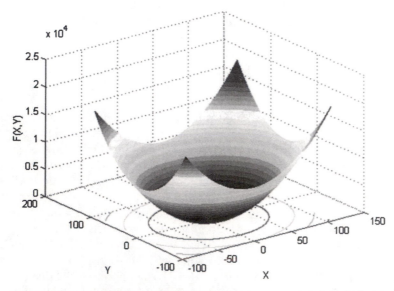

**Figure 6.1** Response Surface Methodology surface with the area under the curve represented as contours in the XY plane.

ogy has the benefit of allowing creation of a surface that visually depicts the interactive behavior of the joints during locomotion. Modifications in the shape of the RSM surface indicate new joint behaviors for a given mode of locomotion (Kurz, Porter, and Steele, 2000; Kurz, Stergiou, and Scott, 2002; Kurz, Stergiou, and Porter, 2003). For example, figure 6.2 depicts two surfaces created for preferred and maximal hopping height conditions where ankle stiffness and knee stiffness are the predictors of leg stiffness.

Although hopping may be classified as a simple form of locomotion, evaluation of the two surfaces suggests that the interactions of the joints are quite complex. The interactions of the joints are not linear. The RSM surfaces in figure 6.2 indicate that as hopping height increased (between experimental conditions), the shape of the RSM surface was modified. During hopping at a preferred height, the surface had a concave-down configuration. However, during hopping at a maximal height, the surface changed to a concave-up configuration. A change in the shape of the surface indicates that the interactive behavior of the joints' stiffness was different between the two conditions. Therefore, hopping height influences the interactive strategies between the two joints for completing the simple mode of locomotion. Changes in the shape of the surface provide a qualitative depiction of the organization of the lower extremity joints for a given criterion (outcome variable).

Theoretically, the area under the surface quantifies the number of possible ways the lower extremity joints can be coupled for a given mode of locomotion. Figure 6.1 displays contour plots in the XY plane. The larger the area covered by the contour plots in the XY plane, the greater the variability in the modeled neuromuscular system during locomotion. Therefore, the area under the surface can provide a theoretical summation of the variability in the system for various modes of locomotion. This information can be used to explore locomotive stability. It has been previously suggested that points of transition from one mode of locomotion to another are characterized by an increased amount of variability (Diedrich and Warren, 1995). It has also been suggested that an increased amount of variability is indicative of an unstable gait pattern. Thus, one can use the area under the surface to gain insight regarding joint behaviors that are stable, as well as where transition points exist as speed of locomotion is increased or decreased.

The absolute minimum or maximum (critical point) of the RSM surface is an attractor that defines the optimal interaction (or preferred mode) of the joints during locomotion (Varberg and Purcell, 1997; Stewart, 1998; Kurz, Porter, and Steele, 2000;

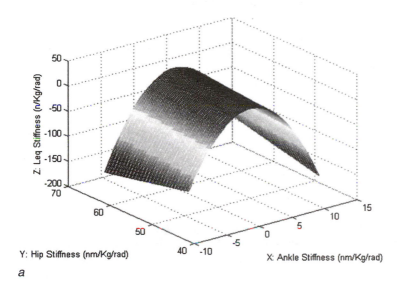

*a*

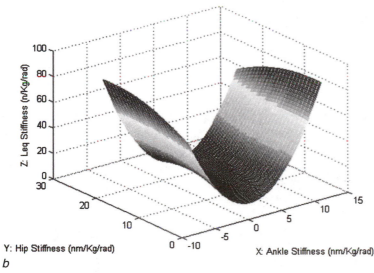

*b*

**Figure 6.2**  Response Surface Methodology surfaces for hip and knee joint interactions: *(a)* preferred hopping height and *(b)* maximal hopping height.

Kurz, Stergiou, and Scott, 2002; Kurz, Stergiou, and Porter, 2003). For example, in figure 6.1, the bottom of the paraboloid would be the attractor for the $x$ and $y$ variables. The **critical point** of the surface defines the preferred interactive behaviors of the respective joints during locomotion. Theoretically, the neuromuscular system seeks the critical point of the surface to maintain a preferred (and possibly stable) gait pattern. Quantifying the joint's behavior at the critical point of the surface can aid in defining the requisites for various modes of locomotion. The critical point of the surface is similar to the mean except that it is influenced by how the joints interact with each other. Therefore, the critical point of the surface can provide information on the behaviors of the joints that one cannot obtain by observing the mean values.

The **gradient vector** of the surface is directed toward the critical point of stability on the surface (Varberg and Purcell, 1997; Stewart, 1998). The gradient vector can be utilized to quantify the changes in behavior of the respective joints that are graphically present in the RSM surface (Box and Draper, 1987; Varberg and Purcell, 1997; Stewart, 1998). The direction of the gradient vector components indicates whether the neuromuscular system is maximizing or minimizing the joint's behavior to achieve a preferred mode of locomotion. The components of the gradient vector quantify the directionality and shape of the various nonlinear terms (i.e., $x^2$ in equation 6.1) in the RSM model. A positive component of the gradient vector indicates that the parabola describing the joint's behavior (i.e., $x^2$ in equation 6.1) is concave up (Varberg and Purcell, 1997; Stewart, 1998). Thus, the values are decreasing as they approach the local minimum of the parabolic term. Alternatively, a negative component of the gradient vector indicates that the parabolic term describing the joint's behavior is concave down (Varberg and Purcell, 1997; Stewart, 1998). Therefore, the values are increasing as they approach the local maximum of the parabolic term. If two components of the gradient vector have the same directionality, then they have a similar behavior as they proceed toward the critical point of the surface (Varberg and Purcell, 1997; Stewart, 1998). Components that share a common directionality are coupled during locomotion. For example, if the $x$ component of the gradient vector is positive and the $y$ component of the gradient vector is positive, then both variables are decreasing in value as they approach the respective local minimum. This indicates that both variables are concurrently minimizing their behavior to achieve a stable locomotive pattern. Thus, the directionality of the gradient vector components provides a means to quantify how the kinematic and kinetic variables are coupled for various modes of locomotion.

The gradient vector of the surface can be used to address which kinematic and kinetic variables determine the selection of a mode of locomotion and the transition from one mode of locomotion to the next. Additionally, the gradient vector of the RSM surface can be used to quantify how the behaviors of the joints influence economy for various modes of locomotion. For example, an RSM surface could be created that models the relationship between economy and the lower extremity joint kinematics. The direction of the gradient vector toward the critical point of the RSM surface will detail the contributions of the respective joint's kinematics for an economical mode of locomotion. Additionally, the directionality of the components of the gradient vector will allow one to quantify which joints have the greatest influence on the shape of the surface. This is because components of the gradient vector that have a larger magnitude have a greater influence on the shape of the RSM surface (Varberg and Purcell, 1997; Stewart, 1998).

One can obtain further information about the behavior of the lower-extremity joints during locomotion by projecting the RSM surface into a two-dimensional plane (Varberg and Purcell, 1997; Stewart, 1998; Kurz, Stergiou, and Millhollin, 2002; figure 6.3). Projections of the RSM surface produce **grid lines** that isolate the behavior of the respective lower-extremity joints. Since the grid line is from the RSM surface, the shape of the grid line is dependent on the behavior of the joint and its interaction with other joints in the lower extremity. Theoretically, variability within the joint can be estimated from the curvature of the grid line (Kurz, Stergiou, and Millhollin, 2002; figure 6.3). Figure 6.3 details the two-dimensional projection of the RSM surface into the YZ plane. The two-dimensional projection of the surface indicates that the behavior of the $x$ variable conforms to a parabola that is concave up. An increased amount of curvature in the projection suggests less variability in the joint's behavior. Conversely, less curvature in the projection suggests increased variability in the joint's behavior. This notion is based on the fact that a parabolic function with less curvature spans a greater range of values (Varberg and Purcell, 1997; Stewart, 1998). An increased range indicates greater variation in the possible solutions of grid line function. Therefore, a grid line with less curvature indicates that the behavior of the joint has a larger amount of variability during locomotion (Kurz, Stergiou, and Millhollin, 2002).

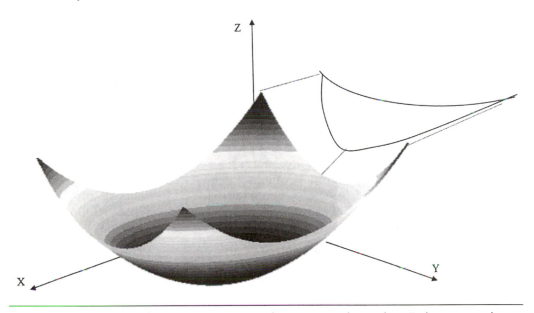

**Figure 6.3**  Response Surface Methodology surface projected into the YZ plane to produce a grid line for the $x$ variable.

Curvature of the grid line for a given joint may be used to classify a stable mode of locomotion. As stated previously, recent investigations have suggested that an increased amount of joint kinematic variability is a precursor to transitioning from one mode of locomotion to the next (Diedrich and Warren, 1995). Typically, variability is reduced during a preferred mode of locomotion (Diedrich and Warren, 1995). However, variability is highest at a point prior to transition from one mode of locomotion to the next. This phenomenon can be observed when humans transition from walking to running (Diedrich and Warren, 1995). Through inspection of the curvature of the respective components of the RSM surface, one can gain insight into how joint variability increases or decreases as the subject transitions from one mode of locomotion to the next. On the basis of this logic, a decreased amount of curvature would suggest an unstable joint's behavior.

The following sections provide additional information on how to create RSM surfaces and quantify the observed changes in the shape of the surface. The examples presented are based on leg and joint stiffness values for 20 subjects completing continuous two-legged hopping at a self-selected frequency. The data set for this example is presented in appendix C. For these representative data, each subject performed 20 consecutive hops. We selected hopping because it can be classified as a simple form of locomotion. Additionally, research has shown that joint stiffness is correlated with changes in leg stiffness (Farley and Morgenroth, 1999). The reader should not be misled into thinking that surfaces can be developed only with kinetic variables. Surfaces can be developed using either kinematic or kinetic variables as long as the variables are related to the criterion. For example, one could create a surface by regressing stride length on ankle, knee, and hip range of motions during gait. The reader should note that RSM models with more than two predictors require projections for visualizing how the joints interact during locomotion. For simplicity, the examples in this chapter focus only on the relationship of two joints for a given criterion. However, the concepts presented can be applied to any $n$th-dimensional surface.

## Fitting the Response Surface Methodology Surface

Response Surface Methodology utilizes a regression model that accounts for the nonlinear properties of the joint and its interaction with other joints that compose the lower extremity (Box and Draper, 1987). A least-squares method is utilized to fit a nonlinear interactive multiple regression model to the respective data (see equation 6.1). The reader may refer to Lay, 2000, and Pedhazur, 1997, for a detailed description of the least-squares method. The parabolic nonlinear terms in the regression equation (equation 6.1) suggest that there is an absolute minimum or maximum (critical point) on the surface where the joint pattern is stable (Box and Draper, 1987; Varberg and Purcell, 1997; Stewart, 1998). This critical point is the collective preferred state of joints' behaviors.

In the development of an RSM model, the predictors used in the regression equation should be centered (mean deviated) about their respective means. Centering the variables eliminates nonessential ill-conditioned multicollinearity in the regression equation caused by the interactive and nonlinear terms of the regression equation (Aiken and West, 1991). The RSM surface should be based on the 15:1 observation-to-variable ratio for a proper fit of a regression equation (Pedhazur, 1997). A lower ratio may result in overfitting of the RSM surface (Pedhazur, 1997). Thus, to investigate the interaction of two predictors (kinematic or kinetic), one should use a minimum of 75 gait cycles to develop the RSM model. If the subjects were pooled together to create the surface, a lower number of gait cycles would be required from each subject.

The direction and sign of the coefficients of the fitted RSM model offer initial insights into neuromuscular strategies. Negative nonlinear terms (i.e., $-b_4x^2$) indicate that the behavior of the joint can be described with a concave-down parabolic function; positive nonlinear terms (i.e., $+b_4x^2$) indicate that a concave-up parabolic function describes the behavior of the joint (Varberg and Purcell, 1997; Stewart, 1998). Differences in concavity between conditions indicate a change in the behavior of the joint. For example, the concave-down shape indicates that the joint is maximizing its behavior for a mode of locomotion (i.e., increasing range of motion or increasing joint torque magnitude); a concave-up shape indicates the joint is minimizing its behavior during locomotion (i.e., decreasing range of motion or decreasing joint torque magnitude, etc.). The larger the coefficient of the nonlinear term, the more steep the slope of the parabolic function. A steeper slope for the coefficient indicates that the behavior of the joint is more constrained and theoretically has less variability in its behavior. Additionally, nonlinear terms with the same sign indicate that the joints are proceeding toward the critical point of the RSM surface in the same direction (Varberg and Purcell, 1997; Stewart, 1998). Therefore, they are coupled as they proceed toward a preferred gait pattern.

### Example 6.1

Based on the joint and leg stiffness data in appendix C, the following RSM function was constructed using a least-squares method:

$$f(x,y) = 22.105 + .124x + 6.451y - 1.216xy - .212x^2 + 1.909y^2 \tag{6.2}$$

where $x$ is the ankle joint stiffness, $y$ is the knee joint stiffness, and leg stiffness was the criterion. The RSM function significantly accounted for 74% of the variance in leg stiffness ($R^2 = .74, F[5,394] = 222.54, p < .0001$).

Evaluation of the nonlinear coefficients of equation 6.2 in this example indicates that the ankle joint stiffness was maximizing while knee joint stiffness was minimizing. This is evident because the directionality of the ankle nonlinear term was negative ($x^2 = -.2$) and the directionality of the knee nonlinear term was positive ($y^2 = +1.9$). Therefore, the behaviors of the ankle and knee were different during hopping. Additionally, it is evident that the nonlinear knee joint term has a larger coefficient than the ankle nonlinear term. Since the coefficient represents the slope of a parabolic function, this suggests that the knee joint function has more curvature than the ankle joint. Figure 6.4 graphically depicts the nonlinear ankle and knee terms. For comparison of the curvature of the two lines, the absolute values of the nonlinear terms are presented. Inspection of the figure confirms the observation that the knee joint nonlinear term has more curvature. Since knee parabolic function is steeper in slope than the ankle parabolic function, the knee tends to gravitate more quickly to the critical point of the RSM surface than the ankle. In figure 6.4, we can graphically observe that the critical point would be where the derivatives of the respective curves are zero. Theoretically, since the knee joint parabolic term has a larger slope, it is more constrained to the critical point of the RSM surface and is not as easily perturbed. The constrained movement pattern suggests that the knee joint is less variable than the ankle during the hopping pattern.

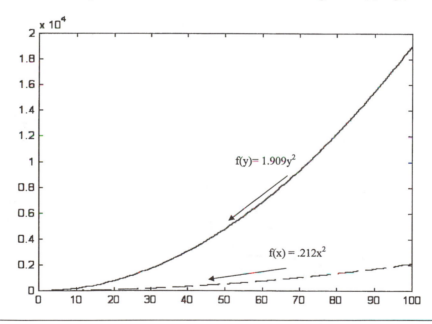

**Figure 6.4**    Plots of the nonlinear terms of the RSM function where $f(y)$ is the knee nonlinear term and $f(x)$ is the ankle nonlinear term.

## Creating a Response Surface Methodology Surface

Software such as MATLAB (MathWorks) can be used to create graphical depictions of the RSM surfaces. To create a surface, the RSM function should be parameterized into a vector function (Varberg and Purcell, 1997; Stewart, 1998):

$$\mathbf{r}(u,v) = x(u,v)\mathbf{i} + y(u,v)\mathbf{j} + z(u,v)\mathbf{k} \qquad (6.3)$$

where $\mathbf{i}$, $\mathbf{j}$, and $\mathbf{k}$ are the respective components of the vector function. Response Surface Methodology functions can be parameterized with spherical, polar, or rectangular coordinate systems.

**Example 6.2**

The RSM function in example 6.1 (equation 6.2) can be parameterized into rectangular coordinates as follows:

$$x = x \quad y = y \quad z = 22.105 + .124x + 6.451y - 1.216xy - .212x^2 + 1.909y^2 \quad (6.4)$$

The parameterized vector function used to develop the RSM surface is then as follows:

$$\mathbf{r}(x,y) = x\mathbf{i} + y\mathbf{j} + (22.105 + .124x + 6.451y - 1.216xy - .212x^2 + 1.909y^2)\mathbf{k} \quad (6.5)$$

where **i, j,** and **k** represent the components of the vector r.

By means of MATLAB's three-dimensional mesh function, one can use the parameterized vector function to create an RSM surface. The mesh function plots the values of the vector function as height values above a rectangular grid and connects the neighboring points into a mesh surface. The surface is typically bounded by the standard deviation of the collected data for each joint. An exaggerated range of values may be necessary to prevent local linearity and allow for observation of the true shape of the surface.

The RSM surface (figure 6.5) from the parameterized equation in example 6.2 (equation 6.5) indicates that the interactive behavior of the ankle and knee joint stiffness results in a saddle configuration. Visually projecting the surface into the XZ plane shows that the surface has the shape of a concave-up parabola. This projection concurs with the result showing that the knee nonlinear term was a concave-up parabola. Visually projecting the surface into the YZ plane shows that the surface has the shape of a concave-down parabola with slight curvature. This projection also concurs with the result indicating that the ankle nonlinear term was a concave-down parabola with a slight curvature. Since the knee joint projection appears to have the greater curvature, it is qualitatively apparent that the knee joint strongly affected the overall shape of the RSM surface. One can graphically display the influence of the various joints on the shape of the surface by projecting the surface into a two-dimensional plane to create grid lines. Projections of the surface can aid in the interpretation of the contributions of the various joints to the shape of the surface. This procedure is detailed later in the chapter. However, at this point the reader can refer to figure 6.7 for the surface projections of figure 6.5.

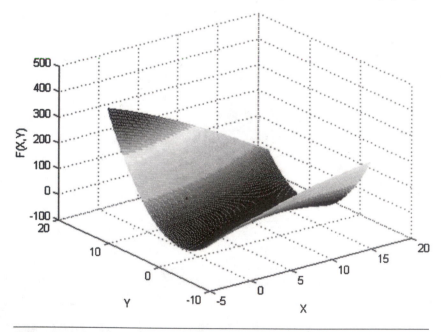

**Figure 6.5**   Response Surface Methodology surface for equation 6.2.

## Area Under the Response Surface Methodology Surface

Theoretically, the area under the RSM surface represents the total number of possible joint interactive patterns for completing the movement pattern (Varberg and

Purcell, 1997; Stewart, 1998). These variations in the joint patterns are necessary to overcome global and local perturbations experienced during the gait cycle. The larger the area under the RSM surface, the greater variations in the neuromuscular responses to local and global perturbations experienced during locomotion. A multiple definite integral can be used to determine the area under the RSM surface where the surface is bounded by the standard deviation of each of the respective joints (equation 6.6).

$$\iint_D f(x,y)\, dA \tag{6.6}$$

## Example 6.3

A definite double integral can be used to find the area under the RSM function detailed in example 6.1 (equation 6.2) that is bounded by $R = \{(x,y) \mid 4.32 \le x \le 7.34,.$

$$\iint_D 22.1 + .12x + 6.45y - 1.22xy - .21x^2 - 1.91y^2\, dx\, dy \tag{6.7}$$

Area under RSM surface = 57.23

The double integral represents the number of possible solutions for completing the gait pattern. The larger the area, the more total variability in the system. Conversely, the smaller the area, the less total variability in the system. Thus, one can use the area under the surface to infer the amount of total variability in the system since it is a collective measure of the total joint variability for a given mode of locomotion.

## *Response Surface Methodology Surface Optimization*

Theoretically, the absolute minimum or maximum of the surface is the critical point where the preferred mode of locomotion occurs on the surface (Varberg and Purcell, 1997; Stewart, 1998). The neuromuscular system is attracted to this point on the surface during locomotion. Variations in the joint's behavior occur when the joint is perturbed from the critical point because of global or local perturbations. Theoretically, once the perturbation subsides, the joint will readily return to the critical point of the surface. The critical point of the surface is evident where the first-order partial derivatives of the RSM surface are equal to zero (Varberg and Purcell, 1997; Stewart, 1998). Matrix algebra can be used to solve for the unknown critical point of the surface.

## Example 6.4

With use of the RSM function detailed in example 6.1 (equation 6.2), the first-order partial derivatives of the RSM surface can be found as follows:

$$\delta f / \delta x = .124 - 1.216y - .424x \tag{6.8}$$

$$\delta f / \delta y = 6.451 - 1.216x + 3.818y \tag{6.9}$$

*(continued)*

Since the first-order partial derivatives represent a system of equations that describe the behavior of the RSM surface, the system can be written in matrix form ($A\mathbf{x} = \mathbf{b}$) such that the unknown critical point values can be solved.

$$\begin{pmatrix} .124 & -.424 & -1.216 \\ 6.451 & 3.818 & -1.216 \end{pmatrix} \begin{pmatrix} x \\ y \end{pmatrix} = \begin{pmatrix} 0 \\ 0 \end{pmatrix} \qquad (6.10)$$

Solving the underdetermined matrix, the x and y critical points of the RSM surface were determined to be as follows:

$$\begin{pmatrix} 1 & 0 & -1.6076 & \vline & 0 \\ 0 & 1 & 2.3978 & \vline & 0 \end{pmatrix}$$

$$y = 1 / 1.6076$$

$$x = -2.3978 / 1.6076$$

$$(6.11)$$

Thus, the optimum point of stability (critical point) for the RSM surface was <–1.49, .622>.

Since the surface is based on the centered data sets (mean deviated), the critical point represents a normalized value for the joint's behavior. To quantify the behavior of the joint in real-world units, the mean of the joint's behavior must be added to the critical point. Therefore, in example 6.4, the optimal point of the RSM surface is <4.32 nm/kg/rad, 2.15 nm/kg/rad >. Note that the critical point of the surface and the mean value are not the same (mean ankle stiffness = 5.83 nm/kg/rad, mean knee stiffness = 1.53 nm/kg/rad). The reason for the difference between the values is that the RSM surface takes into consideration how the joints interact for a preferred mode of locomotion.

Figure 6.6 represents a contour plot of the RSM surface created from equation 6.2 (see example 6.1, p. 168). Figure 6.6 confirms our observation that the RSM surface had a saddle configuration with the optimum point found between the sides of the saddle. Theoretically, the joint pattern is typically varied around this point because of global and local perturbations that occur during locomotion.

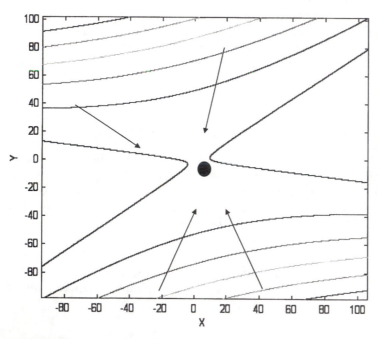

**Figure 6.6** Contour plot of the RSM surface with the critical point identified by the dot. The x- and y-axes are mean deviated joint stiffness.

## Creating Grid Lines From the Response Surface Methodology Surface

From the RSM surface, one can calculate grid lines to isolate the behavior of each joint (Varberg and Purcell, 1997; Stewart, 1998). One calculates grid lines by setting the other joints' variables to zero in the RSM model. The remaining variables in the model are then a projection of the surface into a two-dimensional plane (Varberg and

Purcell, 1997; Stewart, 1998). Projections (grid lines) are used to gain insight into the contribution of each joint to the complex shape of the RSM surface. Such information may help one to interpret which variables in the RSM contribute the most to the changes in the shape of the surface between conditions.

**Example 6.5**

Using the RSM function in example 6.1 (equation 6.2), the grid lines can be calculated as follows:

$$f(x) = 22.105 + .124x - .212x^2 \tag{6.12}$$

$$f(y) = 22.105 + 6.451y + 1.909y^2 \tag{6.13}$$

Figure 6.7 graphically depicts the grid lines in example 6.5. Evaluation of the figure indicates that the ankle grid line function is a concave-down parabola and the knee grid line function is concave up. The fact that the grid line functions are opposite in concavity suggests that the ankle and knee have different neuromuscular strategies during the locomotive pattern. Further inspection of the two grid line functions (figure 6.7) suggests that the knee has qualitatively more curvature than the ankle. Therefore, it can be theoretically suggested that the knee joint stiffness is less variable during the movement pattern and less susceptible to global and local perturbations.

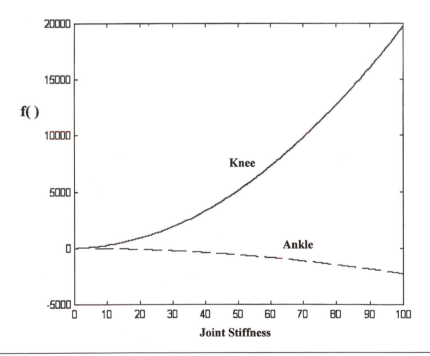

**Figure 6.7**   Response Surface Methodology surface projections. The knee joint is the solid line and the ankle joint is the dashed line.

## Gradient Vectors

Gradient vectors ($\nabla f(x,y)$) for the RSM function are calculated from the second partial derivatives (Varberg and Purcell, 1997; Stewart, 1998). The components of the gradient vector quantify the interactive behavior of each of the respective joints during a preferred mode of locomotion. The gradient vector is directed toward the critical point of the surface where a preferred (and possibly stable) interactive joint pattern exists. Components of the gradient vector with a negative slope indicate that the behavior of the joint is maximizing, and components with positive slopes indicate that the behavior of the joint is minimizing (Varberg and Purcell, 1997; Stewart, 1998). Furthermore, gradient vector components that are larger in magnitude indicate a greater rate of change toward the critical point of the RSM surface (Varberg and Purcell, 1997; Stewart, 1998). The components of the gradient vector should concur with the shape of the grid lines of the RSM surface (Varberg and Purcell, 1997; Stewart, 1998). Therefore, components of the gradient vector that have a larger magnitude have a grid line shape with greater curvature. Additionally, components of the gradient vector that are larger in magnitude have a greater attraction to the critical point and theoretically less variability (based on preceding curvature arguments in this chapter). Additionally, components of the gradient vector that have the same sign are coupled as they proceed toward a point of stability on the RSM surface (Varberg and Purcell, 1997; Stewart, 1998).

## Example 6.6

Using the RSM function in example 6.1 (equation 6.2), the gradient vector of the RSM surface can be found as follows:

$$\delta^2 f / \delta x^2 = -.42 \qquad (6.14)$$

$$\delta^2 f / \delta y^2 = 3.82 \qquad (6.15)$$

Therefore, the gradient vector of the RSM surface is $\nabla f(x,y) = <-.42, 3.82>$.

In example 6.6, the direction of the gradient vector indicates that the ankle and knee are proceeding toward the critical point of the RSM surface in the opposite direction. The ankle is maximizing joint stiffness while the knee joint is minimizing joint stiffness during the movement pattern. Comparison of the magnitudes of the gradient vector components indicates that the knee has a greater attraction to the critical point of the RSM surface. The values of the gradient vector concur with our previous observations of the grid line functions in which it was evident that the ankle had less curvature than the knee (equation 6.12, equation 6.13). Thus, we can suggest that the ankle joint stiffness is more variable than that of the knee during this movement pattern.

### Curvature of Grid Lines

Curvature is a measure of how quickly the grid line changes direction at the critical point (Varberg and Purcell, 1997; Stewart, 1998). One can use curvature of the grid lines to quantify the attraction of the joints to the critical point of the surface. Greater curvature indicates that the parabola has steep slopes and that theoretically local and global perturbations do not easily perturb the joint from the critical point of the surface. A grid line with less curvature indicates that the joint has theoretically more variability in its behavior and is more readily perturbed from the critical point. Curvature of the grid line can be calculated with equation 6.16 (Varberg and Purcell, 1997; Stewart, 1998):

$$\kappa(x) = \frac{|f''(x)|}{\left[1 + (f'(x))^2\right]^{3/2}} \qquad (6.16)$$

where f(x) is the grid line function, f' (x) is the first derivative, and f" (x) is the second derivative. Curvature is calculated at the local minimum or maximum of the respective grid line. The local minimum or maximum occurs when the derivative of the grid line function is equal to zero (Varberg and Purcell, 1997; Stewart, 1998).

## Example 6.7

Using grid line functions in example 6.5 (equations 6.12 and 6.13), the following measures of curvature were calculated for the respective grid lines:

$$\kappa(x) = \frac{|-.424|}{\left[1 + (.124 - .424x)^2\right]^{3/2}} = 0.424$$

where the local maximum for the grid line function was .292.

$$\kappa(y) = \frac{|3.82|}{\left[1 + (6.45 + 3.82y)^2\right]^{3/2}} = 3.82$$

where the local minimum was for the grid line function was −1.69.

In example 6.7 it is apparent that the knee joint has more curvature than the ankle. The greater curvature value suggests that the knee joint variable has a greater attraction to the critical point of the surface. The greater curvature also suggests that the knee joint stiffness has theoretically less variability, since it is a parabola with a steeper slope. The lesser curvature indicates that the ankle stiffness had more variability.

### Summary of Response Surface Methodology

Response Surface Methodology modeling provides useful information on the organization of the neuromuscular system during locomotion. On the basis of the RSM model, one can develop a surface that graphically details the interactive behavior and stability of the joints under a wide range of perturbations. Characteristics of the surface, such as the gradient vector and **grid line curvature,** provide quantitative information about the interactive behavior of the joints during various modes of locomotion. Traditional tools that use the mean cannot provide such important information. The RSM model approach can provide new insights on the requisites for locomotion. Mathematical tools presented in this section can further our scientific understanding of the biological properties that control the selection of various modes of locomotion. We believe that future scientific studies can benefit by using this type of theoretical model that is based on a sound mathematical framework.

# Variability

As discussed in the beginning of the chapter, recent research efforts have used measures of variability to explore the various modes of locomotion (Winter, 1983; Gabel and Nayak, 1984; Diedrich and Warren, 1995; Maki, 1997; Heiderscheit, Hamill, and van Emmerick, 1999; van Emmerick et al., 1999; Hamill et al., 1999; Byrne et al., 2002; Kurz and Stergiou, 2003; Kurz, Stergiou and Blanke, 2003). In the past it has been proposed that the variability in the joint kinematics from one gait cycle to the next is the result of biological noise in the neuromuscular system (Glass and Mackey, 1988). However, recent investigations have determined that these fluctuations are not random noise superimposed on top of the control system. Rather these fluctuations have a deterministic structure (Hausdorff et al., 1995, 1997).

The notion that variability in the locomotive pattern has a biological meaning has stimulated further investigations. Recent researchers have noted that points of transition from one mode of locomotion to another are characterized by an increased amount of kinematic variability (Diedrich and Warren, 1995). Additionally, several studies have indicated that variability may be related to the health and stability of the neuromuscular system (Winter, 1983; Gabel and Nayak, 1984; Heiderscheit, Hamill, and van Emmerik, 1999; Dingwell et al., 1999; Hamill et al., 1999; Kurz and Stergiou, 2003). These investigations have just begun to explore the wealth of information that variability has to offer. Further research is necessary to elucidate the biological meaning of the changes in variability for various modes of locomotion.

Within-subject variability during gait can be defined from the standard deviations of the parameters under study through the use of multiple gait cycles. Many current investigations of variability use the mean ensemble curve to quantify the amount of variability in the neuromuscular system. The larger the standard deviations of the mean ensemble curve, the larger the variability in the system. Traditionally, the coefficient of variation and mean deviation are used to quantify the amount of variability observed in the mean ensemble curve. However, there are times when the absolute measure of variability is a better measure than the relative amount of variability. In this portion of the chapter we present the spanning set as an absolute measure of variability from the mean ensemble curve. The spanning set offers the researcher an alternative method for defining within-subject variability from the mean ensemble curve.

## Creating a Mean Ensemble Curve

The mean ensemble curve for joint angular displacement describes the average joint pattern and the variations in the joint pattern from one trial to the next (figure 6.8). The mean ensemble curve is constructed from multiple time-normalized data series (trials) where a mean and confidence interval (CI) are calculated for every $i$th point of the joint pattern. Equation 6.17 represents the mean at the $i$th point. Equation 6.18 is the CI about each $i$th point.

$$Y = \frac{\sum_{i=1}^{N} x_i}{N} \tag{6.17}$$

$$CI = Y \pm \sqrt{\frac{\sum x_i^2}{N} - Y^2} \tag{6.18}$$

where $N$ is the number of trials averaged and $x$ is the actual value of the time series for each $i$th trial. The CI is used to construct the standard deviation curves about the mean ensemble curve. The distance between the two standard deviation curves represents the variability in the joint pattern (figure 6.8).

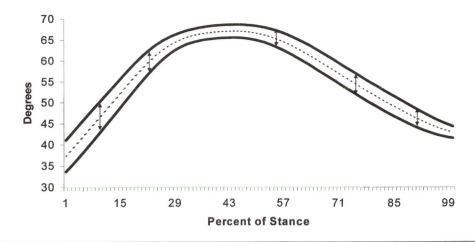

**Figure 6.8** Mean ensemble curve for the knee joint. Dashed line is the mean, and the bold lines are the standard deviation.

## Coefficient of Variation and Mean Deviation

As stated previously, the coefficient of variation (CV; equation 6.19) and the mean deviation (MD; equation 6.20) are typically used to quantify the amount of variability in the joint pattern from the mean ensemble curve:

$$CV = \frac{\frac{1}{N}\sum_{i=1}^{N} S_i}{\frac{1}{N}\sum_{i=1}^{N} |x_i|} \qquad (6.19)$$

$$MD = \frac{\sum_{i=1}^{N} |S_i|}{N} \qquad (6.20)$$

where $S$ is the standard deviation of the entire mean ensemble curve, $x$ is the $i$th point of the mean ensemble curve, and $N$ is the number of points in the mean ensemble curve. Note that the CV is a ratio of the standard deviation of the entire mean ensemble curve over the mean joint pattern. The CV provides a way to quantify the variation in the joint relative to the mean. The MD measure provides an average of the standard deviation of the entire mean ensemble curve.

It is worth mentioning here that there also may be a problem with using CV to compare variability across subjects (equation 6.19). The difficulty in using the CV formula may result from the fact that the mean value of the joint movement pattern is contained in the denominator. A larger denominator influences the magnitude of the CV ratio. For example, the CV was calculated for subject A's and subject B's mean ensemble joint patterns (table 6.1). Although both subjects have the average standard deviation about the mean ensemble curve for their respective joint patterns (5.5°), the CV measure suggests that subject B's joint pattern is more variable than subject A's. This discrepancy is attributable to the different mean joint pattern values. Subject A's larger mean value diminishes the amount of true variability present in the joint pattern because the mean value is located in the denominator of the CV measure (equation 6.19). Therefore, different mean joint pattern magnitudes affect the reliability of the CV measure to quantify variability between subjects.

**Table 6.1   Exemplar Descriptive Statistics for the Mean Ensemble Joint Pattern of Two Subjects**

|            | Mean joint pattern | Standard deviation | Coefficient of variation |
|------------|--------------------|--------------------|--------------------------|
| Subject A  | 60                 | 5.5                | .09                      |
| Subject B  | 45                 | 5.5                | .12                      |

Values are presented in degrees.

Utilizing the CV to determine the amount of variability in a movement pattern from the mean ensemble curve may not be the best mathematical measure. If subjects have different ranges of motion during the gait pattern, the CV will not be able to quantify the true variability in the movement pattern. The spanning set can be used as an alternative measure of variability about the mean ensemble curve. The spanning set provides a measure of variability that is not susceptible to changes in the configuration of the mean curve.

## Spanning Set

The spanning set is composed of vectors that describe the possible linear combinations for a system of equations (Lay, 2000). Essentially, the greater the number of the possible linear combinations of the vectors, the greater the variations in the possible solutions. Graphically, the spanning set can be described as a plane in $\mathbf{R}^n$ (figure 6.9).

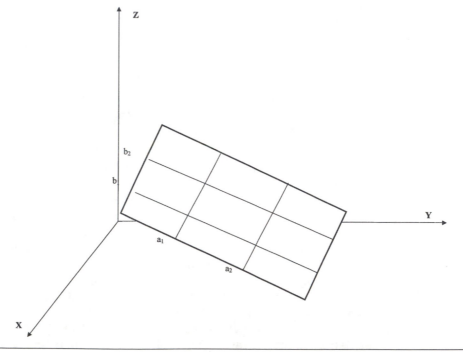

**Figure 6.9**   Plane in $\mathbf{R}^n$ that represents the span of the two vectors, where $a_1$, $a_2$, $b_1$, and $b_2$ represent the components of the respective vectors.

In figure 6.9, the vectors that compose the spanning set can be visualized as the edges of the plane, and the surface between the two vectors contains all possible linear combinations (or variations in the solutions) of the system of equations. The larger the distance between the vectors (or edges of the plane) that define the spanning set, the greater the span of the plane. A larger span between the two vectors suggests greater variability in the possible linear combinations (solutions) of the two vectors.

The standard deviations about the mean ensemble curve can be represented as a spanning set in which the two standard deviation curves represent the two vectors (Kurz and Stergiou, 2003; Kurz, Stergiou, and Blanke, 2003). The greater the distance between the two standard deviation curves of the mean ensemble, the larger the span of the two vectors. Therefore, a larger span between the two vectors would suggest greater variability in the joint pattern.

## Creating Polynomials for the Standard Deviation Curves

The two vectors in the spanning set are based on the standard deviation curves (figure 6.8) of the mean ensemble. Infinite power series expansion formulas (equations 6.21 and 6.22) are used to create polynomials that describe the properties of each standard deviation curve in $\mathbf{P_n}$.

$$p(t) = \sum_{n=o}^{\infty} a_n t^n = a_0 + a_1 t + a_2 t^2 + \dots \tag{6.21}$$

$$g(t) = \sum_{n=o}^{\infty} b_n t^n = b_0 + b_1 t + b_2 t^2 + \dots \tag{6.22}$$

Equations 6.21 and 6.22 consist of a sum of an infinite number of power terms to the 0 degree, 1st degree, 2nd degree, and so on up to the $n$th degree. Differences between the two polynomials will be due to the magnitude of the coefficients (i.e., $a_n$) and which power terms are used to describe the respective standard deviation curves (i.e., $b_2 t^2$ vs. $b_5 t^5$). There are many mathematical methods to determine what infinite power series polynomial best describes the standard deviation curve. In this chapter we use a least-squared method. The least-squared procedure is well documented in statistics and linear algebra textbooks such as Pedhazur's (1997) and Lay's (2000).

Although all terms are expressed in the power expansion series, not all power terms are going to provide the best fit of the standard deviation curve. These terms will be represented as zero in the infinite power series polynomial. One can use a stepwise approach to determine the nonlinear terms used in the polynomial that best describe standard deviation curves.

### Example 6.8

Using the mean ensemble curve data presented in appendix C, with a stepwise approach, we can create a polynomial that best describes the above-the-mean standard deviation curve (equation 6.21) and the below-the-mean standard deviation curve (equation 6.22) for figure 6.10.

$$p(t) = a_0 + a_1 t + a_2 t^2 + a_7 t^7 = 54.76 + .62t - .008t^2 + .000000000000112t^7 \tag{6.23}$$

$$g(t) = b0 + b_1 t + b_2 t^2 + b_4 t^4 + b_7 t^7 = 31.39 + 1.75t - .0249t^2 + .00000109t^4 - .000000000000244t^7 \tag{6.24}$$

The above-the-mean standard deviation polynomial accounted for 99.7% of the variance in the above-the mean standard deviation curve ($R = .99$, $F[3,96] = 1,110.99$, $p < .0001$). The below-the-mean standard deviation polynomial accounted for 99.8% of the variance in the below-the-mean standard deviation curve ($R = .99$, $F[4,95] = 1,4143.10$, $p < .0001$).

(continued)

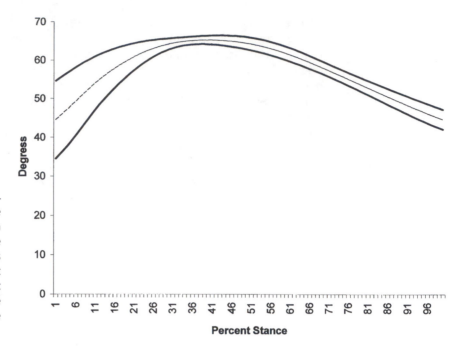

**Figure 6.10** Mean ensemble curve for the knee joint data presented in appendix C, table C.2. The bold line represents the standard deviation about the mean ensemble curve at a 68% confidence interval. The dashed line represents the mean ensemble curve.

## Creating a Spanning Set From the Polynomial

Since $\mathbf{P}_n$ is isomorphic to $\mathbf{R}^{n+1}$, the coefficients from the respective power series polynomials can be used to map to a vector space that defines vectors in the spanning set (equation 6.25).

$$[\mathbf{f}]_B = \begin{pmatrix} a_0 & b_o \\ . & \\ . & . \\ a_n & b_n \end{pmatrix} \tag{6.25}$$

Coordinate mapping is possible because the spanning set forms a basis (linearly independent set). Since $[\mathbf{f}]_B$ was a basis, the coordinate mapping resulted in a 1:1 relationship and accurately described the properties of the polynomials. The two vector spaces $(\mathbf{P}_n$ and $\mathbf{R}^{n+1})$ are different in notation and terminology, but they are indistinguishable in that every vector space calculation in $\mathbf{R}^{n+1}$ is accurately reproduced in $\mathbf{P}_n$ (Lay, 2000). Therefore, the coefficients in $\mathbf{R}^n$ can be used to describe the properties of the standard deviation from the mean ensemble curve.

### Example 6.9

Using the polynomial from example 6.8 (equations 6.23 and 6.24), the spanning set can be defined as follows (equation 6.26):

$$[\mathbf{f}] = \begin{pmatrix} 54.76 & 31.39 \\ .62 & 1.75 \\ -.008 & -.0249 \\ 0 & .00000109 \\ .000000000000112 & -.000000000000244 \end{pmatrix} \tag{6.26}$$

## Determining the Magnitude of the Spanning Set

One determines the magnitude of the spanning set by calculating the norm of the difference between the two vectors of the spanning set (equation 6.27).

$$y = \left\| \mathbf{u} - \mathbf{v} \right\|$$   (6.27)

where **u** represents the vector formed by the coefficients from the first polynomial and **v** represents the vector formed by the coefficients from the second polynomial. The larger the norm of the difference between the two vectors in the spanning set, the greater the span between the two standard deviation curves about the mean ensemble. Therefore, a larger span indicates more variability in the joint pattern (Kurz and Stergiou, 2003; Kurz, Stergiou and Blanke, 2003).

---

### Example 6.10

The norm of the difference between the two vectors of the spanning set from example 6.9 (equation 6.26) can be calculated as follows:

$$u - v = \begin{pmatrix} 54.76 \\ .62 \\ -.008 \\ 0 \\ .00000000000012 \end{pmatrix} - \begin{pmatrix} 31.39 \\ 1.75 \\ -.0249 \\ .00000109 \\ -.000000000000244 \end{pmatrix} = \begin{pmatrix} 23.37 \\ -1.13 \\ .0169 \\ -.00000109 \\ .000000000000356 \end{pmatrix}$$

$$\left\| u - v \right\| = \sqrt{(23.37)^2 + (-1.13)^2 + (.0169)^2 + (-.00000109)^2 + (.0000000000003)^2}$$

$$\left\| u - v \right\| = 547.43$$

---

The magnitude of the spanning set provides a quantitative measure of the distance between the two polynomials that describe the properties of the standard deviation curves. The larger the magnitude of the spanning set, the larger the confidence interval about the mean ensemble. A larger confidence interval would indicate greater variability of the joint movement pattern (Kurz and Stergiou, 2003; Kurz, Stergiou, and Blanke, 2003).

Kurz, Stergiou, and Blanke (2003) evaluated the ability of the CV, MD, and spanning set methodology to detect change in variability during gait, evaluating changes in variability as the subject ran barefoot and with footwear. Variability was based on the standard deviation of the mean ensemble curves for the knee angle during the stance period for each condition. Kurz and colleagues indicated that the spanning set was capable of statistically determining differences in variability between the two conditions. However, CV and MD measures were unable to detect statistical differences between the two conditions. On the basis of the results of this investigation, it was apparent that the spanning set may provide a better measure of differences in variability from the mean ensemble curve. Future researchers on variability based on the mean ensemble curve may want to consider using the spanning set methodology presented in this chapter.

## Summary of Spanning Sets

Within-subject variability during gait can be defined from the standard deviations of the parameters under study using multiple gait cycles. The larger the standard deviations of the mean ensemble curve, the larger the variability in the system. The spanning set provides a novel measure of the variability about the mean ensemble curve. Compared to traditional measures of variability from the mean ensemble curve (i.e., CV and MD), it appears that the spanning set may provide a better measure. Future investigators attempting to link variability in the joint pattern from the mean ensemble curve to movement strategies may want to consider using the spanning set.

## Summary

What variables determine the selection and control of various types of locomotive patterns continues to be a fascinating subject in movement science. To uncover the vast richness in movement strategies, there is a need for new mathematical measures. New mathematical techniques will provide additional scientific information related to gait transitions and neuromuscular pathologies. This chapter presented two novel methods to explore joint strategies and variability for various modes of locomotion.

The RSM model details a surface that describes the nonlinear interactions of the joints during locomotion. Characteristics of the surface, such as the gradient vector and grid line curvature, provide quantitative information about the interactive behavior of the joints for a given mode of locomotion. Analysis of the mean value alone does not provide such scientific information.

Compared to traditional measures of variability of the mean ensemble curve, the spanning set provides an alternative measure of variability. Polynomials are fit to the standard deviations of the mean ensemble curve. The coefficients of the polynomials are used to define the vectors of the spanning set. The greater the distance between the spanning set vectors, the more variability in the mean ensemble curve.

## Work Problems

For the following problems, inspect the nonlinear terms and describe the shape of the RSM surface. Determine which terms are maximizing and minimizing and explain which variables are more constrained to the critical point of the surface.

**1**

$$f(x,y) = -.751 + 1.089x + .576y - .009xy + .148x^2 + .004y^2$$

**2**

$$f(x,y) = .084 + .172x + 6.39y + 1.13xy + .217x^2 - 1.876y^2$$

For the following problems, parameterize the RSM function into a vector function with rectangular coordinates.

**3**

$$f(x,y) = -.751 + 1.089x + .576y - .009xy + .148x^2 + .004y^2$$

**4**

$$f(x,y) = .084 + .172x + 6.39y + 1.13xy + .217x^2 - 1.876y^2$$

For the following problems, calculate the area under the RSM surface for the respective boundaries.

**5**

$$f(x,y) = -.751 + 1.089x + .576y - .009xy + .148x^2 + .004y^2 \text{ bounded by}$$
$$R = \{(x,y) \mid -1.55 \le x \le 1.55, -11.76 \le y \le 11.76\}$$

**6**

$$f(x,y) = .084 + .172x + 6.39y + 1.13xy + .217x^2 - 1.876y^2 \text{ bounded by}$$
$$R = \{(x,y) \mid -1.55 \le x \le 1.55, 1.02 \le y \le 1.02\}$$

For the following problems, calculate the normalized critical point of the RSM surface for the respective variables.

**7**

$$f(x,y) = -.751 + 1.089x + .576y - .009xy + .148x^2 + .004y^2$$

**8**

$$f(x,y) = .084 + .172x + 6.39y + 1.13xy + .217x^2 - 1.876y^2$$

For the following problems, calculate the grid lines of the RSM surface for the respective variables and explain differences in concavity for the $x$ and $y$ variables.

**9**

$$f(x,y) = -.751 + 1.089x + .576y - .009xy + .148x^2 + .004y^2$$

**10**

$$f(x,y) = .084 + .172x + 6.39y + 1.13xy + .217x^2 - 1.876y^2$$

For the following problems, calculate the gradient vector of the RSM and provide conclusions about the behavior of the $x$ and $y$ variables.

**11**

$$f(x,y) = -.751 + 1.089x + .576y - .009xy + .148x^2 + .004y^2$$

**12**

$$f(x,y) = .084 + .172x + 6.39y + 1.13xy + .217x^2 - 1.876y^2$$

For the following problems, calculate the curvature of the grid line and provide conclusions about the amount of variability in the $x$ and $y$ variables.

**13**

$$f(x,y) = -.751 + 1.089x + .576y - .009xy + .148x^2 + .004y^2$$

**14**

$$f(x,y) = .084 + .172x + 6.39y + 1.13xy + .217x^2 - 1.876y^2$$

For the following problems, create a spanning set from the polynomials. The polynomials represent the standard deviation of the mean ensemble curve.

**15**

$$f(x) = -17.16 + 2.10x - .041x^2 + .000000000015x^7 - .00000000000013x^8$$
$$f(x) = -9.305 + 1.85x - .034x^2 + .000000000011x^7 - .000000000000095x^8$$

**16**

$$f(x) = 31.18 - 1.615x - .035x^2 - .0000000000084x^7 + .000000000000061x^8$$
$$f(x) = 37.242 - 1.47x + .0325x^2 - .0000000000066x^7 + .000000000000045x^8$$

For the following problems, determine the magnitude of the spanning set. The polynomials represent the standard deviation curves of the mean ensemble.

**17**

$$f(x) = -17.16 + 2.10x - .041x^2 + .000000000015x^7 - .00000000000013x^8$$
$$f(x) = -9.305 + 1.85x - .034x^2 + .000000000011x^7 - .000000000000095x^8$$

**18**

$$f(x) = 31.18 - 1.615x - .035x^2 - .0000000000084x^7 + .000000000000061x^8$$

$$f(x) = 37.242 - 1.47x + .0325x^2 - .0000000000066x^7 + .000000000000045x^8$$

## Suggested Readings and Other Resources

The majority of the concepts presented in this chapter are based on the foundations of multivariable calculus and linear algebra. For further understanding of these approaches, the reader is directed to the following references:

Box, G.E.P., and N.R. Draper. 1987. *Empirical Model Building and Response Surfaces.* New York: Wiley. Although many of the examples presented in this text are not biologically based, this text will provide the reader with further information on the theoretical foundations for creating and analyzing RSM surfaces.

Kurz, M.J., and N. Stergiou. 2003. The spanning set indicates that variability during the stance period of running is affected by footwear. *Gait and Posture* 17: 132-135. This manuscript uses the spanning set to investigate differences in footwear properties. Such a reference will provide the reader with additional insight on applications of spanning sets for the analysis of variability from the mean ensemble curve.

Kurz, M.J., N. Stergiou, and D. Blanke. 2003. The spanning set defines variability in locomotive patterns. *Medical and Biological Engineering and Computing,* in press. This manuscript compares differences between traditional statistical methods and the spanning set for measuring variability about the mean ensemble curve. This reference will provide additional insights on using the spanning set to analyze variability from the mean ensemble curve.

Lay, D.C. 2000. *Linear Algebra and Its Applications.* New York: Addison-Wesley. This resource provides further logic on how polynomials can be used to create spanning sets.

Varberg, D., and E.J. Purcell. 1997. *Calculus.* New York: Prentice Hall, pp. 698-748, 771-840. Since much of the RSM logic is based on multivariable calculus, this reference provides detailed descriptions (and additional examples) for finding gradient vectors, grid lines, and analysis of concavity for various surfaces. This text also provides mathematical proofs for the use of such tools to explore the characteristics of quadratic surfaces.

## References

Aiken L.S., and S.G. West. 1991. *Multiple Regression: Testing and Interpreting Interactions.* Newbury Park, CA: Sage.

Box, G.E.P., and N.R. Draper. 1987. *Empirical Model Building and Response Surfaces.* New York: Wiley.

Byrne, J.E., N. Stergiou, D. Blanke, J.J. Houser, and M.J. Kurz. 2002. Comparison of gait patterns between young and elderly women: An examination of coordination. *Perceptual and Motor Skills* 94: 265-280.

Diedrich, F.J., and W.H. Warren. 1995. Why change gaits? Dynamics of the walk-run transition. *Journal of Experimental Psychology: Human Perception and Performance* 21(1): 183-202.

Dingwell, J.B., J.S. Ulbrecht, J. Boch, M.B. Becker, J.T. O'Gorman, and P.R. Cavanagh. 1999. Neuropathic gait shows only trends towards increased variability of sagittal plane kinematics during treadmill locomotion. *Gait and Posture* 10: 21-29.

Farley, C.T., J. Glasheen, and T.A. McMahon. 1993. Running springs: Speed and animal size. *Journal of Experimental Biology* 185:71-86.

Farley, C.T., and D.C. Morgenroth. 1999. Leg stiffness primarily depends on ankle stiffness during human hopping. *Journal of Biomechanics* 32(3): 267-273.

Gabel, A., and U.S.L. Nayak. 1984. The effect of age on variability in gait. *Journal of Gerontology* 39(6): 662-666.

Glass, L., and M.C. Mackey. 1988. *From Clocks to Chaos: The Rhythms of Life.* Princeton NJ: Princeton University Press.

Hamill, J., R.E.A. van Emmerik, B.C. Heiderscheit, and L. Li. 1999. A dynamical systems approach to lower extremity running injuries. *Clinical Biomechanics* 14: 297-308.

Hausdorff, J.M., S.L. Mitchell, R. Firtion, C.K. Peng, M.E. Cudkowicz, J.Y. Weit, and A.L. Goldberger. 1997. Altered fractal dynamics of gait: Reduced stride-interval correlations with aging and Huntington's disease. *Journal of Applied Physiology* 82(1): 262-269.

Hausdorff, J.M., P.L. Purdon, C.K. Peng, Z. Ladin, J.Y. Wei, and A.L. Goldberger. 1995. Is walking a random walk? Evidence for long-range correlations in the stride interval of human gait. *Journal of Applied Physiology* 78: 349-358.

Heiderscheit, B.C., J. Hamill, and R.E.A. van Emmerik. 1999. Q-angle influence on the variability of lower extremity coordination during running. *Medicine and Science in Sports and Exercise* 31(9): 1313-1319.

Hill, W.J., and W.G. Hunter. 1966. A review of response surface methodology: A literature survey. *Technometrics* 8(4): 571-590.

Khuri, A.I., and J.A. Cornell. 1987. *Response Surface Design and Analysis.* New York: Marcel Dekker.

Kram, R., and Taylor, C.R. 1990. Energetics of running: A new persepective. *Nature* 346:265-267.

Kurz, M.J., J. Porter, and J.P.H. Steele. 2000. A response surface model approach to predict strategies for adjusting joint stiffness during locomotion. *Proceedings of the 24th Annual Meeting of the American Society of Biomechanics* (pp. 21). Chicago, Illinois.

Kurz, M.J., and N. Stergiou. 2003. The spanning set indicates that variability during the stance period of running is affected by footwear. *Gait and Posture* 17: 132-135.

Kurz, M.J., N. Stergiou, and D. Blanke. 2003. The spanning set defines variability in locomotive patterns. *Medical and Biological Engineering and Computing* 41(2): 211-214.

Kurz, M.J., N. Stergiou, and C. Millhollin. 2002. Response surface curvature suggests that the elderly have altered kinematic variability due to joint interactions. *Journal of Sport and Exercise Psychology* 24: S83.

Kurz, M.J., N. Stergiou, and J. Porter. 2003. Response surface modeling indicates that interactive joint stiffness strategies are dependent on hopping height. In review: *European Journal of Applied Physiology.*

Kurz, M.J., N. Stergiou, and M.M. Scott. 2002. Response surface modeling suggests that kinematic strategies change with aging. *Journal of Sport and Exercise Psychology* 24:S110.

Lay, D.C. 2000. *Linear Algebra and Its Applications.* New York: Addison-Wesley.

Maki, B.E. 1997. Gait changes in older adults: Predictors of falls or indicators of fear? *Journal of the Geriatric Society* 45(4): 313-320.

McClave, J.T., and P.G. Benson. 1988. *Statistics.* San Francisco: Dellen, pp. 647-674.

McMahon, T.A. 1975. Using body size to understand the structural design of animals: quadrupedal locomotion. *Journal of Applied Physiology* 39(4):619-627.

Pedhazur, E.J. 1997. *Multiple Regression in Behavioral Research.* New York: Harcourt College.

Stewart, J. 1998. *Calculus Concepts and Contexts.* New York: Brooks/Cole.

Taylor, C.R., K. Schmidt-Nielsen, J.L. Raab. 1970. Scaling of energetic cost of running to body size in mammals. *American Journal of Physiology* 219(4):1104-1107.

van Emmerik, R.E.A., R.C. Wagenaar, A. Winogrodzka, and E.C. Wolters. 1999. Identification of axial rigidity during locomotion in Parkinson disease. *Archives of Physical Medicine and Rehabilitation* 80: 186-191.

Varberg, D., and E.J. Purcell. 1997. *Calculus.* New York: Prentice Hall.

Winter, D.A. 1983. Biomechanical motor patterns in normal walking. *Journal of Motor Behavior* 15(4): 302-330.

# PART **III**

# Advanced Methods for Data Analysis in Human Movement

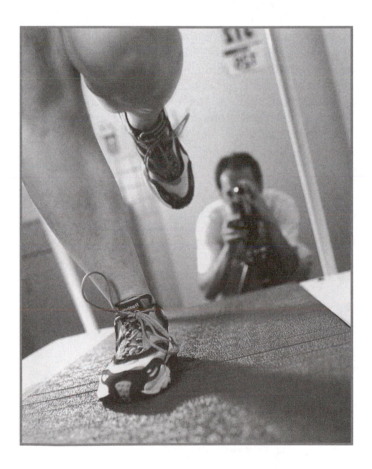

This third section presents methods for analyzing complex data sets using the recent advances in computing power.

In chapter 7, the authors present techniques for the analysis of time series curves. In human movement research, data must be converted to relevant information. These data are typically collected at rates between 60 and 1,000 samples per second. Several seconds of collection can produce a quantity of data that is difficult to deal with. Somehow this large data set must be reduced and interpreted. In this chapter, the authors discuss methods to reduce large quantities of data into a comprehensible set of variables that can be statistically analyzed. Specifically, they discuss how a time series data set can be correlated with another data set (cross-correlation), with itself (autocorrelation), and with known waveforms (i.e., matched filter). They also explain how these techniques can be used to determine differences in phase, extract hidden signals, and reveal frequency content. Examples from a variety of human movements are used to illustrate these methods. Additionally, the authors provide suggestions and recommendations on how cross-correlation can lead to a better understanding of advanced analysis techniques such as Fourier analysis, frequency filtering, and wavelet analysis.

Chapter 8 focuses on bootstrapping. Bootstrapping is a computer-intensive statistical technique that has provided a nonparametric approach to analyze data for which parametric statistics are not appropriate. In addition, bootstrapping provides a new way to explore statistics that traditionally cannot be applied due to a violation of statistical assumptions or a lack of analytic formulas. The authors, in a clear tutorial fashion, present the key idea, computations, and advantages of the use of bootstrapping with human movement data.

The last chapter of the book is about power spectrum analysis and data filtering. Through a series of examples from human movement research, the author illustrates how frequency domain analysis can serve as a valuable tool for distinguishing signal characteristics between experimental groups and conditions. Topics such as data sampling, data filtering, and the differentiation process are discussed in detail. The chapter also covers advanced joint time-frequency domain presentations such as the Wigner function. Lastly, the author offers the reader suggestions and recommendations for improving processing skills and potentially increasing the efficiency of the data analysis procedures.

# CHAPTER 7

# Time Series Analysis: The Cross-Correlation Function

Timothy R. Derrick, PhD, and Joshua M. Thomas, MS

The concept of cross-correlation has been developed in two distinct fields: signal processing and statistics. In the area of signal processing, the cross-correlation function can be used to transform one or more signals so that they can be viewed with an altered perspective. For instance, cross-correlation functions can be used to produce plots that make it easier to identify hidden signals within the data. Cross-correlation functions provide the basis for many more sophisticated signal-processing procedures as well. Digital imaging techniques also rely heavily on cross-correlation procedures, but these methods are not covered in this chapter. In the realm of statistics, **cross-correlation** functions provide a measure of association between signals. The Pearson product-moment correlation coefficient is simply a normalized version of a cross-correlation. When two time series data sets are cross-correlated, a measure of temporal similarity is achieved. The cross-correlation function in its simplest form is easy to use and quite intuitive. This chapter builds on simple cross-correlation procedures to illustrate the wide variety of uses they have in the field of biomechanics and to give the reader an intuitive feel for some more complicated analysis procedures. Concepts from both signal processing and statistics are discussed, and the procedures are applied to several practical problems.

In any discussion of the analysis of data, it is important to be specific about their nature. Biomechanists typically collect time series data that are discrete, equally spaced, and stochastic. *Time series* data change as a function of time. *Discrete* data have been digitized at specific time periods from the continuous biological signal that we are interested in. This chapter assumes that sampling theory restrictions on the digitization process have not been violated (Hamill, Caldwell, and Derrick, 1997). *Equally spaced* indicates that the time between successive digitizations is constant. **Stochastic** refers to the fact that even though successive digitizations of the signal are dependent, they are only partly determined by past values. This is in contrast to a **deterministic** signal, in which case future data points can be exactly predicted from past data points.

## Time Series Analyses

Modern technologies allow biomechanists to collect vast quantities of data in a relatively short period of time. Biological signals such as positions, velocities, accelerations, forces, joint angles, and **electromyography** can be collected at high rates for increasingly longer periods of time. Mass storage devices are still strained to the limits, as they were decades ago, but they are capable of storing thousands of times more data. Rather than letting this space go to waste, we collect additional data to gain a more holistic view of the activity. The days of carrying the data from a biomechanics research project on a floppy disk are gone, and at times a compact disk is now inadequate. This plethora of data presents organizational problems as well as analytical problems.

Although descriptive techniques for analyzing time series data may appear to be basic, they are important for understanding and verifying the quality of the data. The first step in any analysis should be to visually inspect the data. Viewing the raw numbers is usually not practical, but we can rapidly inspect the data in graphical form. Graphing time series data allows one to observe trends and variations, as well as to observe any outliers that are inconsistent with the rest of the data. The outlier may be a true variation in performance or may be an error caused by malfunctioning equipment or unintended movement patterns. It is important to understand one's data in order to make appropriate adjustments.

Once one has an accurate set of curves that one wants to compare, the choice is between three modes of time series analysis typically used by biomechanists. First, pertinent discrete points on the curve could be identified, and their magnitude and/or the time at which the points occur could be noted. Second, the entire curve could be used to calculate a variable such as the average. Third, the curve could be transformed into a different curve, after which one of these three analyses could be applied again. An example is a **differentiation** transformation that one would use to calculate velocity from position data. The change in position is divided by the change in time, and the resultant curve indicates the rate of change of position (or velocity). Peak velocities, time to peak velocities, average velocities, and so on could then be found.

If the timing of the peak value is important, then the time from initiation to the peak value could be recorded, either as an absolute or as a relative value. Relative values may be expressed as a percentage of a complete cycle or a portion of a cycle. For instance, the time to the peak vertical ground reaction force during running is often expressed as a percentage of the stance phase. Other variables could be calculated from the vertical ground reaction force. Differentiating a force curve gives an indication of the rate of loading, while integration results in the calculation of the impulse.

**Discrete point analysis** is certainly not the only way to analyze curves. In fact, there can be a vast amount of information that is ignored during a discrete point analysis. Other techniques analyze the entire curve by condensing it into a single number. For example, rather than finding the maximum value of a particular curve, one could calculate the average value over the entire sampling period. Thus, the entire curve is represented by a single value.

Discrete point analysis has some advantages over whole-curve analysis. It allows the researcher to focus on the pertinent portion of the curve. For instance, if one is considering the potential for injury, it may be useful to know the peak forces involved rather than the average forces. Information at points other than the maximum value may not be important. Figure 7.1 shows the effect of a 10% reduction in stride length on the vertical ground reaction forces during running (Derrick, Caldwell, and Hamill, 2000). The peak impact force decreases from 1.71 body weights (BW) during normal running to 1.55 BW with the reduced stride length. This is a 9.4% reduction in the peak impact force. On the other hand, if the force is averaged over the complete **stance phase,** the value decreases from 1.46 BW to 1.42 BW. This is a reduction of only 2.8%. The portions of the curve other than the peak impact may act to dilute the differences.

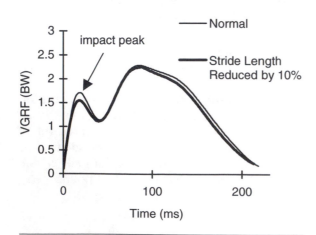

**Figure 7.1** Vertical ground reaction forces (VGRF) at different stride lengths. There is a 9.4% decrease in the impact peak when stride length is reduced by 10%, and there is a 2.8% decrease in the average force. The units are body weights (BW).

Data from Derrick, Caldwell, and Hamill, 2000.

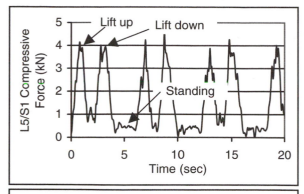

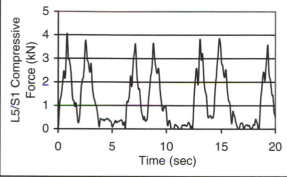

**Figure 7.2**  Pre- and postfatigue compressive forces. The graphs show L5/S1 compressive forces during lifting of a 10-kg crate multiple times *(a)* prefatigue and *(b)* postfatigue. The peak compressive forces decrease 11.3%, and the average forces decrease 18.9%.

In some instances the information before or after the peak values may be important. Figure 7.2 shows the bone-on-bone compressive forces between the fifth lumbar and first sacral vertebrae during lifting of a crate before and after the back muscles became fatigued. The peak compressive force decreased 11.3% after the subject was fatigued (from 2,057 N to 1,826 N), while the average compressive force decreased by 18.9% (from 734 N to 595 N). Some of the difference between pre- and postfatigue occurs when the subject is standing between lifts. It is up to the researcher to determine which method of assessment is more suitable, but a discrete point analysis may not always be the most appropriate technique.

# Defining the Cross-Correlation Function

Cross-correlation is a method by which the degree of similarity between two sets of numbers can be quantified. The process involves two entire curves so that information between peak values is assessed. The procedure is very simple, yet the general concept is used in a variety of advanced analysis techniques. These techniques are all based on the fact that if one carries out a point-by-point multiplication of two data sets, the sum of the products will be a quantification of their relationship (equation 7.1).

$$r_{xy} = \sum_{i=0}^{N-1} x_i y_i \tag{7.1}$$

where N is the number of data points in each data series, $x_i$ is the $i^{th}$ data point of the first data series, $y_i$ is the $i^{th}$ data point of the second data series, and $r_{xy}$ is the correlation.

Readers familiar with matrix notation may note that this cross-correlation is simply the dot product of vectors *x* and *y*. Two time series curves are presented in figure 7.3. Curve 1 does not change, while curve 2 is time shifted in each of the three columns in the figure. Cross-correlation values are given at the bottom of each column. The middle column shows the greatest degree of linear relationship between curve 1 and curve 2. The curves tend to rise and fall at the same time. The cross-correlation value of 2.91 quantifies this similarity. Time shifting curve 2 to the left or to the right tends to reduce the cross-correlation (1.82 and 1.77, respectively). Subtracting the mean value of each time series has the effect of accentuating the cross-correlation values because times at which the curves have the opposite sign reduce the cross-correlation value. For instance, the cross-correlations calculated after removal of the mean values for the three sets of curves in figure 7.3 are 0.47, 1.55, and 0.41 for the left-shifted, normal, and right-shifted curves, respectively. Subtracting the mean value also gives significance to a negative cross-correlation. If the mean values are subtracted, a negative correlation indicates that the two time series have an inverse relationship. As one curve is increasing, the other is decreasing. If the mean values are not subtracted, a negative cross-correlation can be obtained if one of the curves is positive and the other is negative, even if there is a perfect relationship.

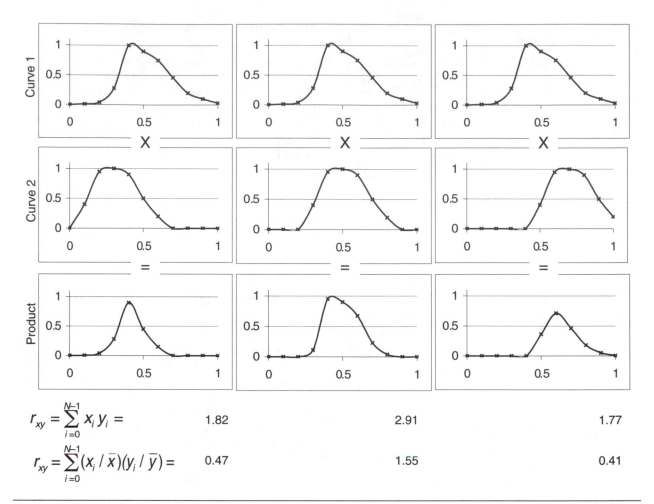

$$r_{xy} = \sum_{i=0}^{N-1} x_i y_i =$$
1.82  2.91  1.77

$$r_{xy} = \sum_{i=0}^{N-1} (x_i / \bar{x})(y_i / \bar{y}) =$$
0.47  1.55  0.41

**Figure 7.3** The effects of time series shifts on the cross-correlation. Numerical values at the bottom of each column represent the cross-correlation of curve 1 and curve 2 without and with removal of the mean values of each curve. The bottom row of curves shows the result of multiplying curve 1 and curve 2.

## Example 7.1

| Time | Curve 1 | Curve 2 |
|------|---------|---------|
| 0.00 | 0.00 | 0.00 |
| 0.10 | 0.01 | 0.40 |
| 0.20 | 0.04 | 0.95 |
| 0.30 | 0.28 | 1.00 |
| 0.40 | 1.00 | 0.90 |
| 0.50 | 0.90 | 0.50 |
| 0.60 | 0.75 | 0.20 |
| 0.70 | 0.46 | 0.00 |
| 0.80 | 0.20 | 0.00 |
| 0.90 | 0.10 | 0.00 |
| 1.00 | 0.03 | 0.00 |

| Curve 1 × Curve 2 |
|-------------------|
| 0.00 |
| 0.00 |
| 0.04 |
| 0.28 |
| 0.90 |
| 0.45 |
| 0.15 |
| 0.00 |
| 0.00 |
| 0.00 |
| 0.00 |
| Sum 1.82 |

Use the values for curve 1 and curve 2 shown in the table to calculate the correlation using equation 7.1.

It is a common practice to shift one of the curves relative to the other as was done with curve 2 of figure 7.3. The number of data points that the signal is shifted is called the **lag** and is denoted by $\ell$.

$$r_{xy}(\ell) = \sum_{i=0}^{N-1} x_i y_i \tag{7.2}$$

Two issues arise with use of this form of the cross-correlation function. Shifting the second of two curves toward the right leaves an unmatched data point at the start of the first curve and another at the end of the second curve. These unmatched data points can then be matched with each other ("wrapped"), or the unmatched data points can be ignored. For instance, if we shift {1,3,5,7} one data point to the right relative to {4,6,8,10}, the matched pairs are 1-6, 3-8, and 5-10. If we wrap the data, we create another matched pair from the last point of the first series and first point of the second series (7-4). If we do not wrap the data, then the 7 and the 4 are ignored and we are left with a shortened series.

The data should be wrapped only if they are circular. For instance, if an activity such as steady state cycling was being investigated and data were collected from top dead center (TDC) of one stroke to TDC of the next stroke of the same leg (a complete cycle), then the data could be considered circular and wrapping may be appropriate. On the other hand, if the stance phase of a running cycle was being studied, the data would represent only a portion of a complete cycle and it would be inappropriate to wrap the data. If the unmatched data points are ignored, then the number of data points that are being cross-correlated will become lower with increasing lags, and this will tend to reduce the sum of the products (equation 7.2) and reduce the number of data points that the correlation is based on.

The second issue arises because the cross-correlation given by equation 7.2 is not unitless. The cross-correlation depends on the units of $x$ and $y$, and therefore it is difficult to compare cross-correlations from different data sets. In order to prevent a reduction in the sum of the products with increasing lags and to make the cross-correlation unitless, most applications "normalize" equation 7.2 by dividing it by the square root of the product of the autocorrelation of $x$ at zero lag and the square root of the autocorrelation of $y$ at zero lag (equation 7.3).

$$\rho_{xy}(\ell) = \frac{r_{xy}(\ell)}{\sqrt{r_{xx}(0)}\sqrt{r_{yy}(0)}} \tag{7.3}$$

The autocorrelation is simply a data series cross-correlated with itself. The autocorrelation of the data series $x$ and a lag of zero would be written $r_{xx}(0)$. Since the denominator represents perfect cross-correlations, it is always greater than the numerator. Thus the value for $\rho$ will never exceed 1.0. When the mean values are subtracted (equation 7.4), this normalized cross-correlation will become negative only when the curves have an inverse relationship. This is the version of the cross-correlation that is most commonly used.

$$\rho_{xy}(\ell) = \frac{\sum_{i=0}^{N-1}\left(x_i - \overline{x}\right) * \left(y_{i-\ell} - \overline{y}\right)}{\sqrt{\sum_{i=0}^{N-1}\left(x_i - \overline{x}\right)^2}\sqrt{\sum_{i=0}^{N-1}\left(y_{i-\ell} - \overline{y}\right)^2}} \tag{7.4}$$

Statisticians will recognize equation 7.4 as the Pearson product-moment correlation coefficient. It is often described as the degree to which $x$ and $y$ vary together divided by the degree to which $x$ and $y$ vary separately. Even with use of the normalized cross-correlation formula, correlations should not be calculated when the lag approaches $N$ unless the data are wrapped. If the data are not wrapped the correlations will be based on too few data points as $\ell$ increases. For instance, if two 25-point series are cross-correlated with a lag of 20, there will be 20 unmatched numbers in each series. This leaves a correlation that is based on only 20% of the original data. A good rule of thumb is to calculate lags only up to $N/2$. In practice, it is often the case that only lags of short duration are of interest.

# Pearson Product-Moment Correlations

The researcher may desire a simple measure of the temporal similarity of two curves. The Pearson product-moment correlation provides such a measure (Derrick, Bates, and Dufek, 1994). This correlation is usually performed with a zero lag; but if appropriate, the peak correlation can also be analyzed when multiple lags are available. Figure 7.4 shows two time series curves along with a scatter plot. The scatter plot

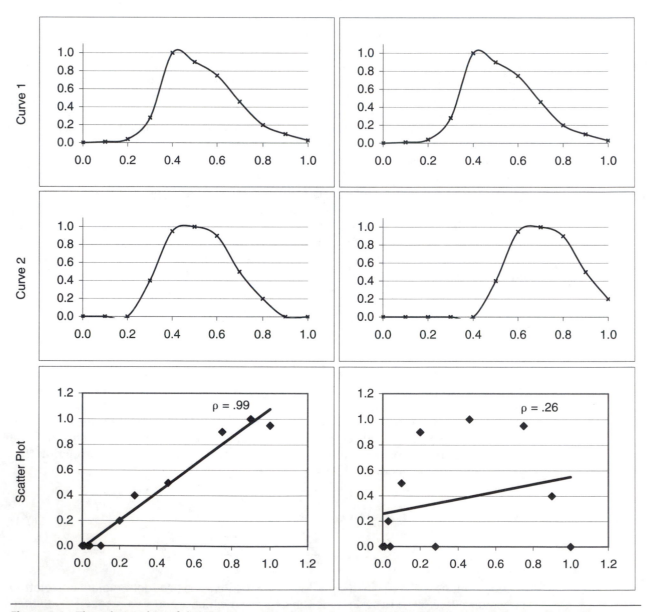

**Figure 7.4**  The relationship of the Pearson product-moment correlation to temporal similarity. Scatter plots show the relationship between the time series curves. The correlation coefficient ρ quantifies the degree of relationship. Curve 1 and curve 2 have good temporal similarity in the left column, but curve 2 has been shifted in the right column in order to portray a disruption in the timing.

shows curve 1 plotted on the horizontal axis with curve 2 plotted on the vertical axis. The linear regression line is also overlaid on the plot. The correlation can be thought of as a measure of how closely the data points line up on the linear regression line. Changes in one of the curves will be matched by changes in the other if the relationship is strong. All of the points will line up on the regression line if the relationship is perfect, and the correlation will be 1.0. The curves from figure 7.3 are reproduced in figure 7.4. In the left column, the peaks occur at the same time and the correlation is .99. The right column shows the relationship when curve 2 is shifted in time so that the peaks no longer line up. The correlation value is reduced to .26. This illustrates the use of the Pearson product-moment correlation as a measure of the temporal similarity of two curves.

Figure 7.5 shows two pairs of curves along with the scatter plots and correlation values. Both sets of curves show a relationship in which the curves begin at zero, rise to peaks that occur at the same time, and then return to the same amplitude. The

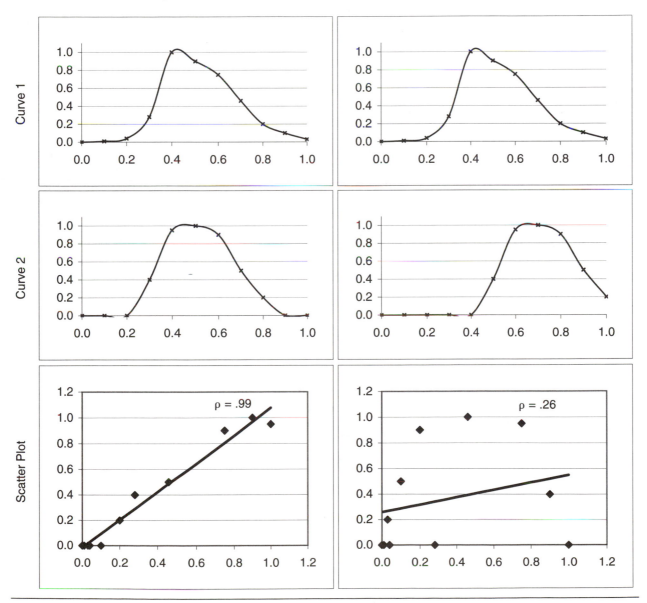

**Figure 7.5** Reductions in the correlation with altered endpoints. Time series curves and scatter plots demonstrate how changing the relationship in a part of the curve affects the correlation coefficient ($\rho$). The left column shows a perfect relationship (curve 1 is a scaled version of curve 2). The right column shows what happens to this relationship if the curves do not end at the same level they start at.

main difference is that the left-hand curves both return to zero while the right-hand curves return to a value somewhat above zero. Although this seems to be an innocuous difference, it does cause the cross-correlation to decrease from 1.00 to .97. This drop occurs because the relationship before the peak value is different from the relationship after the peak value. Separately, the portion before the peak value and the portion after the peak value have nearly perfect relationships ($r \approx 1.00$), but together the relationship is less than perfect.

Stergiou, Bates, and James (1999) used the correlation technique to assess disruptions in the timing between the knee and the subtalar joint during running over an obstacle. Normal running at a self-selected pace produced a correlation of .832 between the knee and the **rearfoot angle** curves. This correlation was reduced to .742 when subjects ran over an obstacle that was 15% of their standing height. The obstacle tended to increase the vertical ground reaction force, which resulted in a change to the shape of the rearfoot curve from unimodal to bimodal (one local minimum to two local minimums). The knee angle curve remained unimodal in both conditions. The change produced greater asynchrony between these joints, which may result in a higher incidence of injury. The change was detected using the Pearson product-moment correlation coefficient because changing the rearfoot curve from unimodal to bimodal made the rearfoot and knee curves less similar.

## Fisher Z-Transformations

Correlation coefficients are not normally distributed. As the correlation increases, the distribution becomes more negatively skewed. The values can be transformed so that statistics that assume a normal distribution can be calculated. The Fisher Z-transformation (an inverse hyperbolic tangent) has been used to accomplish this (Otnes and Enochson, 1978):

$$z = \tanh^{-1}\left[r_{xy}(\ell)\right] = \frac{1}{2}\ln\frac{1+r_{xy}(\ell)}{1-r_{xy}(\ell)} \tag{7.5}$$

where z is the Fisher Z-transformation, $r_{xy}$ is the correlation coefficient and, $\ell$ is the lag.

It can be shown that $z$ is approximately normally distributed with variance

$$\sigma_z^2 = \frac{1}{N-2-\ell} \tag{7.6}$$

where $\sigma_z^2$ is variance of the z-scores, N is the number of data points, and $\ell$ is the lag.

Thus a correlation of .75 would have a z-score of 0.97. Taking the hyperbolic tangent of the z-score will reproduce the correlation coefficient. These z-scores can also be used to average multiple correlations when one is analyzing data across trials or across subjects.

**Example 7.2**

Calculate the z-score for a correlation coefficient of .75.

$$z = \frac{1}{2}\ln\frac{1+r_{xy}(\ell)}{1-r_{xy}(\ell)} = \frac{1}{2}\ln\frac{1+.75}{1-.75} = \frac{1}{2}\ln 7 = \frac{1}{2}(1.946) = 0.97$$

Li and Caldwell (1999) calculated the confidence intervals of the peak cross-correlation to find the **phase shift** between gluteus maximus electromyography (EMG) and the crank angle during cycling. This method allowed the phase shift measure to be objective, in contrast to the traditional subjective threshold identification methods.

The authors calculated the confidence interval (CI) so that standard statistical analysis of the correlation coefficient was possible. They found that the EMG was shifted 20° counterclockwise in the crank cycle as cadence increased from low to high.

# Other Measures of Similarity

Other techniques can be used to measure temporal similarity. For instance, it may be desirable to simply measure the percentage of the time that the signs of the slopes of the curves agree. This measure is sometimes called the slope congruence. Although the resolution is not good, the slope congruence method is able to distinguish the two sets of curves in figure 7.4. The left column produces a slope congruence of 90% while the right column produces a slope congruence of only 70%. These numbers indicate that there are 1 out of the 10 intervals in the left column and 3 out of 10 intervals in the right column in which the slopes of the curves are in the opposite direction. The Pearson product-moment correlation coefficient and the slope congruence methods produce measures of the overall temporal similarity between two curves. For this reason they are not the most sensitive measures if the temporal similarity is important during only a portion of the curves. In this case, one could employ discrete value techniques to measure temporal similarity. For instance, one could calculate the difference in the time to peak values. If the magnitude of the slope difference is important, the curves can be differentiated and then one curve can be subtracted from the other. Local maximum and minimum values would indicate when there is increased disagreement between the slopes of the curves. In this case it is possible to have a peak value even if both curves are in the same direction. One of the curves would be increasing (or decreasing) rapidly while the other is increasing (or decreasing) slowly.

Two curves can have a very high correlation value yet be separated by a large offset. As long as the curves are parallel, the correlation will be high. In order to assess differences in magnitude, one could subtract one of the curves from the other. This produces a difference curve in which peak differences could be identified or average differences calculated. If the curves cross during the period being analyzed, it is more appropriate to calculate the average absolute differences in the curves.

## Example 7.3

Use the values for curve 1 and curve 2 shown in the table to calculate the average difference and the average absolute difference between the curves. Note the substantial difference between these two measures of magnitude differences.

| Time | Curve 1 | Curve 2 | | Difference | Absolute difference |
|---|---|---|---|---|---|
| 0.00 | 0.00 | 0.00 | | 0.00 | 0.00 |
| 0.10 | 0.01 | 0.40 | | −0.39 | 0.79 |
| 0.20 | 0.04 | 0.95 | | −0.91 | 1.86 |
| 0.30 | 0.28 | 1.00 | | −0.72 | 1.72 |
| 0.40 | 1.00 | 0.90 | | 0.10 | 0.80 |
| 0.50 | 0.90 | 0.50 | | 0.40 | 0.10 |
| 0.60 | 0.75 | 0.20 | | 0.55 | 0.35 |
| 0.70 | 0.46 | 0.00 | | 0.46 | 0.46 |
| 0.80 | 0.20 | 0.00 | | 0.20 | 0.20 |
| 0.90 | 0.10 | 0.00 | | 0.10 | 0.10 |
| 1.00 | 0.03 | 0.00 | | 0.03 | 0.03 |
| | | | Average | −0.02 | 0.58 |

# Correlograms

A plot of the cross-correlation values at each lag is called a correlogram. Such plots can be useful for extracting noisy signals, synchronizing signals, or finding hidden frequencies within a signal. Radar and sonar use correlograms to determine the distance of a remote object. A "chirp" signal is transmitted from a radar; it bounces off the object of interest and returns to its original position, where a receiver records the returning signal. We can calculate the distance after measuring the time between transmission and reception. The problem is that a significant amount of noise is introduced to the echoed signal and the original chirp may not be recognizable (figure 7.6). Cross-correlations between the transmitted and echoed signals will peak at the lag in which the signals line up. Figure 7.6 also shows the **correlogram** between a clean chirp signal (transmitted signal) and the same chirp signal with random noise introduced (echoed signal). The noise nearly hides the chirp, yet the correlogram shows a distinct peak cross-correlation at a zero lag. This indicates that the two signals are synchronized. If these signals were from a radar, the peak would occur at a particular lag that could be transformed into a transmission time through multiplying by the sampling period. The distance could be determined through multiplying half of the transmission time by the transmission velocity.

A similar process has been used to measure the conduction velocity in a muscle (Li and Sakamoto, 1997). The propagation wave is measured at more than one location on the muscle, and the delay is determined through cross-correlation of the signals. The peak cross-correlation value occurs at the lag where the signals have the greatest similarity. Conduction velocity can then be calculated from this lag time and from knowledge of the interelectrode distances.

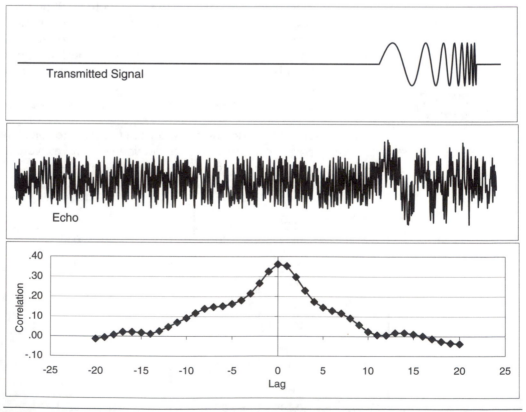

**Figure 7.6**  Correlogram of a chirp. The top curve is a "chirp" signal sent by a radar. The middle curve is the same chirp after it has been echoed off the object of interest and random noise has been introduced. The bottom curve is a correlogram showing the relationship between the signal and the echo at different lags.

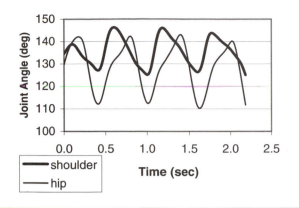

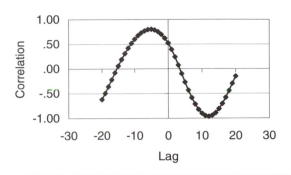

**Figure 7.7** Shoulder-hip angle relationship. Hip and shoulder angles of a dog walking on a treadmill at 1.3 m/s. Approximately three cycles of data are shown.

**Figure 7.8** Shoulder-hip correlogram. A correlogram produced from the cross-correlation between shoulder and hip angles of a dog walking on a treadmill. Peak correlations occur when the shoulder and hip curves line up with each other. The lag at the peak value indicates the phase difference between the shoulder and the hip.

Cross-correlation can provide a cleaner picture of the timing between two signals than discrete point analysis can. Figure 7.7 shows the hip and shoulder angles of a dog walking at 1.3 m/s on a treadmill. These data were hand digitized at a frame rate of 60 Hz. It is difficult to assess the differences in timing between these two joints by evaluating discrete points. The hip seems to reach maximum extension (local maximums) later than the shoulder joint. On the other hand, a look at peak flexion (local minimums) suggests that the hip and shoulder are nearly synchronized. A correlogram evaluates entire curves rather than discrete points on the curve. Figure 7.8 shows the correlogram between the shoulder and hip joint angles. The peak cross-correlation occurs at a lag of –5. This indicates that there is a phase shift of –83 ms (–5/60 s) between the hip and the shoulder joint angles.

## Example 7.4

| lag | r | | Difference | Absolute difference |
|-----|-----|-----|-----|-----|
| –6 | 0.139 | | 0.00 | 0.00 |
| –5 | 0.148 | | –0.39 | 0.79 |
| –4 | 0.152 | | –0.91 | 1.86 |
| –3 | 0.163 | | –0.72 | 1.72 |
| –2 | 0.181 | | 0.10 | 0.80 |
| –1 | 0.215 | | 0.40 | 0.10 |
| 0 | 0.267 | | 0.55 | 0.35 |
| 1 | 0.326 | | 0.46 | 0.46 |
| 2 | 0.362 | | 0.20 | 0.20 |
| 3 | 0.353 | | 0.10 | 0.10 |
| 4 | 0.298 | | 0.03 | 0.03 |
| 5 | 0.229 | Sum | –0.18 | 6.41 |
| 6 | 0.174 | | | |
| 7 | 0.146 | | | |
| 8 | 0.129 | | | |

The data in the table represent a correlogram between two electrodes attached to a nerve fiber. Calculate the nerve conduction velocity if the electrodes are spaced 120 mm apart and each lag represents 1 ms.

There is a lag of 2 between the signals measured at the two electrode sites. Since each lag is 1 ms, the total time between the signals is 2 ms. The propagated wave takes 2 ms to travel 120 mm and thus the velocity is 120 / 2 = 60 mm/ms or 60 m/s.

## Autocorrelograms

A correlogram produced from the autocorrelation is called an autocorrelogram. Figure 7.9 shows correlograms for both the shoulder and the hip joints. These plots show peak correlations at lags of approximately 37 data points. This indicates that as the curve is slid past a copy of itself, the fluctuations in the copy will line up with the original after being shifted 37 data points. The autocorrelogram will continue to have peaks every 37 data points for as long as the lags are calculated. This indicates a cyclical period of 617 ms (37/60 s) or a walking frequency of 1.62 Hz. The same information is obtained using either the shoulder or the hip correlograms. Autocorrelograms can provide much information about a time series, and considerable experience is needed in their interpretation. Such items as nonstationarity, seasonal fluctuation, randomness, alternation, and short-term correlation can be determined from an autocorrelogram of a time series (Chatfield, 1984).

The autocorrelation function has been used to measure conduction velocity in muscles by means of a single channel of data (Spinelli, Felice, Mayosky, Politti, and Valentinuzzi, 2001) as opposed to the two channels required with use of the cross-correlation technique. The difference signal is obtained from two needle electrodes, and autocorrelation techniques are utilized to estimate the conduction velocity in periods as short as 0.3 s.

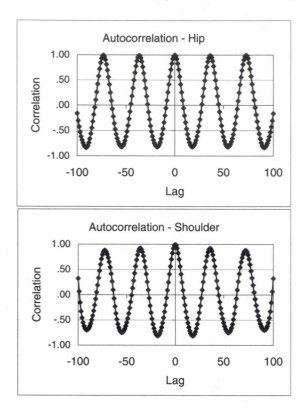

**Figure 7.9** Shoulder and hip autocorrelations. Correlograms produced from the autocorrelation of the shoulder and hip angles of a dog walking on a treadmill. Peak correlations occur when a cycle lines up with a subsequent cycle. The inverse of the time between peak values indicates the cycle frequency.

## Cross-Correlation As a Method for Estimating Spectral Content

Thus far in the chapter, the cross-correlation function has been used as a measure of temporal similarity, as an estimation of the time lag between signals, as a way to find periodicity within a signal, and as a way of discovering signals hidden within a signal (as in the radar example). This last application deserves further exploration. Previous examples correlated two signals that had both been collected from instruments. But it is not necessary for both signals to have been collected from instruments; one of the signals could be created from a function. For instance, the hip joint angle of a dog walking (figure 7.7) could be correlated with a sine wave (equation 7.7) of the same duration. In this case the cross-correlation can indicate how much of the sine wave of frequency $f$ is contained in the signal. The signal could be out of phase with the sine wave, so a complete analysis would also include a cross-correlation between the signal and a cosine wave of frequency $f$.

$$\text{sw}(t) = A\sin(2\pi f t) \tag{7.7}$$

where sw($t$) is the sine wave, $A$ is the amplitude, $f$ is the frequency, and $t$ is the time.

To illustrate this point, a signal h($t$) was created using the following formula:

$$\text{h}(t) = A_1\sin(2\pi f_1 t) + A_2\sin(2\pi f_2 t) + A_3\sin(2\pi f_3 t) \tag{7.8}$$

where $f_1 = 1$, $A_1 = 1$; $f_2 = 3$, $A_2 = 1/3 = 0.33$; and $f_3 = 5$, $A_3 = 1/5 = 0.20$.

These are the first three components of a square wave as calculated by adding sine waves of increasing frequency. The left column of figure 7.10 shows the time series graph of this formula. Cosine and sine waves of increasing frequency were created and graphed in the second column in figure 7.10.

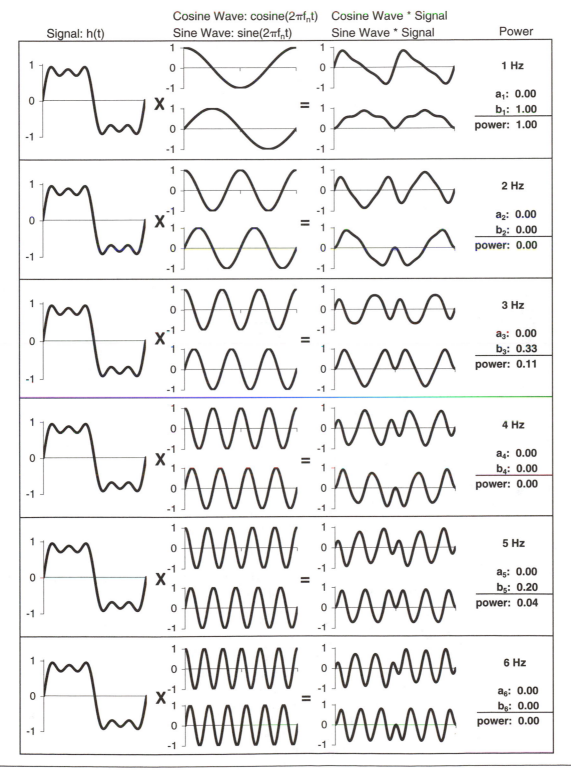

**Figure 7.10**   Calculation of Fourier coefficients using cross-correlation. The signal in the left column $h(t) = \sin(2\pi t)$ $+ 0.3\sin(2\pi 3t) + 0.2\sin(2\pi 5t)$ is cross-correlated with cosine and sine waves to obtain the Fourier coefficients ($a_n$ and $b_n$, respectively). See text for details.

A modified version of equation 7.1 was used to calculate the cross-correlation between the signal and the cosine wave and the sine wave for each frequency ($n$). The cross-correlation was simply divided by the number of data points ($N$) and then multiplied by 2 (equation 7.9). This modification allows the cross-correlation value to correspond to the amplitude of the cosine or sine wave.

$$r_{xy} = \frac{2}{N} \sum_{i=0}^{N-1} x_i y_i \qquad (7.9)$$

Since the signal was composed of 1-, 3-, and 5-Hz sine waves, all of the $a_n$ coefficients (from the cosine wave) are zero, and the only $b_n$ coefficients (from the sine wave) that are non-zero are at 1, 3, and 5 Hz. Note that the $b_n$ coefficients are 1.00, 0.33, and 0.20 for the 1-, 3-, and 5-Hz sine waves, respectively, and that these coefficients correspond to the amplitudes of the sine waves that composed the original signal $h(t)$. It is possible to decompose any signal into constituent sine and cosine waves of varying amplitude.

The $a_n$ and $b_n$ coefficients were calculated by cross-correlating the signal with sine and cosine waves of varying frequency ($n$) using equation 7.9. These coefficients are called **Fourier coefficients** after the French mathematician Jean Baptiste Joseph, Baron de Fourier (1768-1830). By squaring the $a_n$ and $b_n$ coefficients and then adding them together, one obtains the power for the frequency $n$. A plot of the power by frequency is generally known as a **power spectrum.** There are better ways to calculate the Fourier coefficients for a given signal, but this method has the advantage that it is intuitive. Chapter 9 describes power spectrums in more detail. The power spectrum of the signal represented by equation 7.8 is plotted in figure 7.11 for the frequencies from 1 to 6 Hz. The plot indicates the frequencies that are present in the signal. This method may be used for any time-varying signal, not just one composed of sine waves.

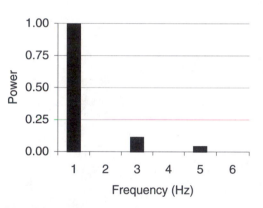

**Figure 7.11** Power spectrum of a complex sine function. The power spectrum is of the signal $h(t) = \sin(2\pi t) + 0.33\sin(2\pi 3t) + 0.2\sin(2\pi 5t)$. The powers were obtained by summing the square of the cosine and sine amplitudes of the Fourier coefficients.

Knowledge of the spectral content of a signal is important because different frequencies may have different biological significance. Shorten and Winslow (1992) used an efficient algorithm called a Fast Fourier Transform (FFT) to calculate the power spectrums from leg accelerations during running. The methods were used to determine which acceleration frequencies were present during running. The authors identified three frequency ranges (4-8 Hz, 12-20 Hz, and 60-90 Hz) in an accelerometer attached to the leg. The low frequencies were the result of muscular activity; the middle frequencies were the result of the foot striking the ground; and the high-frequency range resulted from resonance of the accelerometer attachment. The authors were most interested in the midfrequency range because of the potential for injury caused by the impact.

Other analysis and processing techniques such as data filtering can be built from knowledge of these frequency domain methods. For instance, data filtering can be accomplished through removal of particular frequencies from the data. Electrical equipment can contaminate signals by producing frequencies around 60 Hz. The signal can be transformed into the frequency domain; frequencies around 60 Hz could be eliminated or reduced; and then the signal could be transformed back into the time domain. Fourier filters eliminate the unwanted frequencies in the frequency domain; digital filters eliminate the frequencies in the time domain without transformations.

# Matched Filters

Matched filters can be used whenever a known signal is present within a more complex signal or a corrupted signal. In an earlier example, the chirp from a transmitter was compared to the returning signal that was bounced off an object. The signal of interest, or template, is compared to the complex signal, and a high correlation indicates a "match." The template that is cross-correlated with the complex signal can be a sine wave or a cosine wave, as indicated in the section on spectral content, or it can be any other function. It can even be a portion of a curve. For instance, **motor unit action potential** (MUAP) **decomposition** can utilize cross-correlation techniques to identify the location of individual MUAPs. Low-force electromyographic (EMG) recordings are composed of several motor units, each firing at a relatively consistent

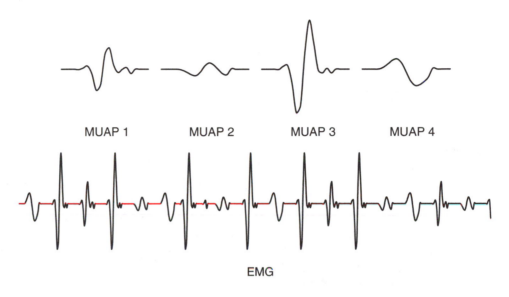

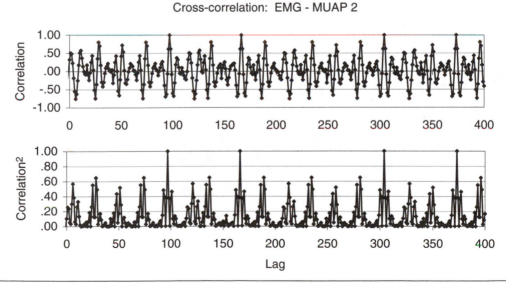

**Figure 7.12**  Identification of waveforms in a complex signal. Schematic of waveforms of four individual motor unit action potentials (MUAP) and an EMG signal created from multiple firings of the individual MUAP waveforms. Correlograms display MUAP 2 cross-correlated with the EMG signal. Firing of this motor unit occurs when the correlation approaches 1.0. The bottom correlogram shows that squaring the correlation coefficient can make identification of the firing easier.

rate. The EMG is digitized at high sampling frequency, and the first clear occurrence of each MUAP is manually extracted and used as a template. Several motor units are typically recorded at the same time, but each has its own unique waveform (see the four MUAP waveform templates at the top of figure 7.12). Thus, cross-correlating each waveform template with the EMG allows one to identify each firing of the motor unit within the EMG. It is then possible to reconstruct the MUAP trains by joining the multiple occurrences of each waveform together. This algorithm is referred to as a **matched filter** because it allows the template to be matched to subsequent occurrences of the same general shape (Karu, 1995).

Figure 7.12 shows a schematic waveform of four individual motor unit action potentials (MUAP). The EMG is composed of multiple firings of these four motor units. It is often useful to calculate statistics such as firing frequency and the variability of the firing frequency of each motor unit (DeLuca, 1983). In order to calculate these statistics, it is necessary to decompose the EMG into the individual MUAP firings. The correlograms in figure 7.12 show the second MUAP cross-correlated with the EMG in an attempt to identify when the second MUAP is firing. Correlations that approach 1.0 indicate a likely match between the template MUAP waveform and a firing of that motor unit in the EMG. Four such firings are indicated by the correlogram. The bottom correlogram shows that the correlations can be squared to give a clearer picture of when the motor unit is firing.

## Summary

The analysis of time series curves consumes a relatively large portion of a research biomechanist's time. To be successful, the researcher must be familiar with both the data that have been collected and the possible analysis techniques. A time series data set can be correlated with another data set (cross-correlation), with itself (autocorrelation), with a function such as a sine wave, or with a template such as a known waveform (matched filter). In its simplest form the cross-correlation measures the similarity between two signals. It can also be used to determine differences in phase, extract hidden signals, and reveal frequency content. The purpose of this chapter was to acquaint the reader with these types of analysis techniques. Understanding the basis for cross-correlation can be useful in itself but can also lead to an intuitive feel for more advanced analysis techniques such as Fourier analysis, frequency filtering, and wavelet analysis.

## Work Problems

**1**

Using the tables in example 7.1, shift curve 2 by 0.2 s and by 0.4 s and recalculate the correlations using equation 7.1.

**2**

Show that the average of three correlation coefficients (.70, .80, and .90) is not .80 when the Fisher Z-transformation is used to convert the coefficients to z-scores prior to taking the average. The average z-score can be transformed back into a correlation by using the following formula:

$$r = \frac{\exp(2z) - 1}{\exp(2z) + 1}$$

where r is the average correlation coefficient and z is the averaged z-score.

# Suggested Readings and Other Resources

### Books and Articles

Karu, Z.Z. (1995). *Signals and systems made ridiculously simple.* Cambridge, MA: ZiZi Press.

Oppenheim, A.V., and Schafer, R.W. (1989). *Discrete-time signal processing.* Englewood Cliffs, NJ: Prentice Hall.

Proakis, J.G., and Manolakis, D.G. (1988). *Introduction to digital signal processing.* New York: Macmillan.

Transnational College of LEX. (1995). *Who is Fourier? A mathematical adventure.* Translated by Alan Gleason. Belmont, MA: Language Research Foundation.

### Web Sites

www.falstad.com/fourier/

www.stats.gla.ac.uk/steps/glossary/time_series.html

www.statsoftinc.com/textbook/sttimser.html

# References

Chatfield, C. (1984). *The analysis of time series.* New York: Chapman & Hall.

DeLuca, C.J. (1983). Myoelectric manifestations of localized muscular fatigue in humans. *Critical Reviews in Biomedical Engineering,* 11(4):251-279.

Derrick, T.R., Bates, B.T., and Dufek, J.S. (1994). Comparative evaluation of time-series data sets using the Pearson product-moment correlation coefficient. *Medicine and Science in Sports and Exercise,* 26(7): 919-928.

Derrick, T.R., Caldwell, G.E., and Hamill, J. (2000). Modeling the stiffness characteristics of the human body while running with various stride lengths. *Journal of Applied Biomechanics,* 16(1):36-51.

Hamill, J., Caldwell, G.E., and Derrick, T.R. (1997). A method for reconstructing digital signals using Shannon's sampling theorem. *Journal of Applied Biomechanics,* 13:226-238.

Karu, Z.Z. (1995). *Signals and systems made ridiculously simple.* Cambridge, MA: ZiZi Press.

Li, L., and Caldwell, G.E. (1999). Coefficient of cross correlation and the time domain correspondence. *Journal of Electromyography and Kinesiology,* 9:385-389.

Li, W., and Sakamoto, K. (1997). Distribution of muscle fiber conduction velocity of M. biceps brachii during voluntary isometric contraction with use of surface array electrodes. *Applied Human Science: Journal of Physiological Anthropology,* 15(1):41-53.

Otnes, R.K., and Enochson, L. (1978). *Applied time series analysis.* New York: Wiley.

Shorten, M.R., and Winslow, D.S. (1992). Spectral analysis of impact shock during running. *Journal of Sport Biomechanics,* 8:288-304.

Spinelli, E., Felice, C.J., Mayosky, M., Politti, J.C., and Valentinuzzi, M.E. (2001). Propagation velocity measurement: Autocorrelation technique applied to the electromyogram. *Medical and Biological Engineering and Computing,* 39(5):590-593.

Stergiou, N., Bates, B.T., and James, S.L. (1999). Asynchrony between subtalar and knee joint function during running. *Medicine and Science in Sports and Exercise,* 31(11):1645-1655.

# CHAPTER 8

# Principles and Applications of Bootstrapping Statistical Analysis

Weimo Zhu, PhD, and Songning Zhang, PhD

Research interests in human movement cover broad areas such as injury mechanisms, injury prevention, and performance enhancement. Research questions may concern injury mechanisms due to impact forces in dynamic movements, efficacy of interventions using shoes for the prevention of running injuries, and variables related to one's jump height. In real-life situations, we rarely try to address these questions by directly studying the population, which is the entire collection of events (e.g., jump heights of every individual in the population), because of the huge size of the population and the high costs that would be associated with such a study. Instead, we draw samples from the population and try to make inferences about the population based on sample data. Because of many factors (e.g., sampling error), however, a sample will not always represent the population. Thus, some kind of quality control of the sample-based inferences that we have made becomes necessary. Like scientists in other disciplines, human movement researchers often use statistical analysis to try to extract information from the sample data so that phenomena they have studied can be better understood at a predetermined confidence level.

Statistical analyses are based on statistics, which are functions of data and which are selected according to certain principles (e.g., likelihood). Most of the currently used "classical" statistics, however, were developed at the end of the 19th century or the early part of the 20th century when computing power was slow and expensive. Pearson's product-moment correlation coefficient, Student's *t*-test, and analysis of variance (ANOVA) are a few familiar examples of these statistics. Prior to data collection, such a statistic is a random quantity having a probability distribution. This means that in order for the investigator to use the statistic to make an inference, there are two requirements.

The first requirement is that a formula to calculate the parameters of the distribution must be readily available. In fact, this is one of the major reasons statisticians in the past devoted most of their time to developing new formulas. Still, because formulas have not been properly developed or because they are too difficult to derive mathematically, only limited statistics are available. For example, we often use only mean, median, and mode to estimate central tendency, although other statistics, such as the trimmed mean, may be more appropriate in certain circumstances. The second requirement is that the distribution assumption(s) about the **sampling distribution** must be known. The most familiar assumption is the "normal" or "Gaussian" distribution, in which data are assumed to be distributed under a bell-shaped curve; the assumptions become much more complex when multivariate statistics are used (Chen and Zhu, 2001). The statistics based on such assumptions are known as *parametric statistics.*

Data that we collect in our practice, however, do not often satisfy the assumptions. For example, the data usually are not normally distributed when the sample size is small or when the original population itself is not normally distributed. Furthermore, studies (DiCiccio and Romano, 1989; Efron, 1987) have shown that only in rare cases do parametric assumptions allow for an exact estimation of sampling distributions, especially when the sample size is small (conventionally $n < 30$).

There are generally three alternatives when parametric statistics are either not appropriate or not available. The first alternative for the researcher is to make statistical inferences using parametric statistics, disregarding the fact that the assumptions of the statistics have been violated and hoping that the results will be robust. In other words, it is still possible to achieve accurate inferences even if the assumptions have been violated. The commonly used *t*-test and ANOVA, for example, are quite robust with respect to violation of the normality assumption. The major drawback of this alternative is that the degree of assumption violation is often difficult to measure, although various robust indexes (see Hill and Dixon, 1982) have been developed.

The second alternative is to apply a transformation in which scores are systematically changed (e.g., when the square root of each of the scores is taken, so that the assumptions can be met). The major shortcomings of this option are that the transformation may not be necessary (e.g., the use of variance-stabilizing transformations has little effect on the significance level and power of the *F*-test [Budescu and Appelbaum, 1981]) and that the transformation may be not helpful at all (e.g., no transformation will make the data more suitable for ANOVA if the means of treatment levels are approximately equal but variances of the error effects are heterogeneous [Kirk, 1982]).

The third alternative is to make inferences using classical nonparametric statistics. The sign test, the matched-pair Wilcoxon test, and the Kruskal-Wallis test are just a few familiar examples in this category. The major drawback of using nonparametric statistics is that much information about data is lost in the computing process (i.e., only the ordering of observations is taken into account, not their values), and thereafter an inference derived from the nonparametric statistic may lack the statistical power needed. Bootstrapping statistics, a relatively new statistical method, provide us with another, new alternative.

# Bootstrapping

**Bootstrapping** was invented by Bradley Efron, a statistician at Stanford University, in 1979. In bootstrapping, subsamples are drawn from the original sample by means of sampling replacement procedures, and statistics of interest (e.g., means) and sampling distributions (e.g., the sampling distribution of the means) are then derived from the subsamples. The term "bootstrap" was derived from the old saying about pulling oneself up by one's own bootstraps, reflecting the fact that the one available sample gives rise to many others.

## *Bootstrapping and Jackknife*

Bootstrapping is also closely associated with the idea of the **jackknife** (Miller, 1974), which is used to reduce the bias of an estimator. Statistics in both techniques are developed based on subsamples from a sample originally drawn from the population. However, jackknife sampling is conducted in a more systematic way: Each observation is omitted, one at a time in turn, to generate a set of separate subsamples, whereas bootstrap sampling is conducted randomly through replacement.

## Bootstrapping and Computers

The development of bootstrapping cannot be separated from the development of modern digital computers. Compared to the computing power available when classical statistics were developed, modern computing power is much faster and cheaper. Today's statisticians can afford more computations on a single problem than the world's yearly total expenditure for statistical computations in the 1920s! The computing involved in bootstrapping is so intensive that it is impossible to implement the idea without the assistance of modern computer power. This is why bootstrapping, together with randomization tests and Monte Carlo tests, is referred to as **"computer-intensive statistics"** in the literature (Diaconis and Efron, 1983; Noreen, 1989). Computer-intensive statistics take advantage of the high-speed computing power of modern digital computers. The payoff for such intensive computation is freedom from two major limiting factors that have dominated classical statistical theory since its beginning: the assumption that the data conform to a bell-shaped curve, and the need to focus on statistical measures whose theoretical properties can be analyzed mathematically (Diaconis and Efron, 1983).

## Bootstrapping and Monte Carlo Simulations

To many, drawing a sample using a computer with some random process may seem like a **Monte Carlo simulation.** Although there are some similarities between the two procedures (e.g., both include steps such as defining a population, sampling with replacement, and conducting evaluation), there is a major difference in the way the population is defined in the two procedures. In Monte Carlo simulations, a hypothesized population is defined based on researchers' best knowledge of the phenomenon, and random samples are then drawn from the population; test statistics and their sampling distributions can be computed and derived from the samples. In bootstrapping, however, a **pseudo-population** is defined based on a sample—hopefully a representative one—drawn from the targeted population. Bootstrap samples are then generated from this pseudo-population; bootstrap sample statistics and the bootstrap sampling distribution of the statistics are derived from the bootstrap samples.

# Bootstrap Samples and Bootstrap Sampling Distributions

Like other commonly used parametric inference statistics, bootstrapping is based on a sampling distribution, more specifically a bootstrap sampling distribution. The **bootstrap sampling distribution,** however, is developed based neither on many repeated samples from the population nor on analytic formulas. To develop a bootstrap sampling distribution, we first draw a sample randomly from a population. Instead of drawing succeeding samples repeatedly from the population or computing parameters of the sampling distribution based on existing analytic formulas as is typical, we redraw many subsamples, which we call **"bootstrap samples,"** randomly *with replacement* from the sample. Thus, "bootstrapping" here actually means redrawing samples randomly from the original sample with replacement. Sample statistics, such as mean and median, are then computed for each bootstrap sample. Distributions of the bootstrap sample statistics are the bootstrapping sampling distributions of the statistics. Statistics, or estimates, derived from the bootstrap sampling distributions are the bootstrap statistics. Thus, both parametric statistics and bootstrap statistics estimate the parameters of a sampling distribution from a sample. Parameter statistics estimate the parameters based on available analytic formulas, which are restricted by related distribution assumptions and availability of analytic formulas; in contrast, bootstrap statistics estimate the parameters based on the bootstrap samples, which are free of these assumptions. Differences in developing the sampling distribution and the bootstrap sampling distribution are illustrated in figure 8.1.

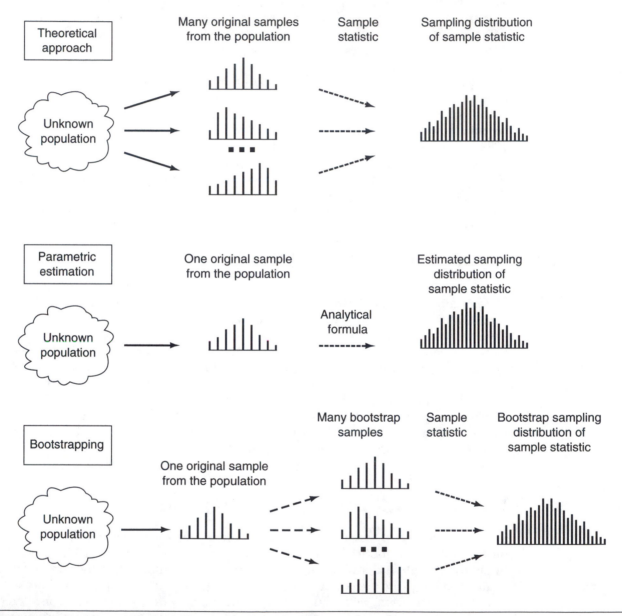

**Figure 8.1**  Differences in developing sampling distributions.
From Zhu, 1997, p. 46.

# How Bootstrapping Works

The idea behind bootstrapping is quite simple. Although it is impossible to get many samples from a population ($F$), it is possible to get repeated samples from a population ($\hat{F}$) whose distribution approximates the population. Given a sample that is drawn randomly from the population, the cumulative distribution of the sample—known also as the "empirical distribution function"—is the optimal estimator of the population. Let $X = (x_1, x_2, \ldots, x_n)$ denote a random sample of size $n$ from a population. If $F$ denotes the cumulative distribution function of the population, then $F(x) = P(X \leq x)$, where $P(X \leq x)$ denotes the probability of yielding a value less than or equal to $x$. This allows the cumulative distribution of the sample denoted by $\hat{F}(x_{(i)})$ to become the maximum likelihood estimate of the population distribution function, $F(x)$:

$$\hat{F}\left(x_{(i)}\right) = \frac{i}{n} = \frac{number\ of\ x_1, x_2, \ldots, x_n \leq x_i}{n}$$

$$(8.1)$$

where $i$ is, in fact, the cumulative frequency and $n$ is the sample size (see Rohatgi, 1984, pp. 234-236 for more details).

Let us use a landing example to illustrate how $\hat{F}(x_{(i)})$ is developed and implemented in determining the central tendency of a population. One of the most important parameters in studying the landing is the peak vertical ground reaction force (GRF) during the landing phase of the jump. To aid in understanding the characteristics of this parameter, a sample of 15 participants was asked to perform the landing. Their GRF scores (N/kg) were recorded and are presented in table 8.1. The maximum value was 75.40 and the minimum value was 33.34, with a mean of 46.38 and a standard deviation of 12.11. After the data were arranged in an order of increasing magnitude yielding $x_{(1)}, x_{(2)}, \ldots, x_{(n)}$, the cumulative distribution of the sample, $\hat{F}(x_{(i)})$, was computed (table 8.1) and plotted (figure 8.2). For example, the cumulative frequency of $x_{(3)}$, that is, value 35.22, was 3, and the corresponding $\hat{F}(x_{(3)})$ was 0.20 (3/15), which meant that for this population, the probability of yielding a value less than or equal to 35.22 was 0.20.

One of the nice features of the cumulative distribution function is that its value has a range from 0 to 1 and each $x$ value has its own corresponding cumulative proportion (CP). For example, because the $\hat{F}(x_{(1)})$ was 0.07 and the $\hat{F}(x_{(2)})$ was 0.13, the CP of the value 33.39, $CP_{(33.39)}$, was in the range $0.07 < CP_{(33.39)} < 0.13$ (see table 8.1). On the basis of the cumulative distribution function of the sample and its corresponding cumulative proportions, we can begin our bootstrap sampling process. First, we generate a uniform random number between 0 and 1 using a random number generator. Then

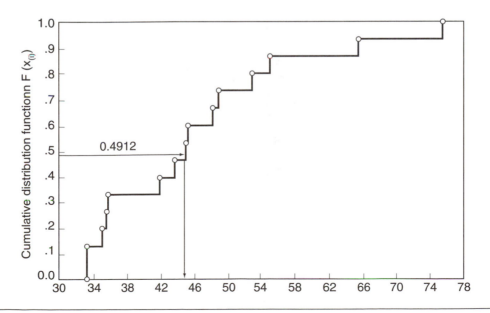

**Figure 8.2**  Cumulative distribution function of the sample.

**Table 8.1 Landing Sample Data and Their Cumulative Distribution Function**

**Raw data** (n=15)

| 33.34 | 33.39 | 35.22 | 35.70 | 35.85 | 41.89 | 43.73 | 45.15 | 45.28 | 48.30 | 48.88 | 53.01 | 55.08 | 65.42 | 75.40 |
|---|---|---|---|---|---|---|---|---|---|---|---|---|---|---|

**Ordered observations** $X_{(1)} \cdots X_{(n)}$

| | 33.34 | 33.39 | 35.22 | 35.70 | 35.85 | 41.89 | 43.73 | 45.15 | 45.28 | 48.30 | 48.88 | 53.01 | 55.08 | 65.42 | 75.40 |
|---|---|---|---|---|---|---|---|---|---|---|---|---|---|---|---|
| Frequency | 1 | 1 | 1 | 1 | 1 | 1 | 1 | 1 | 1 | 1 | 1 | 1 | 1 | 1 | 1 |
| Cumulative frequency | 1 | 2 | 3 | 4 | 5 | 6 | 7 | 8 | 9 | 10 | 11 | 12 | 13 | 14 | 15 |
| $\hat{F}(x_{(i)})$ | 0.07 | 0.13 | 0.20 | 0.27 | 0.33 | 0.40 | 0.47 | 0.53 | 0.60 | 0.67 | 0.73 | 0.80 | 0.87 | 0.93 | 1.00 |

we determine the corresponding value of $x$ according to the cumulative distribution function. Say, for example, the first uniform random number generated is 0.4912, which, according to the cumulative distribution function (table 8.1), falls into the CP range of the value 45.15 ($0.47 < CP_{(45.15)} < 0.53$) (see figure 8.2). The value of 45.15, therefore, is selected, and this value becomes the first observation in the first bootstrap example. We repeat this process 15 times (each bootstrap sample is the same size as the original sample), and the first bootstrap sample with 15 observations is generated. We illustrate 15 such bootstrap samples in table 8.2, including in these samples the frequencies and statistics of resampled observations.

Based on these bootstrap samples, bootstrap sample statistics can be calculated. For example, the mean of the first bootstrap sample is 42.56. The bootstrap estimates can be computed based on the bootstrap sample statistics. The bootstrap estimate ($\hat{\theta}_B$) of the population mean, $\theta$, can be defined as follows:

$$\hat{\theta}_B = \frac{1}{B}\sum_{i=1}^{B}\hat{\theta}_i^*$$

(8.2)

where $B$ is the number of bootstrap samples ($B = 15$ in this case) and $\hat{\theta}^*$ is the bootstrap sample mean. (We used $\theta$ to denote a true parameter of a population and $\hat{\theta}$ to denote an estimate of $\theta$ based on sample data.) The bootstrap estimate of the mean, which is the average of the bootstrap sample means, is 47.31 with an estimated standard error of 4.32 (see table 8.2). Two other bootstrap estimates, median and 5% trimmed mean, which were computed in a similar way, are also summarized in table 8.2. The ability to compute other useful statistics (some of which cannot be computed by parametric statistics) using the same bootstrap samples is, in fact, an advanced feature of bootstrapping. In practice, for example, researchers may have a data set with some outliers. To remove the biased effect, researchers can look at the trimmed means with different degrees of trimness (e.g., 5-50%). With bootstrapping it is easy to compute these useful statistics.

# Practical Issues of Bootstrapping Applications

Bootstrap computing, as illustrated in the preceding section, is straightforward. The procedure is to (1) generate a bootstrap sample randomly with replacement from the original sample, (2) compute bootstrap sample statistics and save them, (3) repeat steps 1 and 2 $B$ times, and (4) compute bootstrap estimates. The amount of computing required by most bootstrap applications, although considerable, is well within the capabilities of today's desktop computers. It is important to be aware of several practical issues related to these steps, namely, original sample and its size, size of bootstrap sample, and computing software.

## *Original Sample and Its Size*

The bootstrapping sample statistic and its sampling distribution, as mentioned earlier, are generated based on a pseudo-population, which is further developed based on a sample drawn from the targeted population. The representation of the sample thus becomes critically important. If the sample drawn is biased, the bootstrap statistics derived will likely also be biased. One simple way to increase the chance of representation of a sample is to increase its size. Yung and Chan (1999), for example, showed that a better resemblance to the population distribution was achieved when the sample size was increased from 31 to 51. The larger sample size, however, does not guarantee that the sample drawn will represent the targeted population (Yung and Chan, 1999). As a general rule, the number 30 is used as a minimum sample size (Chernick, 1999).

# Table 8.2 Frequencies (f) and Statistics of 15 Bootstrap Samples

| Bootstrap sample | 33.34 | 33.39 | 35.22 | 35.70 | 35.85 | 41.89 | 43.73 | 45.15 | 45.28 | 48.30 | 48.88 | 53.01 | 55.08 | 65.42 | 75.40 | Mean | Median | 5% trimmed mean |
|---|---|---|---|---|---|---|---|---|---|---|---|---|---|---|---|---|---|---|
| Data: (f) | 1 | 1 | 1 | 1 | 1 | 1 | 1 | 1 | 1 | 1 | 1 | 1 | 1 | 1 | 1 | | | |
| 1 | 2 | | | 2 | | 1 | 1 | | | | 2 | | | 1 | 1 | 42.56 | 35.70 | 41.25 |
| 2 | 2 | 1 | 5 | 1 | 1 | 1 | 3 | 1 | 2 | 1 | | 1 | | 1 | 1 | 43.81 | 43.73 | 42.63 |
| 3 | 1 | 1 | | 1 | 2 | 1 | 1 | | 1 | 1 | 1 | 1 | | | 3 | 49.10 | 45.28 | 48.51 |
| 4 | 1 | 2 | | 1 | 1 | 1 | 2 | | | | 1 | 1 | 1 | 2 | 1 | 47.95 | 43.73 | 47.24 |
| 5 | 3 | | | 1 | | | 3 | 1 | 3 | | 1 | 2 | 2 | 2 | 2 | 52.29 | 45.28 | 51.93 |
| 6 | 3 | 2 | 1 | 2 | 1 | | | 1 | 1 | | | 1 | 1 | 2 | | 42.57 | 35.70 | 41.82 |
| 7 | | | 2 | | 1 | 2 | 2 | | 1 | | 2 | | 1 | 1 | 3 | 51.15 | 45.28 | 50.69 |
| 8 | 1 | | 2 | 5 | 2 | 1 | | | | 1 | 2 | 1 | 1 | | | 39.66 | 35.70 | 39.27 |
| 9 | | | | 3 | 3 | 1 | 2 | 1 | | | | 2 | 2 | | 2 | 47.40 | 43.73 | 46.49 |
| 10 | 1 | 1 | 3 | 1 | 1 | | | | 2 | | | 3 | | 1 | 2 | 47.31 | 45.28 | 46.53 |
| 11 | 2 | 1 | | 1 | | 1 | 1 | | | 1 | 1 | | 2 | 1 | 5 | 55.85 | 55.08 | 56.01 |
| 12 | | 1 | | | | 2 | 2 | 1 | 2 | 1 | | | 2 | 3 | 1 | 51.36 | 45.28 | 51.03 |
| 13 | 2 | | 1 | | 1 | 1 | | 1 | 2 | | 2 | 1 | 2 | | 2 | 48.47 | 45.28 | 47.81 |
| 14 | 1 | 3 | | | 2 | 2 | | | 1 | 1 | 1 | 3 | | | 1 | 44.39 | 41.89 | 43.28 |
| 15 | 2 | 3 | | 1 | 1 | | 1 | | | 1 | 1 | 3 | 1 | 1 | 1 | 45.73 | 43.73 | 44.77 |
| Bootstrap estimate | | | | | | | | | | | | | | | | 47.31 | 43.38 | 46.62 |
| Standard error | | | | | | | | | | | | | | | | 4.32 | 4.92 | 4.58 |

214

## *Size of Bootstrap Samples*

Generally, 50 to 200 bootstrap samples are adequate (Efron and Tibshirani, 1986). To build bootstrap confidence intervals, however, one should employ at least 1,000 bootstrap samples (Efron, 1988). Using a set of sit-up data, Zhu (1997) examined the effect of increasing the number of bootstrap samples ($B$) on bootstrap statistics by computing the bootstrap statistics when $B = 50$, $B = 200$, $B = 500$, $B = 1,000$, $B = 2,000$, and $B = 5,000$. An overall trend as indicated by the results was that the more bootstrap samples, the better the estimation. Improvement of estimation accuracy, however, was limited, if one considers the degree of improvement from the estimates when $B = 10$ and when $B = 5,000$. Because today's computer power is so cheap and convenient to use, $B = 1,000$ is recommended for standard error estimation, and $B = 10,000$ is recommended for confidence interval estimations (Chernick, 1999).

# Advantages and Limitations of Bootstrapping

Bootstrapping is a statistical approach to making inferences about population parameters, but it differs fundamentally from traditional parametric statistics in that it employs large numbers of repetitive computations to estimate the shape of a statistic's sampling distribution, rather than depending on distribution assumptions and analytic formulas. Thus, the major advantage of bootstrapping over traditional parametric statistics is that sometimes it may be better to draw conclusions about the parameters of a population strictly from the sample at hand (e.g., a small sample) than to make perhaps unrealistic assumptions about the population.

Another major advantage of bootstrapping is that it provides a useful alternative when the parametric statistics are not available. For example, the $x$% **trimmed mean** is a very useful measure of central tendency when sample data come from a long-tailed population (Huber, 1981). The robust estimator is obtained through "trimming" of the data to exclude values that are far from others in order to produce a smaller measurement error. In the past, however, there were no analytic parametric formulas for computing the standard error of this statistic. Thus the accuracy of the estimation could not be determined and application of the statistic was limited. Through use of bootstrapping, as illustrated earlier, the problem is easy to solve. We simply generated bootstrap sampling distributions for the $x$% trimmed means (e.g., 5% trimmed mean in this study) and used the standard deviations of the distributions as the estimated standard error of the statistics (see table 8.2).

Bootstrapping has been applied in a variety of fields including medicine (Koehler and McGovern, 1990), psychophysiology (Wasserman and Bockenholt, 1989), social science (Hand, 1990), and education (Maguire, 1986), as well as to a variety of statistics, such as correlation coefficients (Lunneborg, 1985), factor analysis (Borrello and Thompson, 1989), repeated measures of analysis of variance (Lunneborg and Tousignant, 1985), and covariance structure analysis (Bollen and Stine, 1990). By 1999, more than 600 bootstrap research and application papers had been published (Chernick, 1999).

Bootstrapping was introduced into the field of kinesiology by Zhu (1997) using physical fitness data; and three bootstrapping applications have been reported by Zhu (1994), Zhang, Dufek, and Bates (1994), and Kelley (1997), respectively. Bootstrapping also has great application potential in assessment and data analyses in human movement research because many of our data distributions are skewed (e.g., pull-up data in fitness testing) and many of our researchers use small samples in their studies. For example, the development of prediction equations using regression has been of considerable interest to biomechanical researchers. To determine whether an equation can be applied to other populations or settings or to select a better equation, the researcher often needs to use cross-validation, which requires that a large sample be available. Fortunately, as a later example will show, it is easy to remove this limitation by employing the *K*-fold cross-validation method, a bootstrapping-based approach.

No statistical procedure always yields the correct answers. This is true also for bootstrapping, especially if the resampling scheme does not parallel the structure of the actual sampling mechanism. Recall that bootstrapping sampling distribution is derived from one "original" sample from the population. If this "original" sample is not a representative sample of the population, bootstrapping may give misleading results. Efron and Tibshirani (1993), for example, reported that bootstrapping may fail when the true distribution *F* in the extreme tail is estimated. Also, application of bootstrapping to some statistical procedures, such as correlation coefficients (Rasmussen, 1987) and hypothesis testing (Noreen, 1989), has been questioned. For example, Rasmussen (1987) found that bootstrapping resulted in overly restricted confidence intervals and overly liberal Type I error rates of correlation coefficients. Furthermore, a biased random number generator could cause bootstrapping to fail, so every effort should be made to control and monitor the function of the random number generator (see Ripley, 1987, for related tests).

## Example 8.1

Prediction using linear regression is one of the most popular statistical applications in human movement and biomechanical research (e.g., Burkholder and Leber, 1996; Dapena, McDonald, and Cappaert, 1990; Heil, Derrick, and Whittlesey, 1997) and is usually based on a convenient, small sample. One question often asked about the results of a prediction study is how well the derived prediction model can be generalized when it is applied to a different sample. To estimate the prediction error, the average residual error is often computed (Efron and Tibshirani, 1993):

$$\text{average residual error} = \Sigma(y_i - \hat{y}_i)^2 / n \tag{8.3}$$

where $y_i$ is the observed outcome/dependent measure, $\hat{y}_i$ is the predicted outcome measure, and *n* is the sample size. Two additional error indexes have been used often in exercise science literature: total error and standard error of estimate (SEE; Jackson, 1989; Lohman, 1981). The index of total error is defined as:

$$\text{total error} = \sqrt{\left(\Sigma\left(y_i - \hat{y}_i\right)^2 / n\right)} \tag{8.4}$$

Total error, thus, is the squared root of the average residual error. Standard error of estimate is defined as:

$$\text{SEE} = \sqrt{\left(\Sigma\left(y_i - \hat{y}_i\right)^2 / n - 2\right)} \tag{8.5}$$

The difference between SEE and total error, therefore, is that the factor of degree of freedom is taken into consideration. Note that when there are more than three prediction variables, an additional adjustment may be needed (i.e., replace *n* – 2 with *n* – *p*, where *p* is the number of prediction variables). In practice, SEE is perhaps the most commonly used prediction accuracy index since, like a standard deviation, it can provide an error band around predicted scores (e.g., a 68% confidence band for a prediction when ±1 SEE is applied; see Jackson, 1989, for a more detailed interpretation of SEE).

All these indexes, however, tend to underestimate the true prediction error, since the same data are used to derive and fit the model. In other words, the training, or calibration, sample is exactly the same as the test, or cross-validation, sample. In the literature, underestimates of prediction errors due to such use of the same data have been aptly called "apparent error" estimates (Efron and Tibshirani, 1993).

The ideal way to decrease the apparent error is to estimate the prediction error using a different test sample. Often, however, this is not easy in practice because of costs or logistics.

Another way to reduce the apparent error, when the sample size is large, is to randomly split the data into half, one half to be used for training and the other half to be used for cross-validation. This practice is also not easy to implement since the sample size in biomechanical research is usually small and the data themselves are not truly "crossed" (i.e., the data still came from the same sample).

Fortunately, a method based on the bootstrapping and jackknife idea has been developed (Efron and Tibshirani, 1993). The method is called "K-fold cross-validation." The steps to implement this method are quite straightforward:

1. Split the data randomly into K approximately equal-sized parts.
2. For the kth part, fit the model to the other K – 1 parts of the data, and compute the prediction error of the fitted model when predicting the kth part of the data.
3. Repeat steps 1 and 2 for $k = 1, 2, \ldots, K$ and combine the K estimates of prediction error.

When the sample size is small, n, the sample size, is often used as k, resulting in a "leave-one-out" cross-validation. In the "leave-one-out" cross-validation, one of the cases is first selected and a prediction equation is developed using the rest of the data. The data from the selected case are then applied to the derived prediction equation, and a residual between the observed and predicted score is computed. This process is repeated until all the cases in the sample have been selected. The prediction error can then be computed through averaging the derived residuals.

Let us use a real data set to illustrate how the K-fold cross-validation method works. It is well known that athletes' jump heights are directly related to the impulse they generate during the jumping phase (see, e.g., Dapena, McDonald, and Cappaert, 1990). To verify this relationship in a plyometric setting, a drop jump study was conducted in which 14 participants were asked to drop from a height equal to their maximum vertical jump height and to jump as high as possible. The jump height (cm) was computed from kinematic data (high-speed video) and jump impulse (unit: N/kg/s) is the impulse of the vertical GRF during the jumping phase. The data of jump height and impulse are illustrated in the second and third columns of table 8.3. With use of a linear regression, the relationship between impulse and jump height was statistically modeled, and the derived prediction equation was:

$$\text{predicted jump height} = 13.768 + 4.898(\text{jump impulse}) \qquad (8.6)$$

With use of this equation, the participants' predicted jump height scores were computed; these are shown in the fourth column in table 8.3. In addition, the difference between participants' predicted jump scores and their observed jump height scores was computed; these values are shown in the residual column (fifth column) in table 8.3. Using the information in the residual column and equations 8.3 to 8.5, average residual error, total error, and SEE were computed; these equaled 31.61, 5.62, and 6.07, respectively (see table 8.3). As pointed out earlier, these values would likely be underestimated if equation 8.6 were applied to another sample.

Let us see how the K-fold cross-validation method can help here. Since the sample size is small, we decided to use the "leave-one-out" method. More specifically, in the first-round analysis, we withheld the information for the first participant and derived a linear equation using the rest of the data (n = 13) in columns 2 and 3: predicted jump height score = 12.442 + 5.127(jump impulse). By applying the withheld information, we computed the first participant's predicted jump height score, which equaled 34.00. By computing the difference between this predicted number and the original observed jump height (37.05), we determined the K-fold residual (3.05). We then brought back the information for the first participant, but withheld the information for the second participant. Similarly, after the second prediction equation was derived (intercept = 17.802; slope = 4.218), the second participant's predicted jump score and residual were computed. This process was repeated until all the participants were rotated. One can easily implement the computation using the K-fold functions developed for the

*(continued)*

**Table 8.3   Results of Prediction and *K*-Fold Cross-Validation**

| ID | Prediction using linear regression | | | | | K-fold cross-validation | | | | |
|---|---|---|---|---|---|---|---|---|---|---|
| | Jump height (cm) | Jump impulse (N/kg/sec) | Predicted jump height (cm) | Residual (Res) | Res² | Intercept | Slope | Predicted jump height (cm) | Residual (Res) | Res² |
| 1 | 37.05 | 4.21 | 34.36 | 2.69 | 7.22 | 12.44 | 5.13 | 34.00 | 3.05 | 9.29 |
| 2 | 25.36 | 4.30 | 34.81 | -9.45 | 89.23 | 17.80 | 4.22 | 35.92 | -10.56 | 111.53 |
| 3 | 32.75 | 4.19 | 34.29 | -1.54 | 2.38 | 14.55 | 4.76 | 34.50 | -1.75 | 3.08 |
| 4 | 45.50 | 4.72 | 36.88 | 8.62 | 74.23 | 12.54 | 5.01 | 36.21 | 9.29 | 86.35 |
| 5 | 37.72 | 4.27 | 34.69 | 3.03 | 9.19 | 12.42 | 5.13 | 34.32 | 3.40 | 11.58 |
| 6 | 41.28 | 5.51 | 40.76 | .52 | .27 | 13.97 | 4.85 | 40.68 | .60 | .36 |
| 7 | 36.70 | 4.24 | 34.54 | 2.16 | 4.66 | 12.76 | 5.07 | 34.27 | 2.43 | 5.92 |
| 8 | 42.22 | 6.13 | 43.80 | -1.58 | 2.49 | 12.24 | 5.25 | 44.41 | -2.19 | 4.81 |
| 9 | 31.32 | 4.27 | 34.66 | -3.34 | 11.17 | 15.27 | 4.64 | 35.07 | -3.75 | 14.07 |
| 10 | 24.38 | 4.23 | 34.48 | -10.10 | 102.08 | 18.57 | 4.07 | 35.79 | -11.41 | 130.15 |
| 11 | 42.35 | 6.71 | 46.62 | -4.27 | 18.22 | 4.80 | 6.88 | 50.98 | -8.63 | 74.44 |
| 12 | 42.68 | 4.74 | 36.99 | 5.69 | 32.40 | 13.04 | 4.96 | 36.55 | 6.13 | 37.63 |
| 13 | 47.78 | 5.05 | 38.50 | 9.28 | 86.13 | 14.40 | 4.62 | 37.72 | 10.06 | 101.21 |
| 14 | 36.34 | 4.96 | 38.05 | -1.71 | 2.92 | 13.75 | 4.93 | 38.18 | -1.84 | 3.40 |
| | | | | | Sum = 442.59 | | | | | Sum = 593.82 |
| | | | | | Mean = 31.62 | | | | | Mean = 42.42 |
| | | | | | Total error = $\sqrt{31.62}$ = 5.62 | | | | | Total error = $\sqrt{42.42}$ = 6.51 |
| | | | | | SEE = $\sqrt{(442.59 / (14-2))}$ = 6.07 | | | | | |

*S*-Plus statistical software library (e.g., the "Bootstrap" and "Boot" functions mentioned later in the section on software). The information for the *K*-fold-based intercepts, slopes, predicted scores, and residuals is summarized in columns 7 to 11 in table 8.3, respectively.

Based on the *K*-fold cross-validation residual information, the average residual and total errors were again computed; these equaled 42.42 and 6.51 ($\sqrt{42.42}$), respectively (note: SEE is usually not computed in cross-validation). Comparison of these values with those derived from the linear regression showed that the average residual errors and total errors based on the linear regression underestimated the cross-validation prediction errors by about 25.47% [(42.4155 − 31.6137) / 42.4155 × 100%] and 13.67% [(6.5127 − 5.6226) / 6.5127 × 100%], respectively. With an adjustment of degrees of freedom, the prediction significantly improved (SEE = 6.0731 vs. *K*-fold total error = 6.5127), although it still accounted for about 6.75% (6.5127 − 6.0731 / 6.5127) of the underestimation. The results indicate that the *K*-fold cross-validation is a useful approach to estimate prediction error when a small-sample-based equation is applied to another sample, and this method should be integrated into regression data analysis in biomechanical and other human movement research.

# Summary

The development of bootstrapping, one of the most commonly used computer-intensive statistics, has provided us with not only a nonparametric approach to analyzing data for which parametric statistics are not appropriate, but also a new way to explore those statistics that traditionally cannot be applied due to a violation of statistical assumptions or a lack of analytic formulas. This chapter illustrated the key idea, computations, and advantages of bootstrapping using a set of real data from biomechanics. The major advantage of bootstrapping over traditional statistics is that sometimes it may be better to draw conclusions about the parameters of a population strictly from the sample at hand rather than on the basis of perhaps unrealistic assumptions about the population. Bootstrapping should not be considered preferable to parametric statistics, but rather as an alternative in situations in which parametric statistics are not appropriate or not available.

# Suggested Readings and Other Resources

## Books and Articles

Chernick, M.R. (1999). *Bootstrap methods: A practitioner's guide.* New York: Wiley.

Davison, A.C., and Hinkley, D.V. (1997). *Bootstrap methods and their application.* New York: Cambridge University Press.

Efron, B. (1982). *The jackknife, the bootstrap and other resampling plans.* Philadelphia: Society for Industrial and Applied Mathematics.

Efron, B., and Tibshirani, R. (1993). *An introduction to the bootstrap.* New York: Chapman & Hall.

Hjorth, U. (1994). *Computer intensive statistical methods: Validation model selection and bootstrap.* New York: Chapman & Hall.

Mooney, C.Z., and Duval, R.D. (1993). *Bootstrapping: A nonparametric approach to statistical inference.* Newbury Park, CA: Sage.

Noreen, E.W. (1989). *Computer-intensive methods for testing hypotheses.* New York: Wiley.

Shao, J., and Tu, D. (1995). *The jackknife and bootstrap.* New York: Springer.

Stine, R. (1990). An introduction to bootstrap methods. In J. Fox and J.S. Long (Eds.), *Modern methods of data analysis.* Newbury Park, CA: Sage.

## Bootstrapping Software

Computing language codes for the commonly used statistics have been included in several introductory texts (see Mooney and Duval, 1993; Noreen, 1989). Most commercial statistical software packages, such as SAS, *S*-Plus, and SPSS-X, have the capacity to conduct bootstrapping with some programming. Among such packages, *S*-Plus, published by Insightful, is perhaps the one most commonly used because of the availability of bootstrap-related functions, for example, "Bootstrap" by Efron and Tibshirani (1986), which can be downloaded at http://lib.stat.cmu.edu/DOS/S/, and "Boot," which was developed at the Swiss Federal Institute of Technology (http://statwww.epfl.ch/davison/BMA/library.html) and described by Davison and Hinkley (1997). In addition, several computer programs specializing in bootstrapping, such as BOJA (Boomsma, 1991) and BOOTSTRP (Lunneborg, 1987), have been developed.

## Web Sites

www.insightful.com/Hesterberg/bootstrap. A rich resource for bootstrap application using *S*-Plus software, a data analysis software package developed by Insightful, Inc.

www.resample.com/. A site with rich information on resampling statistics that includes bootstrapping.

www.statistics.com/. A general statistics site with resampling statistics information.

www.uvm.edu/~dhowell/StatPages/Resampling/Resampling.html. Some useful information on bootstrapping and randomization is presented in this rather simple Web site. Includes free software that can be downloaded from the site.

# References

Bollen, K.A., and Stine, R.A. (1990). Direct and indirect effects: Classical and bootstrap estimates of variability. In C.C. Clogg (Ed.), *Sociological methodology 1990* (pp. 115-140). Oxford: Blackwell.

Boomsma, A. (1991). *BOJA: A program for bootstrap and jackknife analysis* [Computer program]. Groningen, The Netherlands: iec ProGAMMA.

Borrello, G.M., and Thompson, B. (1989). A replication bootstrap analysis of the structure underlying perceptions of stereotypic love. *Journal of General Psychology, 116,* 317-327.

Budescu, D.V., and Appelbaum, M.I. (1981). Variance stabilizing transformations and the power of the *F* test. *Journal of Educational Statistics, 6,* 55-74.

Burkholder, T.J., and Leber, R.L. (1996). Stepwise regression is an alternative to splines for fitting noise data. *Journal of Biomechanics, 29*(2), 235-238.

Chen, A., and Zhu, W. (2001). Revisiting the assumptions for inferential statistical analyses: A conceptual guide. *Quest, 53,* 418-439.

Chernick, M.R. (1999). *Bootstrap methods: A practitioner's guide.* New York: Wiley.

Davison, A.C., and Hinkley, D.V. (1997). *Bootstrap methods and their application.* New York: Cambridge University Press.

Dapena, J., McDonald, C., and Cappaert, J. (1990). A regression analysis of high jumping technique. *International Journal of Sport Biomechanics, 6,* 246-261.

Diaconis, P., and Efron, B. (1983). Computer-intensive methods in statistics. *Scientific American, 248*(5), 116-130.

DiCiccio, T.J., and Romano, J.P. (1989). The automatic percentile method: Accurate confidence limits in parametric models. *Canadian Journal of Statistics, 17,* 155-169.

Efron, B. (1979). Bootstrap methods: Another look at the jackknife. *Annals of Statistics, 7,* 1-26.

Efron, B. (1987). Better bootstrap confidence intervals. *Journal of the American Statistical Association, 82,* 171-185.

Efron, B. (1988). Bootstrap confidence intervals: Good or bad? *Psychological Bulletin, 104,* 293-296.

Efron, B., and Tibshirani, R. (1986). Bootstrap methods for standard errors, confidence intervals, and other measures of statistical accuracy. *Statistical Science, 1,* 54-77.

Efron, B., and Tibshirani, R. (1993). *An introduction to the bootstrap.* New York: Chapman & Hall.

Hand, M.L. (1990). A resampling analysis of federal family assistance program quality control data: An application of the Bootstrap. *Evaluation Review, 14,* 391-410.

Heil, D.P., Derrick, T.R., and Whittlesey, S. (1997). The relationship between preferred and optimal positioning during submaximal cycle ergometry. *European Journal of Applied Physiology, 75*(2), 160-165.

Hill, M., and Dixon, W.J. (1982). Robustness in real life: A study of clinical laboratory data. *Biometrics, 38,* 377-396.

Huber, P. (1981). *Robust statistics.* New York: Wiley.

Jackson, A.S. (1989). Application of regression analysis to exercise science. In M.J. Safrit and T.M. Wood (Eds.), *Measurement concepts in physical education and exercise science* (pp. 181-205). Champaign, IL: Human Kinetics.

Kelley, G.A. (1997). Bootstrap procedures for corroborating mean outcomes form meta-analytic data: A brief tutorial. *Measurement in Physical Education and Exercise Science, 1,* 203-212.

Kirk, R.E. (1982). *Experimental design for the behavioral sciences* (2nd ed.). Belmont, CA: Brooks/Cole.

Koehler, K.J., and McGovern, P.G. (1990). An application of the LEP survival model to smoking cessation data. *Statistics in Medicine, 9,* 409-421.

Lohman, T.G. (1981). Skinfolds and body density and their relation to body fatness: A review. *Human Biology, 53*(2), 181-225.

Lunneborg, C.E. (1985). Estimating the correlation coefficient: The bootstrap approach. *Psychological Bulletin, 98,* 209-215.

Lunneborg, C.E. (1987). Bootstrap applications for the behavioral sciences. *Educational and Psychological Measurement, 47,* 627-629.

Lunneborg, C.E., and Tousignant, J.P. (1985). Efron's bootstrap with application to the repeated measures design. *Multivariate Behavioral Research, 20,* 161-178.

Maguire, T.O. (1986). Applications of new directions in statistics to educational research. *Alberta Journal of Educational Research, 2,* 154-171.

Miller, R.G. (1974). The jackknife: A review. *Biometrika, 61,* 1-15.

Noreen, E.W. (1989). *Computer-intensive methods for testing hypotheses.* New York: Wiley.

Rasmussen, J.L. (1987). Estimating correlation coefficients: Bootstrap and parametric approaches. *Psychological Bulletin, 101,* 136-139.

Ripley, B.D. (1987). *Stochastic simulation.* New York: Wiley.

Rohatgi, V.K. (1984). *Statistical inference.* New York: Wiley.

Wasserman, S., and Bockenholt, U. (1989). Bootstrapping: Applications to psychophysiology. *Psychophysiology, 26,* 208-221.

Yung, Y.-F., and Chan, W. (1999). Statistical analysis using bootstrapping: Concepts and Implementation. In R.H. Hoyle (Ed.), *Statistical strategies for small sample research* (pp. 81-105). Thousand Oaks, CA: Sage.

Zhang, S., Dufek, J.S., and Bates, B.T. (1994). Viability of bootstrap techniques in biomechanics research. In W. Herzog, B. Nigg, and T. van den Bogert (Eds.), *Proceedings of Canadian Society for Biomechanics, VIIIth Biennial Conference* (pp. 310-311). Calgary, Canada: University of Calgary.

Zhu, W. (1994). Bootstrap estimation of a population's physical fitness status using small samples. *Medicine and Science in Sports and Exercise, 26*(Suppl. 5), S217.

Zhu, W. (1997). Making bootstrap statistical inferences: A tutorial. *Research Quarterly for Exercise and Sport, 68,* 44-55.

# CHAPTER 9

# Power Spectrum Analysis and Filtering

**Giannis Giakas, PhD**

The analysis of kinematic and kinetic data is usually limited to the identification of local minimum and maximum values and their respective times of occurrence (figure 9.1). This is the typical approach employed to evaluate curves that describe a movement's pattern and is termed **time domain** analysis since the data are represented as a function of time. Such a representation is useful when there are clear differences in the selected parameters, but it is important to note that this representation does not take into account the overall pattern of data. For example, a patient with Parkinson's disease may exhibit similar minimums and maximums in the force curves during walking pre- and postmedication; however, the force patterns before medication could be more oscillated due to the tremor. Conventional time domain analysis would indicate that there is no difference before and after medication, even after the tremor has been reduced. These differences can be revealed through examination of another form of data representation, known as **frequency domain** analysis, that presents the data as a function of frequencies.

Frequency domain analysis is a powerful data analysis tool employed by a variety of disciplines such as mathematics, optics, genetics, and physics (Braun, 2002; Cramer et al., 2001; Lee et al., 2002; Rozan et al., 2000; Sadik and Litwin, 2002; Tavazoie et al., 1999). Through use of this type of analysis, a given set of data is represented-transformed as a function of frequencies. Among a number of frequency domain transforms, the most commonly used in biomechanics is the Fourier transform, invented by the French mathematician Jean Baptiste Joseph, Baron de Fourier, in 1810, who used sums of sine and cosine functions to represent more complex functions employed in solving heat conduction problems. Fourier showed that periodic signals could be represented as weighted sums of harmonically related sinusoids. This work has significantly influenced the development of physics and engineering and will continue to influence these fields in the future. A reference search for the period between January 2000 and May 2002 revealed that the word "Fourier" had appeared in 12,000 academic papers.

The aim of this chapter is to describe frequency domain analysis and demonstrate two major applications in human movement: (1) discrimination between patterns with a similar time domain but different frequency domains and (2) data **filtering.** It is useful to mention that all tools needed to perform the frequency domain analysis and the associated graphics used in this chapter are available from the computing language MATLAB (MathWorks, version 6.5).

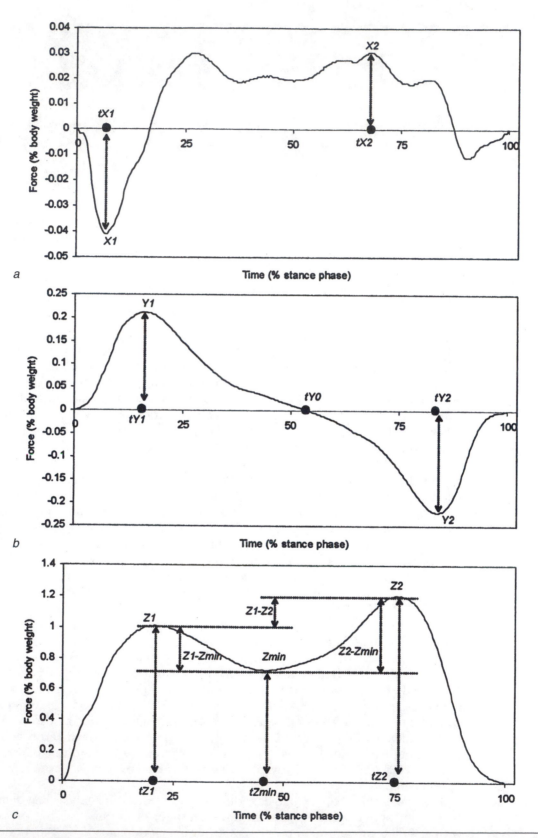

**Figure 9.1** Selected time domain parameters of ground reaction force data during walking of a healthy subject: *(a)* mediolateral, *(b)* anterior-posterior, *(c)* vertical. The analysis usually contains local minimum and maximum values (such as *X1, X2, Y1, Y2, Z1, Zmin,* and *Z2*), their times of occurrence (such as *tX1, tX2, tY1, tY2, tZ1, tZmin,* and *tZ2*), differences between force parameters (such as *Z1-Zmin, Z1-Z2, and Z2-Zmin*), and time of anterior-posterior pattern zero cross (*t0*).

# Time and Frequency Domain Representations:
# A Simple Signal

The following example provides a description of the frequency domain and its relation to the time domain. A pendulum (figure 9.2) of 1-m length completes one full sway (or otherwise cycle) in 1 s around point O. This means that the pendulum sways with a frequency of 1 Hz. The time domain pattern of the movement of point A at the edge of the pendulum is a well-known sinusoid wave (also in figure 9.3a). Its frequency domain pattern is represented by a bar at the frequency of 1 Hz with amplitude of 1 (figure 9.3b). Similarly, a much smaller pendulum with a length of 0.02 m is attached to the first pendulum and sways with a frequency of 15 Hz (figure 9.4a). The frequency domain of the movement of point B relative to A is represented by a bar at 15 Hz with amplitude of 0.02 (figure 9.4b). Examining the movement of B relative to O shows that point B follows a combined pattern that is the sum of the two individual patterns (figure 9.5a). If we try to compare these two signals (figures 9.3a and 9.5a) in terms of selected time domain parameters such as minimum and maximum values, then unfortunately we will find no clear differences. In contrast, there is an obvious difference in the frequency components between the two signals (one of them has an additional low amplitude at 15 Hz). The process of viewing the same signal in two domains is performed with the Fourier transform described later.

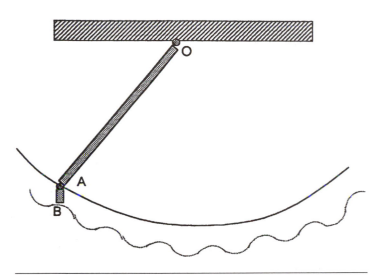

**Figure 9.2** The movement of two points (A and B) on two attached pendulums.

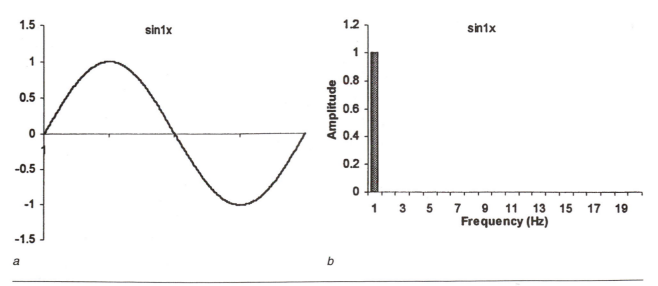

a

b

**Figure 9.3** (a) Time domain and (b) frequency domain patterns of the long pendulum. The pendulum in this figure represents an example of the data we want to collect.

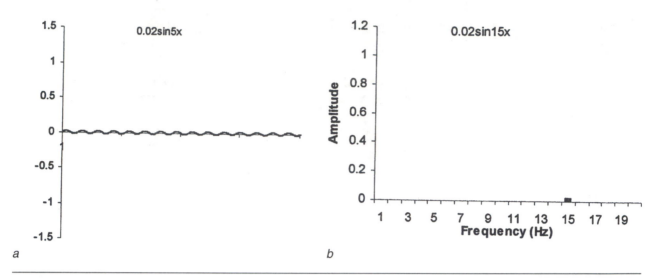

**Figure 9.4**    *(a)* Time domain and *(b)* frequency domain patterns of the short pendulum. The pendulum in this figure represents an example of noisy data superimposed on the signal.

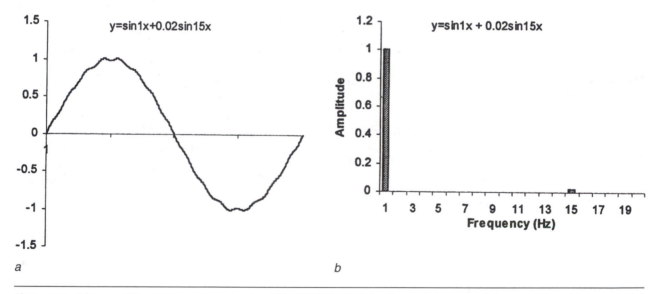

**Figure 9.5**    The combined movement of point B (figure 9.2) when both pendulums from figures 9.3 and 9.4 sway. The *(a)* time domain and *(b)* frequency domain of combined movement are used to demonstrate the superimposed error (low amplitude-high frequency) in measurements (high amplitude-low frequency).

# Frequency Domain Transform and the Discrete Fourier Transform

In human movement, we are used to observing data collected in the time domain as a one-dimensional context. As mentioned earlier, this is not the only way of examining the data. The data can be expressed as a sum of simple sinusoids, each having a specific frequency. This can be achieved by means of a **Fourier transformation,** which is discussed and explained in detail in this chapter. For information on "alternative" frequency domain transformation functions, readers are referred to other sources such as Bogner and Constantinides, 1975; Lynn and Fuerst, 1994; Rabiner and Gold, 1975; and Smith and Mersereau, 1992.

The Fourier transform is basically a mathematical procedure to describe complicated analog signals with the help of simpler functions: the sine and the cosine. In practice, though, before we examine a signal we need to sample it digitally. This requirement has created the discrete Fourier transform (DFT), which is the application of the original "analog" Fourier transform to the "digital" world. The DFT (Cooley and Tukey, 1965) calculates the contribution of sinusoidal components to reconstruct a signal.

The DFT components of a signal, called the *harmonics,* represent specific frequencies. The frequency of the first harmonic (fundamental frequency) is equal to the sampling frequency divided by the number of data. To obtain the frequencies of subsequent harmonics we multiply the fundamental frequency with the harmonic number. The following formula can be used to calculate each harmonic $\alpha_k$ (which is a complex number):

$$\alpha_k = \sum_{n=0}^{N-1} x[n]W^{kn} \tag{9.1}$$

where $W = e^{2j\pi/N}$, $x[n]$ represents the $n^{\text{th}}$ data point, $\pi = 3.1415\ldots$, $N$ is the number of data, $k$ is the harmonic number, and $n$ is the data point index. Using equation 9.1 it is possible to transform the signal from the time domain to the frequency domain through calculation of the Fourier coefficients $\alpha_k$. The standard algorithm for calculating the Fourier coefficients (Cappozzo and Gazzani, 1983) is quite slow (Brigham, 1974), and therefore fast approximations have been developed (www.fftw.org; Frigo and Johnson, 1998). These discrete Fourier approximations use a lower number of calculations, making the process faster, when the number of data can be analyzed in small prime numbers (Rader, 1968). These algorithms are generally called **Fast Fourier Transformations** (FFT). The FFT is mainly applied when the total number of data is a power of 2. However, other fast approximations are also used when the number of data can be analyzed in small prime numbers. If it cannot be analyzed in any prime numbers, then the FFT cannot be applied. In MATLAB the signal can be padded with zeroes until the number of data reaches a power of 2. Using MATLAB, if $x$ is a vector that contains the raw data, then using $\alpha_k = fft(x)$ we can calculate the DFT harmonics. If all harmonics $\alpha_k$ are known, we can synthesize the signal precisely, thus transforming a signal from the frequency domain to the time domain:

$$x[n] = \frac{1}{N}\sum_{k=0}^{N-1} a_k W^{kn} \tag{9.2}$$

where $W = e^{2j\pi/N}$, $\alpha_k$ represents the $k^{\text{th}}$ harmonic, $\pi = 3.1415\ldots$, $N$ is the number of data, $k$ is the harmonic number, and $n$ is the data point index. This process is called inverse DFT. In MATLAB, if $X$ is a vector containing the harmonics $\alpha_k$, the function *ifft(X)* returns the original signal $x$.

## Example 9.1

A useful example is the simple signal $y = 10 \times \cos(2\pi 3x)$. Using the *fft* function of MATLAB we obtain its Fourier transformation (table 9.1). Depending on the number of data used to reconstruct the function, at the frequency of 3 Hz the real value approaches 5. The value of 5 is half the amplitude; the other half is in the mirrored side of the Fourier transform. The imaginary value approaches 0. It is evident that the values from table 9.1, that is, 5 and 0, are not exact. This is the case because the 1,000 or 100,000 data points reconstruct the function $10 \times \cos(2\pi 3x)$ in a digital manner. Practically, the higher the number of data points to $y = 10 \times \cos(2\pi 3x)$ reconstruct the same function, the more accurate the Fourier coefficients. *(continued)*

**Table 9.1　The First 10 Fourier Transform Coefficients of the Signal $y = 10 \times \cos(2\pi 3x)$ Using 1,000 and 100,000 Data Points to Digitally Reconstruct the Analog Function**

| Frequency | 1,000 data points | 100,000 data points |
| --- | --- | --- |
| 0 | −0 .0000 | −0.0000 |
| 1 | 0.0000 + 0.0000i | 0.0000 + 0.0000i |
| 2 | 0.0000 + 0.0000i | 0.0000 + 0.0000i |
| 3 | 4.9991 + 0.0942i | 5.0000 + 0.0009i |
| 4 | 0.0000 + 0.0000i | 0.0000 + 0.0000i |
| 5 | 0.0000 + 0.0000i | 0.0000 + 0.0000i |
| 6 | 0.0000 + 0.0000i | 0.0000 + 0.0000i |
| 7 | 0.0000 + 0.0000i | 0.0000 + 0.0000i |
| 8 | 0.0000 + 0.0000i | 0.0000 + 0.0000i |
| 9 | 0.0000 + 0.0000i | 0.0000 + 0.0000i |
| 10 | 0.0000 + 0.0000i | 0.0000 + 0.0000i |

The first 10 Fourier transform coefficients (including frequency 0, which represents the mean) of the signal $y = 10 \times \cos(2\pi 3x)$. Total time is 1 s. At the frequency of 3 Hz we expect a real value of 5 (this is half the amplitude; the other half is in the mirrored side of the Fourier transform) and an imaginary value of 0. This happens because the 1,000 or 100,000 data points reconstruct the function $y = 10 \times \cos(2\pi 3x)$ in a digital manner. Practically, the higher the number of data points used to reconstruct the same function, the more accurate the Fourier coefficients.

Let's look at the function $y = 4 \times \sin(2\pi 8x)$. The Fourier transform of the sine signal should give us a non-zero amplitude only at the frequency of 8 Hz (table 9.2). Of course this is not true, since the digital sampling of the signal improves dramatically when the same signal is sampled at a higher rate. Theoretically we get the exact values when the sampling frequency is infinite, which means that we have an analog signal.

**Table 9.2　The First 10 Fourier Transform Coefficients of the Signal $y = 4 \times \sin(2\pi 8x)$ Using 1,000 and 100,000 Data Points to Digitally Reconstruct the Analog Function**

| Frequency | 1,000 data points | 100,000 data points |
| --- | --- | --- |
| 0 | −0.0000 | −0.0000 |
| 1 | 0.0000 + 0.0000i | 0.0000 + 0.0000i |
| 2 | 0.0000 + 0.0000i | 0.0000 + 0.0000i |
| 3 | 0.0000 + 0.0000i | 0.0000 + 0.0000i |
| 4 | 0.0000 + 0.0000i | 0.0000 + 0.0000i |
| 5 | 0.0000 + 0.0000i | 0.0000 + 0.0000i |
| 6 | 0.0000 + 0.0000i | 0.0000 + 0.0000i |
| 7 | 0.0000 + 0.0000i | 0.0000 + 0.0000i |
| 8 | 0.1005 - 1.9975i | 0.0010 − 2.0000i |
| 9 | 0.0000 + 0.0000i | 0.0000 + 0.0000i |
| 10 | 0.0000 + 0.0000i | 0.0000 + 0.0000i |

The first 10 Fourier transform coefficients (including frequency 0, which represents the mean) of the signal $y = 4 \times \sin(2\pi 8x)$. Total time is 1s. At the frequency of 8 Hz we expect an imaginary amplitude value of −2 (this is half the amplitude; the other half as a positive number is in the mirrored side of the Fourier transform) and a real value of 0. This happens because the 1,000 or 100,000 data points reconstruct the function $y = 4 \times \sin(2\pi 8x)$ in a digital manner. Practically, the higher the number of data points to reconstruct the same function, the more accurate the Fourier coefficients.

Let's add these two signals together (figure 9.6). The new function will be $y = 10 \times \cos(2\pi 3x) + 4 \times \sin(2\pi 8x)$. The sum of the two functions will result in the sum of their Fourier transforms as well. On the basis of the previous two paragraphs, we will get a real amplitude of 5 at the frequency of 3 Hz and an imaginary amplitude of –2 at the frequency of 8 Hz (table 9.3).

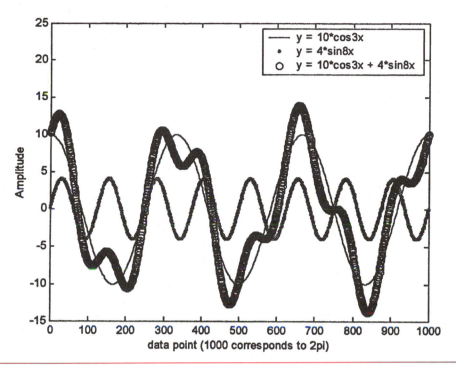

**Figure 9.6**  The time domain of the functions $y = 10 \times \cos(2\pi 3x)$; $y = 4 \times \sin(2\pi 8x)$; and their sum $y = 4 \times \sin(2\pi 8x) + 10 \times \cos(2\pi 3x)$.

| Table 9.3    The First 10 Fourier Transform Coefficients of the Signal $y = 10 \times \cos(2\pi 3x) + 4 \times \sin(2\pi 8x)$ Using 1,000 and 100,000 Data Points to Digitally Reconstruct the Analog Function | | |
|---|---|---|
| **Frequency** | **1,000 data points** | **100,000 data points** |
| 0 | –0.0000 | –0.0000 |
| 1 | 0.0000 + 0.0000i | 0.0000 + 0.0000i |
| 2 | 0.0000 + 0.0000i | 0.0000 + 0.0000i |
| 3 | 4.9991 + 0.0942i | 5.0000 + 0.0009i |
| 4 | 0.0000 + 0.0000i | 0.0000 + 0.0000i |
| 5 | 0.0000 + 0.0000i | 0.0000 + 0.0000i |
| 6 | 0.0000 + 0.0000i | 0.0000 + 0.0000i |
| 7 | 0.0000 + 0.0000i | 0.0000 + 0.0000i |
| 8 | 0.1005 – 1.9975i | 0.0010 – 2.0000i |
| 9 | 0.0000 + 0.0000i | 0.0000 + 0.0000i |
| 10 | 0.0000 + 0.0000i | 0.0000 + 0.0000i |

The first 10 Fourier transform coefficients (including frequency 0, which represents the mean) of the signal $y = 10 \times \cos(2\pi 3x) + 4 \times \sin(2\pi 8x)$. Total time is 1 s. At the frequency of 3 Hz we expect a real amplitude value of 5, and at the frequency of 8 Hz we expect an imaginary amplitude of –2. All other amplitudes should be 0. The signal is reconstructed using 1,000 and 100,000 data points. Refer also to tables 9.1 and 9.2 for simpler functions.

These examples demonstrate the basic concept of the relationship between the two forms of the Fourier transform using simple sinusoids. Moving on to a more complicated signal from a human movement experiment, we will use data presented by Lanshammar (1982; Pezzack et al., 1977). These data have been used in the past for the development of filtering techniques for human movement signals (Giakas et al., 1998; Ismail and Asfour, 1999; Vint and Hinrichs, 1996; Walker, 1998). These data show the angular movement of a rotating arm, indicating the elbow flexion/extension angle (figure 9.7*a*). Using the *fft* function we can calculate the Fourier coefficients and examine separately their real (figure 9.7*b*) and their imaginary part (figure 9.7*c*). The Fourier coefficients are used to examine the contribution that each frequency component makes to the signal. From figure 9.7, *b* and *c,* we can observe that only the initial 9 to 10 Fourier coefficients have significant amplitude; all the remaining coefficients corresponding to the higher frequencies are nearly

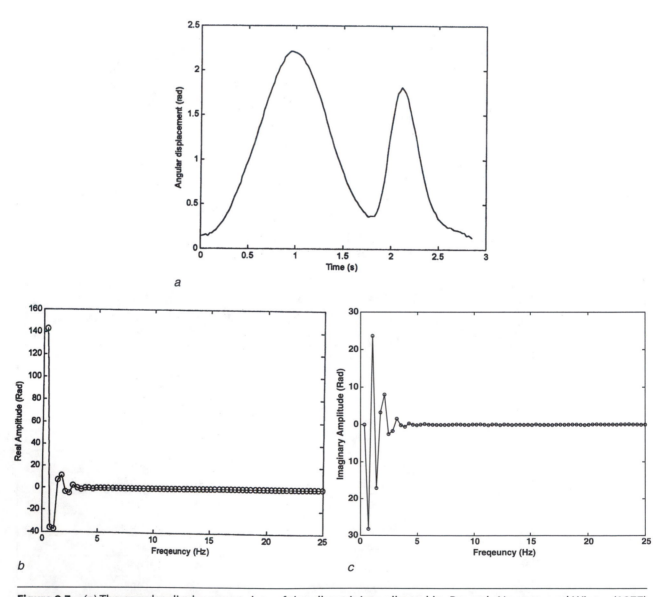

**Figure 9.7** *(a)* The angular displacement data of the elbow joint collected by Pezzack, Norman, and Winter (1977) and modified by Lanshammar (1982), who added white noise to make the signal more representative because the original was very smooth. After the Fourier transform of the signal, the *(b)* real and *(c)* imaginary parts of each Fourier coefficient can be graphically represented.

zero. These amplitudes indicate the contribution of each frequency component to the signal. We determine this by calculating the amplitudes (square root of the sum of squares of the real and imaginary parts) of each frequency or the power spectrum (PS), that is, the square magnitude of the DFT (sum of squares of the real and imaginary parts for each frequency) (figure 9.8a). The power spectrum of the Lanshammar data shows that the lower frequencies have much higher power than the higher frequencies (note that the Y scale of the graph is logarithmic). If we sum the power of all frequencies, the result will be the overall power of the signal. Having that calculated, we could express the power of each frequency as a percentage of the overall power of the signal (table 9.4, PS column). In most kinematic signals, the first harmonic that represents a frequency counts for most of the power of the signal, and in our case it counts for approximately 81%. By progressively adding the power of the subsequent harmonics we advance to higher frequencies that obviously account for a higher percentage of the power of the signal (table 9.4, Cumulative PS column).

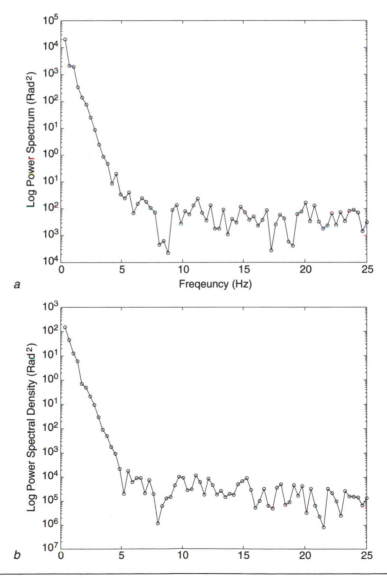

**Figure 9.8** The (a) power spectrum and (b) power spectral density of the Lanshammar data (1982). Both spectrums are generally used to examine the distribution of frequencies to the total power. Usually they are expressed as a percentage of their total sum. Refer to figure 9.7 for the time domain pattern.

**Table 9.4   Contribution of Frequencies to the Overall Power Spectrum (PS) and Power Spectral Density (PSD) of the Data Shown in Figure 9.8**

| Harmonic # | Frequency (Hz) | PS (%) | Cumulative PS (%) | PSD (%) | Cumulative PSD (%) |
|---|---|---|---|---|---|
| 1 | 0.35 | 81.35 | 81.35 | 69.81 | 69.81 |
| 2 | 0.70 | 8.44 | 89.80 | 20.81 | 90.62 |
| 3 | 1.05 | 7.82 | 97.61 | 5.86 | 96.49 |
| 4 | 1.41 | 1.36 | 98.97 | 2.79 | 99.28 |
| 5 | 1.76 | 0.56 | 99.53 | 0.32 | 99.61 |
| 6 | 2.11 | 0.31 | 99.84 | 0.22 | 99.83 |
| 7 | 2.46 | 0.10 | 99.94 | 0.10 | 99.93 |
| 8 | 2.82 | 0.04 | 99.98 | 0.04 | 99.98 |
| 9 | 3.17 | 0.01 | 99.99 | 0.01 | 99.99 |
| 10 | 3.52 | 0.00 | 99.99 | 0.00 | 99.99 |
| All the rest | 25 | 0.0049 | 100 | 0.0046 | 100 |

The frequency of the first harmonic is 0.352 Hz.

Another way of examining the contribution of frequencies to the signal is through the calculation of the **power spectral density** (PSD) of the signal (Thomson, 1982; Welch, 1967). The power spectrum for a given signal is unique; it represents the square magnitude of the signal's Fourier transform. On the other hand, the PSD is an estimate of the power spectrum of a signal if the signal was to be repetitively collected multiple times (infinitely). However this is practically impossible as most biomechanical signals are collected just once (figure 9.8b). Table 9.4 shows the percentage amplitudes of PSD and cumulative PSD of the signal of the initial 10 harmonics in Lanshammar's study. In this particular example, the main difference between PS and PSD is the values of the first two harmonics. The first harmonic has 81.35% of the signal's PS, but 69.81% of the signal's PSD. The second harmonic has 8.44% of the signal's PS, but 20.81% of the signal's PSD. However if we add them, the first two harmonics have a total of 89.80% of the signal's PS and 90.62% of the signal's PSD. This indicates that these two harmonics have a different contribution but have a similar sum; the rest of the harmonics are very similar.

Practically, if we collect the same signal many times, the power spectral density is the mean of the power spectrums. Theoretically, the difference between the power spectrum and power spectral density is caused by random noise. Therefore, if the collected signal is purely deterministic (without noise), then the power spectrum and power spectral density are the same. The power spectrum and power spectral density of kinematic and kinetic data from human movements generally have a high power at the lower frequencies and low power at the higher frequencies, usually considered as added noise (Winter et al., 1974). In figure 9.8, for example, the power spectrum and power spectral density of the Lanshammar data have a very similar pattern. This is not the case for electromyography (EMG) data, however, because they represent a random process that makes their power spectrum "noisy" (figure 9.9) as opposed to the estimated power spectral density, which represents a relatively smooth process. This is the case because the power spectral density estimates a mean value of repeated EMG signals, thus reducing dramatically the noise effect on the power spectrum.

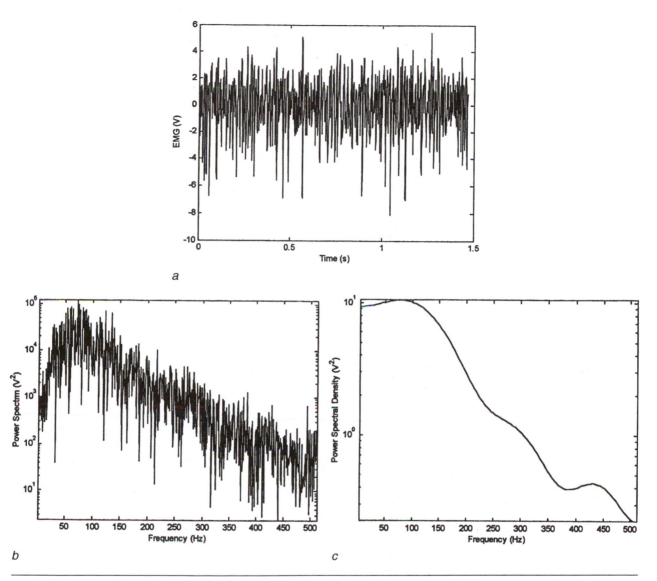

**Figure 9.9** *(a)* An electromyographic signal of the vastus medialis muscle at 90° of knee flexion at 60% of maximum voluntary contraction for 1.5 s. *(b)* Its PS is very noisy as opposed to its *(c)* PSD, which is smoother because it estimates a mean value of repeated PS.

The power spectrum and the power spectral density have been used to differentiate the ground reaction forces of healthy and pathological walking (Giakas and Baltzopoulos, 1997c; Giakas et al., 1996; Stergiou et al., 2002). For example, figure 9.10 presents the mediolateral ground reaction force pattern of two male subjects of similar mass; one is healthy, and the other has anterior knee pain that makes walking "stiff," as demonstrated by the flat pattern of the ground reaction force period. The minimum and maximum values of the two patterns are nearly the same; however, a visual inspection indicates that the pattern of the healthy subject is more oscillated than the anterior knee pain pattern. This is confirmed by the power spectrum values (table 9.5) demonstrating that the healthy subject exhibited more frequency components than the anterior knee pain subject. To quantify the oscillations of the ground reaction force data, the highest frequency needed to split the signal into two parts (e.g., 99% of the signal and 1% of the signal) was used in this example.

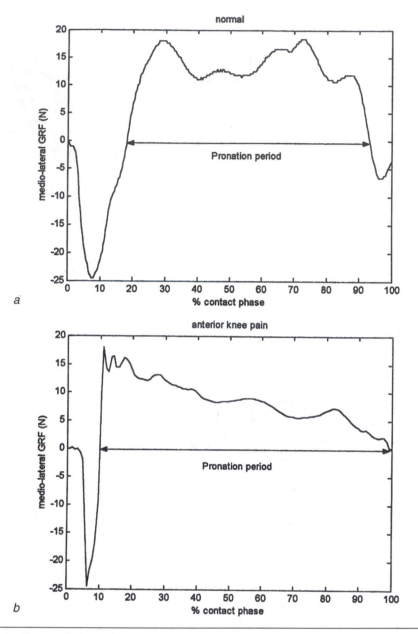

**Figure 9.10** The mediolateral ground reaction force pattern of two male subjects of similar mass. *(a)* The healthy subject exhibits a more oscillated pattern than *(b)* the other subject, who has anterior knee pain.

| Table 9.5 The Cumulative Power Spectral Density Normalized to 100% of Total Power of the Mediolateral Ground Reaction Force During Walking | | |
|---|---|---|
| **Frequency (Hz)** | **Anterior knee pain (%)** | **Healthy (%)** |
| 2 | 77.7585 | 70.2360 |
| 4 | 99.0670 | 95.8599 |
| 6 | 99.2775 | 98.0689 |
| 8 | 99.5952 | 99.2720 |

With a criterion of 99% of total power, the anterior knee pain subject required just 4 Hz while the healthy subject exhibited more oscillation and required 8 Hz.

**Example 9.2**

Table 9.4 provides another example, in which it is evident that the first three harmonics representing 1.05 Hz contain approximately 97% of the signal. Therefore, if we had the criterion of 95% of signal power spectrum (or power spectral density), this would be the frequency of 1.05 Hz for Lanshammar's signal (the frequency of 1.05 Hz contains 97.61% of PS or 97.49% of PSD). A similar signal that has more oscillations or more noise is expected to require a higher frequency in order to contain 95% of the signal.

Using this method we can distinguish patterns that have similar time domain parameters (minimum and maximum values) but different oscillation patterns. Also, we can quantify the relationship between the deterministic signal and the added noise. For example, in figure 9.8, the power spectrum of the Lanshammar data becomes nearly flat after the frequency of 4 to 5 Hz, which is a known pattern for noise.

The power spectrum and power spectral density have also been used to develop a criterion for automatic selection of a cutoff frequency in kinematic data filtering (DAmico and Ferrigno, 1990; DAmico and Ferrigno, 1992; Giakas and Baltzopoulos, 1997a; Jackson, 1979; Winter, 1990; Yu, 1989; Yu et al., 1999). Since this is one of the most important processes of kinematic data analysis in human movement, we discuss it in detail in a subsequent section of this chapter.

Furthermore, the power spectrum and spectral density are widely used in the analysis of EMG signals, especially during fatigue protocols. It has been noticed (DeAngelis et al., 1990; DeLuca, 1984; Linsseen et al., 1993) that the firing rate of the muscles becomes slower as we get fatigued. This is a special case in which the frequency components of a signal change over time. Another section toward the end of this chapter is dedicated to this issue.

Data processing requires that a major assumption be satisfied—that the data have been collected with the appropriate sampling frequency. In most cases the problem is that the equipment used cannot sample with the high sampling frequency required. This results in loss of essential information. The following section is devoted to this important issue.

# Data Sampling

Prior to the analysis of the signals associated with human movement, one must ensure that the data have been collected properly. For example, if we are interested in knowing the progress of the horizontal velocity of a long distance runner during a marathon, then one sample per second will probably be adequate. However, if we are interested in examining the knee joint flexion/extension kinematics during each step for the same athlete, then a far higher sampling rate is required. One may ask "How fast should I sample the data, or in other words, what should be my sampling frequency?" The answer is given by Shannon's sampling theorem (Hamill et al., 1997), which states that a band limited at the frequency of $f$ *(Nyquist frequency)* signal should be captured at a rate of at least $2f$. Observation of the wheels of a bicycle or a car as they rotate provides a practical example of this theorem. Let's assume that human vision samples approximately 14 pictures per second (14 Hz). When the bicycle moves slowly, we can clearly see the wheel rotating forward. This is so because the wheel rotates with a sampling frequency less than 7 Hz. If the wheel is rotated with a higher angular velocity (thus increasing the frequency), for example with 13 Hz, then we will see the wheel rotating backward. The reason is that the sampling frequency of our eyes is not high enough to capture the fast movement of the wheel. If the wheel moves even faster (rotates with 15 Hz), we would perceive a rotation of 1 Hz; however, we would not be able to understand that the wheel has also completed one cycle.

Practically, in human movement this problem exists only with kinematic analysis (i.e., video) because other types of analyses (i.e., force platform, EMG) entail sampling with very high sampling frequencies (>1,000 Hz). Currently, kinematic analysis can be completed with cameras that can sample at a range of 50 to 400 Hz. It is always a good suggestion to use the highest sampling rate without dramatically limiting the field of view of the camera. When the sampling rate is increased to a level higher than the camera's standard sampling rate, the resolution or field of view is decreased accordingly. With the advancement of technology, faster and higher-resolution cameras will be developed, allowing higher sampling frequencies. Very high sampling rates, however, require more storage space. It is expected that technology will solve the problems concerning sampling rates and data storage.

A very high sampling frequency creates a practical difficulty when the actual signal develops slowly such that there is no change in the recorded signal for some frames. For example, the measurement of the center of pressure to assess balance using the force platform is a relatively constant signal. If the sampling frequency is extremely high for this type of movement (>1,000 Hz), then for a subject with good balance there could be periods in which the center of pressure does not change (this could also occur because the analog-to-digital converter has a low resolution or because the force platform ranges have not been set properly). Nevertheless, the driving force for selecting an efficient sampling frequency is the actual movement under investigation, the type of equipment being used, and whether the digitization is manual or automatic. For example, a 50-Hz camera would probably suffice to record walking data to examine knee joint kinematics (Vaughan et al., 1992). However, this frequency will not be enough to investigate the kinematics of more dynamic activities such as javelin throwing (table 9.6) (Bartlett et al., 1996; Brunt et al., 1999; Cavanaugh et al., 1999; Dingwell and Cavanagh, 2001; Ferdjallah et al., 1999; Onell, 2000).

| Table 9.6    Examples With Recommended Minimum Sampling Frequencies | | |
|---|---|---|
| **Movement** | **System** | **Sampling frequency (Hz)** |
| Balance | Video | 50 |
| | Force plates | 50 |
| Walking | Video | 50-100 |
| | Force plates | 200-500 |
| Running-sprinting | Video | 100-200 |
| | Force plates | 1,000 |
| Javelin throwing | Video | 200-300 |
| | Force plates | 1,000 |
| Muscle contraction | EMG | Equal or higher than 1,000 |

Data from Bartlett et al., 1996; Brunt et al., 1999; Cavanaugh et al., 1999; Dingwell and Cavanagh, 2001; Ferdjallah et al., 1999; Onell, 2000.

A sampling frequency lower than the Nyquist frequency will hide important information. Figure 9.11 shows the vertical displacement of a marker positioned at the heel (lower part of the calcaneus) during three overground walking cycles using two sampling frequencies, 120 Hz and 20 Hz. With the lower frequency we were able to

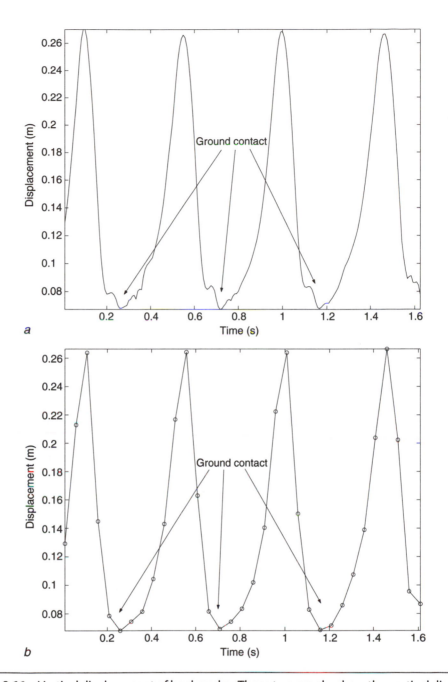

**Figure 9.11** Vertical displacement of heel marker. These two graphs show the vertical displacement of a marker placed at the heel during walking sampled at *(a)* 120 Hz and *(b)* 20 Hz. The circles in *(b)* are the data points, which are connected with straight lines. Notice the small oscillations due to the contact with the ground. These oscillations are lost when we sample at 20 Hz instead of 120 Hz.

capture the main movement patterns; however, the fine oscillations during the contact with the ground were lost.

To estimate the Nyquist frequency, one collects the data with a high sampling rate. Examination of the power spectrum of the signal provides information on the highest frequency in the signal. If, for example, the highest frequency of the signal is 17 Hz and the data were collected with 100 Hz, in the frequency domain analysis we will be able to see some power at the highest frequency of 50 Hz. This is so for two reasons—first, because there is always noise superimposed in any measurement, which is evident in all frequencies, and second, because the signal is digital. The procedure of expressing frequencies normalized to total power, as explained in the previous section, is useful here. If the cumulative power up to the frequency of 17 Hz represents more than 99.99% of the signal, then we can be sure that the signal will be accurately collected even with a sampling frequency of 34 Hz. For example, in table 9.4, the first 3.5 Hz covers more than 99.99% of the signal power. This is approximately the Nyquist frequency. Theoretically, sampling at a rate of twice the Nyquist will result in a valid frequency analysis, but the time domain signal will not necessarily be well represented. To be on the safe side, though, one has to collect data at least four to six times this frequency because this is a post hoc analysis and a new "similar" movement could have significant components in a higher-frequency band.

In this section we discussed only one application of the DFT. Earlier we also mentioned the inverse DFT. The next section describes the importance of inverse DFT, in which the power of unwanted frequency components is assumed to be noise and therefore zeroed. This process, called filtering, remains a complex issue in human movement data analysis.

## Biomechanical Data Filtering

In human movement, the general definition of filtering is the reduction or amplification of certain components of the signal (for a review see Wood, 1982). Any technique or transformation performed in either the time or the frequency domain that has such an effect is called filtering. Another term used is "smoothing," which usually refers to a way of "forcing" the data to fit a polynomial or a spline model (Dohrmann et al., 1988; Pezzack et al., 1977; Vaughan, 1982; Woltring, 1985).

In the frequency domain, filtering changes the values of the Fourier coefficients and therefore the power spectrum also. This is performed in a controlled manner using the value of a "key" frequency. This frequency, known as the cut-off frequency, signals the filter to keep or remove subsequent (or remaining) frequency spectrums. The function used to change the coefficients of specific frequency components is called the *transfer function*. Depending on the type of transfer function, certain frequencies are emphasized and other frequencies are de-emphasized. The transfer function of the filter is characterized by three bands: pass, transition, and stop. The pass band allows all the covered frequency components to remain unchanged. Subsequently, the transition band progressively decreases the power of the frequency components that it covers. Finally, in the stop band, all the remaining frequencies are drastically reduced or eliminated. The order of the filter defines the sharpness of the filter, which is indicated by the width of the transition band. A higher order will sharpen the filter. Figure 9.12 shows the power spectral density of the Lanshammar data (also shown in figure 9.8b, p. 231) upon which is superimposed a transfer function that affects the frequencies above 4.5 Hz. During the filtering process, the power spectral density is multiplied by the transfer function. This results in unaffected frequency components below 4.5 Hz, gradually decreased frequency components in the area of 4.5 Hz→7 Hz, and elimination of components above 7 Hz.

In the previous section we examined the cumulative frequency content of the Lanshammar data (table 9.4) and found that the first three harmonics (or 1.05 Hz) contain approximately 97% of the signal's power spectrum. Let's assume that the collected signal has some low-power random noise superimposed on it and distributed evenly throughout the power spectrum. The lower frequencies shown in figure 9.8a and table 9.4 have very high power relative to the higher frequencies, which means that in the lower-frequency spectrum the predominant part is the actual real signal that Lanshammar aimed to collect (we use this phrasing although the original signal was collected by Pezzack and colleagues [1977]). It is therefore concluded that the higher frequencies are predominantly noise that needs to be removed.

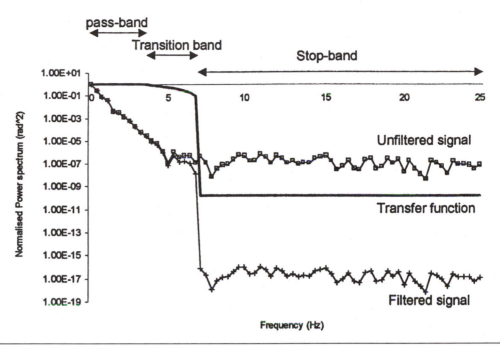

**Figure 9.12**   The three bands of a low pass filter.

## Example 9.3

Let's try to filter the signal with a low pass transfer function that will allow the first three harmonics to pass unaffected and zero all the rest of the harmonics (this is a Fourier filter). In this case the cutoff frequency is 1 Hz. Similarly, allowing more harmonics increases the cut-off frequency (figure 9.13). Practically, the higher the frequency, the more closely the filtered signal approaches the original signal because more frequencies are allowed to pass. It is obvious that if the cutoff frequency reaches its maximum (half the sampling frequency), then the filtered signal will be exactly the same as the original signal.

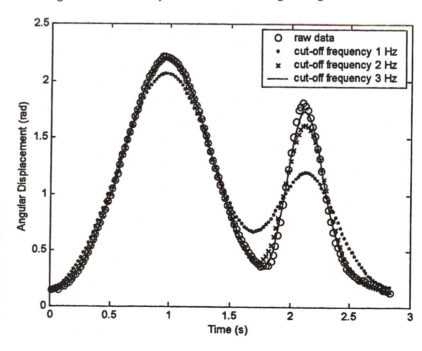

**Figure 9.13**   The Lanshammar data presented in figure 9.8 (p. 231) and filtered here with different cutoff frequencies. A higher cutoff frequency "pushes" the reconstructed signal toward the raw signal. The cutoff frequency of 3 Hz provides an acceptable reconstruction of the angular displacement data.

In this example, the filter we used was very sharp, without a transition band. This means that the power for some (lower) frequencies was multiplied by 1 and the power for the remaining frequencies was multiplied by 0, so there was not a progressive decrease of the power spectrum (multiplication by 0.9, 0.8, 0.7, etc.) around the cut-off frequency. This filter is also called optimal or square because its transfer function creates an orthogonal parallelogram as shown in figure 9.14.

Among the four types of filters, displayed in figure 9.14, those used frequently in human movement are the low pass filter for kinematic data filtering and the band pass filter for EMG data filtering (Lariviere et al., 2001; Winter et al., 1974). Each type is used depending on the frequencies that we want to eliminate. To eliminate the higher frequencies we use a low pass filter; to eliminate the lower frequencies we use a high pass filter; to remove frequency components at the central part of the power spectrum we use a band stop filter; and to keep central frequency components and remove the components at the lowest and the highest bands of the spectrum we use a band pass filter. Figure 9.14 represents these filters graphically. In the low pass and high pass filters, $Fc$ indicates the cut-off frequency. In the band stop and band pass filters, $Fc_1$ and $Fc_2$ define an area in the frequency where specific frequency bands are to be removed or kept.

These types of filters are used depending on the power spectral properties of the signal and the associated noise. If the signal is predominant in the lower frequencies

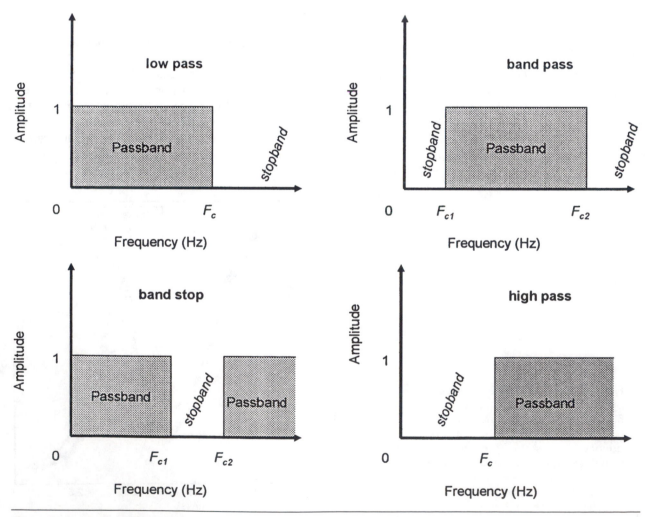

**Figure 9.14** The four filter types.

and the noise is in the higher frequencies (e.g., see figure 9.8*a* for the elbow flexion/extension angular data), then the low pass filter is used. In contrast, the EMG signal is a pseudo-random process in which the lower frequencies are created by the movement of the electrodes. For EMG signals, therefore, a high pass or a band pass filter is used. For example, figure 9.15 shows the EMG activity of the vastus medialis of a healthy subject during a maximum isokinetic concentric knee extension trial. When the vastus medialis is passive, the EMG signal does not fluctuate, making a very thin line pattern (resting period in the graph); when it is active, the EMG signal is highly oscillated (contraction period). During the passive period, the unfiltered signal is not a straight line; instead it has a slow-moving pattern that is noise created by the movement of the wires between the electrodes and the preamplifier. This slow-moving pattern is a low-frequency noise that can be removed through use of a band pass or a high pass filter. We used a band pass filter that kept the frequencies between 15 Hz and 499 Hz (Perry et al., 2001) and removed the unwanted low-frequency components (filtered signal is elevated by 1 V). Now the EMG signal shows a straight line during the resting period, as realistically it should. Most studies in which EMG data are collected use this type of filtering,

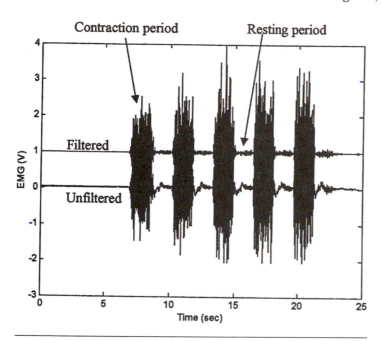

**Figure 9.15**   The electromyographic activity of the vastus medialis of a healthy subject during a maximum isokinetic concentric knee extension trial.

although the cutoff frequency range may vary (Aruin et al., 2001; Lariviere et al., 2001; Madeleine et al., 2001; Perry et al., 2001). The reasons for the variation are differences in the equipment used and the movements examined.

The greatest challenge, though, is the selection of an appropriate cutoff frequency. When we use a high cutoff frequency, the reconstructed-filtered signal is "forced" to get closer to the collected signal because more frequency components are allowed to pass. We do not know, however, if the additional frequency components allowed with a higher cutoff frequency are part of the signal or the noise (or both). On the other hand, a low cutoff frequency could perhaps eliminate useful frequency components that are part of the signal. For example, in figure 9.13, the cutoff frequency of 1 Hz will dramatically eliminate useful signal characteristics, especially in the second curve. Over the last 20 years, numerous investigators (Challis, 1999; DAmico and Ferrigno, 1990, 1992; Giakas and Baltzopoulos, 1997a, 1997b; Hatze, 1981; Woltring, 1986; Yu et al., 1999) have tried to solve this problem, which has no optimal solution because human movement signals have large variability of signal and noise characteristics. Stergiou and colleagues (1999) published a paper that indicates the importance of the appropriate selection of cut-off frequencies. More specifically, Stergiou and his colleagues (1999) examined the asynchrony between subtalar and knee joint angular movement during running. They found that the rearfoot movement is described by a bimodal angle curve as opposed to the unimodal angle curves presented in the literature (Stergiou and Bates, 1997). This observation was possible only after careful selection of an appropriate cut-off frequency.

A number of resources are available on the Internet through the official Web site of the International Society of Biomechanics (www.isbweb.org) to help users with the selection of an appropriate cut-off frequency. Various software programs including generalized cross-validated splines and GGPSA (http://ggiakas.users.uth.gr) provide their own guidelines for users. The latter was built based on the idea that there is not a perfect solution for all data sets. Instead, the user has to make some adjustments with the parameters selected from the software program. This program has a Help file that can be used to make the adjustments.

In addition to the cutoff frequency, there are other filter properties that need to be identified. First, one has to select the type of filter. The most popular filter in human movement research is the Butterworth filter (Winter et al., 1974). This filter shifts the timing of the data, and therefore the same filter is also applied in the reverse direction to remove this time shift. This means that when a filter with a cutoff frequency of $fc$ is applied to a time series $x_1, x_2, \ldots, x_n$ creating the filtered time series $y_1, y_2, \ldots, y_n$, the same filter with the same cutoff frequency $fc$ is applied to the filtered sequence $y_n, y_{n-1}, \ldots, y_2, y_1$. Second, the order of the filter is a property that indicates how sharp the transfer function will be. A high order will have a sharper transfer function than a low order; this is not really a big issue as the second order operates reasonably well for biomechanical data (Winter et al., 1974). The effect of the order is mainly visible in the second derivative (described in the next section). Figure 9.16 shows the calculated acceleration of a free-falling ball as recorded by Vaughan (1982) after filtering with a Butterworth digital filter (cutoff frequency 4 Hz). The acceleration should be equal to the gravitational acceleration ($-9.8$ m/s$^2$), but as one can see, it is not. This graph demonstrates the effects of different orders of the filter. The increase of order from 2 to 4 and then to 6 introduces unwanted oscillations in the acceleration pattern.

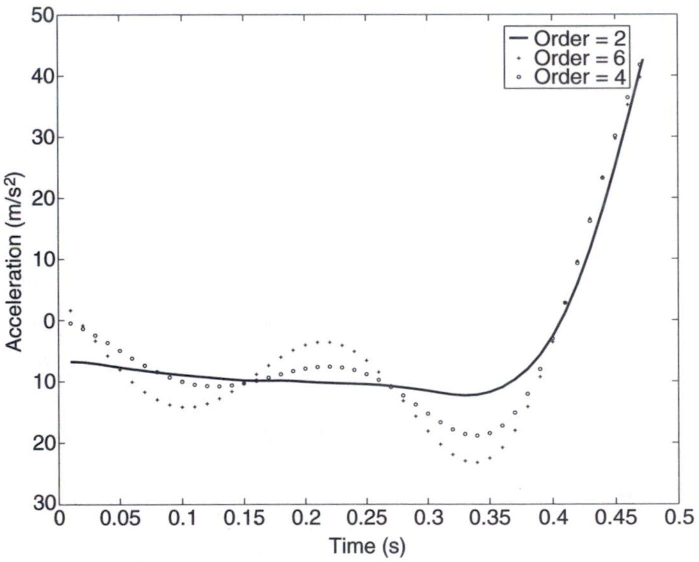

This filter effectively becomes a fourth order because it has to be applied in the forward and the backward sequence of data to avoid a time shift. This type of filtering also creates a problem at the edges of the signal because the calculation of a filtered data point requires previous filtered and future unfiltered data points; at the beginning of the signal, there are no previous filtered data points and at the end of the signal there are no future unfiltered data points. This problem led to the development of various extrapolation techniques such as linear, reverse mirror, linear prediction, and polynomial (Giakas et al., 1998; Smith, 1989; Vint and Hinrichs, 1996). Again, the problem at the edges is mainly visible in the second derivative. Again in figure 9.16, one sees that the problem of a nonconstant acceleration around the area of $-9.84$ m/s$^2$ is more evident at the edges. Especially at the end of the signal, the acceleration has been dramatically distorted.

**Figure 9.16** The calculated acceleration of a free-falling ball as recorded by Vaughan (1982) after filtering with a Butterworth digital filter (cutoff frequency 4 Hz).

An important question that arises from the preceding discussion, especially for kinematic data processing, is the following: If the signal is relatively "smooth" and the power of the signal is much higher than the power of the noise, then do we need to filter the signal? Especially for kinematic data, we are also interested in the higher derivatives such as velocity and acceleration. During the differentiation process, the power of the noise, which is predominant in the higher frequencies, is highly amplified. In contrast, the signal is also amplified but on a much lower scale. Therefore at the higher derivatives, the properties of the signal are masked by the properties of the noise. This process makes filtering an important signal-processing tool.

# The Differentiation Process

The differentiation process is used to calculate the higher derivatives of a signal. The differentiation of a signal can be performed in the time domain using finite differences or in the frequency domain through the analytical differentiation of Fourier series. In table 9.7, we have included the analytical higher derivatives of the basic trigonometric functions. In the analytical differentiation of the sine and cosine functions, the frequency after derivation is multiplied by the amplitude of the function. Therefore the frequency components that are predominant at the higher frequencies will be highly amplified. Since at the higher frequencies of a signal there is noise, unfortunately the differentiation process increases the power of noise much more than the power of the signal.

**Table 9.7  Analytical Calculation of the Derivatives of the Sine and Cosine Function That Are the Basis of the Fourier Transform**

| Signal | 1st derivative | 2nd derivative |
|---|---|---|
| $S = a \times \sin(bx)$ | $S_{(1)} = -b \times a \times \cos(bx)$ | $S_{(2)} = -b \times b \times a \times \sin(bx)$ |
| $S = a \times \cos(bx)$ | $S_{(1)} = b \times a \times \sin(bx)$ | $S_{(2)} = -b \times b \times a \times \cos(bx)$ |

The frequency after derivation is multiplied by the amplitude of the function.

## Example 9.4

In this example, for the sake of simplicity we use only the sine function and the signals described in figure 9.5a (p. 226). Let's assume that the reference signal is described by the function sin(1x) (low frequency-high amplitude) (figure 9.3) and the superimposed noise is described by the function 0.02sin(15x) (high frequency-low amplitude) (figure 9.4). The overall signal (figure 9.5a) therefore is described by the function

$$S = \text{real signal} + \text{superimposed noise} = 1 \times \sin(2\pi x) + 0.02\sin(2\pi 15x) \quad (9.3)$$

Let's examine how the ratio between the amplitude of the signal and the amplitude of noise develops during the differentiation process. For the signal S, the signal-to-noise ratio (SNR) is

$$\text{SNR}_S = [\text{amplitude-power of the signal}] / [\text{amplitude-power of the noise}] = 1 / 0.02 = 50 \quad (9.4)$$

This ratio is a measure of the quality of the signal collected. A higher SNR corresponds to a better quality of the signal. A pure sinusoidal signal without noise has an infinite SNR because the amplitude of noise is zero. Every single measure that we do, though, is susceptible to error, so this is only a theoretical consideration.

In the analytical first derivative of this signal, the new amplitude of each trigonometric function is the product of the amplitude with the frequency. Hence, the first derivative of the signal described, along with its SNR values, is

$$S_{(1)} = \cos(2\pi x) + 15 \times 0.02 \cos(2\pi 15x) = 1 \times \cos(2\pi x) + 0.3 \cos(2\pi 15x) \quad (9.5)$$

Therefore the SNR of the first derivative of signal S is:

$$\text{SNR}_{S(1)} = 1 / 0.3 = 3.33 \quad (9.6)$$

and the second derivative will be:

$$S_{(2)} = -\sin(2\pi x) - 15 \times 0.3 \sin(2\pi 15x) = -1 \times \sin(2\pi x) - 4.5 \cos(2\pi 15x) \quad (9.7)$$

$$\text{SNR}_{S(2)} = 1 / 4.5 = 0.22 \quad (9.8)$$

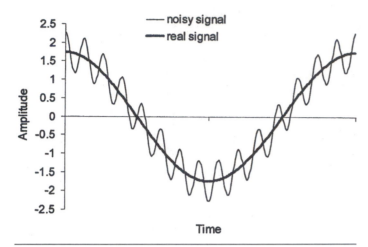

**Figure 9.17** The first derivative (velocity) of the signal presented in figure 9.5 on page 226 (thin line), which contains very low levels of noise, and the reference first derivative (thick line) of the clean signal. Compare the oscillations with figure 9.5. Note the dramatic increase of the noisy component of the signal.

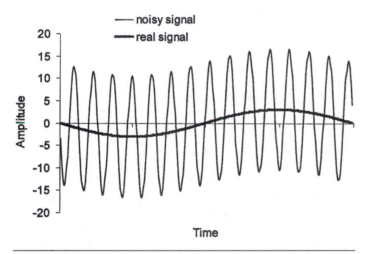

**Figure 9.18** The second derivative (acceleration) of the signal presented in figure 9.5 on page 226 (thin line), and the reference second derivative (thick line) of the clean signal. Compare the oscillations with those in figure 9.5. Note the dramatic increase of the noisy component of the signal.

The SNR values indicate that the quality of the signal has been dramatically reduced, and this is clearly demonstrated by their plot (figures 9.17 and 9.18). Especially in figure 9.18 we can see that the real signal has been masked and that the noise is the most representative part of the overall signal.

The amplification of the higher frequencies creates a difficult problem in human movement research because the signal and noise frequency components are not clearly identified. Returning to the Lanshammar data, the power spectral density of the angular velocity and acceleration demonstrate the problem created by the noise at the higher frequencies of the signal. In figure 9.19*a* the power of the higher frequencies tends to become equal to the power of the lower frequencies, while in figure 9.19*b* the power of the lower frequencies is much smaller than the power of the higher frequencies. These power spectral densities should be graphically examined together with figure 9.8*b* (p. 231), which shows the power spectral density of the displacement data. The result of the high-frequency amplification in the time domain of the acceleration pattern presented in figure 9.19*b* is demonstrated by the noisy acceleration pattern (continuous line) superimposed on the real acceleration pattern (asterisks) in figure 9.20. This figure shows that if we do not filter the data before differentiation, selected time domain parameters in the acceleration pattern will be wrong; for example, at time near 2 s the local minimum would be approximately $-180$ rad/$s^2$ as shown by the continuous line instead of approximately $-95$ rad/$s^2$ as shown by the asterisked line. This problem has been previously addressed by several investigators (Hatze, 1981, 1990; Woltring, 1986, 1995), and is the reason it is necessary to filter the data before differentiation (Giakas and Baltzopoulos, 1997b).

This problem can also affect the calculation of joint moments and powers using the inverse dynamics process. The inverse dynamics process involves the combination of kinetic, kinematic, and anthropometric data (Winter, 1990). The kinematic data used are the second derivatives of the joint rotations. Inaccurate estimations of the acceleration patterns due to inappropriate filtering will surely affect the joint moments and powers (Giakas, 1997).

## Joint Time-Frequency Domain Representations

A major assumption in the previous sections of this chapter was that the signal being examined is a stationary process. A signal is stationary when its statistical and frequency characteristics do not change over time. Although this assumption is true

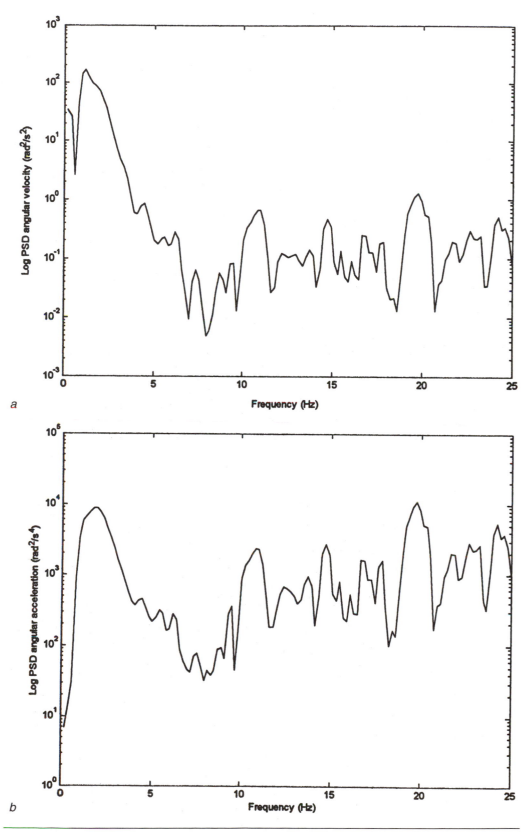

**Figure 9.19** The power spectral density of the *(a)* first and *(b)* second derivatives of the Lanshammar data previously presented in figure 9.7*a*. The power spectral density of the raw Lanshammar data (1982) is also drawn in figure 9.8*b*. The power at the higher-frequency band increases dramatically, especially at the second derivative in figure 9.19*b*.

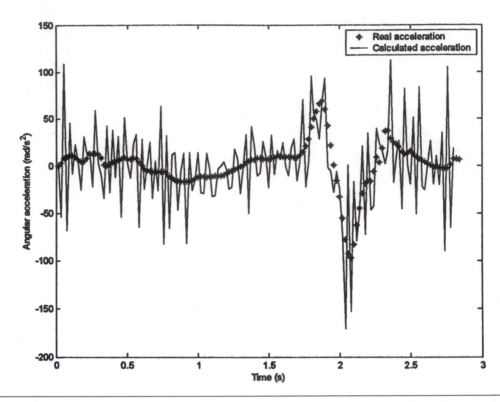

**Figure 9.20**   The real acceleration of the data by Lanshammar (1982) measured with an accelerometer (*) and the calculated second derivative of the angular displacement data (continuous line). These results clearly demonstrate why we need to filter the data if we want to calculate the higher derivatives.

in many cases, especially when the movement examined is short, there are numerous cases in which the assumption is violated. In these cases another approach is used. This approach (called **joint time-frequency domain** analysis) is an extension of the Fourier transform. Joint time-frequency analysis is used to describe signals that change their characteristics over time. These signals are called nonstationary. Music is a characteristic example of a nonstationary signal. When we listen to music, at any given time a range of frequencies is interpreted by our acoustic senses. These frequencies change over time; otherwise we would always hear the same sound. The firing rate during an isometric contraction decreases as the muscle fatigues (DeAngelis et al., 1990; DeLuca, 1984). The graphs in figure 9.21, *b* and *c,* are two 1-s windows of the long EMG signal shown in the graph in figure 9.21*a*. Figure 9.21*b* is the EMG during the initial few seconds of the trial, and figure 9.21*c* shows the EMG during the final period of the trial. The "slowness" of the EMG oscillations at the end of the trial point to a change in the frequency content. The power spectral density of the whole EMG signal (in figure 9.21*a*) shows that the maximum power is in the range of 30 to 60 Hz, as these frequencies contain more than 99% of the total signal power of the EMG (figure 9.22). However, the power spectral density does not provide any temporal information indicating, for example, what the spectrum at the initial, the central, and the final period of the contraction was. It might therefore be inappropriate to use the frequency domain analysis method to characterize these nonstationary signals.

In contrast to frequency domain or time domain in which the frequency or time characteristics of the signals are known, other combined *time-frequency* domain methods have been developed (Allen and Rabiner, 1977; Kayhan, 1991) and applied in

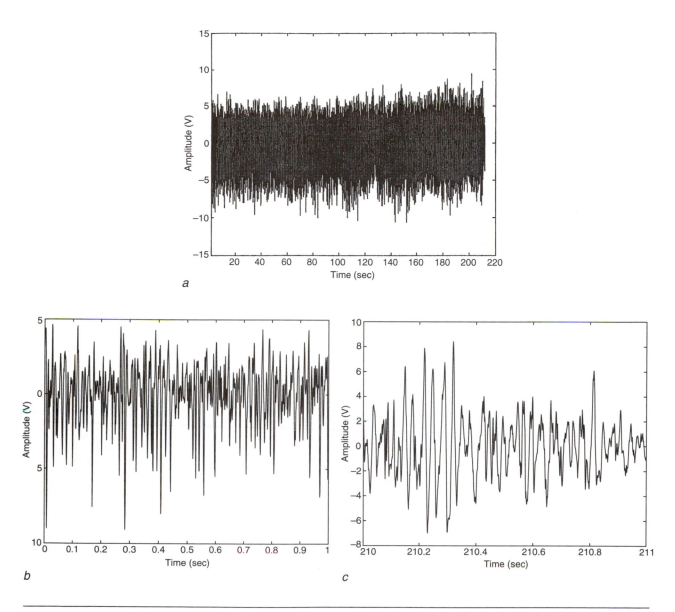

**Figure 9.21**  A fatigue protocol: *(a)* electromyographic signal of an isometric contraction of the vastus medialis at 90° of knee flexion at 60% of maximum voluntary contraction for 210 s. The signal at the initial phase of the recording is highly oscillated *(b)* as opposed to that at the final part of the recording *(c)*.

human movement to estimate the change of frequency as a function of time. A simple time-frequency domain method is the windowed short-time Fourier transform. The windowed short-time Fourier method to estimate the combined time-frequency domain characteristics starts with identification of the size of the window to be used, which controls the time and frequency resolution. This procedure is used in EMG signals to calculate the drifting of the median frequency over time after a prolonged isometric contraction (DeLuca, 1984). In this method the overall signal is divided into smaller continuous or overlapping parts (windows) (figure 9.23). In each window it is assumed that the signal is stationary, and the Fourier transform is applied to calculate the power spectrum. In this way a number of "localized" frequency domain representations will provide a three-dimensional (3-D) time-frequency domain analysis

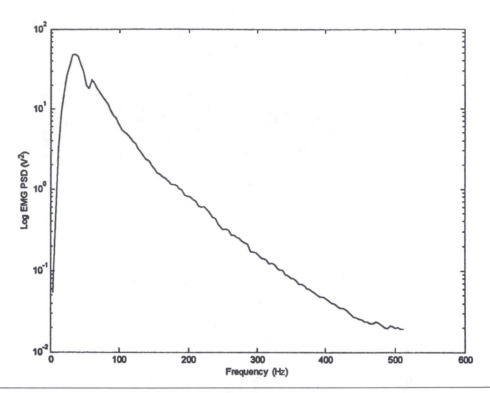

**Figure 9.22** The power spectral density of the electromyographic signal previously presented in figure 9.21*a*. Note that the scale is logarithmic. The range of frequencies between 30 Hz and 60 Hz contains more than 99% of the total signal power. This graph does not provide any temporal information.

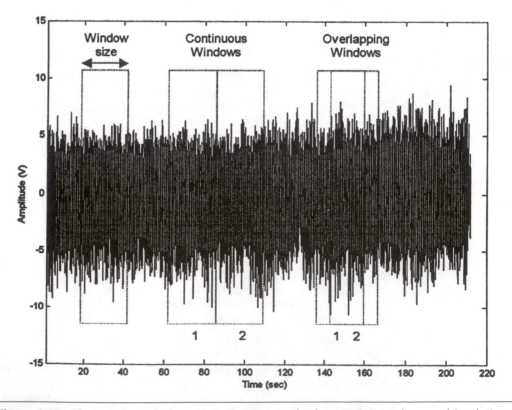

**Figure 9.23** The windowed short-time Fourier method to estimate the combined time-frequency domain characteristics.

(figure 9.24). This 3-D plot indicates the change of frequency content over time. In a fatigue protocol, the firing rate decreases as a function of time. This decrease relates to the decrease of the median frequency (the frequency that splits the power spectrum into two halves) of the power spectrum of each window (DeAngelis et al., 1990; DeLuca, 1984; Linsseen et al., 1993). The median frequency for each window can be plotted against time and can demonstrate the drifting of the power spectrum toward the lower frequencies, which in turn indicates the drifting of the firing rate due to fatigue (figure 9.25).

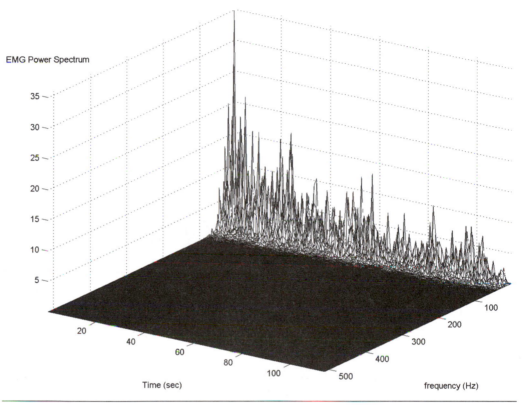

**Figure 9.24**   The time-frequency distribution of the electromyographic signal in figure 9.21 shows the concentration of signal power at the lower frequencies as the muscle fatigued. The power also increases with time. This graph is shown to demonstrate the change of frequencies over time.

Thus, the joint time-frequency domain transform provides information in both the time and the frequency domains. The combined time-frequency domain representation is based on a theory that describes the amount of information we are able to get in both domains simultaneously (Munger, 1998). This theory, also known as the *scaling uncertainty theory,* states that in order for us to obtain more information in the frequency domain, the information in the time domain has to be narrowed and vice versa. For example, in the windowed short-time Fourier transform described in figure 9.23, a 1-s window was used. With an increase in the window size (e.g., to 2 s), the frequency domain representation part will be more accurate than before, at the expense of the time domain representation of the signal (lower resolution).

Time-frequency domain analysis has been used mainly in the analysis of EMG signals to aid in understanding the mechanisms of muscle fatigue during prolonged isometric contractions (Merletti and DeLuca, 1990), as well as in analysis of the center of pressure during prolonged postural sway (Ferdjallah et al., 1999; Schumann et al., 1995). Recently, however, more sophisticated methods of time-frequency domain analysis have also been used to filter nonstationary kinematic data (Giakas et al.,

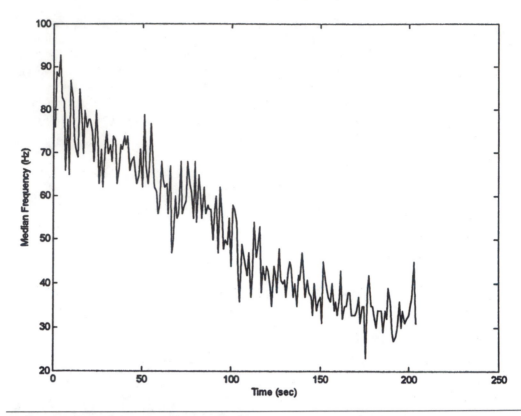

**Figure 9.25** The development of the median frequency as an effect of a decreasing firing rate as the muscle gets fatigued. This graph was produced from figure 9.24 by calculating the median value of every single Fourier transform. It shows the development of the median frequency of the EMG signal as a function of time. The median frequency was decreased from an initial value of approximately 80 Hz to 35 Hz at the end of the trial.

2000; Wachowiak et al., 2000). For example, nonstationary kinematic data are collected when the movement involves impacts such as landing and jumping, or release of an object such as the javelin.

## The Wigner Function

The Wigner distribution $W(t,\omega)$ is a joint time-frequency analysis method to identify signal characteristics in both the time and frequency domains simultaneously (Claasen and Mecklenbrauker, 1980a, 1980b; Cohen, 1989). The equation to calculate the Wigner function of a signal $s$ is:

$$W_s(t,\omega) = \frac{1}{2\pi} \int_{-\infty}^{\infty} s\left(t + \frac{\tau}{2}\right) s^*\left(t - \frac{\tau}{2}\right) e^{-j\omega\tau} \, d\tau \qquad (9.9)$$

where $t$ and $\omega$ are the time and frequency variables, respectively; $s(t)$ is the signal; $s^*$ is its complex conjugate; $\tau$ is a time lag; and $j$ is the imaginary unit. The Wigner distribution possesses maximum time and frequency resolution. Figure 9.26 shows the acceleration time domain pattern (figure 9.26*a*) and combined time-frequency domain analysis using the Wigner function of the effects of a soft (figure 9.26*b*) and a hard mat (figure 9.26*c*) placed on the ground during heel impact after landing from a jump. The highest frequency in the soft mat case was approximately 20 Hz; that for the hard mat was approximately 120 Hz. With the use of this distribution it is possible

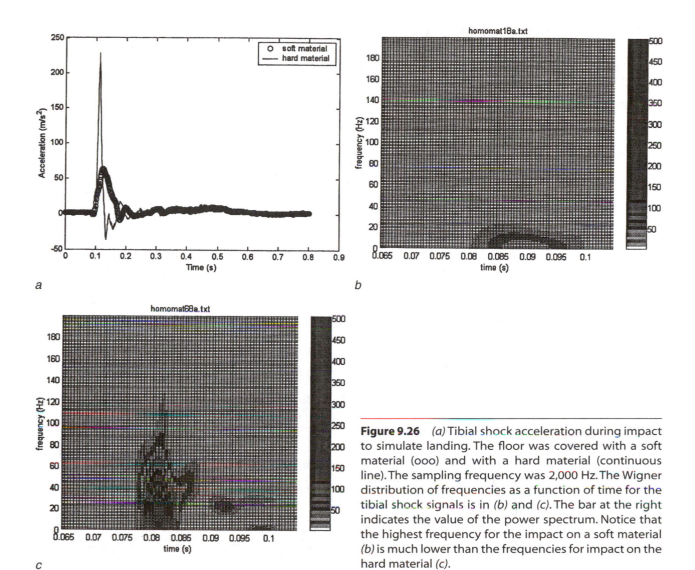

**Figure 9.26** *(a)* Tibial shock acceleration during impact to simulate landing. The floor was covered with a soft material (ooo) and with a hard material (continuous line). The sampling frequency was 2,000 Hz. The Wigner distribution of frequencies as a function of time for the tibial shock signals is in *(b)* and *(c)*. The bar at the right indicates the value of the power spectrum. Notice that the highest frequency for the impact on a soft material *(b)* is much lower than the frequencies for impact on the hard material *(c)*.

to understand not only when the highest peak occurred, but also the absorption of the frequency content as a function of time. The shaded bar at the right of the graphs in figure 9.26, *b* and *c*, indicates the value of the power spectrum

Recently, the Wigner function has been used in the data filtering of kinematic signals that are nonstationary (Giakas et al., 2000). The data are represented in a combined time-frequency domain, and a window that changes the cutoff frequency is applied. As a result, the high-frequency components (e.g., during impact) have a high cutoff frequency, while the lower-frequency components have a low cutoff frequency (Giakas et al., 2000).

In order to filter in the time-frequency plane, we consider a region R and a filtering function $F(t,f)$, which is equal to unity inside the region R and 0 outside. The filtered Wigner function is now given as follows:

$$W'(t,f) = W_s(t,f)\, F(t,f) \qquad (9.10)$$

The difference between this type of filtering and the conventional methods of filtering must be stressed. In the special case in which R is a strip with axes parallel to the time axis, this is a conventional frequency domain filter. However, for more general R, we have filtering in both the time and frequency domains, which is a deeper concept than filtering in frequency only or in time only. The choice of R will depend on the particular signal. The function $f$, which describes the R region, is given by the following equation.

$$f = \pm A_c \pm A_0 \exp[-\frac{(t-t_0)^2}{2\sigma_t^2}]Hz \qquad (9.11)$$

In the following paragraphs we describe all the parameters used in this equation.

$A_c$ is the width of the all-pass frequency band that corresponds to the "background" signal (the signal before or long after the impact). This can be calculated using the power spectrum assessment method (Giakas and Baltzopoulos, 1997a) applied to a relatively large part of the signal outside the impact region. $A_0$ is the frequency increase from $A_c$ to the highest cutoff frequency $(A_c + A_0)$ applied to the signal at the time of the impact $t_0$. We point out that the results depend weakly on this value. For example, if a lower value is chosen, the acceleration may not quite reach its peak value, but the overall signal will still be represented adequately.

The estimation of the center time point of the impact phase $t_0$ is important. This is the point at which the impact shown has the maximum acceleration. If this point is not known from the experiment (e.g., with force plate data), we have to estimate it from the noisy data. We can use several methods for this purpose. As an example, we can calculate finite differences and detect the impact point using simple slope-based criteria; we can then determine the parameter $t_0$ as the first time point in the data sequence that satisfies $[s(t_0) - s(t_{0-n})] > T$, T being a "slope-based" threshold, and $n$ the order of the finite differences. We use the first-derivative criterion; but other more sophisticated criteria, which might also involve higher-order derivatives, should be explored in future work.

$\sigma_t$ is the width of the "impact envelope," which corresponds to the duration of the impact phase. The duration depends on the type of the impact, conditions prior to the impact (e.g., kinematics), material of the bodies, and so on. A high-acceleration impact will require a smaller width value than a low-acceleration impact.

With use of joint time-frequency domain filtering, it is possible to keep the high frequencies during the impact phase and at the same time filter them out of the rest of the signal. The reasonable assumption has been made that in general, high frequencies are associated with noise and are therefore undesirable, except in the impact phase, where the high-frequency content is mainly due to rapid change in the useful signal.

The measured displacement signal is reconstructed back in the time domain using an "inverse" process for the Wigner transform (equation 9.9). The new signal represents the filtered displacement. The energy of the reconstructed signal is normalized with respect to the initial energy (i.e., the reconstructed signal has energy equal to 1) to compensate for the energy loss due to filtering. First-order finite differences are used to calculate the higher derivatives, that is, velocity and acceleration. The problem at the endpoints of the signal was resolved by the use of an appropriate extrapolation procedure (Giakas et al., 1998) before filtering.

This method has been further developed to make the selection of cutoff frequencies an automatic process (Georgakis et al., 2002, 2003). Currently the limitation of the method is its requirements in computing time, which are such that this method is not suitable for very long signals such as EMG data collected during a fatigue protocol. However it is anticipated that the method will be a valuable tool in the near future with the development of computer technology. Figure 9.27a shows displacement data, including an impact, of a marker placed at the tibia. These data were filtered using the method described (figure 9.27, b and c) and also using two conventional methods: the GGPSA method based on the LAMBDA method (DAmico and Ferrigno, 1990) (figure 9.27d), and the well-known cross-validated splines method (Woltring, 1986) (figure 9.27e). The superiority of filtering in the time-frequency domain for this type of signal is clear.

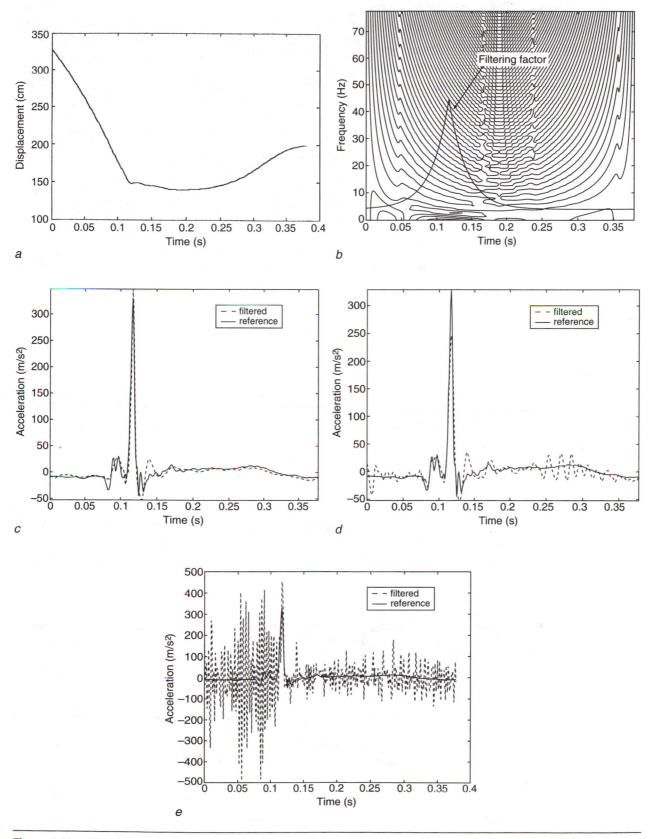

**Figure 9.27** Displacement data including an impact (a) and its Wigner function showing also the filtering performed (b). The reference data (dark line) superimposed on the filtered acceleration (light line) calculated with the Wigner function (c), with the GGPSA (d), and with cross-validated quintic splines (e).

There are also other time-frequency domain transforms that allow time-frequency analysis. These include wavelets (MATLAB includes a dedicated "wavelet" toolbox with many useful functions and a graphics user interface for signal analysis) and various Cohen class time-frequency representations (Cohen, 1989). An excellent reading and software (MATLAB and C/C++) resource is provided on the Internet by the signal-processing group from Rice (www-dsp.rice.edu/); this resource also includes MATLAB code for the Wigner function.

## Summary

Frequency domain analysis has received significant attention during the last few years by human movement researchers and has proved to be a valuable tool to distinguish signal characteristics among experimental groups and/or conditions, but also in data filtering. The purpose of this chapter was to provide an introduction to the reader who is not familiar with this type of processing using several examples from human movement research. We anticipate that this chapter will enable researchers to extend their "processing" skills and potentially improve the efficiency of their data analysis procedures.

## Work Problems

**1**

Create the function $16 \times \cos(2\pi 2x) + 6 \times \sin(2\pi 14x)$ using 1,000 data points (total time would be 1 s). Calculate the discrete Fourier transform of the function and provide the amplitude of the first 20 Fourier coefficients (only for the positive frequencies).

**2**

Create the function $7 \times \cos(2\pi 3x) + 10 \times \sin(2\pi 11x)$ using 100,000 data points (total time would be 1 s). Calculate the discrete Fourier transform of the function and provide the amplitude of the first 20 Fourier coefficients (only for the positive frequencies).

**3**

Calculate the power spectrum and power spectral density of the data provided by Dowling (1985). (The corresponding data can be found at www.unocoe.unomaha.edu/hper/bio/home.htm.) The sampling frequency is 512 Hz.

**4**

Calculate the highest frequency included in 90%, 95%, 99%, and 99.5% of the total signal's power for the Dowling data. The sampling frequency is 512 Hz.

**5**

Calculate the highest frequency included in 90%, 95%, and 99% of the total signal's power for the Pezzack et al. (1977) data. The sampling frequency is 50 Hz.

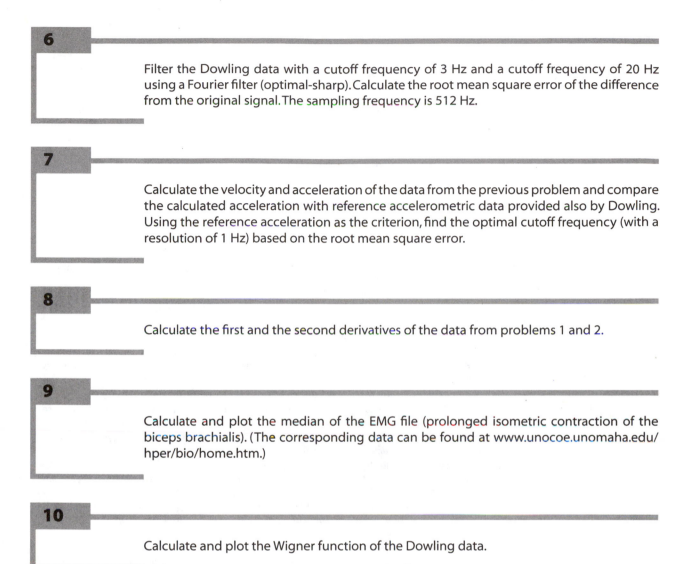

**6**    Filter the Dowling data with a cutoff frequency of 3 Hz and a cutoff frequency of 20 Hz using a Fourier filter (optimal-sharp). Calculate the root mean square error of the difference from the original signal. The sampling frequency is 512 Hz.

**7**    Calculate the velocity and acceleration of the data from the previous problem and compare the calculated acceleration with reference accelerometric data provided also by Dowling. Using the reference acceleration as the criterion, find the optimal cutoff frequency (with a resolution of 1 Hz) based on the root mean square error.

**8**    Calculate the first and the second derivatives of the data from problems 1 and 2.

**9**    Calculate and plot the median of the EMG file (prolonged isometric contraction of the biceps brachialis). (The corresponding data can be found at www.unocoe.unomaha.edu/hper/bio/home.htm.)

**10**    Calculate and plot the Wigner function of the Dowling data.

# Suggested Readings and Other Resources

## Books and Articles

Lynn, P.A., and W. Fuerst. 1994. *Digital signal processing.* New York: Wiley. A general resource for information on signal processing.

Peter T., and J. Williams. 1998. *The Fourier transform in biomedical engineering.* Boston: Birkhauser. For applications of the Fourier transform in biomedical engineering.

Wood, G.A. 1982. Data smoothing and differentiation procedures in biomechanics. In *Exercise and Sport Science Review,* ed. R. Terjung, 308-362. Philadelphia: Franklin Institute Press. For further information on biomechanical data filtering.

## Web Sites

www.dsp.rice.edu. The Web site of Rice's signal-processing group, for further information on time-frequency domain analysis.

www.fftw.org. Web site for FFT approximations, the official Web site for programming. The site includes many different computing language solutions.

www.isbweb.org. The International Society of Biomechanics Web site, listing further resources for data and software on digital filtering specifically for biomechanical data.

www.mathworks.com. The Web site for MATLAB, for help with MATLAB programming language.

# References

Allen, J.B., and L.R. Rabiner. 1977. A unified approach to short-time Fourier analysis and synthesis. *Proceedings of the IEEE* 65: 1558-1564.

Aruin, A.S., T. Ota, and M.L. Latash. 2001. Anticipatory postural adjustments associated with lateral and rotational perturbations during standing. *Journal of Electromyography and Kinesiology* 11: 39-51.

Bartlett, R., E. Muller, S. Lindinger, F. Brunner, and C. Morriss. 1996. Three-dimensional evaluation of the kinematic release parameters for javelin throwers of different skill levels. *Journal of Applied Biomechanics* 12: 58-71.

Bogner, R.E., and A.G. Constantinides. 1975. *Introduction to digital filtering.* Chichester, England: Wiley.

Braun, D. 2002. Quantum chaos and quantum algorithms. *Physical Review A* 65: art. no. 042317, part A.

Brigham, E.O. 1974. *The Fast Fourier transform.* Englewood Cliffs, NJ: Prentice Hall.

Brunt, D., S.M. Liu, M. Trimble, J. Bauer, and M. Short. 1999. Principles underlying the organisation of movement initiation from quiet stance. *Gait and Posture* 10: 121-128.

Cappozzo, A., and F. Gazzani. 1983. Comparative evaluation of techniques for the harmonic analysis of human motion data. *Journal of Biomechanics* 16: 767-776.

Cavanaugh, J.T., M. Shinberg, L. Ray, K.M. Shipp, M. Kuchibhatla, and M. Schenkman. 1999. Kinematic characterization of standing reach: Comparison of younger vs. older subjects. *Clinical Biomechanics* 14: 271-279.

Challis, J. 1999. A procedure for the automatic determination of filter cutoff frequency for the processing of biomechanical data. *Journal of Applied Biomechanics* 15: 303-317.

Claasen, T.A.C.M., and W.F.G. Mecklenbrauker. 1980a. The Wigner distribution: A tool for time-frequency signal analysis (part I). *Philips Journal of Research* 35: 217-250.

Claasen, T.A.C.M., and W.F.G. Mecklenbrauker. 1980b. The Wigner distribution: A tool for time-frequency signal analysis (part II). *Philips Journal of Research* 35: 276-300.

Cohen, L. 1989. Time-frequency distributions—a review. *Proceedings of the IEEE* 77: 941-981.

Cooley, J.W., and J.W. Tukey. 1965. An algorithm for the machine computation of the complex Fourier series. *Mathematics and Computation* 19: 297-301.

Cramer, P., D.A. Bushnell, and R.D. Kornberg. 2001. Structural basis of transcription: RNA polymerase II at 2.8 angstrom resolution. *Science* 292: 1863-1876.

DAmico, M., and G. Ferrigno. 1990. Technique for the evaluation of derivatives from noisy biomechanical displacement data using a model-based-bandwidth-selection procedure. *Medical and Biological Engineering and Computing* 28: 407-415.

DAmico, M., and G. Ferrigno. 1992. Comparison between the more recent techniques for smoothing and derivative assessment in biomechanics. *Medical and Biological Engineering and Computing* 30: 193-204.

DeAngelis, G., L.D. Gilmore, and C.J. DeLuca. 1990. Standardized evaluation of techniques for measuring the spectral compression of the myoelectric signal. *IEEE Transactions in Biomedical Engineering* 37: 844-849.

DeLuca, C.J. 1984. Myoelectric manifestations of localised muscular fatigue in humans. *Critical Reviews in Biomedical Engineering* 11: 251-279.

Dingwell, J.B., and P.R. Cavanagh. 2001. Increased variability of continuous overground walking in neuropathic patients is only indirectly related to sensory loss. *Gait and Posture* 14: 1-10.

Dohrmann, C., H. Busby, and D. Trujillo. 1988. Smoothing noisy data using dynamic programming and generalised cross-validation. *Journal of Biomechanical Engineering* 110: 37-41.

Dowling, J.J. 1985. A modeling strategy for the smoothing of biomechanical data. In *Biomechanics XB.* ed. B. Johnsson, 1163-1167. Champaign, IL: Human Kinetics

Ferdjallah, M., G.F. Harris, and J.J. Wertsch. 1999. Instantaneous postural stability characterization using time-frequency analysis. *Gait and Posture* 10: 129-134.

Frigo, M., and S.G. Johnson. 1998. FFTW: An adaptive software architecture for the FFT. In *International Conference on Acoustics, Speech, and Signal Processing* 3: 1381-1384.

Georgakis, A., L.K. Stergioulas, and G. Giakas. 2003. An automatic algorithm for filtering kinematic signals with impacts in the Wigner representation. *Medical and Biological Engineering and Computing* 40(6): 625-633.

Georgakis, A., L. Stergioulas, and G. Giakas. 2002. Wigner filtering with smooth roll-off boundary for differentiation of noisy non-stationary signals. *Signal Processing* 82 (10): 1411-1415.

Giakas, G. 1997. *Time and frequency domain applications in biomechanics.* PhD thesis, Manchester Metropolitan University.

Giakas, G., and V. Baltzopoulos. 1997a. A comparison of automatic filtering techniques applied to biomechanical walking data. *Journal of Biomechanics* 30: 847-850.

Giakas, G., and V. Baltzopoulos. 1997b. Optimal digital filtering requires a different cut-off frequency strategy for the determination of the higher derivatives. *Journal of Biomechanics* 30: 851-855.

Giakas, G., and V. Baltzopoulos. 1997c. Time and frequency domain analysis of ground reaction forces during walking: An investigation of variability and symmetry. *Gait and Posture* 5: 189-197.

Giakas, G., V. Baltzopoulos, and R.M. Bartlett. 1998. Improved extrapolation techniques in recursive digital filtering: A comparison between least squares and linear prediction. *Journal of Biomechanics* 31: 87-91.

Giakas, G., V. Baltzopoulos, P. Dangerfield, J. Dorgan, and S. Dalmira. 1996. Comparison of gait patterns between healthy and scoliotic patients using time and frequency domain analysis of ground reaction forces. *Spine* 21: 2235-2242.

Giakas, G., L. Stergioulas, and A. Vourdas. 2000. Time-frequency analysis and filtering of kinematic signals with impacts using the Wigner function: Accurate estimation of the second derivative. *Journal of Biomechanics* 33: 567-574.

Hamill, J., G. Caldwell, and T. Derrick. 1997. Reconstructing digital signals using Shannon's sampling theorem. *Journal of Applied Biomechanics* 13: 226-238.

Hatze, H. 1981. The use of optimally regularized Fourier series for estimating higher-order derivatives of noisy biomechanical data. *Journal of Biomechanics* 14: 13-18.

Hatze, H. 1990. Data conditioning and differentiation techniques. In *Biomechanics of human movement: Applications in rehabilitation, sports and ergonomics,* ed. N. Berme and A. Cappozzo, 237-248. Worthington, OH: Bertec Corporation.

Ismail, A.R., and S.S. Asfour. 1999. Discrete wavelet transform: A tool in smoothing kinematic data. *Journal of Biomechanics* 32: 317-321.

Jackson, K. 1979. Fitting mathematical function to biomechanical data. *IEEE Transactions in Biomedical Engineering* 26: 122-124.

Kayhan, A.S. 1991. *Time-varying spectral estimators for non-stationary signals.* PhD thesis, University of Pittsburgh.

Lanshammar, H. 1982. On practical evaluation of differentiation techniques for human gait analysis. *Journal of Biomechanics* 15: 99-105.

Lariviere, C., A.B. Arsenault, D. Gravel, D. Gagnon, and P. Loisel. 2001. Median frequency of the electromyographic signal: Effect of time-window location on brief step contractions. *Journal of Electromyography and Kinesiology* 11: 65-71.

Lee, K.J., M.N. Paley, I.D. Wilkinson, and P.D. Griffiths. 2002. Fast two-dimensional MR imaging by multiple acquisition with micro B-0 array (MAMBA). *Magnetic Resonance Imaging* 20: 119-125.

Linsseen, W., D. Stegeman, E. Joosten, M. Vant'hof, R. Binkhorst, and S. Notermans. 1993. Variability and interrelationship of surface EMG parameters during local muscle fatigue. *Muscle and Nerve* 16: 849-856.

Lynn, P.A., and W. Fuerst. 1994. *Digital signal processing.* New York: Wiley.

Madeleine, P., P. Bajaj, K. Sogaard, and L. Arendt-Nielsen. 2001. Mechanomyography and electromyography force relationships during concentric, isometric and eccentric contractions. *Journal of Electromyography and Kinesiology* 11: 113-121.

Merletti, R., and C.J. DeLuca. 1990. Myoelectric manifestations of muscle fatigue during voluntary and electrically elicited contractions. *Journal of Applied Physiology* 69: 1810-1820.

Munger, P. 1998. Beyond Fourier: The wavelet transform. In *The Fourier transform in biomedical engineering,* ed. T. Peter and J. Williams. Boston: Birkhauser.

Onell, A. 2000. The vertical ground reaction force for analysis of balance? *Gait and Posture* 12: 7-13.

Perry, J.E., B.L. Davis, and M.G. Luciano. 2001. Quantifying muscle activity in non-ambulatory children with spastic cerebral palsy before and after selective dorsal rhizoctomy. *Journal of Electromyography and Kinesiology* 11: 31-37.

Pezzack, J.C., R.W. Norman, and D.A. Winter. 1977. An assessment of derivative determining techniques used for motion analysis. *Journal of Biomechanics* 10: 377-382.

Rabiner, L.R., and B. Gold. 1975. *The ry and applications of digital signal processing.* Englewood Cliffs, NJ: Prentice Hall.

Rader, C.M. 1968. Discrete Fourier transforms when the number of data samples is prime. *Proceedings of the IEEE* 56: 1107-1108.

Rozan, T.F., M.E. Lassman, D.P. Ridge, and G.W. Luther. 2000. Evidence for iron, copper and zinc complexation as multinuclear sulphide clusters in oxic rivers. *Nature* 406: 879-882.

Sadik, A.M., and D. Litwin. 2002. An extrapolation method based on the variable-wavelength interferometry optical Fourier transform technique for fibre birefringence measurement. *Journal of Optics A-Pure and Applied Optics* 4: 135-139.

Schumann, T., M.R. Redfern, J.M. Furman, A. El-Jaroudi, and L.F. Chaparro. 1995. Time-frequency analysis of postural sway. *Journal of Biomechanics* 28: 603-607.

Smith, G. 1989. Padding point extrapolation techniques for the Butterworth digital filter. *Journal of Biomechanics* 22: 967-971.

Smith, M.J.T., and R.M. Mersereau. 1992. *Introduction to digital signal processing.* New York: Wiley.

Stergiou N., and B.T. Bates. 1997. The relationship between subtalar and knee joint function as a possible mechanism for running injuries. *Gait and Posture* 6: 177-185.

Stergiou, N., B.T. Bates, and S.L. James. 1999. Asynchrony between subtalar and knee joint function during running. *Medicine and Science in Sports and Exercise* 31: 1645-1655.

Stergiou, N., G. Giakas, J.E. Byrne, and V. Pomeroy. 2002. Frequency domain characteristics of ground reaction forces during walking of young and elderly females. *Clinical Biomechanics* 17(8): 615-617.

Tavazoie, S., J.D. Hughes, M.J. Campbell, R.J. Cho, and G.M. Church. 1999. Systematic determination of genetic network architecture. *Nature Genetics* 22: 281-285.

Thomson, D.J. 1982. Spectrum estimation and harmonic analysis. *Proceedings of the IEEE* 70: 1055-1096.

Vaughan, C.L. 1982. Smoothing and differentiation of displacement-time data: An application of splines and digital filtering. *International Journal of Biomedical Computing* 13: 375-386.

Vaughan, C., B. Davis, and J. O'Connor. 1992. *Gait analysis laboratory.* Champaign, IL: Human Kinetics.

Vint, P.F., and R.N. Hinrichs. 1996. Endpoint error in smoothing and differentiation raw kinematic data: An evaluation of four popular methods. *Journal of Biomechanics* 29: 1637-1642.

Wachowiak, M.P., G.S. Rash, P.M. Quesada, and A.H. Desoky. 2000. Wavelet-based noise removal for biomechanical signals: A comparative study. *IEEE Transactions in Biomedical Engineering* 47: 360-368.

Walker, J.A. 1998. Estimating velocities and accelerations of animal locomotion: A simulation experiment comparing numerical differentiation algorithms. *Journal of Experimental Biology* 201: 981-995.

Welch, P.D. 1967. The use of Fast Fourier Transform for the estimation of power spectra: A method based on time averaging over short, modified periodograms. *IEEE Transactions in Audio Electroacoustics* 15: 70-73.

Winter, D.A. 1990. *Biomechanics and motor control of human movement* (2nd ed.). Toronto, Ontario: Wiley InterScience.

Winter, D.A., H.G. Sidwall, and D.A. Hobson. 1974. Measurement and reduction of noise in kinematics of locomotion. *Journal of Biomechanics* 7: 157-159.

Woltring, H.J. 1985. On optimal smoothing and derivative estimation from noisy displacement data in biomechanics. *Human Movement Science* 4: 229-245.

Woltring, H.J. 1986. A Fortran package for generalised, cross-validatory spline smoothing and differentiation. *Advances in Engineering Software* 8: 104-107.

Woltring, H.J. 1995. Smoothing and differentiation techniques applied to 3-D data. In *Three-dimensional analysis of human movement,* ed. P. Allard, I.A.F. Stokes, and J.P. Blanchi, 79-99. Champaign, IL: Human Kinetics.

Wood, G.A. 1982. Data smoothing and differentiation procedures in biomechanics. In *Exercise and Sport Science Review,* ed. R. Terjung, 308-362. Philadelphia: Franklin Institute Press.

Yu, B. 1989. Determination of the optimum cut-off frequency in the digital filter data smoothing procedure. In *XII International Congress of Biomechanics,* ed. R. Gregor, R. Zernicke, and W. Whiting. Los Angeles: University of California.

Yu, B., D. Gabriel, L. Noble, and K.N. An. 1999. Estimate of the optimum cut-off frequency for the Butterworth low-pass digital filter. *Journal of Applied Biomechanics* 15: 318-329.

# Solutions to Work Problems

## Chapter 2

1. For part A: Worked-out solutions are provided for the first data point (i.e., 0% stride) only. Answers for all data points are given in Table A.1.

From equation 2.12:

$$M_0 = (11.3 + 11.2 + 10.6) / 3 = 11.0°$$

From equation 2.13:

$$SD_0 = [((11.3 - 11.0)^2 + (11.2 - 11.0)^2 + (10.6 - 11.0)^2) / (3 - 1))]^{1/2} = 0.38°$$

From equation 2.14:

$$CV_0 = (0.38 / 11.0) \times 100 = 3.5\%$$

| Table A.1 | | | |
|---|---|---|---|
| % Stride | $M_i$ | $SD_i$ | $CV_i$ |
| 0 | 11.0 | 0.38 | 3.5 |
| 1 | 12.5 | 0.40 | 3.2 |
| 2 | 14.2 | 0.59 | 4.2 |
| 3 | 16.4 | 0.87 | 5.3 |
| 4 | 19.2 | 1.15 | 6.0 |
| 5 | 22.4 | 1.48 | 6.6 |
| 6 | 29.3 | 5.72 | 19.5 |
| 7 | 30.0 | 2.23 | 7.4 |
| 8 | 34.4 | 2.53 | 7.4 |
| 9 | 39.1 | 2.95 | 7.5 |

Worked-out solutions for part B:

From equation 2.15:

$$SD_{avg} = [(0.38^2 + 0.40^2 + 0.59^2 + 0.87^2 + 1.15^2 + 1.48^2 + 5.72^2 + 2.23^2 + 2.53^2 + 2.95^2) / 10] = 2.40°$$

From equation 2.16:

$$CV_{avg} = 2.40 / [(11.0 + 12.5 + 14.2 + 16.4 + 19.2 + 22.4 + 29.3 + 30.0 + 34.4 + 39.1) / 10] \times 100 = 10.5\%$$

2. For part A: Examples of worked-out solutions are provided for the first data point (i.e., 0% stride) only. Answers for all data points are given in the tables following these examples.

Step 1. Calculate $M_{iy}$ and $M_{ix}$ for each data point.
Example for the ankle ($y$) and knee ($x$) data at 0% stride:
From equation 2.12:

$$M_{0y} = (-46.1 + -45.4 + -47.3) / 3 = -46.3$$

$$M_{0x} = (11.3 + 11.2 + 10.6) / 3 = 11.0$$

Step 2. Calculate $R_{ij}$ for each cycle and data point.
Example for 0% stride, cycle one:
From equation 2.17:

$$R_{0,1} = [(-46.1 - -46.3)^2 + (11.3 - 11.0)^2]^{1/2} = 0.36$$

Step 3. Calculate $RMS_i$ for each data point.
Example for 0% stride:
From equation 2.30:

$$RMS_0 = [(0.36^2 + 0.92^2 + 1.08^2) / 3]^{1/2} = 0.85$$

**Table A.2**

| % Stride | $M_{iy}$ | $M_{ix}$ |
|---|---|---|
| 0 | −46.3 | 11.0 |
| 1 | −45.9 | 12.5 |
| 2 | −45.4 | 14.2 |
| 3 | −44.5 | 16.4 |
| 4 | −43.4 | 19.2 |
| 5 | −41.9 | 22.4 |
| 6 | −40.3 | 29.3 |
| 7 | −38.4 | 30.0 |
| 8 | −36.3 | 34.4 |
| 9 | −33.9 | 39.1 |

**Table A.3**

| | Cycle 1 | Cycle 2 | Cycle 3 | |
|---|---|---|---|---|
| % Stride | $R_{ij}$ | $R_{ij}$ | $R_{ij}$ | $RMS_i$ |
| 0 | 0.36 | 0.92 | 1.08 | 0.85 |
| 1 | 0.10 | 1.08 | 0.98 | 0.84 |
| 2 | 0.46 | 1.37 | 0.82 | 0.96 |
| 3 | 0.92 | 1.56 | 0.72 | 1.13 |
| 4 | 1.35 | 1.91 | 0.50 | 1.38 |
| 5 | 1.89 | 2.22 | 0.42 | 1.70 |
| 6 | 5.21 | 2.24 | 6.27 | 4.88 |
| 7 | 2.83 | 3.12 | 0.40 | 2.44 |
| 8 | 3.30 | 3.63 | 0.48 | 2.85 |
| 9 | 3.82 | 4.19 | 0.61 | 3.29 |

Part B:

Step 1. Calculate $RMS_j$ for each cycle.
From equation 2.18:

$$RMS_1 = [(0.36^2 + 0.10^2 + 0.46^2 + 0.92^2 + 1.35^2 + 1.89^2 + 5.21^2 + 2.83^2 + 3.30^2 + 3.82^2) / 10]^{1/2} = 2.59$$

$$RMS_2 = [(0.92^2 + 1.08^2 + 1.37^2 + 1.56^2 + 1.91^2 + 2.22^2 + 2.24^2 + 3.12^2 + 3.63^2 + 4.19^2) / 10]^{1/2} = 2.46$$

$$RMS_3 = [(1.08^2 + 0.98^2 + 0.82^2 + 0.72^2 + 0.50^2 + 0.42^2 + 6.27^2 + 0.40^2 + 0.48^2 + 0.61^2) / 10]^{1/2} = 2.09$$

Step 2. Calculate $RMS_{avg}$ across all cycles.
From equation 2.19:

$$RMS_{avg} = (2.59 + 2.46 + 2.09) / 3 = 2.38$$

# Chapter 4

1. Shank phase plot

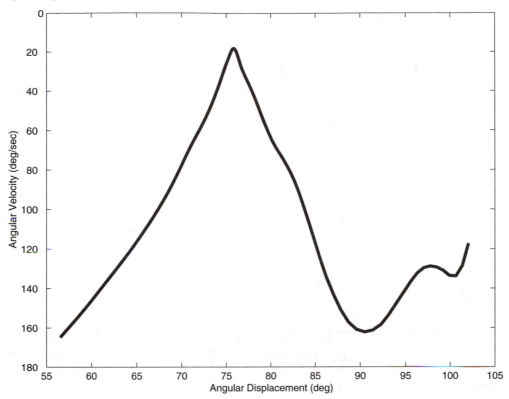

Thigh phase plot

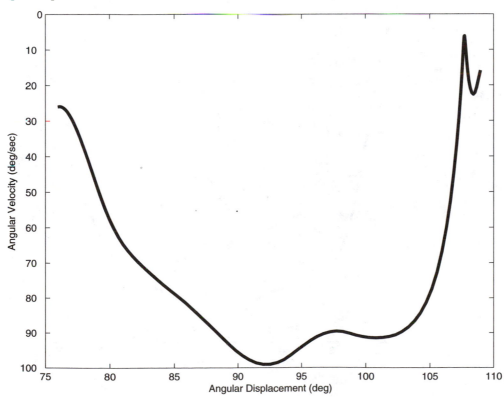

## 2. Shank phase angle

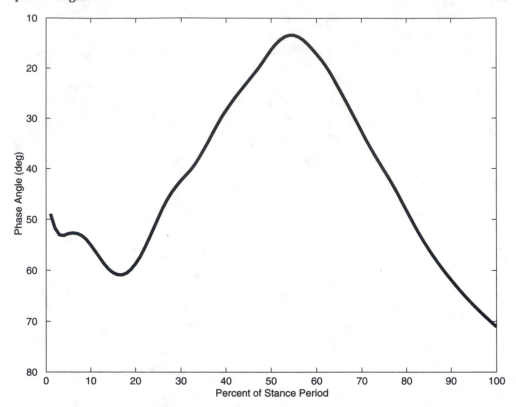

Thigh phase angle

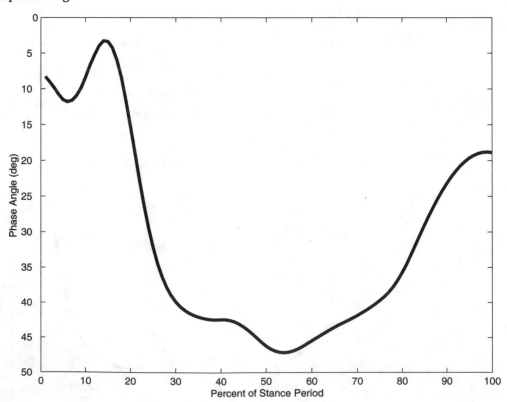

## 3. Relative phase

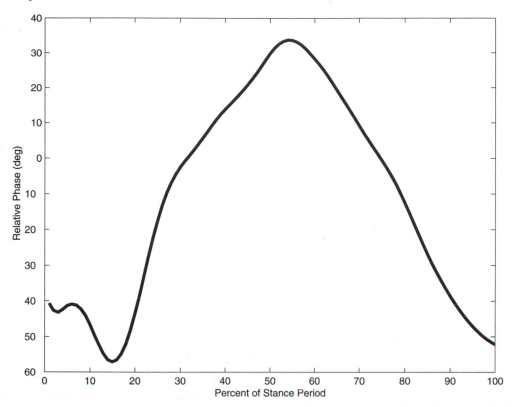

## 4. Shank phase plot

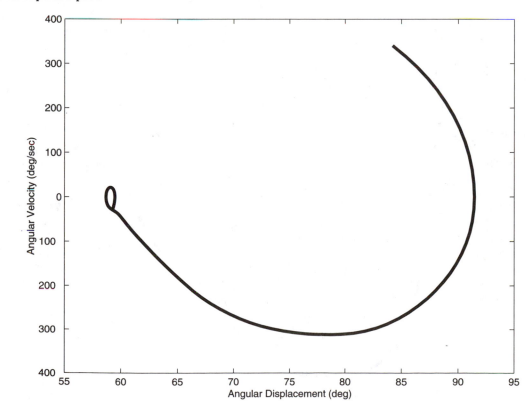

Thigh phase plot

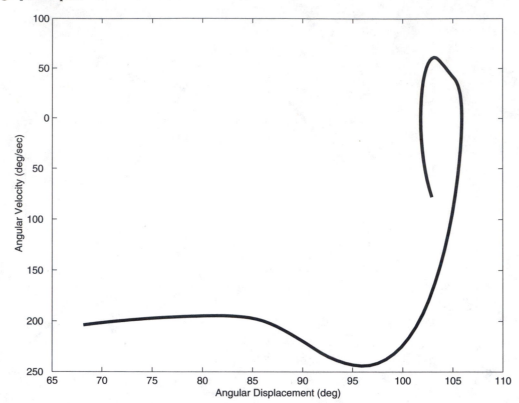

5. Shank phase angle

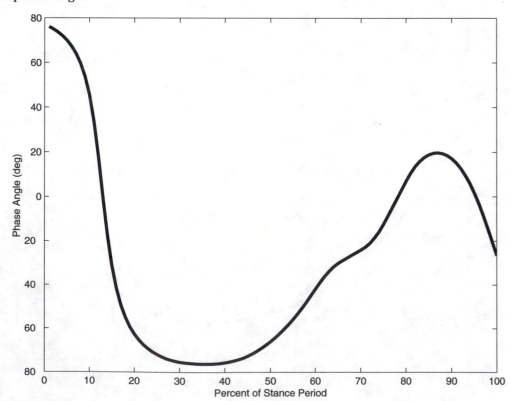

**Thigh phase angle**

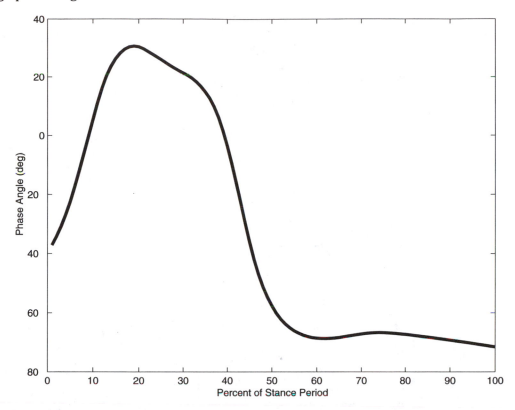

## 6. Running relative phase

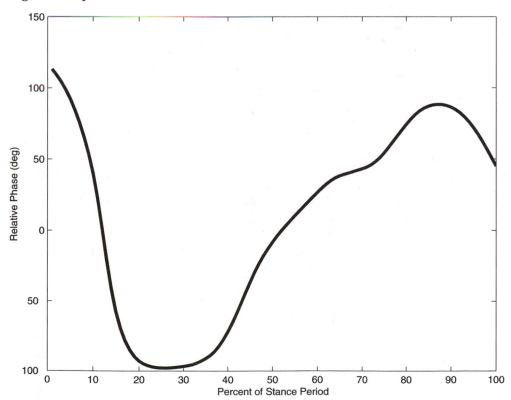

# Chapter 5

1. $\bar{\theta} = 32.6°$; $\hat{\theta} = 29°$; $\bar{\rho} = 0.77$; $s_b = 38.8°$; range = 150°
2. $\bar{\theta} = 175.5°$; $\hat{\theta} = 199°$; $\bar{\rho} = 0.26$; $s_b = 69.9°$; range = 278°

3.

| Table A.4 | |
|---|---|
| **Radial case** | **Axial case** |
| $\bar{\theta} = 131.4°$ | $\bar{\theta}_a = 73.3°$ |
| $\hat{\theta} = 18.5°$ | $\hat{\theta}_a = 70°$ |
| $\bar{\rho} = 0.09$ | $\bar{\rho}_a = 0.54$ |
| $s_b = 77.4°$ | $s_{b,a} = 27.4°$ |

4.

| Table A.5 | | |
|---|---|---|
| | $\bar{\theta}$ | $\bar{\rho}$ |
| S1 | 124.8 | 0.61 |
| S2 | 304.4 | 0.15 |
| S3 | 105.2 | 0.58 |
| S4 | 169.1 | 0.67 |
| S5 | 216.0 | 0.42 |
| S6 | 142.4 | 0.69 |
| S7 | 82.6 | 0.34 |
| S8 | 115.9 | 0.38 |

$\bar{\theta}_{so} = 137.4°$; $\bar{\rho}_{so} = 0.35$; $s_{b,so} = 65.1°$.

5.

| Table A.6 | | |
|---|---|---|
| | $\bar{\theta}$ | $\bar{\rho}$ |
| S1 | 294.6 | 0.87 |
| S2 | 282.7 | 0.98 |
| S3 | 321.9 | 0.95 |
| S4 | 325.1 | 0.74 |
| S5 | 310.7 | 0.98 |
| S6 | 316.6 | 0.82 |
| S7 | 285.0 | 0.98 |
| S8 | 299.7 | 0.99 |
| S9 | 279.9 | 0.90 |
| S10 | 284.4 | 0.95 |

$\bar{\theta}_{so} = 299.3°$; $\bar{\rho}_{so} = 0.88$; $s_{b,so} = 28.2°$.

6. Based on the value of $\bar{\rho} = 0.77$, and an $n$ of 30, Rayleigh's R = 23.1.

   Using Table H in Batschelet, we see the $p$-value < 0.001. Alternately, we can compute the $p$-value from equation 5.12. In this case, $p$-value = $6.3 \times 10_{-10}$. Therefore, we reject the null hypothesis and conclude that the sample is directed (i.e., not uniformly distributed).

7. Based on the value of $\bar{\rho} = 0.26$, and an $n$ of 15, Rayleigh's R = 3.8.

   Using Table H in Batschelet, we see the $p$-value = 0.37. Alternately, we can compute the $p$-value from equation 5.12. In this case, $p$-value = 0.38. Therefore, we fail to reject the null hypothesis and conclude that the sample is not directed (i.e., uniformly distributed).

8. (a) Radial case: Based on the value of $\bar{\rho} = 0.09$, and an $n$ of 38, Rayleigh's R = 3.3

   Using Table H in Batschelet, we see the $p$-value = 0.75. Alternately, we can compute the $p$-value from equation 5.12. In this case, $p$-value = 0.75. Therefore, we fail to reject the null hypothesis and conclude that the sample is not directed (i.e., uniformly distributed). (b) Axial case: Based on the value of $\bar{\rho} = 0.54$, and an $n$ of 38, Rayleigh's R = 20.6. Using Table H in Batschelet, we see the $p$-value < 0.001. Alternately, we can compute the $p$-value from equation 5.12. In this case, $p$-value = $6.2 \times 10^{-6}$. Therefore, we reject the null hypothesis and conclude that the sample is directed (i.e., not uniformly distributed).

9. Based on the value of $\bar{\theta} = 32.6°$ and $\bar{\rho} = 0.77$,

$$V = \bar{\rho}\cos(\bar{\theta} - \theta_0) = 0.77\cos(32.6 - 45) = 0.75$$

$$v = V\sqrt{2n} = 0.75\sqrt{2*30} = 5.82$$

   Since $v = 5.82$ is greater than the critical value of 1.65 (Batschelet, Table I, $n = 30$, $\alpha = .05$), we reject the null hypothesis and conclude that the data are directed. However, we cannot state whether or not the sample of data has a mean direction different from 45°.

10. Since the range of the data is 150°, we can draw a diameter such that the minimum number of points on one side of the diameter ($m$) is 0. Using Table J of Batschelet, we see that $p$-value < 0.001. Plugging the values into equation 5.16 shows that the probability is:

$$p = 2^{1-n}(n - 2m)\frac{n!}{m!(n-m)!} = 2^{1-30}(30 - 0)\frac{30!}{0!(30-0)!} = 5.6x10^{-8}$$

   Therefore, we reject the null hypothesis and conclude that the data are directed, the same result as in work problem 6.

11. We can draw a diameter such that the minimum number of points on one side of the diameter ($m$) is 3. Using Table J of Batschelet, we see that $p$-value = 0.250. Plugging the values into equation 5.16 shows that the probability is:

$$p = 2^{1-n}(n - 2m)\frac{n!}{m!(n-m)!} = 2^{1-15}(15 - 6)\frac{15!}{3!(15-3)!} = .250$$

   Therefore, we fail to reject the null hypothesis and conclude that the data are not directed, the same result as in work problem 7.

12. We can draw a diameter such that the minimum number of points on one side of the diameter ($m$) is 14. Using Table J of Batschelet, we see that $p$-value = 0.90. The probability cannot be computed accurately from equation 6.16, because $m > n/3$. Therefore, we fail to reject the null hypothesis and conclude that the data are not directed, the same result as in work problem 8 for the radial case.

13. After transforming the data to modulo 360 and sorting the data from smallest to largest, the difference between successive data points is:

   $T$ = [3, 0, 1, 8, 8, 5, 0, 1, 2, 0, 2, 12, 1, 12, 1, 6, 3, 1, 3, 9, 5, 2, 2, 2, 210, 9, 4, 23, 19, 6]

The last point, $T_{30} = \theta_1 - \theta_{30} + 360$. After inserting the $T_i$ into equation 5.18, we find that $U = 216$. Since $U$ is greater than 156.7, the critical value listed in Table L of Batschelet ($n = 30$, $\alpha = .05$), we reject the null hypothesis and conclude that the data are directed.

14. After transforming the data to modulo 360 and sorting the data from smallest to largest, the difference between successive data points is:

$$T = [82, 14, 8, 15, 5, 3, 3, 45, 28, 17, 45, 2, 6, 60, 27]$$

Since $U = 143$ is less than 165.6, the critical value listed in Table L of Batschelet ($n = 15$, $\alpha = .05$), we fail to reject the null hypothesis and conclude that the data are not directed.

15. After transforming the data to modulo 360 and sorting the data from smallest to largest, the difference between successive data points is:

$$T = [5, 9, 0, 1, 2, 2, 7, 8, 2, 6, 11, 2, 1, 1, 9, 8, 5, 3, 75, 5, 8, 9, 8, 0, 2, 13, 3, 2, 2, 4, 2, 11, 4, 21, 4, 14, 44, 47]$$

Since $U = 160.2$ is greater than 155.0, the critical value listed in Table L of Batschelet ($n = 35$, $\alpha = .05$), we reject the null hypothesis and conclude that the data are directed. In both work problem 8 (radial case) and work problem 12, we failed to reject the null hypothesis. Thus we can see that the Rao spacing test is better suited to detecting differences from uniformity for bimodal and multimodal data.

16. (a) The confidence interval is determined by inserting the following values into equation 5.19: $n = 30$, $\bar{\rho} = 0.77$, $\chi_1^2 = 3.84$. Using this information, $\delta = 17.8°$. Thus the 95% confidence interval is $32.6 + 17.8 = [14.8, 50.4]$. (b) For the first method, we can simply see if the hypothetical angle lies within the confidence interval. 45° is within the confidence interval, and 57.3° is not. The second method involves computing a test score, using equations 5.20 and 5.21.

| Table A.7 | | |
|---|---|---|
|  | 45° | 57.3° |
| $\bar{C}$ | 0.75 | 0.70 |
| $P_1$ | 2.34 | 8.55 |
| p-value | 0.13 | 0.004 |

(c) Using both the confidence interval method and the hypothesis test, we fail to reject null hypothesis 1 and conclude that the mean direction of the sample is not different than 45°. However, we reject null hypothesis 2 and conclude that the mean direction of the sample is different than 57.3°.

17. First double all angles in the sample. The pertinent values for constructing the 95% confidence interval are: $n = 38$, $\bar{\rho} = 0.54$, $\bar{\theta} = 146.7$, $\chi_1^2 = 3.84$. Using this information, $\delta = 22.8°$. Thus the 95% confidence interval is $146.7 + 22.8 = [123.9, 169.5]$. If we double the hypothetical angle, we see that 150° lies within the confidence interval and conclude that the mean direction of the data is not different than 75°. Note that after computing delta, we can divide all the values by 2 to transform the data into their original values. In this case, $\delta / 2 = 11.4$, and the confidence interval is $73.3 + 11.4 = [61.9, 84.7]$. The hypothetical direction of 75° lies within this confidence interval. The p-value = 0.77.

18. Since the data can be approximated by a von Mises distribution, the parametric two-sample test is the most powerful test. Null hypothesis: The mean

direction of sample 1 is not different from the mean direction of sample 2. The descriptive statistics are as follows:

$$\bar{\theta}_1 = 32.6°, \ \bar{\rho}_1 = 0.77, \ s_{b,1} = 38.8°$$

$$\bar{\theta}_2 = 64.9°, \ \bar{\rho}_2 = 0.84, \ s_{b,2} = 32.7°$$

First observe that $\bar{\rho} > 0.7$ for both samples, which is a necessary criterion to perform the parametric two-sample test. After solving equations 6.22 and 6.23, we find that:

$$\rho_1 = 23.1, \ \rho_2 = 20.9, \ \rho_T = 42.3, \ \bar{\rho} = 0.80$$

Next, we estimate $\tilde{\kappa} = 2.81$. If you come up with a slightly different value by looking up in table 5.4, it has a negligible effect on the outcome. Using equation 5.24, the test statistic $P_2 = 9.50$, resulting in a $p$-value $= 0.003$. Therefore, we reject the null hypothesis and conclude that the mean directions of the two samples are different.

19. Null hypothesis: The mean direction of sample 1 is not different than the mean direction of sample 3. The descriptive statistics are as follows:

$$\bar{\theta}_1 = 32.6°, \ \bar{\rho}_1 = 0.77, \ s_{b,1} = 38.8°$$

$$\bar{\theta}_3 = 36.1°, \ \bar{\rho}_3 = 0.90, \ s_{b,3} = 26.2°$$

After solving equations 5.22 and 5.23, we find that:

$$\rho_1 = 23.1, \ \rho_3 = 22.4, \ \rho_T = 45.5, \ \bar{\rho} = 0.83$$

Next, we estimate $\tilde{\kappa}_T = 3.17$. If you come up with a slightly different value by looking up in table 5.4, it has a negligible effect on the outcome. Using equation 5.24, the test statistic $P_2 = 0.13$, resulting in a $p$-value $= 0.72$. Therefore, we fail to reject the null hypothesis and conclude that the mean directions of the two samples are no different.

20. Null hypothesis: The mean directions of all three samples are equal. The descriptive statistics are as follows:

$$\bar{\theta}_1 = 32.6°, \ \bar{\rho}_1 = 0.77, \ s_{b,1} = 38.8°$$

$$\bar{\theta}_2 = 64.9°, \ \bar{\rho}_2 = 0.84, \ s_{b,2} = 32.7°$$

$$\bar{\theta}_3 = 36.1°, \ \bar{\rho}_3 = 0.90, \ s_{b,3} = 26.2°$$

After solving equations 5.25 and 5.26, we find that:

$$\rho_1 = 23.1, \ \rho_2 = 20.9, \ \rho_3 = 22.4, \ \rho_T = 64.4, \ \bar{\rho} = 0.83$$

Next, we estimate $\tilde{\kappa}_T = 3.21$. If you come up with a slightly different value by looking up in table 5.4, it has a negligible effect on the outcome. Using equation 5.27, the test statistic $P_k = 6.5$, resulting in a $p$-value $= 0.003$. Therefore, we reject the null hypothesis and conclude that the mean directions are not all equal (i.e., at least one is different). We already know that $\theta_1$ is not equal to $\theta_2$, and $\theta_1 = \theta_3$. Therefore, to complete the analysis, the only other test we need to perform is a test to see if $\theta_2 = \theta_3$. Do this on your own.

21. The descriptive statistics are as follows:

$$\bar{\theta}_1 = 53.3°, \ \bar{\rho}_1 = 0.55, \ s_{b,1} = 54.3°$$

$$\bar{\theta}_2 = 123.9°, \ \bar{\rho}_2 = 0.56, \ s_{b,2} = 53.6°$$

Since the $\bar{\rho}$ are no longer $> 0.7$, we can no longer use the parametric two-sample test. Combine and rank order the data, then transform the ranks into uniform scores (equation 5.28). The uniform scores for each sample are:

$B_1$ = [6.5, 13.1, 19.6, 26.2, 32.7, 45.8, 52.4, 58.9, 65.5, 72, 78.5, 85.1, 104.7, 111.3, 124.4, 130.9, 144, 163.6, 170.2, 189.8, 202.9, 222.5, 229.1, 235.6, 242.2, 320.7, 327.3, 333.8, 340.4, 360];

$B_2$ = [39.3, 91.6, 98.2, 117.8, 137.5, 150.5, 157.1, 176.7, 183.3, 196.4, 209.5, 216, 248.7, 255.3, 261.8, 268.4, 274.9, 281.5, 288, 294.5, 301.1, 307.6, 314.2, 346.9, 353.4]

Then compute the test statistic given by equation 5.29. $NP_2 = 54.4$. Since $n_1 + n_2 \geq 20$, the transformed statistic $NP_2^* = 7.8$. Since 7.8 is greater than the critical $\chi_2^2 = 6.0$, we reject the null hypothesis and conclude that the directions of the two samples are different.

22. The descriptive statistics are as follows:

$$\bar{\theta}_1 = 53.3°, \ \bar{\rho}_1 = 0.55, \ s_{b,1} = 54.3°$$

$$\bar{\theta}_2 = 123.9°, \ \bar{\rho}_2 = 0.56, \ s_{b,2} = 53.6°$$

$$\bar{\theta}_3 = 75.5°, \ \bar{\rho}_3 = 0.65, \ s_{b,3} = 48.0°$$

Combine and rank order the data, then transform the ranks into uniform scores. The uniform scores for each sample are given by equation 5.31 (not shown). If the uniform scores are correct, then the vector lengths (equation 5.32) are as follows:

$$\rho_1^2 = 29.0, \ \rho_2^2 = 66.7, \ \rho_3^2 = 23.0$$

Using this information, we can compute the test statistic $NP_k = 9.1$. Since $NP_k$ is less than the critical value of $\chi_4^2 = 9.5$, we fail to reject the null hypothesis and conclude that the directions of the samples are all equal.

23. Use the parametric one-sample test for second-order analysis to see if this sample is directed. After finding $(X_i, Y_i) = (\bar{\rho}_i \cos \bar{\theta}_i, \bar{\rho}_i \sin \bar{\theta}_i)$, use equation 5.35 to find the sums of squares:

$$SSx = 0.40$$

$$SSy = 0.31$$

$$SSxy = 0.29$$

$$(\bar{X}_{so}, \bar{Y}_{so}) = (0.43, -0.77)$$

Then use equation 5.36 to compute the test statistic:

$$PSO_1 = \frac{10^2 - 2*10}{2}\left[\frac{0.43^2 * 0.31 - 2 * 0.43 * (-0.77) * 0.29 + (-0.77)^2 * 0.40}{0.40 * 0.31 - 0.29^2}\right] = 488$$

Since $PSO_1$ is greater than the critical $F_{2,8} = 4.5$, we reject the null hypothesis and conclude that the second-order sample is directed. Alternately, we can compute the $p$-value $< 0.0001$. If you use equation 5.37, you will find that the 95% confidence interval is approximately [283,317]. This confidence interval can be used to see if the sample is different from a hypothetical direction.

24. Using equation 5.38, the following quantities were computed:

$$SSx_1 = 0.40, SSy_1 = 0.31, SSxy_1 = 0.29, (\overline{X}_1, \overline{Y}_1) = (0.43, -0.77)$$

$$SSx_2 = 1.07, SSy_2 = 0.18, SSxy_2 = 0.03, (\overline{X}_2, \overline{Y}_2) = (0.09, -0.73)$$

$$SSx_c = 1.47, SSy_c = 0.49, SSxy_c = 0.32$$

Then use equation 5.39 to compute the test statistic:

$$PSO_2 = \frac{10+10-3}{2\left(\dfrac{1}{10}+\dfrac{1}{10}\right)}\left[\frac{(0.43-0.09)^2 0.49 - 2(0.43-0.09)(-0.77+0.73)0.32 + (-0.77+0.73)^2 1.47}{1.47*0.49 - 0.32^2}\right]$$

$$PSO_2 = 4.66$$

Since $PSO_2$ is greater than the critical $F_{2,17} = 3.59$, we reject the null hypothesis and conclude that the directions of the two samples are different.

25. First rank order according to the mean vector length.

| Table A.8 | | | | | |
|---|---|---|---|---|---|
| Rank | Subject | $\overline{\rho}$ | $\overline{\theta}$ | $X_r$ | $Y_r$ |
| 1 | 4 | 0.74 | 325.1 | 0.82 | −0.57 |
| 2 | 6 | 0.82 | 316.6 | 1.45 | −1.38 |
| 3 | 1 | 0.87 | 294.6 | 1.25 | −2.73 |
| 4 | 9 | 0.90 | 279.9 | 0.68 | −3.94 |
| 5 | 10 | 0.95 | 284.4 | 1.25 | −4.84 |
| 6 | 3 | 0.95 | 321.9 | 4.72 | −3.70 |
| 7 | 5 | 0.98 | 310.7 | 4.56 | −5.31 |
| 8 | 7 | 0.98 | 285.0 | 2.07 | −7.73 |
| 9 | 2 | 0.98 | 282.7 | 1.98 | −8.78 |
| 10 | 8 | 0.99 | 299.7 | 4.95 | −8.69 |
| Mean | | | | 2.37 | −4.77 |

Then use equation 5.40 to compute the test statistic:

$$NPSO_1 = \sqrt{\frac{2.37^2 + (-4.77)^2}{10}} = 1.68$$

Since $NPSO_1$ is greater than the critical value = 1.048 from Table P in Batschelet, we reject the null hypothesis and conclude that the data are directed.

26. First compute the mean center from the combined second-order samples $(\overline{X}_c, \overline{Y}_c) = (0.26, -0.75)$. Relative to this location, compute new headings for each of the mean angles.

Then using equations 5.28 and 5.29, we find that $NP_2 = 13.6$. Since the combined sample has 20 data points, we also compute $NP_2^* = 5.2$. Since $NP_2^*$ is less than the critical $\chi_2^2 = 6.0$, we fail to reject the null hypothesis and conclude that the directions of the two samples are not different. This result is different than

| Table A.9 | | |
| --- | --- | --- |
| Subject | $\theta_1$ | $\theta_2$ |
| 1 | −21.6 | 63.4 |
| 2 | −102.3 | −13.7 |
| 3 | 18.6 | 157.3 |
| 4 | 43.6 | −79.2 |
| 5 | 1.6 | 45.7 |
| 6 | 29.1 | 23.1 |
| 7 | −92.0 | 168.7 |
| 8 | −25.7 | 149.3 |
| 9 | −128.4 | −172.6 |
| 10 | −98.2 | −157.6 |

the conclusion from work problem 24, possibly because the data arise from a von Mises distribution and thus differences can be detected better with a parametric test.

27. The appropriate tests to use are paired-sample tests. In both the parametric and nonparametric cases, create new $(X_i, Y_i)$ pairs from equation 5.41 (the data are too numerous to print here). Then proceed using the parametric one-sample test for second-order analysis.

$$SSx = 5.41$$

$$SSy = 6.53$$

$$SSxy = -4.64$$

$$\left(\overline{X}_{so}, \overline{Y}_{so}\right) = (0.37, -0.23)$$

$$PSO_1 = 8.04$$

Since $PSO_1$ is greater than the critical $F_{2,23} = 3.4$, we reject the null hypothesis and conclude that the directions from sample 2 are different than the directions from sample 3. Alternately, we can compute the $p$-value = 0.002.

For the nonparametric one-sample test for second-order analysis, $(\overline{X}_r, \overline{Y}_r) = (6.89, -3.75)$

$$NPSO_1 = \sqrt{\frac{6.89^2 + (-3.75)^2}{25}} = 1.57$$

Since $NPSO_1$ is greater than the critical value = 1.020 from Table P in Batschelet, we reject the null hypothesis and conclude that the samples are different.

28. The appropriate tests to use are paired-sample tests for second-order analysis. In both the parametric and nonparametric cases, create new $(X_i, Y_i)$ pairs from equation 5.42.

Then proceed using the parametric one-sample test for second-order analysis.

| Table A.10 | | |
|---|---|---|
| Pair | $X_i$ | $Y_i$ |
| 1 | −0.086 | 0.074 |
| 2 | 0.262 | 0.151 |
| 3 | −0.857 | −0.010 |
| 4 | −0.313 | −0.498 |
| 5 | −0.349 | 0.017 |
| 6 | −0.015 | −0.050 |
| 7 | −0.292 | 0.253 |
| 8 | −0.656 | 0.364 |
| 9 | −0.672 | 0.033 |
| 10 | −0.415 | −0.016 |

$$SSx = 1.02$$
$$SSy = 0.47$$
$$SSxy = -0.03$$
$$(\bar{X}_{so}, \bar{Y}_{so}) = (-0.34, -0.03)$$
$$PSO_1 = 4.56$$

Since $PSO_1$ is greater than the critical $F_{2,8}$ = 4.5, we reject the null hypothesis and conclude that the directions from sample 2 are different than the directions from sample 1. Alternately, we can compute the $p$-value = 0.048.

For the nonparametric one-sample test for second-order analysis, $(\bar{X}_r, \bar{Y}_r)$ = (−4.26, 0.38)

$$NPSO_1 = \sqrt{\frac{-4.26^2 + (0.38)^2}{10}} = 1.35$$

Since $NPSO_1$ is greater than the critical value = 1.048 from Table P in Batschelet, we reject the null hypothesis and conclude that the samples are different.

29. For the parametric case, determine the correlations described in equation 5.43:

$$r_{xc} = -0.146, \ r_{xs} = 0.069, \ r_{cs} = -0.938$$

The linear-circular correlation coefficient $R_{x\theta}$ is then computed (equation 5.44):

$$R_{x\theta} = \sqrt{\frac{-0.146^2 + 0.069^2 - 2(-0.146)(0.069)(-0.938)}{1 - (-0.938)^2}} = 0.24$$

The test statistic (equation 5.45) is:

$$R_{LCP} = \frac{(n-3)0.24^2}{1 - 0.24^2} = 1.34$$

Since $R_{LCP}$ is less than $F_{2,22} = 3.44$, we fail to reject the null hypothesis and conclude that the data are not associated.

For the nonparametric case, compute $T_c$ and $T_s$ (equation 5.50) as demonstrated in table 5.10. You should get:

$$T_c = 26.3, \ T_s = 11.9$$

Since $n$ is odd,

$$a_n = \frac{2\sin^4\left(\dfrac{\pi}{25}\right)}{\left(1+\cos\left(\dfrac{\pi}{25}\right)\right)^3} = 6.24x10^{-5}$$

Thus the nonparametric correlation coefficient is:

$$R_{x\theta,np} = 6.24 \times 10^{-5} \ (26.3^2 + 11.9^2) = 0.05$$

and the test statistic:

$$R_{LCNP} = \frac{24\left(26.3^2 + 11.9^2\right)}{25^2\left(25+1\right)} = 1.23$$

Since $R_{LCNP}$ is less than the critical value of 5.85 given in Table X (Batschelet), we fail to reject the null hypothesis and conclude that there is no association between the linear and circular variable.

30. For the parametric case, determine the correlations described in equation 5.43:

$$r_{xc} = -0.690, \ r_{xs} = 0.785, \ r_{cs} = -0.938$$

The linear-circular correlation coefficient is then computed (equation 5.44):

$$R_{x\theta} = 0.80$$

The test statistic (equation 5.45) is:

$$R_{LCP} = 38.2$$

Since $R_{LCP}$ is greater than $F_{2,22} = 3.44$, we reject the null hypothesis and conclude that a change in linear data is associated with a change in circular data. By plotting the data, you can see that an increase in the linear data is related to an increase in the circular data.

For the nonparametric case, compute $T_c$ and $T_s$ (equation 5.50) as demonstrated in table 5.10. You should get:

$$T_c = 16.1, \ T_s = -75.4$$

Since $n$ is odd,

$$a_n = \frac{2\sin^4\left(\dfrac{\pi}{25}\right)}{\left(1+\cos\left(\dfrac{\pi}{25}\right)\right)^3} = 6.24x10^{-5}$$

Thus the nonparametric correlation coefficient is:

$$R_{x\theta,np} = 6.24 \times 10^{-5} \left(16.1^2 + (-75.4)^2\right) = 0.37$$

and the test statistic:

$$R_{LCNP} = 8.8$$

Since $R_{LCNP}$ is greater than the critical value of 5.85 given in Table X (Batschelet), we reject the null hypothesis and conclude that there is an association between the linear and circular variable.

31. For the parametric case, determine the correlations described in equation 5.46:

$$r_{cc} = 0.121, \ r_{sc} = -0.225, \ r_\theta = -0.831$$

$$r_{cs} = -0.046, \ r_{ss} = 0.157, \ r_\phi = -0.938$$

The circular-circular correlation coefficient $R_{x\theta}$ is then computed (equation 5.47):

$$R_{\theta\phi}^2 = 0.10$$

The test statistic (equation 5.45) is:

$$R_{CCP} = nR_{\theta\phi}^2 = 25*0.104 = 2.6$$

Since $R_{CCP}$ is less than $\chi_4^2 = 9.5$, we fail to reject the null hypothesis and conclude that the data are not associated.

For the nonparametric case, rank each circular sample and transform into uniform angles. Then compute $T_{cc}, T_{cs}, T_{sc}, T_{ss}$ (equation 5.55). You should get:

$$T_{cc} = 0.033, \ T_{cs} = -0.028, \ T_{sc} = -0.0175, \ T_{ss} = -0.0617$$

Thus the nonparametric circular-circular correlation coefficient is the larger of the two:

$$\overline{R}_+^2 = \left(T_{cc} + T_{ss}\right)^2 + \left(T_{sc} - T_{cs}\right)^2 = 0.00095$$

$$\overline{R}_-^2 = \left(T_{cc} - T_{ss}\right)^2 + \left(T_{sc} + T_{cs}\right)^2 = 0.011$$

and the test statistic:

$$R_{\theta\phi,np}^2 = 0.011$$

Since $n > 10$, compute the probability using equation 5.58. Since the p-value = 0.95, we fail to reject the null hypothesis and conclude that there is no association between the two circular samples.

32. For the parametric case, determine the correlations described in equation 5.46:

$$r_{cc} = 0.705, \ r_{sc} = -0.564, \ r_\theta = -0.831$$

$$r_{cs} = -0.781, \ r_{ss} = 0.687, \ r_\phi = -0.936$$

The circular-circular correlation coefficient $R_{x\theta}$ is then computed (equation 5.47):

$$R_{\theta\phi}^2 = 0.71$$

The test statistic (equation 5.45) is:

$$R_{CCP} = nR_{\theta\phi}^2 = 25 * 0.71 = 17.8$$

Since $R_{CCP}$ is greater than $\chi_4^2 = 9.5$, we reject the null hypothesis and conclude that the data are associated. By plotting the data, you can see that an increase in the linear data is related to an increase in the circular data.

For the nonparametric case, rank each circular sample and transform into uniform angles. Then compute $T_{cc}$, $T_{cs}$, $T_{sc}$, $T_{ss}$ (equation 5.55). You should get:

$$T_{cc} = 0.288, \ T_{cs} = -0.036, \ T_{sc} = -0.096, \ T_{ss} = 0.176$$

Thus the nonparametric circular-circular correlation coefficient is the larger of the two:

$$\overline{R}_+^2 = \left(T_{cc} + T_{ss}\right)^2 + \left(T_{sc} - T_{cs}\right)^2 = 0.219$$
$$\overline{R}_-^2 = \left(T_{cc} - T_{ss}\right)^2 + \left(T_{sc} + T_{cs}\right)^2 = 0.030$$

and the test statistic:

$$R_{\theta\phi,np}^2 = 0.219$$

Since $n > 10$, compute the probability using equation 5.58. Since the $p$-value = 0.01, we reject the null hypothesis and conclude that there is an association between the two circular samples.

# Chapter 6

1. The $x$ and $y$ variables are concave up. This observation is based on the fact that the $x^2$ and $y^2$ terms are positive. Additionally, the directionality of the nonlinear terms indicates that both $x$ and $y$ variables are minimizing their behavior. Since both terms are concave up, the overall shape of the surface is a paraboloid. The $x$ term is more constrained to the critical point of the surface than the $y$ term. This observation is based on the fact that the $x^2$ term has a larger slope than the $y^2$ term.

2. The $x$ variable is concave up and the $y$ variable is concave down. This observation is based on the fact that the $x^2$ is positive and the $y^2$ term is negative. Additionally, the directionality of the nonlinear terms indicates that the $x$ is minimizing its behavior and $y$ is maximizing its behavior. Since the terms are opposite in concavity, the overall shape of the surface is a saddle. The $y$ term is more constrained to the critical point of the surface than the $x$ term. This observation is based on the fact that the $y^2$ term has a larger slope than the $x^2$ term.

3. $x = x, y = y, z = -.751 + 1.089x + .576y - .009xy + .148x^2 + .004y^2$

   $r(x,y,z) = x\mathbf{i} + y\mathbf{j} + (-.751 + 1.089x + .576y - .009xy + .148x^2 + .004y^2)\mathbf{k}$

4. $x = x, y = y, z = .084 + .172x + 6.39y + 1.13xy + 2.17x^2 - 1.876y^2$

   $r(x,y,z) = x\mathbf{i} + y\mathbf{j} + z(.084 + .172x + 6.39y + 1.13xy + 2.17x^2 - 1.876y^2)\mathbf{k}$

5. $\iint_D -.751 + 1.089x + .576y - .009xy + .148x^2 + .004y^2 \ dxdy$

   Area under surface $= 32.67$.

6. $\iint_D .084 + .172x + 6.39y + 1.13xy + .217x^2 - 1.88y^2 \, dxdy$

Area under surface = 2.48.

7. $\dfrac{\partial f}{\partial x} = 1.089 - .009y + .296x$

$\dfrac{\partial f}{\partial y} = .576 - .009x + .008y$

$\begin{pmatrix} 1.089 & .296 & -.009 & \bigm| & 0 \\ .576 & -.009 & .008 & \bigm| & 0 \end{pmatrix}$

$\begin{pmatrix} 1 & 0 & .0127 & \bigm| & 0 \\ 0 & 1 & -.0771 & \bigm| & 0 \end{pmatrix}$

$y = -78.74, \; x = 6.07$

8. $\dfrac{\partial f}{\partial x} = .172 + 1.13y + .434x$

$\dfrac{\partial f}{\partial y} = 6.39 - 1.13x - 3.75y$

$\begin{pmatrix} .172 & .434 & 1.13 & \bigm| & 0 \\ 6.39 & 1.13 & -3.75 & \bigm| & 0 \end{pmatrix}$

$\begin{pmatrix} 1 & 0 & -1.13 & \bigm| & 0 \\ 0 & 1 & 3.05 & \bigm| & 0 \end{pmatrix}$

$y = .89, x = -2.71$

9. $f(x) = -46.42 + 1.79x + .148x^2$

The grid line function for the $x$ variable is concave up.

$f(y) = -1.91 + .601y + .004y^2$

The grid line function for the $y$ variable is concave up.

10. $f(x) = -30.98 - 2.89x + .217x^2$

The grid line function for the $x$ variable is concave up.

$f(y) = .409 + 7.396y - 1.876y^2$

The grid line function for the $y$ variable is concave down.

11. $\dfrac{\partial f}{\partial x} = 1.089 - .009y + .296x$

$\dfrac{\partial^2 f}{\partial x^2} = .296$

The $x$ variable is minimizing its behavior.

$\dfrac{\partial f}{\partial y} = .576 - .009x + .008y$

$\dfrac{\partial^2 f}{\partial x^2} = .008$

The $y$ variable is minimizing its behavior.

Since the $x$ variable has a larger slope, it is more constrained to the critical point of the surface.

12. $\dfrac{\partial f}{\partial x} = .172 + 1.13y + .434x$

$\dfrac{\partial^2 f}{\partial x^2} = .433$

The $x$ variable is minimizing its behavior.

$\dfrac{\partial f}{\partial y} = 6.39 + 1.13x - 3.75y$

$\dfrac{\partial^2 f}{\partial x^2} = -3.75$

The $y$ variable is maximizing its behavior.

Since the $y$ variable has a larger slope, it is more constrained to the critical point of the surface.

13. $f(x) = -.751 + 1.089x + .148x^2$

$f'(x) = 1.089 + .296x$

$f''(x) = .296$

$f'(0) = -3.68$

$\kappa(x) = \dfrac{|.296|}{\left[1 + (1.089 + .296x)^2\right]^{3/2}} = 0.296$

$f(y) = -.751 + .576y + .004y^2$

$f'(y) = .576 + .008y$

$f''(y) = .008$

$f'(0) = -72$

$\kappa(y) = \dfrac{|.008|}{\left[1 + (.576\ +\ .008)^2\right]^{3/2}} = 0.008$

The $x$ variable has less variability since it has greater curvature.

14. $f(x) = .084 + .172x + .217x^2$

$f'(x) = .172 + .434x$

$f''(x) = .434$

$f'(0) = -.396$

$\kappa(x) = \dfrac{|.434|}{\left[1 + (.172 + .434x)^2\right]^{3/2}} = 0.434$

$f(y) = .084 + 6.39y - 1.876y^2$

$f'(y) = 6.39 - 3.752y$

$f''(y) = -3.752$

$f'(0) = 1.70$

$\kappa(y) = \dfrac{|3.75|}{\left[1 + (6.39\ +\ 3.75)^2\right]^{3/2}} = 3.75$

The $y$ variable has less variability since it has more curvature.

15.
$$[\mathbf{f}] = \begin{pmatrix} -17.16 & -9.305 \\ 2.10 & 1.85 \\ -.041 & -.034 \\ 1.5 \times 10^{-11} & 1.1 \times 10^{-11} \\ -1.3 \times 10^{-13} & -9.5 \times 10^{-13} \end{pmatrix}$$

16. $[\mathbf{f}] = \quad 31.18 \qquad 37.24$
$$\begin{pmatrix} -1.615 & -1.47 \\ -.035 & -.0325 \\ -8.4 \times 10^{-11} & -6.6 \times 10^{-12} \\ 6.1 \times 10^{-14} & 4.5 \times 10^{-14} \end{pmatrix}$$

17.
$$\|\mathbf{u} - \mathbf{v}\| = \begin{pmatrix} -7.855 \\ .25 \\ .007 \\ 4 \times 10^{-12} \\ -3 \times 10^{-14} \end{pmatrix} = \sqrt{61.76} = 7.86$$

18.
$$\|\mathbf{u} - \mathbf{v}\| = \begin{pmatrix} -6.062 \\ -.145 \\ -.0025 \\ -7.74 \times 10^{-11} \\ 1.6 \times 10^{-14} \end{pmatrix} = \sqrt{36.77} = 6.06$$

# Chapter 7

1. 2.91 and 1.77 for the 0.2- and 0.4-s shifts, respectively.

2. Correlations of 0.70, 0.80, and 0.90 convert to $z$-scores of 8.867, 1.099, and 1.472, respectively. The average $z$-score is 1.146, which converts back to a correlation of 0.82, not 0.80.

# Chapter 9

All the solutions are based on MATLAB programming. If you do not have the full MATLAB software package, the student version (which is much cheaper) may also be used. Sometimes the whole solution is given by a MATLAB file. In that case the problem is named problemX.m; for example, problem1.m is the solution for problem 1 and there is a MATLAB file that you can use to see the solution. In other cases, such as for problem 8, an m file is not provided. The answer sometimes is also provided as a graph.

```
1. clear;clc;
   %digital resolution is 1000
   resolution=1000;
   i=1:resolution;
   j=(2*pi)/resolution;
   %The first function
   amplitude=16;
   frequency=2;
   y1=amplitude*cos(frequency*i*j);
   %The second function
   amplitude=6;
   frequency=14;
   y2=amplitude*sin(frequency*i*j);
   y=y1+y2;
   plot(y)
   FFTy=fft(y)'
```

Answer:

1.0e + 003 *

0.0000

0.0000 – 0.0000i

7.9994 – 0.1005i

–0.0000 + 0.0000i

–0.0000 + 0.0000i

–0.0000 + 0.0000i

–0.0000 – 0.0000i

–0.0000 + 0.0000i

0.0000 + 0.0000i

0.0000 + 0.0000i

–0.0000 + 0.0000i

0.0000 + 0.0000i

0.0000 + 0.0000i

0.0000 + 0.0000i

0.2636 + 2.9884i

0.0000 – 0.0000i

–0.0000 – 0.0000i

–0.0000 – 0.0000i

–0.0000 – 0.0000i

–0.0000 – 0.0000i

0.0000 – 0.0000i

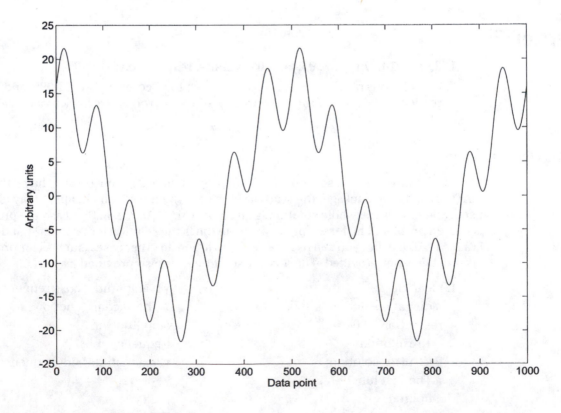

2. By adjusting the resolution, amplitude, and frequency in problem1,

| 1.0e + 005 * | 0.0000 – 0.0000i | 0.0000 + 0.0000i |
|---|---|---|
| –0.0000 | 0.0000 – 0.0000i | 0.0000 + 0.0000i |
| –0.0000 – 0.0000i | 0.0000 – 0.0000i | 0.0000 + 0.0000i |
| –0.0000 – 0.0000i | 0.0000 – 0.0000i | 0.0000 + 0.0000i |
| 3.5000 – 0.0007i | 0.0035 + 5.0000i | 0.0000 + 0.0000i |
| 0.0000 – 0.0000i | 0.0000 + 0.0000i | –0.0000 + 0.0000i |
| 0.0000 – 0.0000i | 0.0000 + 0.0000i | 0.0000 + 0.0000i |
| 0.0000 – 0.0000i | | |

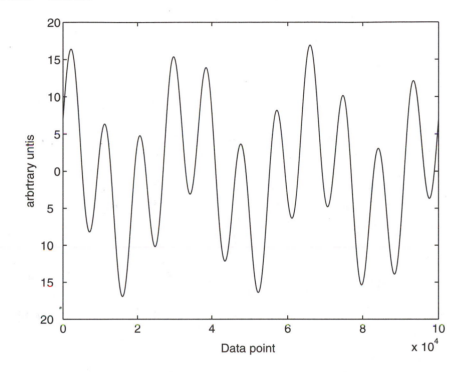

3. 
```
%load the file
disp=load('dowling.dat');
% Sampling frequency is 512
SF=512;
%calculate the PSD of the signal
PSDdisp = sqrt(psd(disp));
ndata=length(PSDdisp);
i=1:ndata;
j=(SF/2)/ndata*i;
figure(1)
semilogy(j,PSDdisp,'b')
xlabel('Frequency (Hz)');
ylabel('Power spectral density (log)');
axis tight
%calculate the FFT of the signal
FFTdisp = fft(disp);
%Calculate the square amplitude of the signal (power spectrum)
PS=FFTdisp.*conj(FFTdisp);
```

```
ndata=round(length(PS)/2);
i=1:ndata;
%the first PS number is the average
PSpositive(i)=PS(i+1);
%construct the frequency axis - up to half the sampling frequency
j=(SF/2)/ndata
figure(2);
semilogy(j,PSpositive,'b')
xlabel('Frequency (Hz)');
ylabel('Power spectrum (log)');
axis tight
```

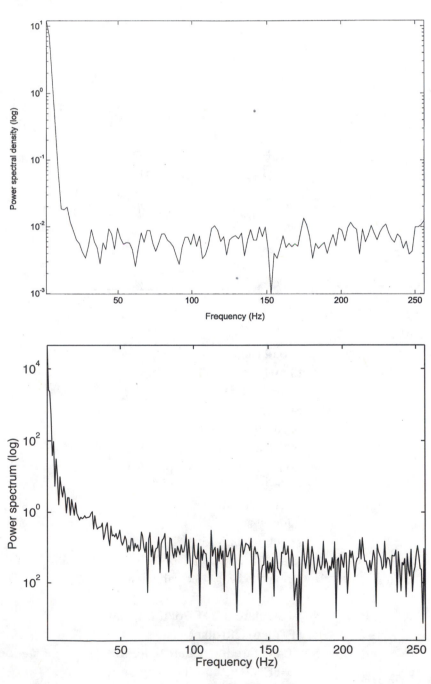

4. %load the file
   disp=load('dowling.dat');
   % Sampling frequency is 512
   SF=512;
   %calculate the FFT of the signal
   FFTdisp = fft(disp);
   %Calculate the square amplitude of the signal (power spectrum)
   PS=FFTdisp.*conj(FFTdisp);
   ndata=round(length(PS)/2);
   i=1:ndata;
   %the first PS number is the average
   PSpositive(i)=PS(i+1);
   %construct the frequency axis - up to half the sampling frequency
   j=(SF/2)/ndata*i'
   %calculate the overall sum of power and the % for each harmonic
   sumPS=sum(PSpositive);
   perc_powerPS=100*PSpositive/sumPS;
   %Find the cumulative sum as a percentage for each harmonic
   cumsumPS = cumsum(perc_powerPS)'

   The cumsumPS data show that:

   | 88.7860% | is on the | 1st | harmonic representing | 0.8533 Hz |
   | 93.8219% | is on the | 2nd | harmonic representing | 1.7067 Hz |
   | 98.4184% | is on the | 3rd | harmonic representing | 2.5600 Hz |
   | 99.4948% | is on the | 4th | harmonic representing | 3.4133 Hz |
   | 99.5721% | is on the | 5th | harmonic representing | 4.2667 Hz |

   Therefore, 90% is included in 1.7 Hz; 95% is included in 2.5 Hz; 99% is included in 3.4 Hz; 99.5% is included in 4.3 Hz.

5. Same as solution to 4, but instead of loading the Lanshammar data, we load the Pezzack data (pezzack.dat). Also, SF = 50.

   The cumsumPS data show that:

   90% is included in the 3rd harmonic representing 1 Hz

   95% is included in the 4th harmonic representing 1.4 Hz

   99% is included in the 6th harmonic representing 2.1 Hz

   99.5% is included in the 7th harmonic representing 2.4 Hz

6. First let's create the filtering function.

   function y = filtering(data, fs, fc, order)
   %Function filtering
   %Low pass butterworth digital filtering
   %Data: input file
   %fs: sampling frequency
   %fc: cutoff frequency
   %Order: order of the filter

   [b,a] = butter(order,2*fc/fs);
   y = filtfilt(b,a,data);

```
data = load('dowling.dat'5);
order = 2;
i = 1:length(data);
filtered_data3 = filtering(data, 512, 3, order);
filtered_data20 = filtering(data, 512, 20, order);

difference3 = abs(data-filtered_data3);
difference20 = abs(data-filtered_data20);

mean3 = mean(difference3)
mean20 = mean(difference20)
```

The results will be
```
mean3 = 0.0458
mean20 = 0.0057
```

7. 
```
data = load('dowling.dat');
realacc = load('dowling.acc');

SF = 512;
order = 2;
i = 1:length(data);
filtered_data3 = filtering(data, 512, 3, order);
filtered_data20 = filtering(data, 512, 20, order);

vel3 = diff(filtered_data3)*512;
vel20 = diff(filtered_data20)*512;

acc3 = diff(vel3)*512;
acc20 = diff(vel20)*512;

%Extend the calculated acceleration by two points to
%Match accelerometer data
ndataacc = length(acc3);
acc3(ndataacc+1) = acc3(ndataacc);
acc3(ndataacc+2) = acc3(ndataacc+1);
acc20(ndataacc+1) = acc20(ndataacc);
acc20(ndataacc+2) = acc20(ndataacc+1);

m = length(acc3);

%Plot graphs
i = 1:m;
figure(1)
plot(i,realacc,'-',i,acc3,'k.');
figure(2)
plot(i,realacc,'-',i,acc20,'k.');

difference3 = abs(realacc-acc3);
difference20 = abs(realacc-acc20);
```

mean3 = mean(difference3)

mean20 = mean(difference20)

The results will be

RMS3 = 26.1268

RMS20 = 15.8847

Best cutoff is around 11 Hz.

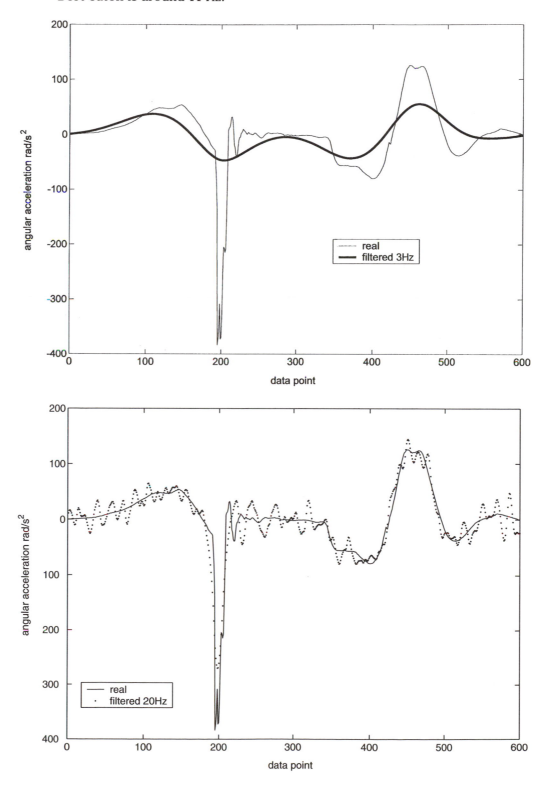

8. Based on the above program the first and second derivatives can be easily
   calculated

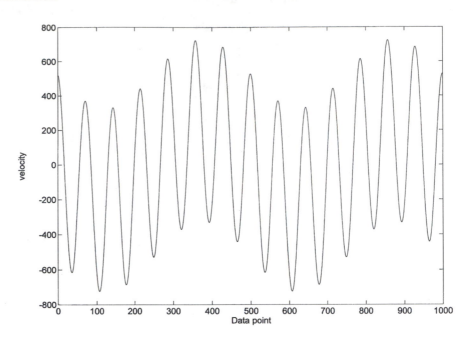

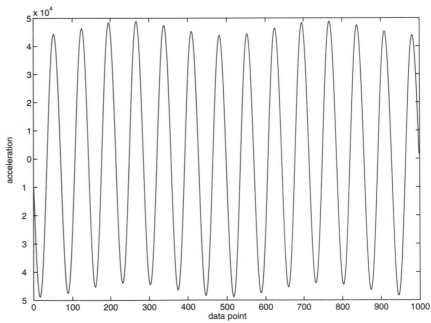

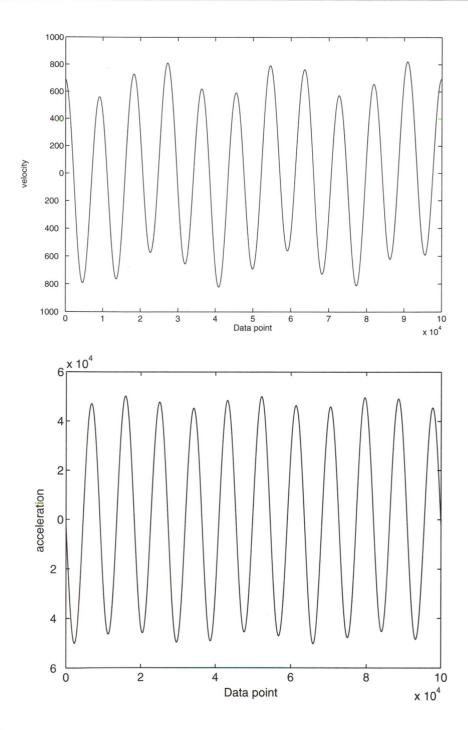

9. clear
   [filename, pathname] = uigetfile('*.*', 'EMG file');
   cd(pathname)
   a = load(filename);
   pause

   windowsize = 1,024;
   ii = 1:windowsize;

   [r,c] = size(a);

```
x = a(:,1);
%Use the hanning window
windowfr = hanning(windowsize);
%Perform the short-time Fourier transform. Also calculate the median fre-
quency.
k = 0;
iii = k*windowsize+ii;
while (k*windowsize+11)<r
k = k+1;
iii = k*windowsize+ii;
if iii<r
signal = x(iii);
Pxx = psd(signal, windowsize, 1,024, windowfr);
plength = length(Pxx);
for p = 1:plength
Pxx_all(k,p) = Pxx(p);
end
median_freq(k) = median(Pxx);
end
end

%Show graphs for the TF distribution.
figure(1)
%Contour(Pxx_all)
mesh(Pxx_all)
%View(-22 , 90);
axis tight
xlabel('frequency (Hz)');
ylabel('Time (sec)');
%zlabel('Amplitude (V)');

%Show graph—the development of median frequency.
figure(2)
plot(median_freq)
xlabel('Time (sec)');
ylabel('Normalised Frequency');
```

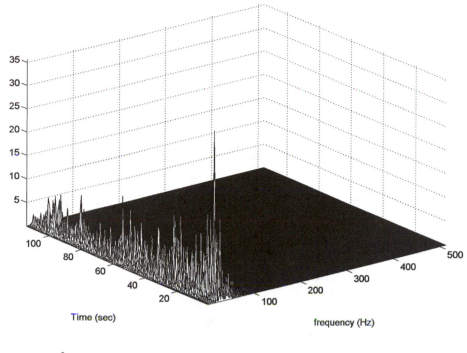

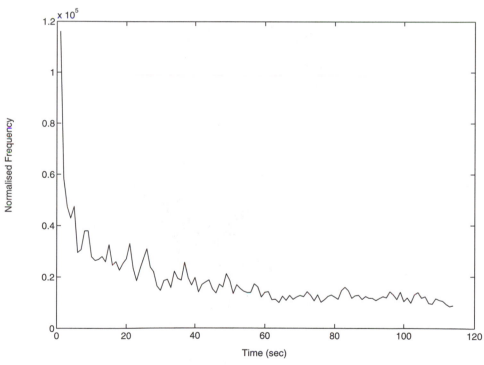

10.

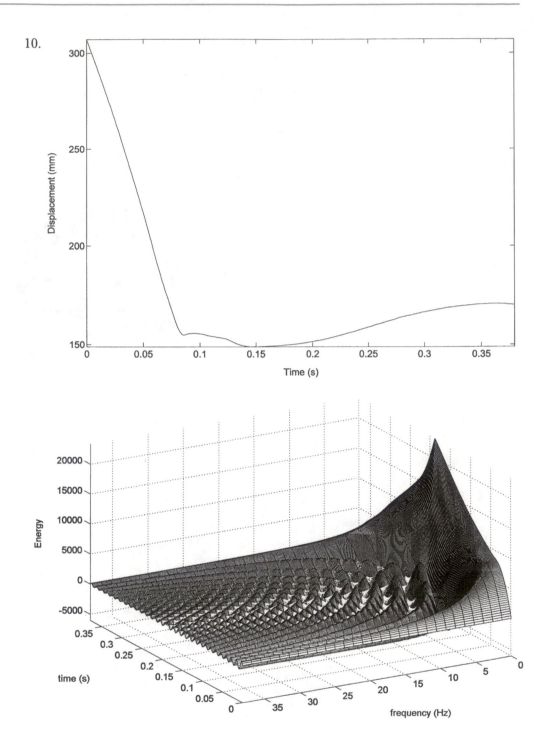

# Data Sets for Chapter 2

# Table B.1 Resultant GRF Data From One Individual for 10 Landing Trials From a Height of 0.45 m

*These data were used to create figures 2.5 and 2.6 and demonstrate equations 2.12-2.16.*
*All GRF data are in Newtons.*

| Time (s) | Trial 1 | Trial 2 | Trial 3 | Trial 4 | Trial 5 | Trial 6 | Trial 7 | Trial 8 | Trial 9 | Trial 10 | M (N) [Eqn 12] | Point-by-point SD (N) [Eqn 13] | Point-by-point SD² (N²) | Point-by-point CV (%) [Eqn 14] | Curve average SD (N) [Eqn 15] | Curve average CV (%) [Eqn 16] |
|---|---|---|---|---|---|---|---|---|---|---|---|---|---|---|---|---|
| 0 | 11 | 13 | 8 | 12 | 12 | 9 | 9 | 10 | 8 | 6 | 9.7 | 2.2 | 4.9 | 22.9 | 274.7 | 24.3 |
| 0.001 | 25 | 61 | 50 | 42 | 28 | 35 | 27 | 51 | 40 | 39 | 39.7 | 11.7 | 135.7 | 29.3 | 274.7 | 24.3 |
| 0.002 | 75 | 118 | 135 | 108 | 83 | 95 | 89 | 116 | 100 | 107 | 102.4 | 18.0 | 324.6 | 17.6 | 274.7 | 24.3 |
| 0.003 | 94 | 109 | 154 | 110 | 117 | 101 | 110 | 123 | 98 | 118 | 113.2 | 17.0 | 288.1 | 15.0 | 274.7 | 24.3 |
| 0.004 | 62 | 111 | 155 | 68 | 101 | 66 | 87 | 127 | 68 | 102 | 94.5 | 30.5 | 929.6 | 32.3 | 274.7 | 24.3 |
| 0.005 | 93 | 217 | 256 | 119 | 150 | 110 | 165 | 232 | 134 | 173 | 164.9 | 55.0 | 3021.9 | 33.3 | 274.7 | 24.3 |
| 0.006 | 187 | 331 | 389 | 221 | 261 | 201 | 271 | 337 | 226 | 274 | 269.8 | 65.4 | 4272.4 | 24.2 | 274.7 | 24.3 |
| 0.007 | 255 | 453 | 550 | 279 | 328 | 243 | 311 | 448 | 261 | 352 | 348.0 | 103.1 | 10635.5 | 29.6 | 274.7 | 24.3 |
| 0.008 | 311 | 659 | 750 | 365 | 468 | 311 | 433 | 657 | 346 | 513 | 481.3 | 159.1 | 25315.5 | 33.1 | 274.7 | 24.3 |
| 0.009 | 421 | 912 | 917 | 513 | 695 | 434 | 645 | 890 | 504 | 742 | 667.2 | 195.6 | 38256.4 | 29.3 | 274.7 | 24.3 |
| 0.010 | 609 | 1158 | 1045 | 726 | 907 | 603 | 876 | 1081 | 712 | 965 | 868.1 | 198.3 | 39305.5 | 22.8 | 274.7 | 24.3 |
| 0.011 | 853 | 1309 | 1087 | 990 | 1088 | 809 | 1112 | 1203 | 948 | 1157 | 1055.6 | 156.0 | 24349.7 | 14.8 | 274.7 | 24.3 |
| 0.012 | 1100 | 1327 | 1032 | 1228 | 1182 | 1003 | 1276 | 1237 | 1141 | 1262 | 1178.7 | 107.5 | 11556.5 | 9.1 | 274.7 | 24.3 |
| 0.013 | 1317 | 1238 | 918 | 1427 | 1170 | 1158 | 1353 | 1196 | 1276 | 1282 | 1233.4 | 138.8 | 19257.1 | 11.3 | 274.7 | 24.3 |
| 0.014 | 1459 | 1060 | 765 | 1544 | 1089 | 1249 | 1341 | 1085 | 1332 | 1217 | 1214.1 | 225.7 | 50934.5 | 18.6 | 274.7 | 24.3 |
| 0.015 | 1480 | 851 | 633 | 1539 | 942 | 1257 | 1225 | 937 | 1286 | 1061 | 1121.2 | 286.8 | 82246.2 | 25.6 | 274.7 | 24.3 |
| 0.016 | 1367 | 678 | 560 | 1419 | 774 | 1202 | 1054 | 821 | 1161 | 882 | 991.6 | 293.3 | 86048.0 | 29.6 | 274.7 | 24.3 |
| 0.017 | 1163 | 563 | 529 | 1225 | 648 | 1117 | 883 | 764 | 1002 | 766 | 865.9 | 252.1 | 63561.6 | 29.1 | 274.7 | 24.3 |
| 0.018 | 934 | 493 | 532 | 1018 | 583 | 1006 | 745 | 741 | 847 | 710 | 761.0 | 189.4 | 35872.1 | 24.9 | 274.7 | 24.3 |
| 0.019 | 734 | 461 | 572 | 839 | 552 | 887 | 648 | 745 | 720 | 677 | 683.4 | 130.6 | 17059.9 | 19.1 | 274.7 | 24.3 |
| 0.020 | 591 | 486 | 666 | 718 | 554 | 794 | 601 | 781 | 644 | 689 | 652.4 | 98.0 | 9601.3 | 15.0 | 274.7 | 24.3 |
| 0.021 | 506 | 551 | 775 | 653 | 600 | 723 | 595 | 835 | 614 | 734 | 658.7 | 104.8 | 10992.0 | 15.9 | 274.7 | 24.3 |
| 0.022 | 468 | 617 | 837 | 618 | 662 | 666 | 599 | 893 | 609 | 773 | 674.3 | 126.3 | 15947.0 | 18.7 | 274.7 | 24.3 |
| 0.023 | 467 | 694 | 868 | 601 | 734 | 647 | 614 | 927 | 618 | 827 | 699.7 | 140.5 | 19753.2 | 20.1 | 274.7 | 24.3 |
| 0.024 | 500 | 770 | 874 | 608 | 814 | 655 | 648 | 916 | 640 | 882 | 730.8 | 139.5 | 19454.1 | 19.1 | 274.7 | 24.3 |
| 0.025 | 552 | 803 | 846 | 634 | 850 | 666 | 684 | 889 | 669 | 891 | 748.4 | 121.0 | 14648.8 | 16.2 | 274.7 | 24.3 |
| 0.026 | 606 | 790 | 810 | 658 | 836 | 681 | 716 | 870 | 699 | 860 | 752.5 | 92.4 | 8531.8 | 12.3 | 274.7 | 24.3 |
| 0.027 | 650 | 759 | 784 | 673 | 804 | 704 | 738 | 864 | 737 | 828 | 754.2 | 67.8 | 4591.1 | 9.0 | 274.7 | 24.3 |
| 0.028 | 686 | 736 | 782 | 689 | 774 | 716 | 733 | 872 | 767 | 811 | 756.6 | 57.2 | 3269.2 | 7.6 | 274.7 | 24.3 |
| 0.029 | 719 | 726 | 811 | 703 | 760 | 710 | 717 | 896 | 764 | 810 | 761.6 | 61.4 | 3767.3 | 8.1 | 274.7 | 24.3 |
| 0.030 | 733 | 730 | 858 | 707 | 759 | 704 | 708 | 940 | 745 | 821 | 770.4 | 78.3 | 6128.2 | 10.2 | 274.7 | 24.3 |
| 0.031 | 722 | 750 | 898 | 699 | 776 | 701 | 707 | 993 | 735 | 842 | 782.3 | 98.8 | 9751.8 | 12.6 | 274.7 | 24.3 |
| 0.032 | 706 | 780 | 929 | 693 | 819 | 701 | 719 | 1035 | 739 | 873 | 799.3 | 114.7 | 13156.6 | 14.4 | 274.7 | 24.3 |
| 0.033 | 700 | 824 | 961 | 695 | 880 | 707 | 742 | 1064 | 756 | 919 | 824.8 | 127.3 | 16196.8 | 15.4 | 274.7 | 24.3 |
| 0.034 | 707 | 890 | 1001 | 703 | 934 | 721 | 771 | 1095 | 778 | 976 | 857.5 | 140.2 | 19667.5 | 16.4 | 274.7 | 24.3 |
| 0.035 | 724 | 948 | 1046 | 716 | 969 | 750 | 802 | 1134 | 809 | 1020 | 891.7 | 149.8 | 22453.4 | 16.8 | 274.7 | 24.3 |
| 0.036 | 749 | 975 | 1131 | 735 | 1004 | 783 | 839 | 1183 | 851 | 1049 | 929.9 | 160.9 | 25879.0 | 17.3 | 274.7 | 24.3 |
| 0.037 | 784 | 1037 | 1304 | 761 | 1047 | 817 | 886 | 1249 | 893 | 1080 | 985.7 | 189.7 | 35969.9 | 19.2 | 274.7 | 24.3 |
| 0.038 | 823 | 1237 | 1500 | 791 | 1107 | 853 | 933 | 1360 | 937 | 1121 | 1066.1 | 241.0 | 58079.4 | 22.6 | 274.7 | 24.3 |
| 0.039 | 857 | 1501 | 1656 | 821 | 1230 | 891 | 970 | 1515 | 993 | 1174 | 1160.9 | 304.8 | 92926.5 | 26.3 | 274.7 | 24.3 |
| 0.040 | 888 | 1686 | 1818 | 852 | 1427 | 937 | 1002 | 1655 | 1043 | 1235 | 1254.4 | 365.1 | 133331.9 | 29.1 | 274.7 | 24.3 |
| 0.041 | 931 | 1855 | 2024 | 891 | 1611 | 990 | 1040 | 1774 | 1075 | 1340 | 1353.0 | 427.6 | 182855.2 | 31.6 | 274.7 | 24.3 |
| 0.042 | 977 | 2103 | 2252 | 941 | 1764 | 1035 | 1087 | 1913 | 1113 | 1496 | 1467.9 | 503.4 | 253418.2 | 34.3 | 274.7 | 24.3 |
| 0.043 | 1005 | 2387 | 2448 | 986 | 1943 | 1064 | 1154 | 2074 | 1164 | 1631 | 1585.6 | 585.6 | 342956.6 | 36.9 | 274.7 | 24.3 |
| 0.044 | 1026 | 2621 | 2599 | 1018 | 2160 | 1086 | 1278 | 2246 | 1229 | 1746 | 1700.7 | 655.1 | 429202.0 | 38.5 | 274.7 | 24.3 |

| | | | | | | | | | | | | | | | | |
|---|---|---|---|---|---|---|---|---|---|---|---|---|---|---|---|---|
| 0.045 | 1057 | 2798 | 2710 | 1053 | 2375 | 1112 | 1454 | 2408 | 1343 | 1886 | 1819.6 | 702.4 | 493423.5 | 38.6 | 274.7 | 24.3 |
| 0.046 | 1096 | 2920 | 2777 | 1092 | 2550 | 1147 | 1606 | 2536 | 1516 | 2036 | 1927.5 | 726.3 | 527504.8 | 37.7 | 274.7 | 24.3 |
| 0.047 | 1146 | 2986 | 2801 | 1151 | 2684 | 1179 | 1734 | 2606 | 1669 | 2180 | 2013.7 | 729.1 | 531520.3 | 36.2 | 274.7 | 24.3 |
| 0.048 | 1248 | 3005 | 2789 | 1276 | 2777 | 1204 | 1893 | 2630 | 1792 | 2317 | 2093.2 | 701.1 | 491549.0 | 33.5 | 274.7 | 24.3 |
| 0.049 | 1401 | 2975 | 2746 | 1454 | 2826 | 1224 | 2070 | 2633 | 1944 | 2428 | 2170.1 | 645.5 | 416689.4 | 29.7 | 274.7 | 24.3 |
| 0.051 | 1526 | 2901 | 2679 | 1602 | 2842 | 1273 | 2237 | 2613 | 2119 | 2490 | 2228.1 | 582.8 | 339712.7 | 26.2 | 274.7 | 24.3 |
| 0.052 | 1622 | 2805 | 2593 | 1725 | 2826 | 1377 | 2382 | 2565 | 2292 | 2518 | 2270.6 | 513.8 | 263983.3 | 22.6 | 274.7 | 24.3 |
| 0.053 | 1741 | 2693 | 2490 | 1878 | 2777 | 1491 | 2490 | 2496 | 2445 | 2529 | 2303.0 | 436.0 | 190060.8 | 18.9 | 274.7 | 24.3 |
| 0.054 | 1889 | 2561 | 2378 | 2051 | 2706 | 1587 | 2555 | 2412 | 2558 | 2515 | 2321.2 | 359.7 | 129348.3 | 15.5 | 274.7 | 24.3 |
| 0.055 | 2041 | 2425 | 2268 | 2205 | 2623 | 1695 | 2585 | 2318 | 2629 | 2473 | 2326.0 | 293.5 | 86125.9 | 12.6 | 274.7 | 24.3 |
| 0.056 | 2177 | 2293 | 2158 | 2318 | 2526 | 1821 | 2588 | 2216 | 2666 | 2410 | 2317.3 | 247.1 | 61078.7 | 10.7 | 274.7 | 24.3 |
| 0.057 | 2277 | 2161 | 2050 | 2390 | 2422 | 1943 | 2565 | 2113 | 2674 | 2335 | 2293.1 | 230.5 | 53110.3 | 10.1 | 274.7 | 24.3 |
| 0.058 | 2331 | 2029 | 1946 | 2420 | 2320 | 2055 | 2518 | 2014 | 2655 | 2251 | 2253.9 | 238.3 | 56802.9 | 10.6 | 274.7 | 24.3 |
| 0.059 | 2351 | 1899 | 1846 | 2413 | 2216 | 2156 | 2457 | 1918 | 2616 | 2161 | 2203.3 | 259.6 | 67407.3 | 11.8 | 274.7 | 24.3 |
| 0.060 | 2345 | 1772 | 1751 | 2383 | 2107 | 2226 | 2390 | 1827 | 2563 | 2067 | 2142.9 | 287.0 | 82343.1 | 13.4 | 274.7 | 24.3 |
| 0.061 | 2312 | 1650 | 1662 | 2334 | 2000 | 2257 | 2315 | 1739 | 2500 | 1970 | 2073.9 | 312.0 | 97314.0 | 15.0 | 274.7 | 24.3 |
| 0.062 | 2259 | 1529 | 1578 | 2270 | 1897 | 2264 | 2233 | 1652 | 2426 | 1873 | 1998.1 | 332.2 | 110359.4 | 16.6 | 274.7 | 24.3 |
| 0.063 | 2196 | 1418 | 1499 | 2199 | 1797 | 2254 | 2147 | 1571 | 2345 | 1782 | 1920.8 | 346.8 | 120275.5 | 18.1 | 274.7 | 24.3 |
| 0.064 | 2123 | 1328 | 1427 | 2122 | 1703 | 2224 | 2059 | 1500 | 2257 | 1694 | 1843.7 | 352.5 | 124232.5 | 19.1 | 274.7 | 24.3 |
| 0.065 | 2041 | 1253 | 1358 | 2035 | 1615 | 2177 | 1967 | 1433 | 2163 | 1607 | 1764.9 | 350.1 | 122579.1 | 19.8 | 274.7 | 24.3 |
| 0.066 | 1953 | 1186 | 1297 | 1943 | 1534 | 2120 | 1876 | 1368 | 2068 | 1529 | 1687.4 | 342.9 | 117571.1 | 20.3 | 274.7 | 24.3 |
| 0.067 | 1867 | 1133 | 1244 | 1854 | 1461 | 2056 | 1784 | 1308 | 1972 | 1461 | 1614.1 | 330.3 | 109129.3 | 20.5 | 274.7 | 24.3 |
| 0.068 | 1781 | 1094 | 1195 | 1762 | 1392 | 1986 | 1692 | 1255 | 1873 | 1395 | 1542.6 | 313.0 | 97992.4 | 20.3 | 274.7 | 24.3 |
| 0.069 | 1695 | 1054 | 1146 | 1665 | 1327 | 1914 | 1606 | 1204 | 1773 | 1335 | 1471.9 | 295.0 | 87020.8 | 20.0 | 274.7 | 24.3 |
| 0.070 | 1613 | 1010 | 1100 | 1569 | 1267 | 1842 | 1526 | 1160 | 1675 | 1286 | 1404.8 | 277.0 | 76733.6 | 19.7 | 274.7 | 24.3 |
| 0.071 | 1533 | 978 | 1063 | 1479 | 1211 | 1767 | 1451 | 1122 | 1582 | 1239 | 1342.5 | 256.6 | 65858.5 | 19.1 | 274.7 | 24.3 |
| 0.072 | 1456 | 955 | 1029 | 1393 | 1155 | 1692 | 1383 | 1086 | 1495 | 1186 | 1283.2 | 235.9 | 55629.0 | 18.4 | 274.7 | 24.3 |
| 0.073 | 1385 | 928 | 1000 | 1314 | 1104 | 1620 | 1323 | 1056 | 1416 | 1137 | 1228.2 | 217.4 | 47266.1 | 17.7 | 274.7 | 24.3 |
| 0.074 | 1321 | 898 | 977 | 1243 | 1062 | 1550 | 1268 | 1033 | 1345 | 1095 | 1179.1 | 199.7 | 39894.1 | 16.9 | 274.7 | 24.3 |
| 0.075 | 1261 | 882 | 955 | 1181 | 1025 | 1482 | 1219 | 1009 | 1281 | 1054 | 1135.0 | 182.0 | 33127.5 | 16.0 | 274.7 | 24.3 |
| 0.076 | 1211 | 870 | 933 | 1130 | 994 | 1418 | 1176 | 985 | 1226 | 1018 | 1095.9 | 166.0 | 27544.5 | 15.1 | 274.7 | 24.3 |
| 0.077 | 1165 | 852 | 915 | 1086 | 970 | 1359 | 1134 | 968 | 1176 | 992 | 1061.5 | 151.1 | 22844.7 | 14.2 | 274.7 | 24.3 |
| 0.078 | 1123 | 841 | 897 | 1048 | 952 | 1302 | 1091 | 953 | 1127 | 972 | 1030.7 | 135.4 | 18322.1 | 13.1 | 274.7 | 24.3 |
| 0.079 | 1087 | 839 | 877 | 1019 | 934 | 1251 | 1050 | 934 | 1084 | 951 | 1002.6 | 121.4 | 14735.5 | 12.1 | 274.7 | 24.3 |
| 0.080 | 1058 | 827 | 856 | 992 | 916 | 1209 | 1012 | 915 | 1046 | 932 | 976.2 | 112.1 | 12572.3 | 11.5 | 274.7 | 24.3 |
| 0.081 | 1027 | 809 | 836 | 964 | 898 | 1171 | 978 | 900 | 1007 | 914 | 950.4 | 104.4 | 10907.2 | 11.0 | 274.7 | 24.3 |
| 0.082 | 992 | 798 | 813 | 939 | 878 | 1135 | 951 | 884 | 970 | 896 | 925.6 | 97.2 | 9451.0 | 10.5 | 274.7 | 24.3 |
| 0.083 | 963 | 785 | 791 | 918 | 855 | 1103 | 928 | 863 | 939 | 879 | 902.3 | 92.2 | 8499.1 | 10.2 | 274.7 | 24.3 |
| 0.084 | 940 | 765 | 773 | 892 | 838 | 1073 | 902 | 842 | 909 | 867 | 880.1 | 88.4 | 7810.6 | 10.0 | 274.7 | 24.3 |
| 0.085 | 914 | 751 | 756 | 865 | 823 | 1045 | 876 | 826 | 882 | 857 | 859.5 | 83.8 | 7021.4 | 9.7 | 274.7 | 24.3 |
| 0.086 | 891 | 746 | 739 | 846 | 805 | 1018 | 857 | 806 | 862 | 843 | 841.4 | 79.2 | 6278.3 | 9.4 | 274.7 | 24.3 |
| 0.087 | 872 | 735 | 723 | 831 | 786 | 991 | 844 | 788 | 850 | 828 | 824.8 | 76.1 | 5791.6 | 9.2 | 274.7 | 24.3 |
| 0.088 | 853 | 718 | 713 | 813 | 769 | 963 | 828 | 775 | 840 | 814 | 808.6 | 72.4 | 5243.7 | 9.0 | 274.7 | 24.3 |
| 0.089 | 837 | 709 | 707 | 796 | 750 | 936 | 812 | 762 | 828 | 798 | 793.5 | 67.8 | 4599.8 | 8.5 | 274.7 | 24.3 |
| 0.090 | 826 | 702 | 701 | 783 | 732 | 913 | 796 | 746 | 817 | 780 | 779.7 | 64.3 | 4138.3 | 8.3 | 274.7 | 24.3 |
| 0.091 | 813 | 691 | 698 | 766 | 719 | 894 | 778 | 734 | 804 | 767 | 766.6 | 61.1 | 3735.6 | 8.0 | 274.7 | 24.3 |
| 0.092 | 798 | 686 | 699 | 745 | 709 | 876 | 760 | 727 | 788 | 756 | 754.4 | 56.4 | 3181.6 | 7.5 | 274.7 | 24.3 |
| 0.093 | 785 | 687 | 702 | 727 | 697 | 858 | 744 | 719 | 773 | 741 | 743.3 | 51.3 | 2634.7 | 6.9 | 274.7 | 24.3 |
| 0.094 | 777 | 685 | 704 | 713 | 688 | 841 | 730 | 709 | 761 | 727 | 733.2 | 47.7 | 2276.3 | 6.5 | 274.7 | 24.3 |
| 0.095 | 765 | 680 | 708 | 696 | 680 | 823 | 715 | 702 | 748 | 717 | 723.6 | 44.1 | 1940.6 | 6.1 | 274.7 | 24.3 |
| 0.096 | 750 | 682 | 715 | 681 | 675 | 805 | 702 | 697 | 735 | 713 | 715.4 | 39.6 | 1571.9 | 5.5 | 274.7 | 24.3 |
| 0.097 | 741 | 687 | 721 | 669 | 671 | 791 | 692 | 692 | 722 | 709 | 709.5 | 36.8 | 1355.5 | 5.2 | 274.7 | 24.3 |
| 0.098 | 734 | 688 | 726 | 659 | 669 | 780 | 682 | 689 | 711 | 707 | 704.4 | 35.6 | 1269.8 | 5.1 | 274.7 | 24.3 |
| 0.099 | 723 | 686 | 734 | 648 | 669 | 769 | 674 | 687 | 699 | 706 | 699.5 | 35.3 | 1245.0 | 5.0 | 274.7 | 24.3 |
| 0.100 | 714 | 690 | 743 | 640 | 669 | 758 | 669 | 685 | 688 | 704 | 696.1 | 35.5 | 1258.6 | 5.1 | 274.7 | 24.3 |

## Table B.2 Right Ankle and Knee Angle Data for One Individual for 10 Consecutive Running Strides (Cycles)

***These data were used to create figures 2.7, 2.8, and 2.9 and demonstrate equations 2.17-2.35.***
***All joint angle values are in degrees (ankle: + is dorsiflexion; knee: + is flexion).***

| | Ankle(y-variable) | | | | | | | | | | | | | | |
|---|---|---|---|---|---|---|---|---|---|---|---|---|---|---|---|
| %Stride | Cycle 1 | Cycle 2 | Cycle 3 | Cycle 4 | Cycle 5 | Cycle 6 | Cycle 7 | Cycle 8 | Cycle 9 | Cycle 10 | $M_{yi}$ | $SD_{yi}$ (population) | $SD_{yi}$ (sample) | $y_{pi}$ [Eqn 34] | $y_{mi}$ [Eqn 35] |
| 0 | -46.1 | -45.4 | -47.3 | -46.1 | -47.8 | -46.6 | -48.1 | -48.3 | -47.9 | -46.5 | -47.00 | 0.94 | 0.99 | -46.06 | -47.94 |
| 1 | -46.0 | -44.9 | -46.8 | -45.7 | -47.7 | -46.8 | -47.9 | -47.9 | -47.8 | -46.4 | -46.79 | 0.99 | 1.05 | -45.80 | -47.78 |
| 2 | -45.8 | -44.2 | -46.1 | -45.3 | -47.5 | -46.8 | -47.6 | -47.5 | -47.5 | -46.0 | -46.42 | 1.08 | 1.14 | -45.34 | -47.50 |
| 3 | -45.2 | -43.3 | -45.1 | -44.6 | -46.9 | -46.5 | -47.1 | -46.6 | -46.9 | -45.3 | -45.75 | 1.19 | 1.25 | -44.56 | -46.93 |
| 4 | -44.4 | -42.0 | -43.7 | -43.7 | -46.1 | -45.8 | -46.2 | -45.4 | -46.0 | -44.2 | -44.75 | 1.32 | 1.39 | -43.43 | -46.06 |
| 5 | -43.3 | -40.4 | -42.1 | -42.5 | -44.9 | -44.8 | -45.1 | -44.0 | -44.9 | -42.9 | -43.49 | 1.47 | 1.55 | -42.02 | -44.96 |
| 6 | -42.0 | -38.5 | -40.3 | -41.2 | -43.6 | -43.7 | -43.8 | -42.3 | -43.6 | -41.3 | -42.01 | 1.65 | 1.74 | -40.36 | -43.67 |
| 7 | -40.4 | -36.4 | -38.4 | -39.6 | -42.0 | -42.3 | -42.2 | -40.3 | -42.1 | -39.4 | -40.32 | 1.86 | 1.96 | -38.46 | -42.18 |
| 8 | -38.7 | -33.9 | -36.2 | -37.8 | -40.3 | -40.6 | -40.5 | -38.1 | -40.4 | -37.3 | -38.38 | 2.08 | 2.20 | -36.30 | -40.47 |
| 9 | -36.7 | -31.2 | -33.8 | -35.7 | -38.3 | -38.7 | -38.5 | -35.6 | -38.6 | -35.0 | -36.20 | 2.32 | 2.44 | -33.89 | -38.52 |
| 10 | -34.5 | -28.3 | -31.3 | -33.4 | -36.0 | -36.4 | -36.2 | -32.9 | -36.5 | -32.3 | -33.77 | 2.55 | 2.69 | -31.22 | -36.32 |
| 11 | -32.0 | -25.1 | -28.5 | -30.9 | -33.5 | -33.9 | -33.7 | -29.9 | -34.2 | -29.5 | -31.01 | 2.77 | 2.92 | -28.33 | -33.87 |
| 12 | -29.2 | -21.8 | -25.6 | -28.1 | -30.6 | -31.0 | -31.0 | -26.8 | -31.7 | -26.4 | -28.22 | 2.95 | 3.11 | -25.27 | -31.16 |
| 13 | -26.2 | -18.5 | -22.5 | -25.2 | -27.6 | -27.9 | -28.0 | -23.5 | -29.0 | -23.3 | -25.17 | 3.08 | 3.25 | -22.09 | -28.25 |
| 14 | -23.2 | -15.2 | -19.3 | -22.2 | -24.4 | -24.5 | -25.0 | -20.2 | -26.2 | -20.1 | -22.02 | 3.16 | 3.33 | -18.86 | -25.18 |
| 15 | -20.0 | -12.1 | -16.2 | -19.1 | -21.1 | -21.1 | -21.9 | -17.0 | -23.4 | -16.9 | -18.86 | 3.18 | 3.35 | -15.68 | -22.04 |
| 16 | -16.9 | -9.2 | -13.2 | -16.1 | -17.8 | -17.7 | -18.8 | -13.8 | -20.5 | -13.8 | -15.79 | 3.14 | 3.31 | -12.64 | -18.93 |
| 17 | -14.0 | -6.6 | -10.3 | -13.3 | -14.7 | -14.5 | -15.9 | -10.9 | -17.7 | -11.0 | -12.88 | 3.06 | 3.23 | -9.81 | -15.94 |
| 18 | -11.2 | -4.3 | -7.7 | -10.7 | -11.8 | -11.5 | -13.2 | -8.2 | -15.1 | -8.4 | -10.22 | 2.95 | 3.11 | -7.27 | -13.17 |
| 19 | -8.9 | -2.4 | -5.5 | -8.4 | -9.2 | -8.8 | -10.8 | -5.9 | -12.7 | -6.2 | -7.89 | 2.80 | 2.96 | -5.08 | -10.69 |
| 20 | -6.9 | -1.0 | -3.6 | -6.5 | -7.1 | -6.5 | -8.8 | -4.0 | -10.7 | -4.4 | -5.94 | 2.64 | 2.79 | -3.30 | -8.59 |
| 21 | -5.4 | -0.0 | -2.2 | -5.0 | -5.4 | -4.6 | -7.2 | -2.5 | -9.0 | -3.1 | -4.44 | 2.47 | 2.60 | -1.97 | -6.90 |
| 22 | -4.3 | 0.4 | -1.2 | -3.9 | -4.3 | -3.2 | -6.1 | -1.5 | -7.7 | -2.3 | -3.41 | 2.29 | 2.41 | -1.13 | -5.70 |
| 23 | -3.8 | 0.3 | -0.8 | -3.3 | -3.7 | -2.4 | -5.6 | -1.1 | -6.9 | -2.0 | -2.91 | 2.11 | 2.22 | -0.80 | -5.01 |
| 24 | -3.9 | -0.3 | -0.8 | -3.2 | -3.6 | -2.1 | -5.5 | -1.2 | -6.6 | -2.2 | -2.95 | 1.93 | 2.03 | -1.02 | -4.88 |
| 25 | -4.5 | -1.6 | -1.4 | -3.6 | -4.1 | -2.3 | -6.0 | -1.9 | -6.9 | -3.1 | -3.56 | 1.77 | 1.86 | -1.79 | -5.32 |
| 26 | -5.7 | -3.4 | -2.6 | -4.6 | -5.2 | -3.2 | -7.1 | -3.1 | -7.7 | -4.6 | -4.73 | 1.62 | 1.71 | -3.10 | -6.35 |
| 27 | -7.5 | -5.7 | -4.3 | -6.2 | -6.9 | -4.7 | -8.7 | -5.0 | -9.0 | -6.6 | -6.45 | 1.51 | 1.59 | -4.94 | -7.96 |
| 28 | -9.7 | -8.5 | -6.5 | -8.2 | -9.0 | -6.8 | -10.7 | -7.4 | -10.7 | -9.2 | -8.68 | 1.43 | 1.51 | -7.25 | -10.11 |
| 29 | -12.4 | -11.6 | -9.1 | -10.7 | -11.6 | -9.3 | -13.2 | -10.2 | -12.9 | -12.2 | -11.34 | 1.38 | 1.45 | -9.96 | -12.71 |
| 30 | -15.3 | -14.9 | -12.0 | -13.6 | -14.6 | -12.3 | -16.0 | -13.4 | -15.4 | -15.5 | -14.31 | 1.34 | 1.41 | -12.97 | -15.65 |
| 31 | -18.5 | -18.3 | -15.2 | -16.6 | -17.7 | -15.6 | -19.0 | -16.7 | -18.1 | -19.0 | -17.47 | 1.30 | 1.37 | -16.17 | -18.77 |
| 32 | -21.6 | -21.6 | -18.4 | -19.8 | -20.9 | -18.9 | -22.0 | -20.1 | -20.9 | -22.3 | -20.66 | 1.25 | 1.32 | -19.40 | -21.91 |
| 33 | -24.6 | -24.6 | -21.5 | -22.8 | -24.1 | -22.2 | -24.9 | -23.4 | -23.6 | -25.5 | -23.73 | 1.19 | 1.26 | -22.54 | -24.92 |
| 34 | -27.4 | -27.3 | -24.5 | -25.7 | -26.9 | -25.3 | -27.6 | -26.4 | -26.2 | -28.3 | -26.55 | 1.11 | 1.17 | -25.45 | -27.66 |
| 35 | -29.8 | -29.6 | -27.1 | -28.2 | -29.4 | -28.1 | -29.9 | -29.1 | -28.5 | -30.6 | -29.03 | 1.02 | 1.07 | -28.00 | -30.04 |
| 36 | -31.9 | -31.5 | -29.2 | -30.2 | -31.5 | -30.5 | -31.9 | -31.3 | -30.4 | -32.4 | -31.08 | 0.92 | 0.96 | -30.16 | -31.99 |
| 37 | -33.5 | -32.9 | -30.9 | -31.9 | -33.0 | -32.3 | -33.3 | -33.0 | -31.9 | -33.6 | -32.65 | 0.82 | 0.86 | -31.83 | -33.47 |
| 38 | -34.7 | -33.9 | -32.2 | -33.0 | -34.0 | -33.7 | -34.3 | -34.2 | -33.0 | -34.3 | -33.72 | 0.74 | 0.78 | -32.98 | -34.45 |
| 39 | -35.4 | -34.3 | -32.8 | -33.6 | -34.4 | -34.5 | -34.8 | -34.9 | -33.7 | -34.2 | -34.26 | 0.71 | 0.74 | -33.56 | -34.97 |
| 40 | -35.7 | -34.2 | -33.0 | -33.6 | -34.1 | -34.7 | -34.8 | -35.0 | -33.8 | -33.6 | -34.27 | 0.76 | 0.80 | -33.51 | -35.02 |
| 41 | -35.4 | -33.7 | -32.6 | -33.1 | -33.2 | -34.4 | -34.3 | -34.5 | -33.5 | -32.3 | -33.71 | 0.91 | 0.96 | -32.80 | -34.61 |
| 42 | -34.7 | -32.5 | -31.7 | -32.1 | -31.6 | -33.4 | -33.2 | -33.4 | -32.7 | -30.3 | -32.56 | 1.15 | 1.22 | -31.40 | -33.71 |
| 43 | -33.3 | -30.8 | -30.3 | -30.4 | -29.4 | -31.8 | -31.6 | -31.6 | -31.3 | -27.6 | -30.78 | 1.48 | 1.56 | -29.31 | -32.26 |
| 44 | -31.3 | -28.4 | -28.3 | -28.1 | -26.4 | -29.4 | -29.3 | -29.0 | -29.3 | -24.1 | -28.35 | 1.86 | 1.96 | -26.49 | -30.21 |
| 45 | -28.6 | -25.2 | -25.7 | -25.2 | -22.7 | -26.3 | -26.4 | -25.7 | -26.8 | -19.8 | -25.22 | 2.29 | 2.41 | -22.93 | -27.51 |
| 46 | -25.1 | -21.2 | -22.4 | -21.7 | -18.4 | -22.3 | -22.8 | -21.5 | -23.5 | -14.8 | -21.39 | 2.72 | 2.87 | -18.66 | -24.11 |
| 47 | -20.8 | -16.5 | -18.6 | -17.6 | -13.6 | -17.7 | -18.5 | -16.7 | -19.6 | -9.3 | -16.88 | 3.13 | 3.30 | -13.75 | -20.01 |
| 48 | -15.8 | -11.2 | -14.0 | -13.0 | -8.4 | -12.3 | -13.6 | -11.2 | -15.1 | -3.5 | -11.81 | 3.45 | 3.63 | -8.37 | -15.26 |
| 49 | -10.3 | -5.6 | -9.0 | -8.1 | -3.1 | -6.7 | -8.3 | -5.3 | -10.1 | 2.4 | -6.40 | 3.63 | 3.82 | -2.78 | -10.03 |
| 50 | -4.5 | -0.0 | -3.7 | -3.1 | 2.1 | -0.9 | -2.8 | 0.6 | -4.9 | 7.8 | -0.93 | 3.65 | 3.85 | 2.72 | -4.58 |
| 51 | 1.1 | 5.1 | 1.7 | 1.8 | 7.0 | 4.6 | 2.5 | 6.2 | 0.3 | 12.7 | 4.28 | 3.53 | 3.72 | 7.81 | 0.76 |
| 52 | 6.1 | 9.4 | 6.6 | 6.3 | 11.3 | 9.5 | 7.3 | 11.1 | 5.0 | 16.6 | 8.92 | 3.29 | 3.47 | 12.21 | 5.63 |
| 53 | 10.2 | 12.6 | 10.8 | 10.2 | 15.1 | 13.7 | 11.3 | 15.1 | 8.9 | 19.5 | 12.74 | 3.01 | 3.18 | 15.75 | 9.73 |
| 54 | 13.3 | 14.7 | 14.0 | 13.4 | 18.1 | 17.0 | 14.4 | 17.9 | 11.9 | 21.5 | 15.61 | 2.77 | 2.92 | 18.38 | 12.84 |
| 55 | 15.2 | 15.6 | 16.2 | 15.6 | 20.4 | 19.3 | 16.5 | 19.7 | 13.8 | 22.4 | 17.47 | 2.63 | 2.77 | 20.01 | 14.84 |
| 56 | 16.0 | 15.4 | 17.2 | 17.0 | 21.9 | 20.8 | 17.6 | 20.4 | 14.6 | 22.3 | 18.31 | 2.65 | 2.80 | 20.97 | 15.66 |
| 57 | 15.7 | 14.2 | 17.1 | 17.5 | 22.7 | 21.3 | 17.6 | 20.0 | 14.4 | 21.3 | 18.16 | 2.86 | 3.02 | 21.03 | 15.30 |
| 58 | 14.3 | 12.1 | 16.0 | 17.1 | 22.7 | 20.9 | 16.6 | 18.7 | 13.2 | 19.4 | 17.08 | 3.22 | 3.39 | 20.30 | 13.87 |
| 59 | 12.1 | 9.3 | 14.0 | 15.8 | 21.9 | 19.7 | 14.7 | 16.4 | 11.1 | 16.7 | 15.15 | 3.63 | 3.83 | 18.78 | 11.53 |
| 60 | 9.1 | 6.1 | 11.2 | 13.8 | 20.4 | 17.6 | 12.1 | 13.4 | 8.4 | 13.2 | 12.53 | 4.01 | 4.23 | 16.54 | 8.51 |
| 61 | 5.7 | 2.8 | 7.9 | 11.2 | 18.1 | 14.7 | 8.9 | 9.9 | 5.4 | 9.4 | 9.40 | 4.27 | 4.50 | 13.67 | 5.13 |
| 62 | 2.2 | -0.2 | 4.5 | 8.1 | 15.2 | 11.2 | 5.4 | 6.2 | 2.3 | 5.5 | 6.05 | 4.31 | 4.54 | 10.36 | 1.74 |
| 63 | -1.0 | -2.8 | 1.1 | 5.0 | 11.8 | 7.5 | 2.0 | 2.7 | -0.5 | 1.9 | 2.76 | 4.11 | 4.33 | 6.87 | -1.34 |
| 64 | -3.7 | -4.8 | -1.8 | 2.0 | 8.2 | 3.7 | -1.1 | -0.5 | -2.8 | -1.1 | -0.18 | 3.66 | 3.86 | 3.48 | -3.84 |
| 65 | -5.6 | -6.0 | -4.0 | -0.7 | 4.6 | 0.3 | -3.6 | -3.0 | -4.4 | -3.4 | -2.56 | 3.03 | 3.19 | 0.46 | -5.59 |
| 66 | -6.8 | -6.5 | -5.4 | -2.8 | 1.4 | -2.5 | -5.3 | -4.7 | -5.3 | -4.8 | -4.26 | 2.30 | 2.43 | -1.96 | -6.57 |
| 67 | -7.3 | -6.4 | -6.1 | -4.2 | -1.2 | -4.5 | -6.3 | -5.6 | -5.6 | -5.5 | -5.25 | 1.60 | 1.68 | -3.66 | -6.85 |
| 68 | -7.1 | -5.7 | -6.0 | -4.9 | -3.0 | -5.8 | -6.5 | -5.8 | -5.3 | -5.5 | -5.57 | 1.02 | 1.07 | -4.55 | -6.58 |
| 69 | -6.4 | -4.8 | -5.4 | -5.1 | -4.1 | -6.2 | -6.2 | -5.4 | -4.6 | -5.0 | -5.32 | 0.71 | 0.75 | -4.61 | -6.03 |
| 70 | -5.3 | -3.6 | -4.5 | -4.7 | -4.5 | -6.0 | -5.4 | -4.6 | -3.6 | -4.1 | -4.63 | 0.73 | 0.77 | -3.91 | -5.36 |
| 71 | -4.1 | -2.3 | -3.3 | -4.0 | -4.2 | -5.2 | -4.3 | -3.6 | -2.5 | -3.1 | -3.66 | 0.85 | 0.89 | -2.81 | -4.51 |
| 72 | -2.8 | -1.1 | -2.0 | -3.0 | -3.5 | -4.2 | -3.1 | -2.4 | -1.4 | -1.9 | -2.54 | 0.91 | 0.96 | -1.62 | -3.45 |
| 73 | -1.5 | -0.0 | -0.8 | -1.9 | -2.5 | -2.9 | -1.8 | -1.2 | -0.3 | -0.8 | -1.38 | 0.89 | 0.94 | -0.49 | -2.27 |
| 74 | -0.4 | 0.9 | 0.2 | -0.8 | -1.4 | -1.6 | -0.6 | -0.1 | 0.7 | 0.2 | -0.31 | 0.79 | 0.83 | 0.48 | -1.09 |
| 75 | 0.5 | 1.6 | 1.1 | 0.3 | -0.3 | -0.5 | 0.4 | 0.7 | 1.4 | 0.9 | 0.62 | 0.63 | 0.67 | 1.25 | -0.01 |
| 76 | 1.2 | 2.1 | 1.6 | 1.2 | 0.8 | 0.5 | 1.2 | 1.4 | 1.9 | 1.5 | 1.34 | 0.45 | 0.47 | 1.78 | 0.89 |
| 77 | 1.6 | 2.3 | 1.7 | 1.9 | 1.6 | 1.3 | 1.8 | 1.8 | 2.2 | 1.9 | 1.80 | 0.28 | 0.30 | 2.09 | 1.52 |
| 78 | 1.8 | 2.2 | 1.6 | 2.4 | 2.2 | 1.7 | 2.1 | 1.9 | 2.3 | 1.9 | 1.99 | 0.25 | 0.26 | 2.24 | 1.74 |
| 79 | 1.7 | 1.8 | 1.0 | 2.5 | 2.5 | 1.8 | 2.0 | 1.7 | 2.0 | 1.6 | 1.87 | 0.41 | 0.43 | 2.28 | 1.46 |
| 80 | 1.2 | 1.1 | 0.1 | 2.4 | 2.5 | 1.5 | 1.7 | 1.1 | 1.4 | 1.0 | 1.40 | 0.65 | 0.69 | 2.05 | 0.75 |
| 81 | 0.4 | -0.0 | -1.2 | 1.8 | 2.2 | 0.7 | 0.9 | 0.2 | 0.5 | -0.0 | 0.55 | 0.93 | 0.98 | 1.47 | -0.38 |
| 82 | -0.8 | -1.6 | -3.0 | 0.9 | 1.6 | -0.5 | -0.4 | -1.2 | -0.9 | -1.5 | -0.74 | 1.23 | 1.29 | 0.49 | -1.96 |
| 83 | -2.5 | -3.6 | -5.2 | -0.5 | 0.6 | -2.2 | -2.1 | -3.0 | -2.8 | -3.5 | -2.48 | 1.54 | 1.62 | -0.94 | -4.02 |
| 84 | -4.7 | -6.1 | -7.9 | -2.4 | -0.8 | -4.5 | -4.3 | -5.3 | -5.1 | -5.9 | -4.71 | 1.85 | 1.95 | -2.86 | -6.56 |
| 85 | -7.4 | -9.1 | -10.9 | -4.9 | -2.8 | -7.3 | -7.1 | -8.1 | -7.9 | -8.9 | -7.43 | 2.14 | 2.25 | -5.29 | -9.57 |
| 86 | -10.7 | -12.5 | -14.2 | -7.9 | -5.2 | -10.7 | -10.3 | -11.2 | -11.1 | -12.2 | -10.60 | 2.38 | 2.50 | -8.22 | -12.97 |
| 87 | -14.3 | -16.1 | -17.9 | -11.4 | -8.2 | -14.4 | -14.0 | -14.8 | -14.6 | -15.9 | -14.14 | 2.54 | 2.68 | -11.60 | -16.69 |
| 88 | -18.2 | -20.0 | -21.6 | -15.2 | -11.7 | -18.4 | -18.0 | -18.6 | -18.3 | -19.8 | -17.96 | 2.62 | 2.76 | -15.35 | -20.58 |
| 89 | -22.2 | -23.9 | -25.3 | -19.3 | -15.5 | -22.5 | -22.2 | -22.5 | -22.1 | -23.8 | -21.93 | 2.60 | 2.74 | -19.33 | -24.52 |
| 90 | -26.2 | -27.7 | -28.9 | -23.5 | -19.6 | -26.5 | -26.4 | -26.4 | -25.9 | -27.7 | -25.89 | 2.48 | 2.61 | -23.41 | -28.37 |
| 91 | -30.0 | -31.4 | -32.3 | -27.7 | -23.8 | -30.4 | -30.4 | -30.2 | -29.6 | -31.4 | -29.71 | 2.29 | 2.42 | -27.42 | -32.00 |
| 92 | -33.5 | -34.8 | -35.4 | -31.6 | -27.9 | -34.0 | -34.1 | -33.7 | -33.0 | -34.8 | -33.29 | 2.05 | 2.16 | -31.23 | -35.34 |
| 93 | -36.6 | -38.0 | -38.1 | -35.2 | -31.8 | -37.2 | -37.5 | -37.0 | -36.1 | -37.9 | -36.54 | 1.79 | 1.89 | -34.75 | -38.33 |
| 94 | -39.3 | -40.7 | -40.5 | -38.4 | -35.4 | -40.1 | -40.5 | -39.8 | -38.8 | -40.6 | -39.41 | 1.53 | 1.61 | -37.88 | -40.94 |
| 95 | -41.5 | -42.9 | -42.4 | -41.1 | -38.5 | -42.5 | -43.0 | -42.3 | -41.1 | -42.8 | -41.79 | 1.29 | 1.36 | -40.50 | -43.08 |
| 96 | -43.1 | -44.6 | -43.8 | -43.3 | -41.0 | -44.3 | -44.9 | -44.2 | -42.9 | -44.4 | -43.64 | 1.09 | 1.15 | -42.54 | -44.73 |
| 97 | -44.3 | -45.9 | -44.8 | -45.0 | -43.0 | -45.8 | -46.3 | -45.7 | -44.4 | -45.6 | -45.09 | 0.95 | 1.00 | -44.14 | -46.04 |
| 98 | -45.3 | -47.1 | -45.8 | -46.7 | -44.9 | -47.2 | -47.7 | -47.1 | -45.7 | -46.6 | -46.39 | 0.86 | 0.91 | -45.53 | -47.26 |
| 99 | -46.4 | -48.4 | -46.8 | -48.4 | -46.9 | -48.6 | -49.1 | -48.5 | -47.2 | -47.7 | -47.79 | 0.87 | 0.92 | -46.92 | -48.66 |

| | Knee (x-variable) | | | | | | | | | | | | | | |
|---|---|---|---|---|---|---|---|---|---|---|---|---|---|---|---|
| %Stride | Cycle 1 | Cycle 2 | Cycle 3 | Cycle 4 | Cycle 5 | Cycle 6 | Cycle 7 | Cycle 8 | Cycle 9 | Cycle 10 | $M_{xi}$ | $SD_{xi}$ (population) | $SD_{xi}$ (sample) | $x_{pi}$ [Eqn 32 34] | $y_{mi}$ [Eqn 33 35] |
| 0 | 11.3 | 11.2 | 10.6 | 10.1 | 9.0 | 8.6 | 11.6 | 9.8 | 10.6 | 11.1 | 10.39 | 0.94 | 0.99 | 11.33 | 9.45 |
| 1 | 12.5 | 12.9 | 12.1 | 11.6 | 10.2 | 9.7 | 12.7 | 11.2 | 11.8 | 12.3 | 11.71 | 1.02 | 1.07 | 12.72 | 10.69 |
| 2 | 14.0 | 14.9 | 13.8 | 13.2 | 11.6 | 11.1 | 14.1 | 12.8 | 13.3 | 13.9 | 13.26 | 1.11 | 1.17 | 14.38 | 12.15 |
| 3 | 15.8 | 17.4 | 16.0 | 15.3 | 13.5 | 12.9 | 16.0 | 15.0 | 15.1 | 15.9 | 15.29 | 1.23 | 1.29 | 16.52 | 14.07 |
| 4 | 18.3 | 20.5 | 18.8 | 17.8 | 15.9 | 15.3 | 18.3 | 17.6 | 17.5 | 18.5 | 17.84 | 1.37 | 1.44 | 19.21 | 16.47 |
| 5 | 21.1 | 24.0 | 22.0 | 20.7 | 18.7 | 18.2 | 21.1 | 20.8 | 20.2 | 21.5 | 20.84 | 1.54 | 1.62 | 22.38 | 19.30 |
| 6 | 24.4 | 28.0 | 25.6 | 23.9 | 22.0 | 21.5 | 24.2 | 24.5 | 23.3 | 25.0 | 24.23 | 1.74 | 1.83 | 25.97 | 22.50 |
| 7 | 28.0 | 32.4 | 29.6 | 27.5 | 25.6 | 25.2 | 27.8 | 28.5 | 26.7 | 28.9 | 28.01 | 1.96 | 2.06 | 29.97 | 26.05 |
| 8 | 32.1 | 37.1 | 33.9 | 31.5 | 29.6 | 29.2 | 31.7 | 32.9 | 30.4 | 33.1 | 32.15 | 2.19 | 2.31 | 34.35 | 29.96 |
| 9 | 36.5 | 42.3 | 38.5 | 35.7 | 33.9 | 33.7 | 35.9 | 37.7 | 34.4 | 37.7 | 36.63 | 2.44 | 2.57 | 39.08 | 34.19 |
| 10 | 41.2 | 47.6 | 43.4 | 40.2 | 38.6 | 38.5 | 40.5 | 42.8 | 38.7 | 42.6 | 41.41 | 2.68 | 2.83 | 44.09 | 38.73 |
| 11 | 46.2 | 53.1 | 48.5 | 44.9 | 43.6 | 43.6 | 45.3 | 48.2 | 43.2 | 47.7 | 46.42 | 2.91 | 3.06 | 49.32 | 43.51 |
| 12 | 51.3 | 58.7 | 53.8 | 49.8 | 48.7 | 48.9 | 50.3 | 53.6 | 47.8 | 53.0 | 51.59 | 3.09 | 3.26 | 54.68 | 48.51 |
| 13 | 56.6 | 64.2 | 59.1 | 54.8 | 54.1 | 54.4 | 55.4 | 59.1 | 52.6 | 58.3 | 56.85 | 3.23 | 3.40 | 60.08 | 53.63 |
| 14 | 61.9 | 69.5 | 64.4 | 59.8 | 59.5 | 60.0 | 60.5 | 64.5 | 57.4 | 63.6 | 62.09 | 3.31 | 3.49 | 65.40 | 58.79 |
| 15 | 67.0 | 74.5 | 69.5 | 64.7 | 64.8 | 65.5 | 65.5 | 69.8 | 62.2 | 68.7 | 67.21 | 3.33 | 3.51 | 70.54 | 63.88 |
| 16 | 72.0 | 79.1 | 74.3 | 69.4 | 70.0 | 70.8 | 70.3 | 74.8 | 66.8 | 73.6 | 72.01 | 3.31 | 3.49 | 75.41 | 68.79 |
| 17 | 76.6 | 83.3 | 78.9 | 73.8 | 74.8 | 75.8 | 74.7 | 79.4 | 71.2 | 78.1 | 76.65 | 3.24 | 3.41 | 79.89 | 73.42 |
| 18 | 80.7 | 87.0 | 82.9 | 77.9 | 79.3 | 80.4 | 78.8 | 83.6 | 75.2 | 82.2 | 80.79 | 3.14 | 3.31 | 83.93 | 77.66 |
| 19 | 84.4 | 90.1 | 86.5 | 81.5 | 83.2 | 84.5 | 82.5 | 87.3 | 78.9 | 85.8 | 84.45 | 3.01 | 3.17 | 87.46 | 81.44 |
| 20 | 87.5 | 92.6 | 89.5 | 84.6 | 86.5 | 88.0 | 85.6 | 90.4 | 82.1 | 88.8 | 87.56 | 2.87 | 3.03 | 90.43 | 84.68 |
| 21 | 90.1 | 94.4 | 91.9 | 87.2 | 89.2 | 90.9 | 88.1 | 92.9 | 84.7 | 91.2 | 90.07 | 2.72 | 2.87 | 92.80 | 87.35 |
| 22 | 92.0 | 95.7 | 93.8 | 89.2 | 91.2 | 93.2 | 90.1 | 94.8 | 86.8 | 93.0 | 91.97 | 2.57 | 2.71 | 94.53 | 89.40 |
| 23 | 93.2 | 96.2 | 94.9 | 90.6 | 92.6 | 94.8 | 91.4 | 96.0 | 88.3 | 94.1 | 93.19 | 2.41 | 2.54 | 95.60 | 90.78 |
| 24 | 93.8 | 96.0 | 95.4 | 91.2 | 93.2 | 95.7 | 92.0 | 96.5 | 89.1 | 94.4 | 93.72 | 2.26 | 2.38 | 95.98 | 91.46 |
| 25 | 93.6 | 95.1 | 95.2 | 91.2 | 93.1 | 95.9 | 92.0 | 96.2 | 89.2 | 94.0 | 93.54 | 2.12 | 2.24 | 95.66 | 91.42 |
| 26 | 92.7 | 93.4 | 94.3 | 90.4 | 92.3 | 95.4 | 91.2 | 95.2 | 88.7 | 92.7 | 92.64 | 2.00 | 2.11 | 94.63 | 90.64 |
| 27 | 91.1 | 91.1 | 92.7 | 89.0 | 90.7 | 94.0 | 89.8 | 93.5 | 87.5 | 90.7 | 91.01 | 1.90 | 2.01 | 92.91 | 89.10 |
| 28 | 88.8 | 88.1 | 90.4 | 86.8 | 88.5 | 92.0 | 87.7 | 91.0 | 85.7 | 88.0 | 88.69 | 1.83 | 1.93 | 90.52 | 86.85 |
| 29 | 85.9 | 84.6 | 87.5 | 84.0 | 85.6 | 89.2 | 84.9 | 87.9 | 83.3 | 84.6 | 85.74 | 1.79 | 1.89 | 87.54 | 83.95 |
| 30 | 82.5 | 80.6 | 84.0 | 80.7 | 82.1 | 85.9 | 81.6 | 84.2 | 80.4 | 80.7 | 82.26 | 1.77 | 1.87 | 84.03 | 80.49 |
| 31 | 78.6 | 76.4 | 80.1 | 76.9 | 78.2 | 82.0 | 78.0 | 80.0 | 77.1 | 76.5 | 78.37 | 1.75 | 1.85 | 80.12 | 76.62 |
| 32 | 74.6 | 72.0 | 75.9 | 72.8 | 74.0 | 77.7 | 74.0 | 75.6 | 73.5 | 72.0 | 74.21 | 1.73 | 1.82 | 75.93 | 72.48 |
| 33 | 70.3 | 67.5 | 71.5 | 68.6 | 69.7 | 73.3 | 69.9 | 71.1 | 69.7 | 67.4 | 69.90 | 1.70 | 1.79 | 71.59 | 68.20 |
| 34 | 66.1 | 63.2 | 67.0 | 64.4 | 65.3 | 68.7 | 65.7 | 66.5 | 65.8 | 62.9 | 65.56 | 1.66 | 1.75 | 67.22 | 63.90 |
| 35 | 61.9 | 58.9 | 62.6 | 60.2 | 60.9 | 64.2 | 61.6 | 62.0 | 61.9 | 58.6 | 61.29 | 1.62 | 1.71 | 62.91 | 59.68 |
| 36 | 57.9 | 54.8 | 58.3 | 56.2 | 56.7 | 59.8 | 57.6 | 57.7 | 58.1 | 54.4 | 57.15 | 1.58 | 1.67 | 58.73 | 55.56 |
| 37 | 54.0 | 50.8 | 54.2 | 52.3 | 52.6 | 55.6 | 53.8 | 53.5 | 54.4 | 50.3 | 53.15 | 1.56 | 1.64 | 54.70 | 51.59 |
| 38 | 50.2 | 47.0 | 50.2 | 48.5 | 48.7 | 51.6 | 50.1 | 49.6 | 50.7 | 46.4 | 49.30 | 1.54 | 1.63 | 50.85 | 47.76 |
| 39 | 46.6 | 43.4 | 46.5 | 44.9 | 44.9 | 47.7 | 46.5 | 45.7 | 47.2 | 42.7 | 45.60 | 1.55 | 1.63 | 47.15 | 44.06 |
| 40 | 43.1 | 39.9 | 42.8 | 41.5 | 41.3 | 44.0 | 43.1 | 42.1 | 43.8 | 39.0 | 42.05 | 1.56 | 1.64 | 43.61 | 40.49 |
| 41 | 39.7 | 36.4 | 39.4 | 38.2 | 37.8 | 40.4 | 39.7 | 38.6 | 40.5 | 35.5 | 38.60 | 1.59 | 1.67 | 40.19 | 37.02 |
| 42 | 36.4 | 33.1 | 36.0 | 34.9 | 34.5 | 37.0 | 36.5 | 35.2 | 37.2 | 32.0 | 35.27 | 1.62 | 1.70 | 36.89 | 33.66 |
| 43 | 33.1 | 29.9 | 32.8 | 31.9 | 31.2 | 33.6 | 33.3 | 31.9 | 34.1 | 28.6 | 32.05 | 1.65 | 1.74 | 33.70 | 30.39 |
| 44 | 30.0 | 26.9 | 29.7 | 28.9 | 28.1 | 30.4 | 30.2 | 28.8 | 31.0 | 25.3 | 28.92 | 1.69 | 1.78 | 30.61 | 27.23 |
| 45 | 26.9 | 23.9 | 26.8 | 26.0 | 25.0 | 27.3 | 27.2 | 25.8 | 28.1 | 22.1 | 25.91 | 1.71 | 1.81 | 27.62 | 24.19 |
| 46 | 24.0 | 21.1 | 23.9 | 23.3 | 22.1 | 24.2 | 24.3 | 22.9 | 25.2 | 19.1 | 23.02 | 1.73 | 1.82 | 24.75 | 21.29 |
| 47 | 21.2 | 18.5 | 21.3 | 20.7 | 19.3 | 21.3 | 21.5 | 20.1 | 22.5 | 16.3 | 20.27 | 1.73 | 1.82 | 22.00 | 18.55 |
| 48 | 18.6 | 16.1 | 18.7 | 18.3 | 16.6 | 18.6 | 18.9 | 17.6 | 19.9 | 13.7 | 17.70 | 1.71 | 1.80 | 19.41 | 15.99 |
| 49 | 16.2 | 13.9 | 16.4 | 16.1 | 14.2 | 16.0 | 16.5 | 15.2 | 17.5 | 11.4 | 15.33 | 1.66 | 1.75 | 16.99 | 13.66 |
| 50 | 14.0 | 11.9 | 14.3 | 14.0 | 11.9 | 13.7 | 14.3 | 13.1 | 15.3 | 9.5 | 13.18 | 1.59 | 1.68 | 14.77 | 11.59 |
| 51 | 12.0 | 10.2 | 12.4 | 12.2 | 9.9 | 11.6 | 12.3 | 11.2 | 13.2 | 7.8 | 11.29 | 1.50 | 1.58 | 12.79 | 9.79 |
| 52 | 10.4 | 8.8 | 10.7 | 10.7 | 8.2 | 9.8 | 10.6 | 9.6 | 11.5 | 6.6 | 9.68 | 1.39 | 1.46 | 11.07 | 8.29 |
| 53 | 9.0 | 7.7 | 9.4 | 9.4 | 6.8 | 8.3 | 9.2 | 8.4 | 10.0 | 5.7 | 8.38 | 1.26 | 1.33 | 9.64 | 7.12 |
| 54 | 7.9 | 7.0 | 8.3 | 8.3 | 5.7 | 7.1 | 8.2 | 7.4 | 8.8 | 5.2 | 7.40 | 1.13 | 1.19 | 8.53 | 6.27 |
| 55 | 7.2 | 6.6 | 7.6 | 7.6 | 4.9 | 6.3 | 7.4 | 6.8 | 8.0 | 5.0 | 6.75 | 1.01 | 1.06 | 7.76 | 5.74 |
| 56 | 6.8 | 6.6 | 7.2 | 7.1 | 4.5 | 5.9 | 7.0 | 6.5 | 7.5 | 5.2 | 6.44 | 0.91 | 0.96 | 7.35 | 5.52 |
| 57 | 6.8 | 6.9 | 7.2 | 7.0 | 4.4 | 5.8 | 6.9 | 6.6 | 7.4 | 5.6 | 6.45 | 0.86 | 0.91 | 7.31 | 5.59 |
| 58 | 7.1 | 7.6 | 7.4 | 7.1 | 4.6 | 6.0 | 7.1 | 7.0 | 7.6 | 6.4 | 6.78 | 0.88 | 0.92 | 7.66 | 5.90 |
| 59 | 7.6 | 8.6 | 8.1 | 7.4 | 5.1 | 6.5 | 7.6 | 7.7 | 8.3 | 7.3 | 7.41 | 0.95 | 1.00 | 8.36 | 6.46 |
| 60 | 8.5 | 9.8 | 9.0 | 8.1 | 5.8 | 7.3 | 8.3 | 8.6 | 9.2 | 8.5 | 8.31 | 1.07 | 1.12 | 9.38 | 7.25 |
| 61 | 9.7 | 11.3 | 10.2 | 8.9 | 6.7 | 8.3 | 9.4 | 9.9 | 10.5 | 9.9 | 9.46 | 1.21 | 1.28 | 10.68 | 8.25 |
| 62 | 11.1 | 13.0 | 11.6 | 10.0 | 7.8 | 9.5 | 10.6 | 11.3 | 12.0 | 11.4 | 10.82 | 1.38 | 1.45 | 12.20 | 9.44 |
| 63 | 12.6 | 14.8 | 13.2 | 11.3 | 9.1 | 10.8 | 12.1 | 12.9 | 13.7 | 13.0 | 12.34 | 1.53 | 1.61 | 13.87 | 10.82 |
| 64 | 14.4 | 16.7 | 14.9 | 12.7 | 10.6 | 12.3 | 13.6 | 14.6 | 15.5 | 14.7 | 13.98 | 1.65 | 1.74 | 15.63 | 12.33 |
| 65 | 16.1 | 18.5 | 16.6 | 14.3 | 12.1 | 13.9 | 15.3 | 16.3 | 17.3 | 16.3 | 15.68 | 1.74 | 1.83 | 17.41 | 13.94 |
| 66 | 17.9 | 20.2 | 18.3 | 15.9 | 13.8 | 15.6 | 16.9 | 18.1 | 19.0 | 18.0 | 17.37 | 1.77 | 1.87 | 19.14 | 15.60 |
| 67 | 19.6 | 21.8 | 19.9 | 17.5 | 15.5 | 17.2 | 18.5 | 19.7 | 20.6 | 19.5 | 19.00 | 1.74 | 1.84 | 20.74 | 17.26 |
| 68 | 21.2 | 23.2 | 21.4 | 19.1 | 17.2 | 18.8 | 20.0 | 21.3 | 22.1 | 20.9 | 20.51 | 1.66 | 1.75 | 22.17 | 18.86 |
| 69 | 22.5 | 24.4 | 22.6 | 20.5 | 18.9 | 20.3 | 21.4 | 22.6 | 23.3 | 22.1 | 21.86 | 1.53 | 1.61 | 23.39 | 20.34 |
| 70 | 23.7 | 25.4 | 23.6 | 21.9 | 20.4 | 21.7 | 22.5 | 23.7 | 24.3 | 23.1 | 23.01 | 1.36 | 1.43 | 24.37 | 21.66 |
| 71 | 24.6 | 26.1 | 24.3 | 23.0 | 21.8 | 22.8 | 23.4 | 24.6 | 25.0 | 23.8 | 23.94 | 1.17 | 1.24 | 25.11 | 22.77 |
| 72 | 25.3 | 26.5 | 24.9 | 23.9 | 23.0 | 23.7 | 24.1 | 25.2 | 25.6 | 24.3 | 24.64 | 0.99 | 1.04 | 25.63 | 23.65 |
| 73 | 25.7 | 26.8 | 25.1 | 24.5 | 23.9 | 24.4 | 24.6 | 25.6 | 25.9 | 24.7 | 25.12 | 0.82 | 0.87 | 25.95 | 24.30 |
| 74 | 25.9 | 26.9 | 25.2 | 25.0 | 24.6 | 24.9 | 24.9 | 25.8 | 26.0 | 24.8 | 25.40 | 0.69 | 0.73 | 26.09 | 24.71 |
| 75 | 26.0 | 26.8 | 25.2 | 25.2 | 25.0 | 25.2 | 25.0 | 25.8 | 26.0 | 24.8 | 25.49 | 0.60 | 0.63 | 26.09 | 24.89 |
| 76 | 25.9 | 26.6 | 24.9 | 25.2 | 25.3 | 25.2 | 24.9 | 25.6 | 25.8 | 24.7 | 25.42 | 0.55 | 0.58 | 25.96 | 24.87 |
| 77 | 25.7 | 26.3 | 24.6 | 25.0 | 25.3 | 25.2 | 24.7 | 25.3 | 25.5 | 24.4 | 25.20 | 0.53 | 0.56 | 25.73 | 24.67 |
| 78 | 25.3 | 25.8 | 24.1 | 24.7 | 25.2 | 24.9 | 24.4 | 25.0 | 25.2 | 24.0 | 24.86 | 0.53 | 0.56 | 25.39 | 24.33 |
| 79 | 24.9 | 25.3 | 23.6 | 24.2 | 24.9 | 24.6 | 23.9 | 24.5 | 24.7 | 23.6 | 24.40 | 0.55 | 0.58 | 24.95 | 23.86 |
| 80 | 24.3 | 24.6 | 22.9 | 23.7 | 24.4 | 24.1 | 23.4 | 23.9 | 24.1 | 23.0 | 23.84 | 0.57 | 0.60 | 24.41 | 23.28 |
| 81 | 23.7 | 23.9 | 22.2 | 23.0 | 23.9 | 23.6 | 22.7 | 23.2 | 23.5 | 22.4 | 23.19 | 0.59 | 0.62 | 23.78 | 22.61 |
| 82 | 22.9 | 23.1 | 21.3 | 22.2 | 23.3 | 22.9 | 21.9 | 22.5 | 22.8 | 21.6 | 22.46 | 0.61 | 0.65 | 23.07 | 21.84 |
| 83 | 22.1 | 22.2 | 20.5 | 21.4 | 22.5 | 22.1 | 21.1 | 21.8 | 22.0 | 20.9 | 21.65 | 0.64 | 0.67 | 22.29 | 21.02 |
| 84 | 21.2 | 21.3 | 19.5 | 20.5 | 21.7 | 21.3 | 20.2 | 20.9 | 21.2 | 20.0 | 20.78 | 0.66 | 0.69 | 21.44 | 20.13 |
| 85 | 20.3 | 20.3 | 18.6 | 19.6 | 20.8 | 20.4 | 19.2 | 20.1 | 20.3 | 19.1 | 19.86 | 0.68 | 0.71 | 20.53 | 19.18 |
| 86 | 19.2 | 19.3 | 17.6 | 18.6 | 19.8 | 19.5 | 18.1 | 19.2 | 19.4 | 18.2 | 18.89 | 0.69 | 0.73 | 19.58 | 18.20 |
| 87 | 18.2 | 18.2 | 16.5 | 17.6 | 18.8 | 18.6 | 17.1 | 18.5 | 18.5 | 17.2 | 17.88 | 0.70 | 0.74 | 18.58 | 17.18 |
| 88 | 17.1 | 17.1 | 15.5 | 16.5 | 17.7 | 17.6 | 16.0 | 17.2 | 17.5 | 16.3 | 16.84 | 0.70 | 0.74 | 17.55 | 16.14 |
| 89 | 16.0 | 16.0 | 14.4 | 15.4 | 16.5 | 16.6 | 15.0 | 16.2 | 16.5 | 15.3 | 15.79 | 0.71 | 0.74 | 16.49 | 15.08 |
| 90 | 14.8 | 14.9 | 13.4 | 14.3 | 15.3 | 15.7 | 13.9 | 15.2 | 15.5 | 14.3 | 14.72 | 0.70 | 0.74 | 15.42 | 14.02 |
| 91 | 13.7 | 13.7 | 12.3 | 13.2 | 14.1 | 14.7 | 12.9 | 14.2 | 14.5 | 13.3 | 13.66 | 0.70 | 0.74 | 14.36 | 12.96 |
| 92 | 12.6 | 12.6 | 11.4 | 12.2 | 12.9 | 13.8 | 11.9 | 13.2 | 13.4 | 12.4 | 12.62 | 0.70 | 0.73 | 13.32 | 11.93 |
| 93 | 11.5 | 11.6 | 10.4 | 11.1 | 11.7 | 12.9 | 10.9 | 12.2 | 12.5 | 11.5 | 11.64 | 0.69 | 0.73 | 12.33 | 10.94 |
| 94 | 10.6 | 10.6 | 9.6 | 10.1 | 10.6 | 12.1 | 10.1 | 11.2 | 11.6 | 10.8 | 10.73 | 0.70 | 0.73 | 11.43 | 10.04 |
| 95 | 9.9 | 9.9 | 9.0 | 9.3 | 9.6 | 11.4 | 9.4 | 10.5 | 10.9 | 10.4 | 10.02 | 0.71 | 0.75 | 10.73 | 9.31 |
| 96 | 9.5 | 9.5 | 8.7 | 8.8 | 8.9 | 10.9 | 8.9 | 10.0 | 10.4 | 10.3 | 9.58 | 0.75 | 0.79 | 10.32 | 8.83 |
| 97 | 9.4 | 9.2 | 8.6 | 8.4 | 8.3 | 10.7 | 8.7 | 9.6 | 10.1 | 10.5 | 9.34 | 0.82 | 0.86 | 10.16 | 8.52 |
| 98 | 9.3 | 9.1 | 8.5 | 8.1 | 7.9 | 10.5 | 8.5 | 9.4 | 9.9 | 10.8 | 9.19 | 0.93 | 0.98 | 10.13 | 8.26 |
| 99 | 9.2 | 8.8 | 8.5 | 7.7 | 7.4 | 10.3 | 8.3 | 9.1 | 9.6 | 11.1 | 8.99 | 1.07 | 1.13 | 10.07 | 7.92 |

# Table B.3 Example Data for Calculating the Curve Average Variability (RMS_avg) for the Angle-Angle Diagram

Equations 2.17-2.19 are demonstrated using the ankle and knee angle data given in table B.2.

| %Stride | Cycle 1 $R_{ij}$ [Eqn 17] | Cycle 2 $R_{ij}$ [Eqn 17] | Cycle 3 $R_{ij}$ [Eqn 17] | Cycle 4 $R_{ij}$ [Eqn 17] | Cycle 5 $R_{ij}$ [Eqn 17] | Cycle 6 $R_{ij}$ [Eqn 17] | Cycle 7 $R_{ij}$ [Eqn 17] | Cycle 8 $R_{ij}$ [Eqn 17] | Cycle 9 $R_{ij}$ [Eqn 17] | Cycle 10 $R_{ij}$ [Eqn 17] |
|---|---|---|---|---|---|---|---|---|---|---|
| 0 | 1.2976 | 1.8026 | 0.3739 | 0.9720 | 1.5507 | 1.8414 | 1.5733 | 1.3833 | 0.9331 | 0.8063 |
| 1 | 1.1248 | 2.2627 | 0.3425 | 1.0584 | 1.7488 | 2.0066 | 1.5250 | 1.2591 | 0.9784 | 0.7631 |
| 2 | 0.9389 | 2.7544 | 0.6060 | 1.1223 | 1.9502 | 2.2315 | 1.4895 | 1.1207 | 1.0390 | 0.7461 |
| 3 | 0.7555 | 3.2559 | 0.9835 | 1.1198 | 2.1453 | 2.4629 | 1.4897 | 0.9354 | 1.1509 | 0.7728 |
| 4 | 0.5577 | 3.8263 | 1.3851 | 1.0603 | 2.3364 | 2.7139 | 1.5297 | 0.7203 | 1.3123 | 0.8329 |
| 5 | 0.3535 | 4.4534 | 1.7924 | 0.9701 | 2.5359 | 2.9656 | 1.6267 | 0.4831 | 1.5295 | 0.9208 |
| 6 | 0.1590 | 5.1424 | 2.1662 | 0.8928 | 2.7452 | 3.2192 | 1.7572 | 0.3281 | 1.8364 | 1.0533 |
| 7 | 0.1152 | 5.8900 | 2.4971 | 0.8556 | 2.9606 | 3.4587 | 1.9384 | 0.4814 | 2.2148 | 1.2235 |
| 8 | 0.3153 | 6.6865 | 2.7841 | 0.9125 | 3.1819 | 3.6720 | 2.1442 | 0.8442 | 2.6826 | 1.4236 |
| 9 | 0.5213 | 7.5007 | 3.0164 | 1.0690 | 3.3962 | 3.8416 | 2.3682 | 1.2545 | 3.2135 | 1.6443 |
| 10 | 0.7207 | 8.2753 | 3.1966 | 1.2838 | 3.5796 | 3.9350 | 2.5976 | 1.6860 | 3.8170 | 1.8707 |
| 11 | 0.8837 | 8.9814 | 3.3364 | 1.5436 | 3.7085 | 3.9525 | 2.8311 | 2.1035 | 4.4558 | 2.0870 |
| 12 | 1.0160 | 9.5330 | 3.4279 | 1.8155 | 3.7372 | 3.8451 | 3.0334 | 2.4760 | 5.1079 | 2.2631 |
| 13 | 1.0990 | 9.8989 | 3.4828 | 2.0931 | 3.6700 | 3.6159 | 3.2160 | 2.7876 | 5.7335 | 2.3930 |
| 14 | 1.1483 | 10.0402 | 3.5108 | 2.3382 | 3.5017 | 3.2752 | 3.3579 | 3.0289 | 6.2995 | 2.4616 |
| 15 | 1.1489 | 9.9500 | 3.4998 | 2.5648 | 3.2330 | 2.8400 | 3.4779 | 3.2108 | 6.7637 | 2.4800 |
| 16 | 1.1205 | 9.6574 | 3.4535 | 2.7508 | 2.9011 | 2.3396 | 3.5552 | 3.3238 | 7.1144 | 2.4407 |
| 17 | 1.0761 | 9.1840 | 3.3789 | 2.8781 | 2.5430 | 1.8284 | 3.5940 | 3.3938 | 7.3241 | 2.3641 |
| 18 | 1.0234 | 8.5735 | 3.2797 | 2.9674 | 2.1882 | 1.3213 | 3.5948 | 3.4307 | 7.4095 | 2.2507 |
| 19 | 0.9812 | 7.8590 | 3.1630 | 2.9991 | 1.8583 | 0.8945 | 3.5611 | 3.4452 | 7.3739 | 2.1084 |
| 20 | 0.9456 | 7.0707 | 3.0436 | 2.9776 | 1.5795 | 0.7030 | 3.5063 | 3.4519 | 7.2427 | 1.9453 |
| 21 | 0.9263 | 6.2220 | 2.9260 | 2.9123 | 1.3445 | 0.8814 | 3.4244 | 3.4351 | 7.0154 | 1.7613 |
| 22 | 0.9217 | 5.3272 | 2.8288 | 2.8031 | 1.1562 | 1.2700 | 3.3238 | 3.4108 | 6.7173 | 1.5363 |
| 23 | 0.9315 | 4.3901 | 2.7497 | 2.6639 | 0.9985 | 1.7287 | 3.2045 | 3.3591 | 6.3564 | 1.2798 |
| 24 | 0.9526 | 3.4362 | 2.7073 | 2.5137 | 0.8624 | 2.1929 | 3.0633 | 3.2823 | 5.9382 | 0.9682 |
| 25 | 0.9806 | 2.4703 | 2.6932 | 2.3548 | 0.7365 | 2.6650 | 2.9029 | 3.1735 | 5.4570 | 0.5973 |
| 26 | 1.0086 | 1.5281 | 2.7148 | 2.1989 | 0.6221 | 3.0977 | 2.7227 | 3.0365 | 4.9096 | 0.1602 |
| 27 | 1.0324 | 0.7238 | 2.7638 | 2.0584 | 0.5203 | 3.4926 | 2.5230 | 2.8540 | 4.3023 | 0.3470 |
| 28 | 1.0477 | 0.6310 | 2.8229 | 1.9433 | 0.4248 | 3.8051 | 2.3032 | 2.6292 | 3.6319 | 0.8845 |
| 29 | 1.0464 | 1.2157 | 2.8839 | 1.8509 | 0.3599 | 4.0094 | 2.0651 | 2.3850 | 2.9194 | 1.4489 |
| 30 | 1.0494 | 1.7610 | 2.9105 | 1.7718 | 0.3168 | 4.1114 | 1.8261 | 2.1038 | 2.1814 | 1.9767 |
| 31 | 1.0288 | 2.1729 | 2.9011 | 1.6963 | 0.3084 | 4.0730 | 1.5871 | 1.8113 | 1.4602 | 2.4330 |
| 32 | 1.0041 | 2.4105 | 2.8366 | 1.6279 | 0.3381 | 3.9163 | 1.3700 | 1.5100 | 0.7836 | 2.7978 |
| 33 | 0.9843 | 2.5116 | 2.7208 | 1.5484 | 0.3994 | 3.6830 | 1.1935 | 1.2197 | 0.2491 | 3.0387 |
| 34 | 0.9950 | 2.5185 | 2.5623 | 1.4637 | 0.4817 | 3.3829 | 1.0376 | 0.9573 | 0.4372 | 3.1480 |
| 35 | 1.0251 | 2.4593 | 2.3853 | 1.3705 | 0.5643 | 3.0660 | 0.9449 | 0.7380 | 0.8507 | 3.1597 |
| 36 | 1.0886 | 2.3846 | 2.1915 | 1.2840 | 0.6294 | 2.7550 | 0.9021 | 0.5927 | 1.1683 | 3.0876 |
| 37 | 1.2008 | 2.3211 | 2.0067 | 1.1836 | 0.6687 | 2.4866 | 0.9117 | 0.5431 | 1.4022 | 2.9911 |
| 38 | 1.3391 | 2.2650 | 1.8259 | 1.0800 | 0.6879 | 2.2595 | 0.9700 | 0.5802 | 1.5761 | 2.9202 |
| 39 | 1.5217 | 2.2243 | 1.6538 | 0.9723 | 0.7010 | 2.0975 | 1.0476 | 0.6631 | 1.7019 | 2.9341 |
| 40 | 1.7585 | 2.1961 | 1.4882 | 0.8573 | 0.7619 | 1.9868 | 1.1472 | 0.7463 | 1.7853 | 3.0982 |
| 41 | 2.0416 | 2.1743 | 1.3085 | 0.7326 | 0.9326 | 1.9280 | 1.2512 | 0.8054 | 1.8662 | 3.4569 |
| 42 | 2.3851 | 2.1432 | 1.1144 | 0.6066 | 1.2330 | 1.9038 | 1.3460 | 0.8163 | 1.9597 | 4.0061 |
| 43 | 2.7676 | 2.1050 | 0.9106 | 0.4535 | 1.6594 | 1.8720 | 1.4586 | 0.7732 | 2.1043 | 4.7544 |
| 44 | 3.1620 | 2.0500 | 0.8023 | 0.2727 | 2.1505 | 1.8257 | 1.5794 | 0.6521 | 2.3310 | 5.6399 |
| 45 | 3.5432 | 1.9864 | 0.9695 | 0.1341 | 2.6533 | 1.7128 | 1.7230 | 0.4691 | 2.6578 | 6.6376 |
| 46 | 3.8548 | 1.8936 | 1.3946 | 0.3943 | 3.1068 | 1.5430 | 1.8731 | 0.2075 | 3.0743 | 7.6470 |
| 47 | 4.0515 | 1.8078 | 1.9446 | 0.8065 | 3.4226 | 1.3080 | 2.0294 | 0.2392 | 3.5303 | 8.5700 |
| 48 | 4.0944 | 1.7399 | 2.4437 | 1.3023 | 3.5449 | 1.0240 | 2.1652 | 0.6447 | 3.9653 | 9.2713 |
| 49 | 3.9390 | 1.6827 | 2.7969 | 1.8292 | 3.4768 | 0.7311 | 2.2130 | 1.1222 | 4.3030 | 9.6037 |
| 50 | 3.6605 | 1.5828 | 2.9429 | 2.2865 | 3.2727 | 0.4712 | 2.1911 | 1.5682 | 4.4683 | 9.5193 |
| 51 | 3.3020 | 1.3733 | 2.8495 | 2.6119 | 3.0140 | 0.4065 | 2.0851 | 1.9249 | 4.4655 | 9.0540 |
| 52 | 2.9161 | 0.9897 | 2.5596 | 2.7616 | 2.8424 | 0.5976 | 1.8829 | 2.1922 | 4.3465 | 8.2678 |
| 53 | 2.6019 | 0.6530 | 2.2066 | 2.6960 | 2.8286 | 0.9364 | 1.6475 | 2.3294 | 4.1447 | 7.2945 |
| 54 | 2.4028 | 1.0102 | 1.8682 | 2.4312 | 3.0146 | 1.3753 | 1.4124 | 2.3281 | 3.9482 | 6.2601 |
| 55 | 2.3174 | 1.8940 | 1.5784 | 2.0176 | 3.4193 | 1.8948 | 1.1616 | 2.2508 | 3.8436 | 5.2228 |
| 56 | 2.3563 | 2.9374 | 1.3875 | 1.4749 | 4.0676 | 2.5038 | 0.9426 | 2.0896 | 3.8255 | 4.2245 |
| 57 | 2.5221 | 4.0283 | 1.2884 | 0.8588 | 4.9382 | 3.1939 | 0.7379 | 1.8654 | 3.8990 | 3.2753 |
| 58 | 2.7937 | 5.0641 | 1.2792 | 0.2937 | 5.9960 | 3.9126 | 0.5824 | 1.5929 | 4.0208 | 2.3671 |
| 59 | 3.1125 | 5.9496 | 1.3405 | 0.6361 | 7.1376 | 4.5905 | 0.4735 | 1.2839 | 4.1517 | 1.4974 |
| 60 | 3.4423 | 6.5762 | 1.4612 | 1.2791 | 8.2310 | 5.1365 | 0.4576 | 0.9533 | 4.2154 | 0.7422 |
| 61 | 3.7123 | 6.8220 | 1.6204 | 1.8514 | 9.1131 | 5.4198 | 0.5429 | 0.6509 | 4.1376 | 0.4213 |
| 62 | 3.8352 | 6.6538 | 1.7533 | 2.2531 | 9.6133 | 5.3627 | 0.6826 | 0.4990 | 3.8996 | 0.7674 |
| 63 | 3.7598 | 6.1262 | 1.8187 | 2.4693 | 9.5786 | 4.9633 | 0.8381 | 0.5484 | 3.5089 | 1.0648 |
| 64 | 3.5149 | 5.3310 | 1.8054 | 2.5018 | 9.0155 | 4.2500 | 0.9902 | 0.6760 | 2.9771 | 1.1471 |
| 65 | 3.1087 | 4.4358 | 1.6790 | 2.3673 | 8.0029 | 3.3581 | 1.0829 | 0.7730 | 2.4179 | 1.0582 |
| 66 | 2.6110 | 3.6362 | 1.4705 | 2.1268 | 6.7030 | 2.5158 | 1.1188 | 0.8095 | 1.9497 | 0.8312 |
| 67 | 2.1037 | 3.0515 | 1.2158 | 1.8499 | 5.3467 | 1.9007 | 1.1134 | 0.7960 | 1.6633 | 0.5623 |
| 68 | 1.6348 | 2.7339 | 0.9545 | 1.5693 | 4.1410 | 1.6821 | 1.0643 | 0.7670 | 1.5712 | 0.3717 |
| 69 | 1.2456 | 2.6151 | 0.7387 | 1.3395 | 3.2061 | 1.7523 | 1.0004 | 0.7337 | 1.5996 | 0.3806 |
| 70 | 0.9546 | 2.5699 | 0.5917 | 1.1566 | 2.5861 | 1.8835 | 0.9176 | 0.6792 | 1.6264 | 0.5078 |
| 71 | 0.7694 | 2.5097 | 0.5501 | 1.0297 | 2.2114 | 1.9389 | 0.8337 | 0.6301 | 1.5960 | 0.6186 |
| 72 | 0.6590 | 2.3689 | 0.5698 | 0.9188 | 1.9503 | 1.8663 | 0.7514 | 0.5708 | 1.4965 | 0.6902 |
| 73 | 0.5903 | 2.1700 | 0.5863 | 0.7924 | 1.6747 | 1.6875 | 0.6752 | 0.4904 | 1.3385 | 0.7161 |
| 74 | 0.5482 | 1.9249 | 0.5795 | 0.6425 | 1.3565 | 1.4258 | 0.6053 | 0.4078 | 1.1365 | 0.7298 |
| 75 | 0.5245 | 1.6558 | 0.5474 | 0.4697 | 0.9919 | 1.1234 | 0.5447 | 0.3107 | 0.9297 | 0.7351 |
| 76 | 0.5188 | 1.4018 | 0.5333 | 0.2891 | 0.5987 | 0.8154 | 0.4987 | 0.2166 | 0.7246 | 0.7581 |
| 77 | 0.5241 | 1.1691 | 0.6245 | 0.2307 | 0.2480 | 0.5263 | 0.4712 | 0.1392 | 0.5485 | 0.7825 |
| 78 | 0.5326 | 0.9864 | 0.8592 | 0.4035 | 0.3502 | 0.2927 | 0.4642 | 0.1319 | 0.4056 | 0.8212 |
| 79 | 0.5319 | 0.8691 | 1.1984 | 0.6724 | 0.7747 | 0.1967 | 0.4931 | 0.1955 | 0.3078 | 0.8608 |
| 80 | 0.5172 | 0.8437 | 1.6158 | 0.9771 | 1.2638 | 0.2768 | 0.5290 | 0.2840 | 0.2594 | 0.9123 |
| 81 | 0.5012 | 0.9108 | 2.0673 | 1.3030 | 1.8230 | 0.4002 | 0.5841 | 0.3753 | 0.2762 | 0.9972 |
| 82 | 0.4858 | 1.0640 | 2.5378 | 1.6434 | 2.4741 | 0.4853 | 0.6515 | 0.4642 | 0.3415 | 1.1137 |
| 83 | 0.4666 | 1.2758 | 2.9816 | 1.9774 | 3.1973 | 0.5454 | 0.7052 | 0.5410 | 0.4287 | 1.2733 |
| 84 | 0.4370 | 1.5073 | 3.3752 | 2.2870 | 3.9742 | 0.5730 | 0.7510 | 0.6099 | 0.5261 | 1.4476 |
| 85 | 0.4025 | 1.7237 | 3.6871 | 2.5490 | 4.7542 | 0.5894 | 0.7835 | 0.6617 | 0.6153 | 1.6307 |
| 86 | 0.3588 | 1.8933 | 3.8817 | 2.7239 | 5.4607 | 0.6353 | 0.7929 | 0.6958 | 0.6852 | 1.7858 |
| 87 | 0.3276 | 1.9932 | 3.9484 | 2.8003 | 6.0199 | 0.7211 | 0.8002 | 0.7156 | 0.7180 | 1.8825 |
| 88 | 0.3133 | 2.0136 | 3.8682 | 2.7631 | 6.3678 | 0.8488 | 0.8167 | 0.7185 | 0.7240 | 1.9297 |
| 89 | 0.3185 | 1.9636 | 3.6614 | 2.6199 | 6.4588 | 0.9929 | 0.8669 | 0.7170 | 0.7260 | 1.9098 |
| 90 | 0.3132 | 1.8553 | 3.3303 | 2.3805 | 6.2897 | 1.1261 | 0.9415 | 0.7072 | 0.7511 | 1.8313 |
| 91 | 0.2877 | 1.7084 | 2.9267 | 2.0857 | 5.9010 | 1.2439 | 1.0328 | 0.6923 | 0.8028 | 1.7098 |
| 92 | 0.2205 | 1.5540 | 2.4591 | 1.7581 | 5.3532 | 1.3441 | 1.1244 | 0.6842 | 0.8760 | 1.5564 |
| 93 | 0.1378 | 1.4088 | 1.9861 | 1.4321 | 4.7026 | 1.4297 | 1.2085 | 0.6641 | 0.9583 | 1.3729 |
| 94 | 0.1671 | 1.2674 | 1.5313 | 1.1507 | 4.0193 | 1.4881 | 1.2742 | 0.6531 | 1.0354 | 1.2071 |
| 95 | 0.3243 | 1.1137 | 1.1620 | 0.9506 | 3.3591 | 1.5214 | 1.3388 | 0.6433 | 1.0975 | 1.0738 |
| 96 | 0.5459 | 0.9681 | 0.9168 | 0.8781 | 2.7823 | 1.5296 | 1.3764 | 0.6440 | 1.1006 | 1.0527 |
| 97 | 0.8111 | 0.8381 | 0.8246 | 0.9574 | 2.3164 | 1.5136 | 1.4073 | 0.6599 | 1.0455 | 1.2403 |
| 98 | 1.1015 | 0.7191 | 0.9020 | 1.1575 | 1.9988 | 1.4919 | 1.4313 | 0.6984 | 0.9585 | 1.6207 |
| 99 | 1.3977 | 0.6127 | 1.1343 | 1.4262 | 1.8645 | 1.4828 | 1.4624 | 0.7526 | 0.8726 | 2.0779 |

| %Stride | Cycle 1 $R_{ij}^2$ | Cycle 2 $R_{ij}^2$ | Cycle 3 $R_{ij}^2$ | Cycle 4 $R_{ij}^2$ | Cycle 5 $R_{ij}^2$ | Cycle 6 $R_{ij}^2$ | Cycle 7 $R_{ij}^2$ | Cycle 8 $R_{ij}^2$ | Cycle 9 $R_{ij}^2$ | Cycle 10 $R_{ij}^2$ |
|---|---|---|---|---|---|---|---|---|---|---|
| 0 | 1.6837 | 3.2495 | 0.1398 | 0.9447 | 2.4045 | 3.3906 | 2.4751 | 1.9135 | 0.8707 | 0.6501 |
| 1 | 1.2652 | 5.1197 | 0.1173 | 1.1203 | 3.0582 | 4.0263 | 2.3255 | 1.5854 | 0.9573 | 0.5824 |
| 2 | 0.8815 | 7.5867 | 0.3673 | 1.2596 | 3.8031 | 4.9794 | 2.2187 | 1.2560 | 1.0796 | 0.5567 |
| 3 | 0.5708 | 10.6009 | 0.9673 | 1.2538 | 4.6024 | 6.0657 | 2.2191 | 0.8749 | 1.3247 | 0.5972 |
| 4 | 0.3110 | 14.6409 | 1.9186 | 1.1243 | 5.4588 | 7.3651 | 2.3399 | 0.5189 | 1.7222 | 0.6937 |
| 5 | 0.1250 | 19.8327 | 3.2127 | 0.9411 | 6.4307 | 8.7949 | 2.6463 | 0.2334 | 2.3393 | 0.8478 |
| 6 | 0.0253 | 26.4440 | 4.6923 | 0.7971 | 7.5360 | 10.3634 | 3.0876 | 0.1077 | 3.3723 | 1.1095 |
| 7 | 0.0133 | 34.6920 | 6.2353 | 0.7321 | 8.7651 | 11.9623 | 3.7573 | 0.2317 | 4.9051 | 1.4968 |
| 8 | 0.0994 | 44.7094 | 7.7515 | 0.8327 | 10.1246 | 13.4839 | 4.5976 | 0.7128 | 7.1965 | 2.0266 |
| 9 | 0.2717 | 56.2602 | 9.0984 | 1.1427 | 11.5343 | 14.7582 | 5.6084 | 1.5737 | 10.3267 | 2.7039 |
| 10 | 0.5194 | 68.4810 | 10.2181 | 1.6482 | 12.8133 | 15.4843 | 6.7476 | 2.8427 | 14.5694 | 3.4996 |
| 11 | 0.7808 | 80.6652 | 11.1313 | 2.3827 | 13.7526 | 15.6222 | 8.0151 | 4.4246 | 19.8544 | 4.3556 |
| 12 | 1.0323 | 90.8789 | 11.7502 | 3.2962 | 13.9670 | 14.7846 | 9.2014 | 6.1304 | 26.0903 | 5.1217 |
| 13 | 1.2078 | 97.9873 | 12.1297 | 4.3812 | 13.4687 | 13.0748 | 10.3429 | 7.7705 | 32.8727 | 5.7266 |
| 14 | 1.3185 | 100.8049 | 12.3256 | 5.4672 | 12.2619 | 10.7269 | 11.2753 | 9.1741 | 39.6837 | 6.0595 |
| 15 | 1.3201 | 99.0020 | 12.2488 | 6.5782 | 10.4525 | 8.0657 | 12.0961 | 10.3095 | 45.7480 | 6.1504 |
| 16 | 1.2556 | 93.2660 | 11.9266 | 7.5667 | 8.4165 | 5.4739 | 12.6393 | 11.0473 | 50.6143 | 5.9568 |
| 17 | 1.1580 | 84.3459 | 11.4170 | 8.2837 | 6.4670 | 3.3432 | 12.9172 | 11.5178 | 53.6422 | 5.5890 |
| 18 | 1.0474 | 73.5043 | 10.7566 | 8.8056 | 4.7880 | 1.7458 | 12.9225 | 11.7699 | 54.9014 | 5.0655 |
| 19 | 0.9628 | 61.7635 | 10.0044 | 8.9949 | 3.4532 | 0.8001 | 12.6812 | 11.8692 | 54.3750 | 4.4455 |
| 20 | 0.8942 | 49.9948 | 9.2637 | 8.8662 | 2.4949 | 0.4942 | 12.2939 | 11.9158 | 52.4571 | 3.7842 |
| 21 | 0.8581 | 38.7128 | 8.5615 | 8.4815 | 1.8077 | 0.7769 | 11.7267 | 11.7997 | 49.2153 | 3.1020 |
| 22 | 0.8495 | 28.3796 | 8.0020 | 7.8575 | 1.3368 | 1.6128 | 11.0476 | 11.6338 | 45.1215 | 2.3601 |
| 23 | 0.8676 | 19.2725 | 7.5610 | 7.0966 | 0.9970 | 2.9884 | 10.2690 | 11.2833 | 40.4043 | 1.6379 |
| 24 | 0.9074 | 11.8075 | 7.3297 | 6.3185 | 0.7438 | 4.8086 | 9.3835 | 10.7738 | 35.2628 | 0.9374 |
| 25 | 0.9615 | 6.1022 | 7.2531 | 5.5450 | 0.5424 | 7.1025 | 8.4267 | 10.0710 | 29.7786 | 0.3568 |
| 26 | 1.0173 | 2.3351 | 7.3702 | 4.8353 | 0.3870 | 9.5960 | 7.4132 | 9.2202 | 24.1040 | 0.0257 |
| 27 | 1.0659 | 0.5239 | 7.6385 | 4.2369 | 0.2707 | 12.1986 | 6.3656 | 8.1454 | 18.5095 | 0.1204 |
| 28 | 1.0978 | 0.3982 | 7.9687 | 3.7765 | 0.1804 | 14.4786 | 5.3045 | 6.9129 | 13.1907 | 0.7823 |
| 29 | 1.0949 | 1.4780 | 8.3169 | 3.4259 | 0.1295 | 16.0749 | 4.2644 | 5.6882 | 8.5226 | 2.0994 |
| 30 | 1.1013 | 3.1011 | 8.4712 | 3.1394 | 0.1004 | 16.9033 | 3.3347 | 4.4260 | 4.7583 | 3.9074 |
| 31 | 1.0584 | 4.7215 | 8.4163 | 2.8773 | 0.0951 | 16.5895 | 2.5190 | 3.2808 | 2.1322 | 5.9193 |
| 32 | 1.0083 | 5.8106 | 8.0463 | 2.6500 | 0.1143 | 15.3378 | 1.8768 | 2.2800 | 0.6141 | 7.8277 |
| 33 | 0.9689 | 6.3082 | 7.4028 | 2.3977 | 0.1596 | 13.5642 | 1.4245 | 1.4876 | 0.0621 | 9.2338 |
| 34 | 0.9900 | 6.3427 | 6.5651 | 2.1425 | 0.2320 | 11.4443 | 1.0766 | 0.9164 | 0.1911 | 9.9100 |
| 35 | 1.0509 | 6.0483 | 5.6895 | 1.8783 | 0.3185 | 9.4005 | 0.8929 | 0.5446 | 0.7237 | 9.9838 |
| 36 | 1.1850 | 5.6865 | 4.8026 | 1.6486 | 0.3962 | 7.5899 | 0.8138 | 0.3513 | 1.3650 | 9.5331 |
| 37 | 1.4420 | 5.3876 | 4.0269 | 1.4008 | 0.4472 | 6.1832 | 0.8312 | 0.2949 | 1.9661 | 8.9469 |
| 38 | 1.7932 | 5.1301 | 3.3341 | 1.1665 | 0.4732 | 5.1052 | 0.9409 | 0.3366 | 2.4842 | 8.5278 |
| 39 | 2.3155 | 4.9477 | 2.7350 | 0.9453 | 0.4914 | 4.3994 | 1.0975 | 0.4397 | 2.8965 | 8.6088 |
| 40 | 3.0922 | 4.8230 | 2.2149 | 0.7350 | 0.5805 | 3.9474 | 1.3160 | 0.5570 | 3.1874 | 9.5989 |
| 41 | 4.1680 | 4.7275 | 1.7121 | 0.5367 | 0.8697 | 3.7173 | 1.5655 | 0.6486 | 3.4828 | 11.9499 |
| 42 | 5.6887 | 4.5932 | 1.2419 | 0.3679 | 1.5203 | 3.6246 | 1.8117 | 0.6663 | 3.8405 | 16.0491 |
| 43 | 7.6593 | 4.4310 | 0.8293 | 0.2056 | 2.7536 | 3.5042 | 2.1274 | 0.5978 | 4.4280 | 22.6040 |
| 44 | 9.9980 | 4.2025 | 0.6437 | 0.0743 | 4.6244 | 3.3332 | 2.4946 | 0.4252 | 5.4334 | 31.8083 |
| 45 | 12.5545 | 3.9459 | 0.9400 | 0.0180 | 7.0400 | 2.9337 | 2.9689 | 0.2201 | 7.0641 | 44.0576 |
| 46 | 14.8597 | 3.5856 | 1.9448 | 0.1555 | 9.6524 | 2.3810 | 3.5086 | 0.0431 | 9.4512 | 58.4759 |
| 47 | 16.4147 | 3.2683 | 3.7814 | 0.6504 | 11.7142 | 1.7110 | 4.1186 | 0.0572 | 12.4630 | 73.4445 |
| 48 | 16.7638 | 3.0272 | 5.9714 | 1.6960 | 12.5664 | 1.0485 | 4.6881 | 0.4157 | 15.7232 | 85.9574 |
| 49 | 15.5156 | 2.8313 | 7.8227 | 3.3459 | 12.0878 | 0.5346 | 4.8974 | 1.2594 | 18.5162 | 92.2319 |
| 50 | 13.3996 | 2.5053 | 8.6607 | 5.2280 | 10.7103 | 0.2221 | 4.8009 | 2.4593 | 19.9660 | 90.6172 |
| 51 | 10.9034 | 1.8859 | 8.1196 | 6.8221 | 9.0842 | 0.1652 | 4.3476 | 3.7052 | 19.9408 | 81.9752 |
| 52 | 8.5034 | 0.9794 | 6.5516 | 7.6265 | 8.0791 | 0.3571 | 3.5453 | 4.8057 | 18.8921 | 68.3557 |
| 53 | 6.7696 | 0.4265 | 4.8691 | 7.2685 | 8.0007 | 0.8768 | 2.7142 | 5.4259 | 17.1784 | 53.2095 |
| 54 | 5.7735 | 1.0206 | 3.4901 | 5.9105 | 9.0879 | 1.8915 | 1.9950 | 5.4200 | 15.5886 | 39.1890 |
| 55 | 5.3705 | 3.5874 | 2.4915 | 4.0706 | 11.6919 | 3.5902 | 1.3494 | 5.0659 | 14.7733 | 27.2777 |
| 56 | 5.5520 | 8.6283 | 1.9250 | 2.1752 | 16.5450 | 6.2692 | 0.8885 | 4.3665 | 14.6347 | 17.8462 |
| 57 | 6.3608 | 16.2272 | 1.6600 | 0.7375 | 24.3857 | 10.2013 | 0.5446 | 3.4797 | 15.2020 | 10.7278 |
| 58 | 7.8047 | 25.6451 | 1.6364 | 0.0862 | 35.9525 | 15.3081 | 0.3392 | 2.5373 | 16.1666 | 5.6033 |
| 59 | 9.6874 | 35.3980 | 1.7970 | 0.4046 | 50.9447 | 21.0730 | 0.2242 | 1.6484 | 17.2367 | 2.2422 |
| 60 | 11.8493 | 43.2459 | 2.1351 | 1.6361 | 67.7490 | 26.3841 | 0.2094 | 0.9088 | 17.7696 | 0.5508 |
| 61 | 13.7814 | 46.5394 | 2.6258 | 3.4277 | 83.0487 | 29.3746 | 0.2948 | 0.4237 | 17.1199 | 0.1775 |
| 62 | 14.7091 | 44.2732 | 3.0740 | 5.0765 | 92.4153 | 28.7590 | 0.4660 | 0.2490 | 15.2066 | 0.5889 |
| 63 | 14.1362 | 37.5303 | 3.3078 | 6.0973 | 91.7489 | 24.6347 | 0.7023 | 0.3008 | 12.3127 | 1.1338 |
| 64 | 12.3548 | 28.4198 | 3.2595 | 6.2591 | 81.2785 | 18.0622 | 0.9804 | 0.4569 | 8.8631 | 1.3159 |
| 65 | 9.6638 | 19.6766 | 2.8189 | 5.6039 | 64.0460 | 11.2769 | 1.1726 | 0.5976 | 5.8462 | 1.1199 |
| 66 | 6.8172 | 13.2222 | 2.1625 | 4.5235 | 44.9309 | 6.3291 | 1.2517 | 0.6553 | 3.8012 | 0.6910 |
| 67 | 4.4255 | 9.3117 | 1.4782 | 3.4222 | 28.5871 | 3.6127 | 1.2396 | 0.6336 | 2.7667 | 0.3162 |
| 68 | 2.6726 | 7.4740 | 0.9110 | 2.4626 | 17.1479 | 2.8294 | 1.1327 | 0.5883 | 2.4687 | 0.1381 |
| 69 | 1.5515 | 6.8390 | 0.5457 | 1.7942 | 10.2793 | 3.0704 | 1.0007 | 0.5383 | 2.5586 | 0.1448 |
| 70 | 0.9113 | 6.6042 | 0.3501 | 1.3377 | 6.6879 | 3.5477 | 0.8419 | 0.4613 | 2.6453 | 0.2579 |
| 71 | 0.5919 | 6.2984 | 0.3027 | 1.0602 | 4.8903 | 3.7593 | 0.6950 | 0.3971 | 2.5474 | 0.3827 |
| 72 | 0.4343 | 5.6115 | 0.3247 | 0.8441 | 3.8037 | 3.4832 | 0.5646 | 0.3258 | 2.2396 | 0.4764 |
| 73 | 0.3484 | 4.7088 | 0.3437 | 0.6279 | 2.8046 | 2.8476 | 0.4560 | 0.2405 | 1.7917 | 0.5128 |
| 74 | 0.3005 | 3.7052 | 0.3358 | 0.4128 | 1.8400 | 2.0328 | 0.3663 | 0.1663 | 1.2916 | 0.5326 |
| 75 | 0.2751 | 2.7416 | 0.2996 | 0.2206 | 0.9838 | 1.2620 | 0.2966 | 0.0965 | 0.8644 | 0.5404 |
| 76 | 0.2691 | 1.9651 | 0.2845 | 0.0836 | 0.3584 | 0.6649 | 0.2487 | 0.0469 | 0.5251 | 0.5748 |
| 77 | 0.2747 | 1.3667 | 0.3900 | 0.0532 | 0.0615 | 0.2770 | 0.2221 | 0.0194 | 0.3009 | 0.6124 |
| 78 | 0.2837 | 0.9730 | 0.7382 | 0.1628 | 0.1226 | 0.0857 | 0.2155 | 0.0174 | 0.1645 | 0.6743 |
| 79 | 0.2830 | 0.7552 | 1.4360 | 0.4521 | 0.6001 | 0.0387 | 0.2431 | 0.0382 | 0.0947 | 0.7409 |
| 80 | 0.2675 | 0.7119 | 2.6109 | 0.9548 | 1.5971 | 0.0766 | 0.2798 | 0.0807 | 0.0673 | 0.8324 |
| 81 | 0.2512 | 0.8296 | 4.2735 | 1.6978 | 3.3234 | 0.1602 | 0.3411 | 0.1409 | 0.0763 | 0.9944 |
| 82 | 0.2360 | 1.1322 | 6.4403 | 2.7008 | 6.1212 | 0.2355 | 0.4245 | 0.2155 | 0.1166 | 1.2404 |
| 83 | 0.2177 | 1.6277 | 8.8897 | 3.9101 | 10.2229 | 0.2975 | 0.4973 | 0.2927 | 0.1838 | 1.6214 |
| 84 | 0.1910 | 2.2718 | 11.3918 | 5.2303 | 15.7942 | 0.3284 | 0.5639 | 0.3720 | 0.2768 | 2.0956 |
| 85 | 0.1620 | 2.9713 | 13.5949 | 6.4974 | 22.6020 | 0.3474 | 0.6139 | 0.4378 | 0.3786 | 2.6591 |
| 86 | 0.1287 | 3.5847 | 15.0678 | 7.4197 | 29.8189 | 0.4036 | 0.6287 | 0.4842 | 0.4695 | 3.1892 |
| 87 | 0.1073 | 3.9727 | 15.5895 | 7.8419 | 36.2387 | 0.5200 | 0.6403 | 0.5121 | 0.5156 | 3.5439 |
| 88 | 0.0981 | 4.0548 | 14.9633 | 7.6345 | 40.5489 | 0.7205 | 0.6670 | 0.5162 | 0.5242 | 3.7237 |
| 89 | 0.1014 | 3.8558 | 13.4058 | 6.8641 | 41.7166 | 0.9859 | 0.7514 | 0.5141 | 0.5270 | 3.6472 |
| 90 | 0.0981 | 3.4421 | 11.0912 | 5.6669 | 39.5597 | 1.2680 | 0.8865 | 0.5002 | 0.5641 | 3.3538 |
| 91 | 0.0828 | 2.9188 | 8.5653 | 4.3502 | 34.8213 | 1.5474 | 1.0667 | 0.4793 | 0.6445 | 2.9233 |
| 92 | 0.0486 | 2.4150 | 6.0473 | 3.0908 | 28.6564 | 1.8065 | 1.2644 | 0.4681 | 0.7673 | 2.4223 |
| 93 | 0.0190 | 1.9846 | 3.9445 | 2.0509 | 22.1143 | 2.0439 | 1.4606 | 0.4410 | 0.9183 | 1.8849 |
| 94 | 0.0279 | 1.6064 | 2.3450 | 1.3242 | 16.1549 | 2.2145 | 1.6236 | 0.4265 | 1.0721 | 1.4571 |
| 95 | 0.1052 | 1.2403 | 1.3501 | 0.9036 | 11.2839 | 2.3148 | 1.7924 | 0.4139 | 1.2044 | 1.1532 |
| 96 | 0.2980 | 0.9373 | 0.8405 | 0.7711 | 7.7411 | 2.3396 | 1.8946 | 0.4148 | 1.2114 | 1.1082 |
| 97 | 0.6579 | 0.7025 | 0.6800 | 0.9167 | 5.3659 | 2.2910 | 1.9806 | 0.4354 | 1.0930 | 1.5384 |
| 98 | 1.2133 | 0.5172 | 0.8137 | 1.3397 | 3.9951 | 2.2258 | 2.0487 | 0.4877 | 0.9187 | 2.6265 |
| 99 | 1.9536 | 0.3754 | 1.2865 | 2.0339 | 3.4764 | 2.1987 | 2.1385 | 0.5664 | 0.7614 | 4.3175 |
| RMS$_j$ = [Eqn 18] | 1.8095 | 4.1886 | 2.2930 | 1.7751 | 3.7519 | 2.5185 | 1.8051 | 1.6335 | 3.3569 | 3.2586 |
| | | | | RMS$_{avg}$ = [Eqn 19] | 2.6391 | | | | | |

**Table B.4  Example Data for Calculating the Values Needed to Plot the Curve Average Variability Band for the Angle-Angle Diagrams**

*Equations 2.21-2.29 are demonstrated using the ankle and knee angle data given in table B.2.*

| | Knee (x-variable) | | | | | | | | | |
|---|---|---|---|---|---|---|---|---|---|---|
| %Stride | Cycle 1 $Rx_{ij}^2$ [Eqn 21] | Cycle 2 $Rx_{ij}^2$ [Eqn 21] | Cycle 3 $Rx_{ij}^2$ [Eqn 21] | Cycle 4 $Rx_{ij}^2$ [Eqn 21] | Cycle 5 $Rx_{ij}^2$ [Eqn 21] | Cycle 6 $Rx_{ij}^2$ [Eqn 21] | Cycle 7 $Rx_{ij}^2$ [Eqn 21] | Cycle 8 $Rx_{ij}^2$ [Eqn 21] | Cycle 9 $Rx_{ij}^2$ [Eqn 21] | Cycle 10 $Rx_{ij}^2$ [Eqn 21] |
| 0 | 0.8987 | 0.7022 | 0.0282 | 0.0686 | 1.8360 | 3.2005 | 1.3642 | 0.3158 | 0.0353 | 0.4330 |
| 1 | 0.6754 | 1.5173 | 0.1168 | 0.0220 | 2.2446 | 4.0248 | 1.0441 | 0.2583 | 0.0125 | 0.3992 |
| 2 | 0.4706 | 2.7423 | 0.2767 | 0.0029 | 2.7027 | 4.8576 | 0.7327 | 0.1971 | 0.0000 | 0.3795 |
| 3 | 0.3014 | 4.4058 | 0.5461 | 0.0017 | 3.2077 | 5.5743 | 0.4476 | 0.1163 | 0.0228 | 0.3956 |
| 4 | 0.1681 | 6.8121 | 0.9025 | 0.0049 | 3.7636 | 6.3001 | 0.2025 | 0.0400 | 0.1296 | 0.4356 |
| 5 | 0.0734 | 9.9919 | 1.3712 | 0.0253 | 4.4058 | 6.9643 | 0.0445 | 0.0001 | 0.3709 | 0.4914 |
| 6 | 0.0213 | 14.0325 | 1.8934 | 0.0864 | 5.0805 | 7.5845 | 0.0006 | 0.0467 | 0.8538 | 0.5868 |
| 7 | 0.0001 | 19.0183 | 2.4367 | 0.2294 | 5.8033 | 8.1168 | 0.0671 | 0.2314 | 1.6615 | 0.7242 |
| 8 | 0.0052 | 24.8801 | 2.9860 | 0.4928 | 6.5638 | 8.4797 | 0.2421 | 0.6209 | 2.9653 | 0.9178 |
| 9 | 0.0237 | 31.5395 | 3.5194 | 0.9293 | 7.2576 | 8.6671 | 0.5098 | 1.2232 | 4.8136 | 1.1364 |
| 10 | 0.0433 | 38.3409 | 3.9681 | 1.5326 | 7.8849 | 8.5147 | 0.8427 | 2.0506 | 7.3333 | 1.3971 |
| 11 | 0.0566 | 44.9168 | 4.3765 | 2.3348 | 8.2254 | 8.0542 | 1.2499 | 3.0346 | 10.4846 | 1.6693 |
| 12 | 0.0640 | 50.2256 | 4.6959 | 3.2870 | 8.1396 | 7.1449 | 1.6719 | 4.0683 | 14.1602 | 1.8961 |
| 13 | 0.0543 | 53.6849 | 4.9151 | 4.3806 | 7.6895 | 5.8709 | 2.1404 | 5.0941 | 18.1732 | 2.0938 |
| 14 | 0.0416 | 54.7008 | 5.0895 | 5.4476 | 6.8330 | 4.4268 | 2.5728 | 5.9341 | 22.1276 | 2.1786 |
| 15 | 0.0296 | 53.1149 | 5.0986 | 6.5127 | 5.6739 | 2.9309 | 2.9998 | 6.6461 | 25.6238 | 2.2141 |
| 16 | 0.0164 | 49.5898 | 5.0266 | 7.4420 | 4.4437 | 1.6589 | 3.3782 | 7.1396 | 28.4942 | 2.1668 |
| 17 | 0.0088 | 44.3023 | 4.8224 | 8.0883 | 3.3270 | 0.7123 | 3.6634 | 7.4857 | 30.2940 | 2.0621 |
| 18 | 0.0040 | 38.1553 | 4.5241 | 8.5439 | 2.3809 | 0.1545 | 3.8534 | 7.7117 | 31.1699 | 1.9238 |
| 19 | 0.0014 | 31.6069 | 4.1698 | 8.6907 | 1.6589 | 0.0018 | 3.9521 | 7.8512 | 31.0026 | 1.7477 |
| 20 | 0.0006 | 25.1502 | 3.8220 | 8.5556 | 1.1556 | 0.2162 | 3.9402 | 7.9806 | 30.1950 | 1.5500 |
| 21 | 0.0002 | 19.0620 | 3.4820 | 8.2025 | 0.7992 | 0.7500 | 3.8181 | 7.9863 | 28.6653 | 1.3363 |
| 22 | 0.0000 | 13.5792 | 3.2220 | 7.6452 | 0.5700 | 1.5750 | 3.5910 | 7.9806 | 26.6772 | 1.0506 |
| 23 | 0.0001 | 8.8209 | 2.9929 | 6.9696 | 0.4096 | 2.6896 | 3.2761 | 7.8400 | 24.3049 | 0.7569 |
| 24 | 0.0007 | 5.0445 | 2.8764 | 6.2700 | 0.2959 | 4.0240 | 2.8696 | 7.5955 | 21.6597 | 0.4436 |
| 25 | 0.0013 | 2.2680 | 2.8090 | 5.5413 | 0.2061 | 5.6454 | 2.4149 | 7.2146 | 18.6970 | 0.1648 |
| 26 | 0.0041 | 0.5837 | 2.8359 | 4.8224 | 0.1414 | 7.3658 | 1.9488 | 6.7288 | 15.4921 | 0.0071 |
| 27 | 0.0069 | 0.0028 | 2.9344 | 4.1494 | 0.0943 | 9.1991 | 1.4811 | 6.0664 | 12.2290 | 0.0824 |
| 28 | 0.0151 | 0.3684 | 3.0381 | 3.5608 | 0.0562 | 10.8439 | 1.0547 | 5.2579 | 8.9820 | 0.4858 |
| 29 | 0.0250 | 1.3971 | 3.1613 | 3.0346 | 0.0369 | 12.0965 | 0.6757 | 4.4437 | 6.0123 | 1.2814 |
| 30 | 0.0445 | 2.7192 | 3.1720 | 2.5889 | 0.0286 | 12.8953 | 0.3832 | 3.5759 | 3.5306 | 2.3994 |
| 31 | 0.0724 | 4.0441 | 3.0941 | 2.1934 | 0.0259 | 12.9528 | 0.1689 | 2.7523 | 1.6926 | 3.6902 |
| 32 | 0.1190 | 4.9952 | 2.9070 | 1.8632 | 0.0342 | 12.3552 | 0.0462 | 2.0022 | 0.5550 | 4.9952 |
| 33 | 0.1892 | 5.5460 | 2.5760 | 1.5750 | 0.0552 | 11.3232 | 0.0012 | 1.3806 | 0.0506 | 6.1256 |
| 34 | 0.2894 | 5.7696 | 2.1845 | 1.3271 | 0.0973 | 9.9730 | 0.0219 | 0.8987 | 0.0520 | 6.9274 |
| 35 | 0.4045 | 5.6835 | 1.7849 | 1.1109 | 0.1552 | 8.5615 | 0.0936 | 0.5417 | 0.3919 | 7.4748 |
| 36 | 0.5402 | 5.4990 | 1.4280 | 0.9312 | 0.2256 | 7.2092 | 0.2162 | 0.2970 | 0.8930 | 7.7562 |
| 37 | 0.6972 | 5.3130 | 1.1130 | 0.7656 | 0.3080 | 6.0762 | 0.3782 | 0.1482 | 1.4520 | 7.9806 |
| 38 | 0.8446 | 5.1121 | 0.8817 | 0.6100 | 0.3982 | 5.1031 | 0.5761 | 0.0620 | 2.0136 | 8.2426 |
| 39 | 0.9722 | 4.9462 | 0.7157 | 0.4679 | 0.4816 | 4.3514 | 0.7850 | 0.0185 | 2.5472 | 8.6084 |
| 40 | 1.0899 | 4.8224 | 0.6147 | 0.3318 | 0.5565 | 3.7404 | 1.0080 | 0.0019 | 3.0068 | 9.1567 |
| 41 | 1.1578 | 4.7263 | 0.5565 | 0.2061 | 0.6147 | 3.2616 | 1.2232 | 0.0006 | 3.4447 | 9.9477 |
| 42 | 1.1816 | 4.5925 | 0.5580 | 0.1109 | 0.6610 | 2.8798 | 1.3853 | 0.0053 | 3.8298 | 10.9098 |
| 43 | 1.1772 | 4.4310 | 0.5852 | 0.0342 | 0.6972 | 2.5122 | 1.5252 | 0.0110 | 4.1820 | 12.1452 |
| 44 | 1.1236 | 4.2025 | 0.6400 | 0.0009 | 0.7396 | 2.1904 | 1.6129 | 0.0169 | 4.4944 | 13.3956 |
| 45 | 1.0692 | 3.9442 | 0.7293 | 0.0154 | 0.7850 | 1.8333 | 1.6487 | 0.0185 | 4.7263 | 14.5619 |
| 46 | 0.9841 | 3.5645 | 0.8317 | 0.0853 | 0.8612 | 1.4689 | 1.6180 | 0.0190 | 4.8929 | 15.5078 |
| 47 | 0.8761 | 3.1471 | 0.9526 | 0.1989 | 0.9683 | 1.1151 | 1.5525 | 0.0180 | 4.9997 | 16.0320 |
| 48 | 0.7885 | 2.6634 | 1.0568 | 0.3576 | 1.1278 | 0.7709 | 1.4593 | 0.0149 | 4.9195 | 16.0160 |
| 49 | 0.6939 | 2.1815 | 1.1513 | 0.5373 | 1.3156 | 0.4665 | 1.3294 | 0.0115 | 4.6786 | 15.2647 |
| 50 | 0.6095 | 1.6623 | 1.1896 | 0.7237 | 1.5858 | 0.2216 | 1.1896 | 0.0080 | 4.3293 | 13.8853 |
| 51 | 0.5198 | 1.1859 | 1.1903 | 0.8668 | 1.8769 | 0.0681 | 1.0630 | 0.0048 | 3.8064 | 11.9578 |
| 52 | 0.4474 | 0.7588 | 1.1213 | 0.9582 | 2.2144 | 0.0046 | 0.9004 | 0.0021 | 3.2002 | 9.6541 |
| 53 | 0.3652 | 0.4067 | 1.0248 | 0.9649 | 2.5751 | 0.0132 | 0.7522 | 0.0002 | 2.5578 | 7.2506 |
| 54 | 0.2886 | 0.1519 | 0.8915 | 0.8840 | 2.9474 | 0.0744 | 0.5978 | 0.0004 | 1.9578 | 4.9899 |
| 55 | 0.2176 | 0.0153 | 0.7491 | 0.7217 | 3.3398 | 0.1677 | 0.4469 | 0.0034 | 1.4508 | 3.0713 |
| 56 | 0.1556 | 0.0258 | 0.6186 | 0.5034 | 3.7500 | 0.2814 | 0.3064 | 0.0109 | 1.0702 | 1.6295 |
| 57 | 0.1058 | 0.2192 | 0.5129 | 0.2737 | 4.2345 | 0.4106 | 0.1834 | 0.0238 | 0.8303 | 0.6853 |
| 58 | 0.0707 | 0.6351 | 0.4461 | 0.0858 | 4.8272 | 0.5702 | 0.0882 | 0.0441 | 0.7139 | 0.1790 |
| 59 | 0.0488 | 1.3087 | 0.4173 | 0.0009 | 5.5790 | 0.7744 | 0.0266 | 0.0724 | 0.7157 | 0.0059 |
| 60 | 0.0397 | 2.2539 | 0.4281 | 0.0649 | 6.5418 | 1.0480 | 0.0010 | 0.1104 | 0.8323 | 0.0417 |
| 61 | 0.0425 | 3.4192 | 0.4749 | 0.2969 | 7.7167 | 1.3969 | 0.0079 | 0.1601 | 1.0590 | 0.1765 |
| 62 | 0.0576 | 4.7528 | 0.5627 | 0.6821 | 9.0715 | 1.8303 | 0.0400 | 0.2210 | 1.3926 | 0.3250 |
| 63 | 0.0883 | 6.0873 | 0.6677 | 1.1727 | 10.4581 | 2.2889 | 0.0800 | 0.2885 | 1.8147 | 0.4318 |
| 64 | 0.1369 | 7.1824 | 0.7744 | 1.6384 | 11.6964 | 2.7225 | 0.1296 | 0.3600 | 2.2201 | 0.4761 |
| 65 | 0.2070 | 7.9242 | 0.8556 | 2.0306 | 12.4962 | 3.0450 | 0.1722 | 0.4422 | 2.5440 | 0.4422 |
| 66 | 0.2830 | 8.2484 | 0.8874 | 2.2741 | 12.6594 | 3.1969 | 0.2007 | 0.5069 | 2.6962 | 0.3624 |
| 67 | 0.3600 | 8.0656 | 0.8281 | 2.2801 | 12.0409 | 3.0976 | 0.2209 | 0.5329 | 2.6569 | 0.2500 |
| 68 | 0.4199 | 7.4420 | 0.7022 | 2.0794 | 17.7060 | 2.7956 | 0.2323 | 0.5446 | 2.3963 | 0.1354 |
| 69 | 0.4462 | 6.5434 | 0.5300 | 1.7477 | 8.8328 | 2.3165 | 0.2520 | 0.5300 | 2.0392 | 0.0433 |
| 70 | 0.4343 | 5.5178 | 0.3238 | 1.3248 | 6.6616 | 1.7716 | 0.2510 | 0.4610 | 1.6104 | 0.0015 |
| 71 | 0.4122 | 4.5454 | 0.1537 | 0.9370 | 4.5710 | 1.2724 | 0.2581 | 0.3869 | 1.1925 | 0.0190 |
| 72 | 0.3709 | 3.6062 | 0.0437 | 0.6100 | 2.8258 | 0.8299 | 0.2611 | 0.3014 | 0.8446 | 0.0906 |
| 73 | 0.3215 | 2.8460 | 0.0003 | 0.3636 | 1.4957 | 0.4942 | 0.2632 | 0.2088 | 0.5730 | 0.1963 |
| 74 | 0.2820 | 2.2530 | 0.0286 | 0.2016 | 0.6545 | 0.2591 | 0.2591 | 0.1376 | 0.3733 | 0.3238 |
| 75 | 0.2520 | 1.7742 | 0.1142 | 0.1142 | 0.2007 | 0.1142 | 0.2480 | 0.0795 | 0.2520 | 0.4330 |
| 76 | 0.2352 | 1.4280 | 0.2352 | 0.0650 | 0.0210 | 0.0342 | 0.2352 | 0.0420 | 0.1640 | 0.5402 |
| 77 | 0.2294 | 1.1428 | 0.3856 | 0.0445 | 0.0119 | 0.0026 | 0.2218 | 0.0193 | 0.1149 | 0.6100 |
| 78 | 0.2333 | 0.9274 | 0.5580 | 0.0350 | 0.0918 | 0.0053 | 0.2089 | 0.0086 | 0.0858 | 0.6675 |
| 79 | 0.2381 | 0.7534 | 0.7259 | 0.0331 | 0.2190 | 0.0317 | 0.2134 | 0.0034 | 0.0718 | 0.6922 |
| 80 | 0.2372 | 0.6194 | 0.9082 | 0.0372 | 0.3564 | 0.0713 | 0.2144 | 0.0022 | 0.0661 | 0.6939 |
| 81 | 0.2362 | 0.4984 | 1.0692 | 0.0416 | 0.4844 | 0.1267 | 0.2343 | 0.0021 | 0.0708 | 0.6956 |
| 82 | 0.2314 | 0.3982 | 1.2522 | 0.0480 | 0.6257 | 0.1772 | 0.2798 | 0.0050 | 0.0847 | 0.6708 |
| 83 | 0.2172 | 0.3204 | 1.4019 | 0.0548 | 0.7327 | 0.2362 | 0.3295 | 0.0092 | 0.1063 | 0.6464 |
| 84 | 0.1910 | 0.2470 | 1.5450 | 0.0590 | 0.8046 | 0.2884 | 0.4007 | 0.0216 | 0.1421 | 0.5975 |
| 85 | 0.1616 | 0.1954 | 1.6848 | 0.0666 | 0.8501 | 0.3387 | 0.4872 | 0.0408 | 0.1954 | 0.5446 |
| 86 | 0.1246 | 0.1467 | 1.7876 | 0.0767 | 0.8336 | 0.4007 | 0.5580 | 0.0692 | 0.2632 | 0.4858 |
| 87 | 0.0888 | 0.1076 | 1.8550 | 0.0912 | 0.7885 | 0.4733 | 0.6273 | 0.1076 | 0.3341 | 0.4251 |
| 88 | 0.0557 | 0.0708 | 1.8879 | 0.1050 | 0.6823 | 0.5715 | 0.6626 | 0.1490 | 0.4045 | 0.3528 |
| 89 | 0.0269 | 0.0416 | 1.8934 | 0.1197 | 0.5388 | 0.7123 | 0.6823 | 0.1971 | 0.4816 | 0.2873 |
| 90 | 0.0081 | 0.0196 | 1.8496 | 0.1444 | 0.3721 | 0.8496 | 0.6561 | 0.2401 | 0.5625 | 0.2209 |
| 91 | 0.0004 | 0.0049 | 1.7689 | 0.1764 | 0.2116 | 1.1025 | 0.6084 | 0.2704 | 0.6241 | 0.1444 |
| 92 | 0.0028 | 0.0001 | 1.6205 | 0.2144 | 0.0767 | 1.3386 | 0.5520 | 0.2884 | 0.6675 | 0.0692 |
| 93 | 0.0112 | 0.0021 | 1.4544 | 0.2767 | 0.0055 | 1.5977 | 0.4844 | 0.2746 | 0.6956 | 0.0135 |
| 94 | 0.0205 | 0.0087 | 1.2551 | 0.3520 | 0.0267 | 1.7868 | 0.4268 | 0.2467 | 0.7169 | 0.0075 |
| 95 | 0.0199 | 0.0126 | 1.0388 | 0.4654 | 0.1816 | 1.8818 | 0.4048 | 0.2041 | 0.7256 | 0.1168 |
| 96 | 0.0075 | 0.0138 | 0.8234 | 0.6422 | 0.5104 | 1.8621 | 0.4037 | 0.1433 | 0.6800 | 0.4825 |
| 97 | 0.0002 | 0.0153 | 0.6170 | 0.9130 | 1.0353 | 1.7596 | 0.4206 | 0.0885 | 0.5875 | 1.2690 |
| 98 | 0.0164 | 0.0187 | 0.4367 | 1.2742 | 1.7631 | 1.6543 | 0.4460 | 0.0442 | 0.4778 | 2.5799 |
| 99 | 0.0548 | 0.0250 | 0.2905 | 1.6952 | 2.6700 | 1.6028 | 0.4692 | 0.0159 | 0.3795 | 4.3098 |
| $RMS_{xj}$ = [Eqn 23] | 0.5192 | 2.8884 | 1.2781 | 1.2768 | 1.6688 | 1.7852 | 0.9578 | 1.2202 | 2.3083 | 1.7428 |

| | Ankle (y-variable) | | | | | | | | | |
| | Cycle 1 $Ry_{ij}^2$ [Eqn 22] | Cycle 2 $Ry_{ij}^2$ [Eqn 22] | Cycle 3 $Ry_{ij}^2$ [Eqn 22] | Cycle 4 $Ry_{ij}^2$ [Eqn 22] | Cycle 5 $Ry_{ij}^2$ [Eqn 22] | Cycle 6 $Ry_{ij}^2$ [Eqn 22] | Cycle 7 $Ry_{ij}^2$ [Eqn 22] | Cycle 8 $Ry_{ij}^2$ [Eqn 22] | Cycle 9 $Ry_{ij}^2$ [Eqn 22] | Cycle 10 $Ry_{ij}^2$ [Eqn 22] |
| %Stride | | | | | | | | | | |
|---|---|---|---|---|---|---|---|---|---|---|
| 0 | 0.7850 | 2.5472 | 0.1116 | 0.8761 | 0.5685 | 0.1901 | 1.1109 | 1.5977 | 0.8354 | 0.2172 |
| 1 | 0.5898 | 3.6024 | 0.0005 | 1.0983 | 0.8136 | 0.0014 | 1.2814 | 1.3271 | 0.9448 | 0.1832 |
| 2 | 0.4109 | 4.8444 | 0.0906 | 1.2566 | 1.1004 | 0.1218 | 1.4860 | 1.0588 | 1.0795 | 0.1772 |
| 3 | 0.2694 | 6.1951 | 0.4212 | 1.2522 | 1.3948 | 0.4914 | 1.7716 | 0.7586 | 1.3019 | 0.2016 |
| 4 | 0.1429 | 7.8288 | 1.0161 | 1.1194 | 1.6952 | 1.0650 | 2.1374 | 0.4789 | 1.5926 | 0.2581 |
| 5 | 0.0515 | 9.8408 | 1.8414 | 0.9159 | 2.0249 | 1.8306 | 2.6018 | 0.2333 | 1.9684 | 0.3564 |
| 6 | 0.0040 | 12.4115 | 2.7989 | 0.7106 | 2.4555 | 2.7789 | 3.0871 | 0.0610 | 2.5186 | 0.5227 |
| 7 | 0.0123 | 15.6737 | 3.7986 | 0.5027 | 2.9618 | 3.8455 | 3.6902 | 0.0004 | 3.2436 | 0.7726 |
| 8 | 0.0943 | 19.8292 | 4.7655 | 0.3399 | 3.5608 | 5.0042 | 4.3556 | 0.0918 | 4.2313 | 1.1088 |
| 9 | 0.2480 | 24.7208 | 5.5790 | 0.2134 | 4.2766 | 6.0910 | 5.0986 | 0.3505 | 5.5131 | 1.5675 |
| 10 | 0.4761 | 30.1401 | 6.2500 | 0.1156 | 4.9284 | 6.9696 | 5.9049 | 0.7921 | 7.2361 | 2.1025 |
| 11 | 0.7242 | 35.7484 | 6.7548 | 0.0480 | 5.5272 | 7.5680 | 6.7652 | 1.3900 | 9.3697 | 2.6863 |
| 12 | 0.9683 | 40.6534 | 7.0543 | 0.0092 | 5.8274 | 7.6397 | 7.5295 | 2.0621 | 11.9301 | 3.2256 |
| 13 | 1.1535 | 44.3023 | 7.2146 | 0.0006 | 5.7792 | 7.2039 | 8.2025 | 2.6765 | 14.6996 | 3.6328 |
| 14 | 1.2769 | 46.1041 | 7.2361 | 0.0196 | 5.4289 | 6.3001 | 8.7025 | 3.2400 | 17.5561 | 3.8809 |
| 15 | 1.2905 | 45.8871 | 7.1503 | 0.0655 | 4.7786 | 5.1348 | 9.0963 | 3.6634 | 20.1242 | 3.9363 |
| 16 | 1.2392 | 43.6762 | 6.9001 | 0.1248 | 3.9728 | 3.8150 | 9.2611 | 3.9077 | 22.1201 | 3.7900 |
| 17 | 1.1492 | 40.0436 | 6.5946 | 0.1954 | 3.1400 | 2.6309 | 9.2538 | 4.0321 | 23.3482 | 3.5269 |
| 18 | 1.0435 | 35.3490 | 6.2325 | 0.2616 | 2.4072 | 1.5914 | 9.0691 | 4.0582 | 23.7315 | 3.1418 |
| 19 | 0.9614 | 30.1566 | 5.8346 | 0.3042 | 1.7943 | 0.7983 | 8.7291 | 4.0180 | 23.3724 | 2.6978 |
| 20 | 0.8935 | 24.8445 | 5.4416 | 0.3106 | 1.3393 | 0.2780 | 8.3537 | 3.9352 | 22.2621 | 2.2342 |
| 21 | 0.8579 | 19.6508 | 5.0796 | 0.2790 | 1.0084 | 0.0270 | 7.9085 | 3.8134 | 20.5500 | 1.7657 |
| 22 | 0.8495 | 14.8003 | 4.7800 | 0.2122 | 0.7668 | 0.0378 | 7.4566 | 3.6531 | 18.4443 | 1.3095 |
| 23 | 0.8675 | 10.4516 | 4.5681 | 0.1270 | 0.5874 | 0.2988 | 6.9929 | 3.4433 | 16.0994 | 0.8810 |
| 24 | 0.9067 | 6.7630 | 4.4532 | 0.0485 | 0.4479 | 0.7846 | 6.5139 | 3.1783 | 13.6030 | 0.4939 |
| 25 | 0.9602 | 3.8342 | 4.4441 | 0.0037 | 0.3363 | 1.4571 | 6.0118 | 2.8564 | 11.0816 | 0.1919 |
| 26 | 1.0132 | 1.7514 | 4.5343 | 0.0129 | 0.2456 | 2.2302 | 5.4644 | 2.4913 | 8.6119 | 0.0186 |
| 27 | 1.0590 | 0.5211 | 4.7041 | 0.0876 | 0.1765 | 2.9995 | 4.8845 | 2.0791 | 6.2805 | 0.0381 |
| 28 | 1.0826 | 0.0298 | 4.9306 | 0.2158 | 0.1243 | 3.6347 | 4.2498 | 1.6551 | 4.2087 | 0.2965 |
| 29 | 1.0700 | 0.0809 | 5.1556 | 0.3914 | 0.0927 | 3.9784 | 3.5888 | 1.2446 | 2.5103 | 0.8179 |
| 30 | 1.0568 | 0.3819 | 5.2992 | 0.5506 | 0.0718 | 4.0080 | 2.9515 | 0.8501 | 1.2277 | 1.5080 |
| 31 | 0.9860 | 0.6773 | 5.3223 | 0.6839 | 0.0692 | 3.6367 | 2.3501 | 0.5285 | 0.4396 | 2.2290 |
| 32 | 0.8893 | 0.8154 | 5.1393 | 0.7868 | 0.0801 | 2.9825 | 1.8306 | 0.2777 | 0.0590 | 2.8325 |
| 33 | 0.7797 | 0.7621 | 4.8268 | 0.8227 | 0.1043 | 2.2410 | 1.4233 | 0.1069 | 0.0114 | 3.1082 |
| 34 | 0.7006 | 0.5730 | 4.3807 | 0.8154 | 0.1347 | 1.4714 | 1.0547 | 0.0177 | 0.1391 | 2.9825 |
| 35 | 0.6464 | 0.3648 | 3.9046 | 0.7674 | 0.1632 | 0.8391 | 0.7992 | 0.0029 | 0.3318 | 2.5091 |
| 36 | 0.6448 | 0.1875 | 3.3746 | 0.7174 | 0.1706 | 0.3807 | 0.5975 | 0.0543 | 0.4720 | 1.7769 |
| 37 | 0.7448 | 0.0745 | 2.9138 | 0.6352 | 0.1391 | 0.1069 | 0.4529 | 0.1467 | 0.5141 | 0.9663 |
| 38 | 0.9487 | 0.0180 | 2.4524 | 0.5565 | 0.0751 | 0.0021 | 0.3648 | 0.2746 | 0.4706 | 0.2852 |
| 39 | 1.3433 | 0.0015 | 2.0192 | 0.4775 | 0.0098 | 0.0480 | 0.3125 | 0.4212 | 0.3493 | 0.0004 |
| 40 | 2.0022 | 0.0006 | 1.6002 | 0.4032 | 0.0240 | 0.2070 | 0.3080 | 0.5550 | 0.1806 | 0.4422 |
| 41 | 3.0102 | 0.0012 | 1.1556 | 0.3306 | 0.2550 | 0.4556 | 0.3422 | 0.6480 | 0.0380 | 2.0022 |
| 42 | 4.5071 | 0.0007 | 0.6839 | 0.2571 | 0.8593 | 0.7448 | 0.4264 | 0.6610 | 0.0106 | 5.1393 |
| 43 | 6.4821 | 0.0000 | 0.2440 | 0.1714 | 2.0564 | 0.9920 | 0.6022 | 0.5868 | 0.2460 | 10.4588 |
| 44 | 8.8744 | 0.0000 | 0.0037 | 0.0734 | 3.8848 | 1.1428 | 0.8817 | 0.4083 | 0.9390 | 18.4127 |
| 45 | 11.4853 | 0.0017 | 0.2107 | 0.0026 | 6.2550 | 1.1004 | 1.3202 | 0.2016 | 2.3378 | 29.4958 |
| 46 | 13.8756 | 0.0210 | 1.1130 | 0.0702 | 8.7912 | 0.9120 | 1.8906 | 0.0240 | 4.5582 | 42.9680 |
| 47 | 15.5386 | 0.1212 | 2.8288 | 0.4515 | 10.7459 | 0.5958 | 2.5661 | 0.0392 | 7.4633 | 57.4124 |
| 48 | 15.9752 | 0.3637 | 4.9146 | 1.3384 | 11.4386 | 0.2776 | 3.2289 | 0.4008 | 10.8037 | 69.9414 |
| 49 | 14.8217 | 0.6498 | 6.6714 | 2.8086 | 10.7722 | 0.0681 | 3.5679 | 1.2479 | 13.8377 | 76.9673 |
| 50 | 12.7901 | 0.8430 | 7.4711 | 4.5043 | 9.1245 | 0.0005 | 3.6112 | 2.4513 | 15.6367 | 76.7319 |
| 51 | 10.3836 | 0.7000 | 6.9293 | 5.9554 | 7.2073 | 0.0971 | 3.2846 | 3.7004 | 16.1344 | 70.0174 |
| 52 | 8.0559 | 0.2206 | 5.4303 | 6.6683 | 5.8646 | 0.3525 | 2.6449 | 4.8035 | 15.6919 | 58.7016 |
| 53 | 6.4044 | 0.0198 | 3.8443 | 6.3036 | 5.4256 | 0.8636 | 1.9620 | 5.4256 | 14.6207 | 45.9589 |
| 54 | 5.4850 | 0.8686 | 2.5985 | 5.0266 | 6.1405 | 1.8171 | 1.3971 | 5.4196 | 13.6309 | 34.1991 |
| 55 | 5.1529 | 3.5721 | 1.7424 | 3.3489 | 8.3521 | 3.4225 | 0.9025 | 5.0625 | 13.3225 | 24.2064 |
| 56 | 5.3963 | 8.6025 | 1.3064 | 1.6718 | 12.7949 | 5.9878 | 0.5822 | 4.3556 | 13.5645 | 16.2167 |
| 57 | 6.2550 | 16.0080 | 1.1470 | 0.4638 | 20.1511 | 9.7906 | 0.3612 | 3.4559 | 14.3717 | 10.0426 |
| 58 | 7.7340 | 25.0100 | 1.1903 | 0.0004 | 31.1252 | 14.7379 | 0.2510 | 2.4932 | 15.4528 | 5.4242 |
| 59 | 9.6385 | 34.0893 | 1.3797 | 0.4037 | 45.3656 | 20.2986 | 0.1977 | 1.5760 | 16.5210 | 2.2362 |
| 60 | 11.8095 | 40.9920 | 1.7069 | 1.5713 | 61.2072 | 25.3361 | 0.2084 | 0.7983 | 16.9373 | 0.5091 |
| 61 | 13.7389 | 43.1202 | 2.1509 | 3.1308 | 75.3320 | 27.9778 | 0.2869 | 0.2636 | 16.0609 | 0.0001 |
| 62 | 14.6514 | 39.5203 | 2.5113 | 4.3944 | 83.3438 | 26.9286 | 0.4260 | 0.0280 | 13.8140 | 0.2639 |
| 63 | 14.0480 | 31.4930 | 2.6402 | 4.9246 | 81.2908 | 22.3459 | 0.6223 | 0.0123 | 10.4980 | 0.7020 |
| 64 | 12.2179 | 21.2374 | 2.4851 | 4.6207 | 69.5821 | 15.3397 | 0.8508 | 0.0969 | 6.6430 | 0.8398 |
| 65 | 9.4567 | 11.7524 | 1.9633 | 3.5733 | 51.5498 | 8.2319 | 1.0004 | 0.1554 | 3.3021 | 0.6776 |
| 66 | 6.5342 | 4.9738 | 1.2751 | 2.2494 | 32.2715 | 3.1322 | 1.0510 | 0.1484 | 1.1050 | 0.3286 |
| 67 | 4.0655 | 1.2461 | 0.6501 | 1.1421 | 16.5462 | 0.5151 | 1.0187 | 0.1007 | 1.0198 | 0.0662 |
| 68 | 2.2527 | 0.0320 | 0.2088 | 0.3833 | 6.4420 | 0.0338 | 0.9004 | 0.0436 | 0.0724 | 0.0027 |
| 69 | 1.1052 | 0.2956 | 0.0157 | 0.0465 | 1.4465 | 0.7539 | 0.7487 | 0.0083 | 0.5194 | 0.1016 |
| 70 | 0.4771 | 1.0864 | 0.0263 | 0.0129 | 0.0263 | 1.7761 | 0.5909 | 0.0002 | 1.0349 | 0.2563 |
| 71 | 0.1798 | 1.7530 | 0.1490 | 0.1232 | 0.3192 | 2.4869 | 0.4369 | 0.0102 | 1.3549 | 0.3636 |
| 72 | 0.0635 | 2.0053 | 0.2810 | 0.2342 | 0.9779 | 2.6533 | 0.3035 | 0.0244 | 1.3950 | 0.3858 |
| 73 | 0.0269 | 1.8628 | 0.3434 | 0.2643 | 1.3089 | 2.3534 | 0.1928 | 0.0317 | 1.2187 | 0.3166 |
| 74 | 0.0186 | 1.4522 | 0.3072 | 0.2112 | 1.1855 | 1.7737 | 0.1073 | 0.0287 | 0.9183 | 0.2088 |
| 75 | 0.0231 | 0.9674 | 0.1854 | 0.1064 | 0.7831 | 1.1478 | 0.0486 | 0.0170 | 0.6124 | 0.1074 |
| 76 | 0.0339 | 0.5371 | 0.0492 | 0.0185 | 0.3374 | 0.6306 | 0.0135 | 0.0049 | 0.3610 | 0.0345 |
| 77 | 0.0453 | 0.2239 | 0.0043 | 0.0087 | 0.0496 | 0.2744 | 0.0002 | 0.0001 | 0.1859 | 0.0024 |
| 78 | 0.0504 | 0.0456 | 0.1802 | 0.1278 | 0.0308 | 0.0804 | 0.0066 | 0.0087 | 0.0787 | 0.0068 |
| 79 | 0.0448 | 0.0018 | 0.7101 | 0.4190 | 0.3811 | 0.0070 | 0.0297 | 0.0349 | 0.0229 | 0.0487 |
| 80 | 0.0303 | 0.0925 | 1.7027 | 0.9175 | 1.2407 | 0.0053 | 0.0655 | 0.0785 | 0.0012 | 0.1385 |
| 81 | 0.0150 | 0.3311 | 3.2044 | 1.6562 | 2.8390 | 0.0334 | 0.1069 | 0.1387 | 0.0055 | 0.2988 |
| 82 | 0.0046 | 0.7340 | 5.1881 | 2.6528 | 5.4955 | 0.0583 | 0.1447 | 0.2105 | 0.0320 | 0.5696 |
| 83 | 0.0005 | 1.3073 | 7.4878 | 3.8554 | 9.4902 | 0.0613 | 0.1678 | 0.2834 | 0.0775 | 0.9749 |
| 84 | 0.0000 | 2.0248 | 9.8468 | 5.1713 | 14.9896 | 0.0400 | 0.1632 | 0.3504 | 0.1347 | 1.4981 |
| 85 | 0.0004 | 2.7759 | 11.9101 | 6.4308 | 21.7520 | 0.0086 | 0.1267 | 0.3970 | 0.1833 | 2.1144 |
| 86 | 0.0041 | 3.4381 | 13.2802 | 7.3430 | 28.9853 | 0.0029 | 0.0707 | 0.4150 | 0.2063 | 2.7034 |
| 87 | 0.0185 | 3.8652 | 13.7344 | 7.7507 | 35.4501 | 0.0467 | 0.0130 | 0.4045 | 0.1815 | 3.1188 |
| 88 | 0.0424 | 3.9840 | 13.0755 | 7.5295 | 39.8666 | 0.1490 | 0.0044 | 0.3672 | 0.1197 | 3.3709 |
| 89 | 0.0745 | 3.8142 | 11.5125 | 6.7444 | 41.1779 | 0.2735 | 0.0692 | 0.3170 | 0.0454 | 3.3599 |
| 90 | 0.0900 | 3.4225 | 9.2416 | 5.5225 | 39.1876 | 0.3844 | 0.2304 | 0.2601 | 0.0016 | 3.1329 |
| 91 | 0.0824 | 2.9139 | 6.7964 | 4.1738 | 34.6097 | 0.4449 | 0.4583 | 0.2089 | 0.0205 | 2.7789 |
| 92 | 0.0458 | 2.4149 | 4.4268 | 2.8764 | 28.5797 | 0.4679 | 0.7123 | 0.1798 | 0.0999 | 2.3532 |
| 93 | 0.0077 | 1.9825 | 2.4901 | 1.7742 | 22.1088 | 0.4462 | 0.9761 | 0.1665 | 0.2228 | 1.8714 |
| 94 | 0.0074 | 1.5977 | 1.0899 | 0.9722 | 16.1283 | 0.4277 | 1.1968 | 0.1798 | 0.3552 | 1.4496 |
| 95 | 0.0853 | 1.2277 | 0.3114 | 0.4382 | 11.1022 | 0.4330 | 1.3877 | 0.2098 | 0.4789 | 1.0363 |
| 96 | 0.2905 | 0.9235 | 0.0172 | 0.1289 | 7.2307 | 0.4775 | 1.4908 | 0.2714 | 0.5314 | 0.6257 |
| 97 | 0.6577 | 0.6872 | 0.0630 | 0.0037 | 4.3306 | 0.5314 | 1.5600 | 0.3469 | 0.5055 | 0.2694 |
| 98 | 1.1968 | 0.4984 | 0.3770 | 0.0655 | 2.2320 | 0.5715 | 1.6028 | 0.4436 | 0.4409 | 0.0467 |
| 99 | 1.8989 | 0.3505 | 0.9960 | 0.3387 | 0.8064 | 0.5960 | 1.6693 | 0.5506 | 0.3819 | 0.0077 |
| $RMSy_j =$ [Eqn 24] | 1.7334 | 3.0335 | 1.9038 | 1.2333 | 3.3604 | 1.7766 | 1.5300 | 1.0860 | 2.4372 | 2.7535 |
| $RMS_j =$ [Eqn 25] | 1.8095 | 4.1886 | 2.2930 | 1.7751 | 3.7519 | 2.5185 | 1.8051 | 1.6335 | 3.3569 | 3.2586 |
| $RMS_{avg} =$ [Eqn 19] | 2.6391 | | | | | | | | | |
| $Ø_j =$ [Eqn 26] | 73.3 | 46.4 | 56.1 | 44.0 | 63.6 | 44.9 | 58.0 | 41.7 | 46.6 | 57.7 |
| $Ø_{avg} =$ [Eqn 27] | 53.22 | | | | | | | | | |
| $RMSx_{avg} =$ [Eqn 28] | 1.5803 | | | | | | | | | |
| $RMSy_{avg} =$ [Eqn 29] | 2.1136 | | | | | | | | | |

# Table B.5 Example Data for Calculating the Point-by-Point Variability (SD$_R$) for the Angle-Angle Diagram

*Equations 2.30-2.31 are demonstrated using the ankle and knee angle data given in table B.2.*

| %Stride | Cycle 1 $R_{ij}$ [Eqn 17] | Cycle 2 $R_{ij}$ [Eqn 17] | Cycle 3 $R_{ij}$ [Eqn 17] | Cycle 4 $R_{ij}$ [Eqn 17] | Cycle 5 $R_{ij}$ [Eqn 17] | Cycle 6 $R_{ij}$ [Eqn 17] | Cycle 7 $R_{ij}$ [Eqn 17] | Cycle 8 $R_{ij}$ [Eqn 17] | Cycle 9 $R_{ij}$ [Eqn 17] | Cycle 10 $R_{ij}$ [Eqn 17] |
|---|---|---|---|---|---|---|---|---|---|---|
| 0 | 1.2976 | 1.8026 | 0.3739 | 0.9720 | 1.5507 | 1.8414 | 1.5733 | 1.3833 | 0.9331 | 0.8063 |
| 1 | 1.1248 | 2.2627 | 0.3425 | 1.0584 | 1.7488 | 2.0066 | 1.5250 | 1.2591 | 0.9784 | 0.7631 |
| 2 | 0.9389 | 2.7544 | 0.6060 | 1.1223 | 1.9502 | 2.2315 | 1.4895 | 1.1207 | 1.0390 | 0.7461 |
| 3 | 0.7555 | 3.2559 | 0.9835 | 1.1198 | 2.1453 | 2.4629 | 1.4897 | 0.9354 | 1.1509 | 0.7728 |
| 4 | 0.5577 | 3.8263 | 1.3851 | 1.0603 | 2.3364 | 2.7139 | 1.5297 | 0.7203 | 1.3123 | 0.8329 |
| 5 | 0.3535 | 4.4534 | 1.7924 | 0.9701 | 2.5359 | 2.9656 | 1.6267 | 0.4831 | 1.5295 | 0.9208 |
| 6 | 0.1590 | 5.1424 | 2.1662 | 0.8928 | 2.7452 | 3.2192 | 1.7572 | 0.3281 | 1.8364 | 1.0533 |
| 7 | 0.1152 | 5.8900 | 2.4971 | 0.8556 | 2.9606 | 3.4587 | 1.9384 | 0.4814 | 2.2148 | 1.2235 |
| 8 | 0.3153 | 6.6865 | 2.7841 | 0.9125 | 3.1819 | 3.6720 | 2.1442 | 0.8442 | 2.6826 | 1.4236 |
| 9 | 0.5213 | 7.5007 | 3.0164 | 1.0690 | 3.3962 | 3.8416 | 2.3682 | 1.2545 | 3.2135 | 1.6443 |
| 10 | 0.7207 | 8.2753 | 3.1966 | 1.2838 | 3.5796 | 3.9350 | 2.5976 | 1.6860 | 3.8170 | 1.8707 |
| 11 | 0.8837 | 8.9814 | 3.3364 | 1.5436 | 3.7085 | 3.9525 | 2.8311 | 2.1035 | 4.4558 | 2.0870 |
| 12 | 1.0160 | 9.5330 | 3.4279 | 1.8155 | 3.7372 | 3.8451 | 3.0334 | 2.4760 | 5.1079 | 2.2631 |
| 13 | 1.0990 | 9.8989 | 3.4828 | 2.0931 | 3.6700 | 3.6159 | 3.2160 | 2.7876 | 5.7335 | 2.3930 |
| 14 | 1.1483 | 10.0402 | 3.5108 | 2.3382 | 3.5017 | 3.2752 | 3.3579 | 3.0289 | 6.2995 | 2.4616 |
| 15 | 1.1489 | 9.9500 | 3.4998 | 2.5648 | 3.2330 | 2.8400 | 3.4779 | 3.2108 | 6.7637 | 2.4800 |
| 16 | 1.1205 | 9.6574 | 3.4535 | 2.7508 | 2.9011 | 2.3396 | 3.5552 | 3.3238 | 7.1144 | 2.4407 |
| 17 | 1.0761 | 9.1840 | 3.3789 | 2.8781 | 2.5430 | 1.8284 | 3.5940 | 3.3938 | 7.3241 | 2.3641 |
| 18 | 1.0234 | 8.5735 | 3.2797 | 2.9674 | 2.1882 | 1.3213 | 3.5948 | 3.4307 | 7.4095 | 2.2507 |
| 19 | 0.9812 | 7.8590 | 3.1630 | 2.9991 | 1.8583 | 0.8945 | 3.5611 | 3.4452 | 7.3739 | 2.1084 |
| 20 | 0.9456 | 7.0707 | 3.0436 | 2.9776 | 1.5795 | 0.7030 | 3.5063 | 3.4519 | 7.2427 | 1.9453 |
| 21 | 0.9263 | 6.2220 | 2.9260 | 2.9123 | 1.3445 | 0.8814 | 3.4244 | 3.4351 | 7.0154 | 1.7613 |
| 22 | 0.9217 | 5.3272 | 2.8288 | 2.8031 | 1.1562 | 1.2700 | 3.3238 | 3.4108 | 6.7173 | 1.5363 |
| 23 | 0.9315 | 4.3901 | 2.7497 | 2.6639 | 0.9985 | 1.7287 | 3.2045 | 3.3591 | 6.3564 | 1.2798 |
| 24 | 0.9526 | 3.4362 | 2.7073 | 2.5137 | 0.8624 | 2.1929 | 3.0633 | 3.2823 | 5.9382 | 0.9682 |
| 25 | 0.9806 | 2.4703 | 2.6932 | 2.3548 | 0.7365 | 2.6650 | 2.9029 | 3.1735 | 5.4570 | 0.5973 |
| 26 | 1.0086 | 1.5281 | 2.7148 | 2.1989 | 0.6221 | 3.0977 | 2.7227 | 3.0365 | 4.9096 | 0.1602 |
| 27 | 1.0324 | 0.7238 | 2.7638 | 2.0584 | 0.5203 | 3.4926 | 2.5230 | 2.8540 | 4.3023 | 0.3470 |
| 28 | 1.0477 | 0.6310 | 2.8229 | 1.9433 | 0.4248 | 3.8051 | 2.3032 | 2.6292 | 3.6319 | 0.8845 |
| 29 | 1.0464 | 1.2157 | 2.8839 | 1.8509 | 0.3599 | 4.0094 | 2.0651 | 2.3850 | 2.9194 | 1.4489 |
| 30 | 1.0494 | 1.7610 | 2.9105 | 1.7718 | 0.3168 | 4.1114 | 1.8261 | 2.1038 | 2.1814 | 1.9767 |
| 31 | 1.0288 | 2.1729 | 2.9011 | 1.6963 | 0.3084 | 4.0730 | 1.5871 | 1.8113 | 1.4602 | 2.4330 |
| 32 | 1.0041 | 2.4105 | 2.8366 | 1.6279 | 0.3381 | 3.9163 | 1.3700 | 1.5100 | 0.7836 | 2.7978 |
| 33 | 0.9843 | 2.5116 | 2.7208 | 1.5484 | 0.3994 | 3.6830 | 1.1935 | 1.2197 | 0.2491 | 3.0387 |
| 34 | 0.9950 | 2.5185 | 2.5623 | 1.4637 | 0.4817 | 3.3829 | 1.0376 | 0.9573 | 0.4372 | 3.1480 |
| 35 | 1.0251 | 2.4593 | 2.3853 | 1.3705 | 0.5643 | 3.0660 | 0.9449 | 0.7380 | 0.8507 | 3.1597 |
| 36 | 1.0886 | 2.3846 | 2.1915 | 1.2840 | 0.6294 | 2.7550 | 0.9021 | 0.5927 | 1.1683 | 3.0876 |
| 37 | 1.2008 | 2.3211 | 2.0067 | 1.1836 | 0.6687 | 2.4866 | 0.9117 | 0.5431 | 1.4022 | 2.9911 |
| 38 | 1.3391 | 2.2650 | 1.8259 | 1.0800 | 0.6879 | 2.2595 | 0.9700 | 0.5802 | 1.5761 | 2.9202 |
| 39 | 1.5217 | 2.2243 | 1.6538 | 0.9723 | 0.7010 | 2.0975 | 1.0476 | 0.6631 | 1.7019 | 2.9341 |
| 40 | 1.7585 | 2.1961 | 1.4882 | 0.8573 | 0.7619 | 1.9868 | 1.1472 | 0.7463 | 1.7853 | 3.0982 |
| 41 | 2.0416 | 2.1743 | 1.3085 | 0.7326 | 0.9326 | 1.9280 | 1.2512 | 0.8054 | 1.8662 | 3.4569 |
| 42 | 2.3851 | 2.1432 | 1.1144 | 0.6066 | 1.2330 | 1.9038 | 1.3460 | 0.8163 | 1.9597 | 4.0061 |
| 43 | 2.7676 | 2.1050 | 0.9106 | 0.4535 | 1.6594 | 1.8720 | 1.4586 | 0.7732 | 2.1043 | 4.7544 |
| 44 | 3.1620 | 2.0500 | 0.8023 | 0.2727 | 2.1505 | 1.8257 | 1.5794 | 0.6521 | 2.3310 | 5.6399 |
| 45 | 3.5432 | 1.9864 | 0.9695 | 0.1341 | 2.6533 | 1.7128 | 1.7230 | 0.4691 | 2.6578 | 6.6376 |
| 46 | 3.8548 | 1.8936 | 1.3946 | 0.3943 | 3.1068 | 1.5430 | 1.8731 | 0.2075 | 3.0743 | 7.6470 |
| 47 | 4.0515 | 1.8078 | 1.9446 | 0.8065 | 3.4226 | 1.3080 | 2.0294 | 0.2392 | 3.5303 | 8.5700 |
| 48 | 4.0944 | 1.7399 | 2.4437 | 1.3023 | 3.5449 | 1.0240 | 2.1652 | 0.6447 | 3.9653 | 9.2713 |
| 49 | 3.9390 | 1.6827 | 2.7969 | 1.8292 | 3.4768 | 0.7311 | 2.2130 | 1.1222 | 4.3030 | 9.6037 |
| 50 | 3.6605 | 1.5828 | 2.9429 | 2.2865 | 3.2727 | 0.4712 | 2.1911 | 1.5682 | 4.4683 | 9.5193 |
| 51 | 3.3020 | 1.3733 | 2.8495 | 2.6119 | 3.0140 | 0.4065 | 2.0851 | 1.9249 | 4.4655 | 9.0540 |
| 52 | 2.9161 | 0.9897 | 2.5596 | 2.7616 | 2.8424 | 0.5976 | 1.8829 | 2.1922 | 4.3465 | 8.2678 |
| 53 | 2.6019 | 0.6530 | 2.2066 | 2.6960 | 2.8286 | 0.9364 | 1.6475 | 2.3294 | 4.1447 | 7.2945 |
| 54 | 2.4028 | 1.0102 | 1.8682 | 2.4312 | 3.0146 | 1.3753 | 1.4124 | 2.3281 | 3.9482 | 6.2601 |
| 55 | 2.3174 | 1.8940 | 1.5784 | 2.0176 | 3.4193 | 1.8948 | 1.1616 | 2.2508 | 3.8436 | 5.2228 |
| 56 | 2.3563 | 2.9374 | 1.3875 | 1.4749 | 4.0676 | 2.5038 | 0.9426 | 2.0896 | 3.8255 | 4.2245 |
| 57 | 2.5221 | 4.0283 | 1.2884 | 0.8588 | 4.9382 | 3.1939 | 0.7379 | 1.8654 | 3.8990 | 3.2753 |
| 58 | 2.7937 | 5.0641 | 1.2792 | 0.2937 | 5.9960 | 3.9126 | 0.5824 | 1.5929 | 4.0208 | 2.3671 |
| 59 | 3.1125 | 5.9496 | 1.3405 | 0.6361 | 7.1376 | 4.5905 | 0.4735 | 1.2839 | 4.1517 | 1.4974 |
| 60 | 3.4423 | 6.5762 | 1.4612 | 1.2791 | 8.2310 | 5.1365 | 0.4576 | 0.9533 | 4.2154 | 0.7422 |
| 61 | 3.7123 | 6.8220 | 1.6204 | 1.8514 | 9.1131 | 5.4198 | 0.5429 | 0.6509 | 4.1376 | 0.4213 |
| 62 | 3.8352 | 6.6538 | 1.7533 | 2.2531 | 9.6133 | 5.3627 | 0.6826 | 0.4990 | 3.8996 | 0.7674 |
| 63 | 3.7598 | 6.1262 | 1.8187 | 2.4693 | 9.5786 | 4.9633 | 0.8381 | 0.5484 | 3.5089 | 1.0648 |
| 64 | 3.5149 | 5.3310 | 1.8054 | 2.5018 | 9.0155 | 4.2500 | 0.9902 | 0.6760 | 2.9771 | 1.1471 |
| 65 | 3.1087 | 4.4358 | 1.6790 | 2.3673 | 8.0029 | 3.3581 | 1.0829 | 0.7730 | 2.4179 | 1.0582 |
| 66 | 2.6110 | 3.6362 | 1.4705 | 2.1268 | 6.7030 | 2.5158 | 1.1188 | 0.8095 | 1.9497 | 0.8312 |
| 67 | 2.1037 | 3.0515 | 1.2158 | 1.8499 | 5.3467 | 1.9007 | 1.1134 | 0.7960 | 1.6633 | 0.5623 |
| 68 | 1.6348 | 2.7339 | 0.9545 | 1.5693 | 4.1410 | 1.6821 | 1.0643 | 0.7670 | 1.5712 | 0.3717 |
| 69 | 1.2456 | 2.6151 | 0.7387 | 1.3395 | 3.2061 | 1.7523 | 1.0004 | 0.7337 | 1.5996 | 0.3806 |
| 70 | 0.9546 | 2.5699 | 0.5917 | 1.1566 | 2.5861 | 1.8835 | 0.9176 | 0.6792 | 1.6264 | 0.5078 |
| 71 | 0.7694 | 2.5097 | 0.5501 | 1.0297 | 2.2114 | 1.9389 | 0.8337 | 0.6301 | 1.5960 | 0.6186 |
| 72 | 0.6590 | 2.3689 | 0.5698 | 0.9188 | 1.9503 | 1.8663 | 0.7514 | 0.5708 | 1.4965 | 0.6902 |
| 73 | 0.5903 | 2.1700 | 0.5863 | 0.7924 | 1.6747 | 1.6875 | 0.6752 | 0.4904 | 1.3385 | 0.7161 |
| 74 | 0.5482 | 1.9249 | 0.5795 | 0.6425 | 1.3565 | 1.4258 | 0.6053 | 0.4078 | 1.1365 | 0.7298 |
| 75 | 0.5245 | 1.6558 | 0.5474 | 0.4697 | 0.9919 | 1.1234 | 0.5447 | 0.3107 | 0.9297 | 0.7351 |
| 76 | 0.5188 | 1.4018 | 0.5333 | 0.2891 | 0.5987 | 0.8154 | 0.4987 | 0.2166 | 0.7246 | 0.7581 |
| 77 | 0.5241 | 1.1691 | 0.6245 | 0.2307 | 0.2480 | 0.5263 | 0.4712 | 0.1392 | 0.5485 | 0.7825 |
| 78 | 0.5326 | 0.9864 | 0.8592 | 0.4035 | 0.3502 | 0.2927 | 0.4642 | 0.1319 | 0.4056 | 0.8212 |
| 79 | 0.5319 | 0.8691 | 1.1984 | 0.6724 | 0.7747 | 0.1967 | 0.4931 | 0.1955 | 0.3078 | 0.8608 |
| 80 | 0.5172 | 0.8437 | 1.6158 | 0.9771 | 1.2638 | 0.2768 | 0.5290 | 0.2840 | 0.2594 | 0.9123 |
| 81 | 0.5012 | 0.9108 | 2.0673 | 1.3030 | 1.8230 | 0.4002 | 0.5841 | 0.3753 | 0.2762 | 0.9972 |
| 82 | 0.4858 | 1.0640 | 2.5378 | 1.6434 | 2.4741 | 0.4853 | 0.6515 | 0.4642 | 0.3415 | 1.1137 |
| 83 | 0.4666 | 1.2758 | 2.9816 | 1.9774 | 3.1973 | 0.5454 | 0.7052 | 0.5410 | 0.4287 | 1.2733 |
| 84 | 0.4370 | 1.5073 | 3.3752 | 2.2870 | 3.9742 | 0.5730 | 0.7510 | 0.6099 | 0.5261 | 1.4476 |
| 85 | 0.4025 | 1.7237 | 3.6871 | 2.5490 | 4.7542 | 0.5894 | 0.7835 | 0.6617 | 0.6153 | 1.6307 |
| 86 | 0.3588 | 1.8933 | 3.8817 | 2.7239 | 5.4607 | 0.6353 | 0.7929 | 0.6958 | 0.6852 | 1.7858 |
| 87 | 0.3276 | 1.9932 | 3.9484 | 2.8003 | 6.0199 | 0.7211 | 0.8002 | 0.7156 | 0.7180 | 1.8825 |
| 88 | 0.3133 | 2.0136 | 3.8682 | 2.7631 | 6.3678 | 0.8488 | 0.8167 | 0.7185 | 0.7240 | 1.9297 |
| 89 | 0.3185 | 1.9636 | 3.6614 | 2.6199 | 6.4588 | 0.9929 | 0.8669 | 0.7170 | 0.7260 | 1.9098 |
| 90 | 0.3132 | 1.8553 | 3.3303 | 2.3805 | 6.2897 | 1.1261 | 0.9415 | 0.7072 | 0.7511 | 1.8313 |
| 91 | 0.2877 | 1.7084 | 2.9267 | 2.0857 | 5.9010 | 1.2439 | 1.0328 | 0.6923 | 0.8028 | 1.7098 |
| 92 | 0.2205 | 1.5540 | 2.4591 | 1.7581 | 5.3532 | 1.3441 | 1.1244 | 0.6842 | 0.8760 | 1.5564 |
| 93 | 0.1378 | 1.4088 | 1.9861 | 1.4321 | 4.7026 | 1.4297 | 1.2085 | 0.6641 | 0.9583 | 1.3729 |
| 94 | 0.1671 | 1.2674 | 1.5313 | 1.1507 | 4.0193 | 1.4881 | 1.2742 | 0.6531 | 1.0354 | 1.2071 |
| 95 | 0.3243 | 1.1137 | 1.1620 | 0.9506 | 3.3591 | 1.5214 | 1.3388 | 0.6433 | 1.0975 | 1.0738 |
| 96 | 0.5459 | 0.9681 | 0.9168 | 0.8781 | 2.7823 | 1.5296 | 1.3764 | 0.6440 | 1.1006 | 1.0527 |
| 97 | 0.8111 | 0.8381 | 0.8246 | 0.9574 | 2.3164 | 1.5136 | 1.4073 | 0.6599 | 1.0455 | 1.2403 |
| 98 | 1.1015 | 0.7191 | 0.9020 | 1.1575 | 1.9988 | 1.4919 | 1.4313 | 0.6984 | 0.9585 | 1.6207 |
| 99 | 1.3977 | 0.6127 | 1.1343 | 1.4262 | 1.8645 | 1.4828 | 1.4624 | 0.7526 | 0.8726 | 2.0779 |

| %Stride | Cycle 1 $R_{ij}^2$ | Cycle 2 $R_{ij}^2$ | Cycle 3 $R_{ij}^2$ | Cycle 4 $R_{ij}^2$ | Cycle 5 $R_{ij}^2$ | Cycle 6 $R_{ij}^2$ | Cycle 7 $R_{ij}^2$ | Cycle 8 $R_{ij}^2$ | Cycle 9 $R_{ij}^2$ | Cycle 10 $R_{ij}^2$ | $RMS_i$ [Eqn 30] | $RMS_i=SD_R$ (population) [Eqn 31] | $RMS_i=SD_R$ (sample) [Eqn 31] |
|---|---|---|---|---|---|---|---|---|---|---|---|---|---|
| 0 | 1.6837 | 3.2495 | 0.1398 | 0.9447 | 2.4045 | 3.3906 | 2.4751 | 1.9135 | 0.8707 | 0.6501 | 1.3313 | 1.3313 | 1.4033 |
| 1 | 1.2652 | 5.1197 | 0.1173 | 1.1203 | 3.0582 | 4.0263 | 2.3255 | 1.5854 | 0.9573 | 0.5824 | 1.4198 | 1.4198 | 1.4966 |
| 2 | 0.8815 | 7.5867 | 0.3673 | 1.2596 | 3.8031 | 4.9794 | 2.2187 | 1.2560 | 1.0796 | 0.5567 | 1.5488 | 1.5488 | 1.6326 |
| 3 | 0.5708 | 10.6009 | 0.9673 | 1.2538 | 4.6024 | 6.0657 | 2.2191 | 0.8749 | 1.3247 | 0.5972 | 1.7052 | 1.7052 | 1.7974 |
| 4 | 0.3110 | 14.6409 | 1.9186 | 1.1243 | 5.4588 | 7.3651 | 2.3399 | 0.5189 | 1.7222 | 0.6937 | 1.8998 | 1.8998 | 2.0026 |
| 5 | 0.1250 | 19.8327 | 3.2127 | 0.9411 | 6.4307 | 8.7949 | 2.6463 | 0.2334 | 2.3393 | 0.8478 | 2.1308 | 2.1308 | 2.2461 |
| 6 | 0.0253 | 26.4440 | 4.6923 | 0.7971 | 7.5360 | 10.3634 | 3.0876 | 0.1077 | 3.3723 | 1.1095 | 2.3987 | 2.3987 | 2.5284 |
| 7 | 0.0133 | 34.6920 | 6.2353 | 0.7321 | 8.7651 | 11.9623 | 3.7573 | 0.2317 | 4.9051 | 1.4968 | 2.6980 | 2.6980 | 2.8439 |
| 8 | 0.0994 | 44.7094 | 7.7515 | 0.8327 | 10.1246 | 13.4839 | 4.5976 | 0.7128 | 7.1965 | 2.0266 | 3.0255 | 3.0255 | 3.1891 |
| 9 | 0.2717 | 56.2602 | 9.0984 | 1.1427 | 11.5343 | 14.7582 | 5.6084 | 1.5737 | 10.3267 | 2.7039 | 3.3657 | 3.3657 | 3.5477 |
| 10 | 0.5194 | 68.4810 | 10.2181 | 1.6482 | 12.8133 | 15.4843 | 6.7476 | 2.8427 | 14.5694 | 3.4996 | 3.6990 | 3.6990 | 3.8991 |
| 11 | 0.7808 | 80.6652 | 11.1313 | 2.3827 | 13.7526 | 15.6222 | 8.0151 | 4.4246 | 19.8544 | 4.3556 | 4.0123 | 4.0123 | 4.2293 |
| 12 | 1.0323 | 90.8789 | 11.7502 | 3.2962 | 13.9670 | 14.7846 | 9.2014 | 6.1304 | 26.0903 | 5.1217 | 4.2691 | 4.2691 | 4.5000 |
| 13 | 1.2078 | 97.9873 | 12.1297 | 4.3812 | 13.4687 | 13.0748 | 10.3429 | 7.7705 | 32.8727 | 5.7266 | 4.4605 | 4.4605 | 4.7018 |
| 14 | 1.3185 | 100.8049 | 12.3256 | 5.4672 | 12.2619 | 10.7269 | 11.2753 | 9.1741 | 39.6837 | 6.0595 | 4.5727 | 4.5727 | 4.8201 |
| 15 | 1.3201 | 99.0020 | 12.2488 | 6.5782 | 10.4525 | 8.0657 | 12.0961 | 10.3095 | 45.7480 | 6.1504 | 4.6040 | 4.6040 | 4.8531 |
| 16 | 1.2556 | 93.2660 | 11.9266 | 7.5667 | 8.4165 | 5.4739 | 12.6393 | 11.0473 | 50.6143 | 5.9568 | 4.5625 | 4.5625 | 4.8093 |
| 17 | 1.1580 | 84.3459 | 11.4170 | 8.2837 | 6.4670 | 3.3432 | 12.9172 | 11.5178 | 53.6422 | 5.5890 | 4.4574 | 4.4574 | 4.6985 |
| 18 | 1.0474 | 73.5043 | 10.7566 | 8.8056 | 4.7880 | 1.7458 | 12.9225 | 11.7699 | 54.9014 | 5.0655 | 4.3047 | 4.3047 | 4.5376 |
| 19 | 0.9628 | 61.7635 | 10.0044 | 8.9949 | 3.4532 | 0.8001 | 12.6812 | 11.8692 | 54.3750 | 4.4455 | 4.1152 | 4.1152 | 4.3378 |
| 20 | 0.8942 | 49.9948 | 9.2637 | 8.8662 | 2.4949 | 0.4942 | 12.2939 | 11.9158 | 52.4571 | 3.7842 | 3.9046 | 3.9046 | 4.1158 |
| 21 | 0.8581 | 38.7128 | 8.5615 | 8.4815 | 1.8077 | 0.7769 | 11.7267 | 11.7997 | 49.2153 | 3.1020 | 3.6748 | 3.6748 | 3.8736 |
| 22 | 0.8495 | 28.3796 | 8.0020 | 7.8575 | 1.3368 | 1.6128 | 11.0476 | 11.6338 | 45.1215 | 2.3601 | 3.4380 | 3.4380 | 3.6240 |
| 23 | 0.8676 | 19.2725 | 7.5610 | 7.0966 | 0.9970 | 2.9884 | 10.2690 | 11.2833 | 40.4043 | 1.6379 | 3.1996 | 3.1996 | 3.3727 |
| 24 | 0.9074 | 11.8075 | 7.3297 | 6.3185 | 0.7438 | 4.8086 | 9.3835 | 10.7738 | 35.2628 | 0.9374 | 2.9711 | 2.9711 | 3.1318 |
| 25 | 0.9615 | 6.1022 | 7.2531 | 5.5450 | 0.5424 | 7.1025 | 8.4267 | 10.0710 | 29.7786 | 0.3568 | 2.7593 | 2.7593 | 2.9086 |
| 26 | 1.0173 | 2.3351 | 7.3702 | 4.8353 | 0.3870 | 9.5960 | 7.4132 | 9.2202 | 24.1040 | 0.0257 | 2.5750 | 2.5750 | 2.7142 |
| 27 | 1.0659 | 0.5239 | 7.6385 | 4.2369 | 0.2707 | 12.1986 | 6.3656 | 8.1454 | 18.5095 | 0.1204 | 2.4305 | 2.4305 | 2.5620 |
| 28 | 1.0978 | 0.3982 | 7.9687 | 3.7765 | 0.1804 | 14.4786 | 5.3045 | 6.9129 | 13.1907 | 0.7823 | 2.3257 | 2.3257 | 2.4515 |
| 29 | 1.0949 | 1.4780 | 8.3169 | 3.4259 | 0.1295 | 16.0749 | 4.2644 | 5.6882 | 8.5226 | 2.0994 | 2.2604 | 2.2604 | 2.3827 |
| 30 | 1.1013 | 3.1011 | 8.4712 | 3.1394 | 0.1004 | 16.9033 | 3.3347 | 4.4260 | 4.7583 | 3.9074 | 2.2191 | 2.2191 | 2.3391 |
| 31 | 1.0584 | 4.7215 | 8.4163 | 2.8773 | 0.0951 | 16.5895 | 2.5190 | 3.2808 | 2.1322 | 5.9193 | 2.1820 | 2.1820 | 2.3000 |
| 32 | 1.0083 | 5.8106 | 8.0463 | 2.6500 | 0.1143 | 15.3378 | 1.8768 | 2.2800 | 0.6141 | 7.8277 | 2.1346 | 2.1346 | 2.2501 |
| 33 | 0.9689 | 6.3082 | 7.4028 | 2.3977 | 0.1596 | 13.5642 | 1.4245 | 1.4876 | 0.0621 | 9.2338 | 2.0739 | 2.0739 | 2.1860 |
| 34 | 0.9900 | 6.3427 | 6.5651 | 2.1425 | 0.2320 | 11.4443 | 1.0766 | 0.9164 | 0.1911 | 9.9100 | 1.9953 | 1.9953 | 2.1032 |
| 35 | 1.0509 | 6.0483 | 5.6895 | 1.8783 | 0.3185 | 9.4005 | 0.8929 | 0.5446 | 0.7237 | 9.9838 | 1.9113 | 1.9113 | 2.0147 |
| 36 | 1.1850 | 5.6865 | 4.8026 | 1.6486 | 0.3962 | 7.5899 | 0.8138 | 0.3513 | 1.3650 | 9.5331 | 1.8268 | 1.8268 | 1.9256 |
| 37 | 1.4420 | 5.3876 | 4.0269 | 1.4008 | 0.4472 | 6.1832 | 0.8312 | 0.2949 | 1.9661 | 8.9469 | 1.7586 | 1.7586 | 1.8537 |
| 38 | 1.7932 | 5.1301 | 3.3341 | 1.1665 | 0.4732 | 5.1052 | 0.9409 | 0.3366 | 2.4842 | 8.5278 | 1.7115 | 1.7115 | 1.8041 |
| 39 | 2.3155 | 4.9477 | 2.7350 | 0.9453 | 0.4914 | 4.3994 | 1.0975 | 0.4397 | 2.8965 | 8.6088 | 1.6993 | 1.6993 | 1.7912 |
| 40 | 3.0922 | 4.8230 | 2.2149 | 0.7350 | 0.5805 | 3.9474 | 1.3160 | 0.5570 | 3.1874 | 9.5989 | 1.7336 | 1.7336 | 1.8273 |
| 41 | 4.1680 | 4.7275 | 1.7121 | 0.5367 | 0.8697 | 3.7173 | 1.5655 | 0.6486 | 3.4828 | 11.9499 | 1.8270 | 1.8270 | 1.9258 |
| 42 | 5.6887 | 4.5932 | 1.2419 | 0.3679 | 1.5203 | 3.6246 | 1.8117 | 0.6663 | 3.8405 | 16.0491 | 1.9851 | 1.9851 | 2.0924 |
| 43 | 7.6593 | 4.4310 | 0.8293 | 0.2056 | 2.7536 | 3.5042 | 2.1274 | 0.5978 | 4.4280 | 22.6040 | 2.2168 | 2.2168 | 2.3367 |
| 44 | 9.9980 | 4.2025 | 0.6437 | 0.0743 | 4.6244 | 3.3332 | 2.4946 | 0.4252 | 5.4334 | 31.8083 | 2.5107 | 2.5107 | 2.6465 |
| 45 | 12.5545 | 3.9459 | 0.9400 | 0.0180 | 7.0400 | 2.9337 | 2.9689 | 0.2201 | 7.0641 | 44.0576 | 2.8591 | 2.8591 | 3.0137 |
| 46 | 14.8597 | 3.5856 | 1.9448 | 0.1555 | 9.6524 | 2.3810 | 3.5086 | 0.0431 | 9.4512 | 58.4759 | 3.2258 | 3.2258 | 3.4003 |
| 47 | 16.4147 | 3.2683 | 3.7814 | 0.6504 | 11.7142 | 1.7110 | 4.1186 | 0.0572 | 12.4630 | 73.4445 | 3.5724 | 3.5724 | 3.7657 |
| 48 | 16.7638 | 3.0272 | 5.9714 | 1.6960 | 12.5664 | 1.0485 | 4.6881 | 0.4157 | 15.7232 | 85.9574 | 3.8452 | 3.8452 | 4.0532 |
| 49 | 15.5156 | 2.8313 | 7.8227 | 3.3459 | 12.0878 | 0.5346 | 4.8974 | 1.2594 | 18.5162 | 92.2319 | 3.9880 | 3.9880 | 4.2037 |
| 50 | 13.3996 | 2.5053 | 8.6607 | 5.2280 | 10.7103 | 0.2221 | 4.8009 | 2.4593 | 19.9660 | 90.6172 | 3.9821 | 3.9821 | 4.1975 |
| 51 | 10.9034 | 1.8859 | 8.1196 | 6.8221 | 9.0842 | 0.1652 | 4.3476 | 3.7052 | 19.9408 | 81.9752 | 3.8334 | 3.8334 | 4.0408 |
| 52 | 8.5034 | 0.9794 | 6.5516 | 7.6265 | 8.0791 | 0.3571 | 3.5453 | 4.8057 | 18.8921 | 68.3557 | 3.5735 | 3.5735 | 3.7668 |
| 53 | 6.7696 | 0.4265 | 4.8691 | 7.2685 | 8.0007 | 0.8768 | 2.7142 | 5.4259 | 17.1784 | 53.2095 | 3.2671 | 3.2671 | 3.4438 |
| 54 | 5.7735 | 1.0206 | 3.4901 | 5.9105 | 9.0879 | 1.8915 | 1.9950 | 5.4200 | 15.5886 | 39.1890 | 2.9894 | 2.9894 | 3.1511 |
| 55 | 5.3705 | 3.5874 | 2.4915 | 4.0706 | 11.6919 | 3.5902 | 1.3494 | 5.0659 | 14.7733 | 27.2777 | 2.8155 | 2.8155 | 2.9678 |
| 56 | 5.5520 | 8.6283 | 1.9250 | 2.1752 | 16.5450 | 6.2692 | 0.8885 | 4.3665 | 14.6347 | 17.8462 | 2.8077 | 2.8077 | 2.9596 |
| 57 | 6.3608 | 16.2272 | 1.6600 | 0.7375 | 24.3857 | 10.2013 | 0.5446 | 3.4797 | 15.2020 | 10.7278 | 2.9921 | 2.9921 | 3.1539 |
| 58 | 7.8047 | 25.6451 | 1.6364 | 0.0862 | 35.9525 | 15.3081 | 0.3392 | 2.5373 | 16.1666 | 5.6033 | 3.3329 | 3.3329 | 3.5131 |
| 59 | 9.6874 | 35.3980 | 1.7970 | 0.4046 | 50.9447 | 21.0730 | 0.2242 | 1.6484 | 17.2367 | 2.2422 | 3.7504 | 3.7504 | 3.9533 |
| 60 | 11.8493 | 43.2459 | 2.1351 | 1.6361 | 67.7490 | 26.3841 | 0.2094 | 0.9088 | 17.7696 | 0.5508 | 4.1526 | 4.1526 | 4.3772 |
| 61 | 13.7814 | 46.5394 | 2.6258 | 3.4277 | 83.0487 | 29.3746 | 0.2948 | 0.4237 | 17.1199 | 0.1775 | 4.4364 | 4.4364 | 4.6763 |
| 62 | 14.7091 | 44.2732 | 3.0740 | 5.0765 | 92.4153 | 28.7590 | 0.4660 | 0.2490 | 15.2066 | 0.5889 | 4.5257 | 4.5257 | 4.7705 |
| 63 | 14.1362 | 37.5303 | 3.3078 | 6.0973 | 91.7489 | 24.6347 | 0.7023 | 0.3008 | 12.3127 | 1.1338 | 4.3807 | 4.3807 | 4.6177 |
| 64 | 12.3548 | 28.4198 | 3.2595 | 6.2591 | 81.2785 | 18.0622 | 0.9804 | 0.4569 | 8.8631 | 1.3159 | 4.0156 | 4.0156 | 4.2328 |
| 65 | 9.6638 | 19.6766 | 2.8189 | 5.6039 | 64.0460 | 11.2769 | 1.1726 | 0.5976 | 5.8462 | 1.1199 | 3.4903 | 3.4903 | 3.6791 |
| 66 | 6.8172 | 13.2222 | 2.1625 | 4.5235 | 44.9309 | 6.3291 | 1.2517 | 0.6553 | 3.8012 | 0.6910 | 2.9049 | 2.9049 | 3.0620 |
| 67 | 4.4255 | 9.3117 | 1.4782 | 3.4222 | 28.5871 | 3.6127 | 1.2396 | 0.6336 | 2.7667 | 0.3162 | 2.3621 | 2.3621 | 2.4898 |
| 68 | 2.6726 | 7.4740 | 0.9110 | 2.4626 | 17.1479 | 2.8294 | 1.1327 | 0.5883 | 2.4687 | 0.1381 | 1.9449 | 1.9449 | 2.0501 |
| 69 | 1.5515 | 6.8390 | 0.5457 | 1.7942 | 10.2793 | 3.0704 | 1.0007 | 0.5383 | 2.5586 | 0.1448 | 1.6829 | 1.6829 | 1.7740 |
| 70 | 0.9113 | 6.6042 | 0.3501 | 1.3377 | 6.6879 | 3.5477 | 0.8419 | 0.4613 | 2.6453 | 0.2579 | 1.5377 | 1.5377 | 1.6209 |
| 71 | 0.5919 | 6.2984 | 0.3027 | 1.0602 | 4.8903 | 3.7593 | 0.6950 | 0.3971 | 2.5474 | 0.3827 | 1.4465 | 1.4465 | 1.5248 |
| 72 | 0.4343 | 5.6115 | 0.3247 | 0.8441 | 3.8037 | 3.4832 | 0.5646 | 0.3258 | 2.2396 | 0.4764 | 1.3457 | 1.3457 | 1.4184 |
| 73 | 0.3484 | 4.7088 | 0.3437 | 0.6279 | 2.8046 | 2.8476 | 0.4560 | 0.2405 | 1.7917 | 0.5128 | 1.2117 | 1.2117 | 1.2772 |
| 74 | 0.3005 | 3.7052 | 0.3358 | 0.4128 | 1.8400 | 2.0328 | 0.3663 | 0.1663 | 1.2916 | 0.5326 | 1.0481 | 1.0481 | 1.1047 |
| 75 | 0.2751 | 2.7416 | 0.2996 | 0.2206 | 0.9838 | 1.2620 | 0.2966 | 0.0965 | 0.8644 | 0.5404 | 0.8707 | 0.8707 | 0.9178 |
| 76 | 0.2691 | 1.9651 | 0.2845 | 0.0836 | 0.3584 | 0.6649 | 0.2487 | 0.0469 | 0.5251 | 0.5748 | 0.7086 | 0.7086 | 0.7469 |
| 77 | 0.2747 | 1.3667 | 0.3900 | 0.0532 | 0.0615 | 0.2770 | 0.2221 | 0.0194 | 0.3009 | 0.6124 | 0.5981 | 0.5981 | 0.6305 |
| 78 | 0.2837 | 0.9730 | 0.7382 | 0.1628 | 0.1226 | 0.0857 | 0.2155 | 0.0174 | 0.1645 | 0.6743 | 0.5863 | 0.5863 | 0.6180 |
| 79 | 0.2830 | 0.7552 | 1.4360 | 0.4521 | 0.6001 | 0.0387 | 0.2431 | 0.0382 | 0.0947 | 0.7409 | 0.6843 | 0.6843 | 0.7213 |
| 80 | 0.2675 | 0.7119 | 2.6109 | 0.9548 | 1.5971 | 0.0766 | 0.2798 | 0.0807 | 0.0673 | 0.8324 | 0.8648 | 0.8648 | 0.9116 |
| 81 | 0.2512 | 0.8296 | 4.2735 | 1.6978 | 3.3234 | 0.1602 | 0.3411 | 0.1409 | 0.0763 | 0.9944 | 1.0995 | 1.0995 | 1.1589 |
| 82 | 0.2360 | 1.1322 | 6.4403 | 2.7008 | 6.1212 | 0.2355 | 0.4245 | 0.2155 | 0.1166 | 1.2404 | 1.3734 | 1.3734 | 1.4477 |
| 83 | 0.2177 | 1.6277 | 8.8897 | 3.9101 | 10.2229 | 0.2975 | 0.4973 | 0.2927 | 0.1838 | 1.6214 | 1.6662 | 1.6662 | 1.7563 |
| 84 | 0.1910 | 2.2718 | 11.3918 | 5.2303 | 15.7942 | 0.3284 | 0.5639 | 0.3720 | 0.2768 | 2.0956 | 1.9625 | 1.9625 | 2.0687 |
| 85 | 0.1620 | 2.9713 | 13.5949 | 6.4974 | 22.6020 | 0.3474 | 0.6139 | 0.4378 | 0.3786 | 2.6591 | 2.2420 | 2.2420 | 2.3632 |
| 86 | 0.1287 | 3.5847 | 15.0678 | 7.4197 | 29.8189 | 0.4036 | 0.6287 | 0.4842 | 0.4695 | 3.1892 | 2.4738 | 2.4738 | 2.6076 |
| 87 | 0.1073 | 3.9727 | 15.5895 | 7.8419 | 36.2387 | 0.5200 | 0.6403 | 0.5121 | 0.5156 | 3.5439 | 2.6359 | 2.6359 | 2.7785 |
| 88 | 0.0981 | 4.0548 | 14.9633 | 7.6345 | 40.5489 | 0.7205 | 0.6670 | 0.5162 | 0.5242 | 3.7237 | 2.7102 | 2.7102 | 2.8568 |
| 89 | 0.1014 | 3.8558 | 13.4058 | 6.8641 | 41.7166 | 0.9859 | 0.7514 | 0.5141 | 0.5270 | 3.6472 | 2.6902 | 2.6902 | 2.8357 |
| 90 | 0.0981 | 3.4421 | 11.0912 | 5.6669 | 39.5597 | 1.2680 | 0.8865 | 0.5002 | 0.5641 | 3.3538 | 2.5774 | 2.5774 | 2.7168 |
| 91 | 0.0828 | 2.9188 | 8.5653 | 4.3502 | 34.8213 | 1.5474 | 1.0667 | 0.4793 | 0.6445 | 2.9233 | 2.3958 | 2.3958 | 2.5254 |
| 92 | 0.0486 | 2.4150 | 6.0473 | 3.0908 | 28.6564 | 1.8065 | 1.2644 | 0.4681 | 0.7673 | 2.4223 | 2.1676 | 2.1676 | 2.2849 |
| 93 | 0.0190 | 1.9846 | 3.9445 | 2.0509 | 22.1143 | 2.0439 | 1.4606 | 0.4410 | 0.9183 | 1.8849 | 1.9199 | 1.9199 | 2.0238 |
| 94 | 0.0279 | 1.6064 | 2.3450 | 1.3242 | 16.1549 | 2.2145 | 1.6236 | 0.4265 | 1.0721 | 1.4571 | 1.6808 | 1.6808 | 1.7718 |
| 95 | 0.1052 | 1.2403 | 1.3501 | 0.9036 | 11.2839 | 2.3148 | 1.7924 | 0.4139 | 1.2044 | 1.1532 | 1.4752 | 1.4752 | 1.5550 |
| 96 | 0.2980 | 0.9373 | 0.8405 | 0.7711 | 7.7411 | 2.3396 | 1.8946 | 0.4148 | 1.2114 | 1.1082 | 1.3250 | 1.3250 | 1.3967 |
| 97 | 0.6579 | 0.7025 | 0.6800 | 0.9167 | 5.3659 | 2.2910 | 1.9806 | 0.4354 | 1.0930 | 1.5384 | 1.2515 | 1.2515 | 1.3192 |
| 98 | 1.2133 | 0.5172 | 0.8137 | 1.3397 | 3.9951 | 2.2258 | 2.0487 | 0.4877 | 0.9187 | 2.6265 | 1.2723 | 1.2723 | 1.3411 |
| 99 | 1.9536 | 0.3754 | 1.2865 | 2.0339 | 3.4764 | 2.1987 | 2.1385 | 0.5664 | 0.7614 | 4.3175 | 1.3823 | 1.3823 | 1.4571 |

# APPENDIX C

# Data Sets for Chapter 4 Work Problems

| Table C.1    Exemplar Leg and Joint Stiffness Values From Continuous Two-Legged Hopping | | |
|---|---|---|
| Leg stiffness (f(x,y)) (nm/kg/rad) | Ankle stiffness (x) (nm/kg/rad) | Knee stiffness (y) (nm/kg/rad) |
| 15.23 | 4.4 | 0.64 |
| 13.36 | 5.06 | 0.48 |
| 15.72 | 4.86 | 0.35 |
| 14.73 | 4.89 | 0.56 |
| 16.29 | 5.82 | 0.28 |
| 18.58 | 6.14 | 1.11 |
| 18.13 | 5.68 | 0.77 |
| 16.48 | 5.19 | 0.37 |
| 17.48 | 5.78 | 0.67 |
| 17.81 | 6.01 | 1.25 |
| 17.96 | 6.22 | 0.69 |
| 17.6 | 5.77 | 0.65 |
| 17.27 | 5.84 | 0.38 |
| 19.02 | 5.88 | 0.77 |
| 18.38 | 5.89 | 0.57 |
| 23.3 | 6.77 | 1.88 |
| 18.97 | 6.22 | 0.97 |
| 21.53 | 6.89 | 1.88 |
| 22.48 | 6.57 | 2.09 |
| 22.6 | 6.94 | 2.81 |
| 22.93 | 8.53 | 3.07 |
| 20.26 | 5.6 | 0.56 |
| 22.22 | 6.82 | 2.64 |
| 19.48 | 5.8 | 1.42 |
| 22.26 | 8 | 2.89 |

*(continued)*

Table C.1 *(continued)*

| Leg stiffness (f(x,y)) (nm/kg/rad) | Ankle stiffness (x) (nm/kg/rad) | Knee stiffness (y) (nm/kg/rad) |
|---|---|---|
| 19.56 | 6.51 | 1.39 |
| 19.66 | 5.67 | 1.05 |
| 23.56 | 9.23 | 2.67 |
| 20.74 | 7.09 | 2.21 |
| 16 | 5.58 | 1.1 |
| 29.91 | 6.85 | 2.19 |
| 25.97 | 3.54 | 2.02 |
| 39.68 | 8.43 | 4.07 |
| 36.3 | 6.98 | 2.29 |
| 37 | 6.83 | 3.07 |
| 34.19 | 6.25 | 2.1 |
| 34.67 | 6.79 | 2.97 |
| 39.1 | 7.58 | 3.36 |
| 30.95 | 6.42 | 2.55 |
| 39.51 | 7.26 | 2.01 |
| 43.37 | 8.04 | 3.51 |
| 35.71 | 7.2 | 2.93 |
| 33.52 | 7.08 | 1.91 |
| 38.26 | 8.24 | 3.69 |
| 30.3 | 5.96 | 1.79 |
| 15.5 | 3.93 | 0.4 |
| 16.22 | 4.8 | 0.72 |
| 14.13 | 3.24 | 0.66 |
| 12.16 | 2.67 | 0.64 |
| 14.46 | 2.79 | 0.52 |
| 12.52 | 2.53 | 0.37 |
| 13.81 | 3.5 | 0.12 |
| 15.6 | 4.17 | 1.02 |
| 13.68 | 3.38 | 0.91 |
| 15.27 | 3.55 | 0.81 |
| 14.86 | 3.69 | 0.91 |
| 17.34 | 4.24 | 1.17 |
| 18.18 | 4.73 | 1.28 |
| 20.62 | 5.14 | 1.96 |
| 18.6 | 5.15 | 1.71 |
| 15.23 | 4.4 | 0.64 |
| 13.36 | 5.06 | 0.48 |
| 15.72 | 4.86 | 0.35 |
| 14.73 | 4.89 | 0.56 |
| 16.29 | 5.82 | 0.28 |
| 18.58 | 6.14 | 1.11 |

| Leg stiffness (f(x,y)) (nm/kg/rad) | Ankle stiffness (x) (nm/kg/rad) | Knee stiffness (y) (nm/kg/rad) |
|---|---|---|
| 18.13 | 5.68 | 0.77 |
| 16.48 | 5.19 | 0.37 |
| 17.48 | 5.78 | 0.67 |
| 17.81 | 6.01 | 1.25 |
| 17.96 | 6.22 | 0.69 |
| 17.6 | 5.77 | 0.65 |
| 17.27 | 5.84 | 0.38 |
| 19.02 | 5.88 | 0.77 |
| 18.38 | 5.89 | 0.57 |
| 23.3 | 6.77 | 1.88 |
| 18.97 | 6.22 | 0.97 |
| 21.53 | 6.89 | 1.88 |
| 22.48 | 6.57 | 2.09 |
| 22.6 | 6.94 | 2.81 |
| 22.93 | 8.53 | 3.07 |
| 20.26 | 5.6 | 0.56 |
| 22.22 | 6.82 | 2.64 |
| 19.48 | 5.8 | 1.42 |
| 22.26 | 8 | 2.89 |
| 19.56 | 6.51 | 1.39 |
| 19.66 | 5.67 | 1.05 |
| 23.56 | 9.23 | 2.67 |
| 20.74 | 7.09 | 2.21 |
| 16 | 5.58 | 1.1 |
| 29.91 | 6.85 | 2.19 |
| 25.97 | 3.54 | 2.02 |
| 39.68 | 8.43 | 4.07 |
| 36.3 | 6.98 | 2.29 |
| 37 | 6.83 | 3.07 |
| 34.19 | 6.25 | 2.1 |
| 34.67 | 6.79 | 2.97 |
| 39.1 | 7.58 | 3.36 |
| 30.95 | 6.42 | 2.55 |
| 39.51 | 7.26 | 2.01 |
| 43.37 | 8.04 | 3.51 |
| 35.71 | 7.2 | 2.93 |
| 33.52 | 7.08 | 1.91 |
| 38.26 | 8.24 | 3.69 |
| 30.3 | 5.96 | 1.79 |
| 15.5 | 3.93 | 0.4 |
| 16.22 | 4.8 | 0.72 |

*(continued)*

**Table C.1** *(continued)*

| Leg stiffness (f(x,y)) (nm/kg/rad) | Ankle stiffness (x) (nm/kg/rad) | Knee stiffness (y) (nm/kg/rad) |
|---|---|---|
| 14.13 | 3.24 | 0.66 |
| 12.16 | 2.67 | 0.64 |
| 14.46 | 2.79 | 0.52 |
| 12.52 | 2.53 | 0.37 |
| 13.81 | 3.5 | 0.12 |
| 15.6 | 4.17 | 1.02 |
| 13.68 | 3.38 | 0.91 |
| 15.27 | 3.55 | 0.81 |
| 14.86 | 3.69 | 0.91 |
| 17.34 | 4.24 | 1.17 |
| 18.18 | 4.73 | 1.28 |
| 20.62 | 5.14 | 1.96 |
| 18.6 | 5.15 | 1.71 |
| 15.23 | 4.4 | 0.64 |
| 13.36 | 5.06 | 0.48 |
| 15.72 | 4.86 | 0.35 |
| 14.73 | 4.89 | 0.56 |
| 16.29 | 5.82 | 0.28 |
| 18.58 | 6.14 | 1.11 |
| 18.13 | 5.68 | 0.77 |
| 16.48 | 5.19 | 0.37 |
| 17.48 | 5.78 | 0.67 |
| 17.81 | 6.01 | 1.25 |
| 17.96 | 6.22 | 0.69 |
| 17.6 | 5.77 | 0.65 |
| 17.27 | 5.84 | 0.38 |
| 19.02 | 5.88 | 0.77 |
| 18.38 | 5.89 | 0.57 |
| 23.3 | 6.77 | 1.88 |
| 18.97 | 6.22 | 0.97 |
| 21.53 | 6.89 | 1.88 |
| 22.48 | 6.57 | 2.09 |
| 22.6 | 6.94 | 2.81 |
| 22.93 | 8.53 | 3.07 |
| 20.26 | 5.6 | 0.56 |
| 22.22 | 6.82 | 2.64 |
| 19.48 | 5.8 | 1.42 |
| 22.26 | 8 | 2.89 |
| 19.56 | 6.51 | 1.39 |
| 19.66 | 5.67 | 1.05 |
| 23.56 | 9.23 | 2.67 |

| Leg stiffness (f(x,y)) (nm/kg/rad) | Ankle stiffness (x) (nm/kg/rad) | Knee stiffness (y) (nm/kg/rad) |
|---|---|---|
| 20.74 | 7.09 | 2.21 |
| 16 | 5.58 | 1.1 |
| 29.91 | 6.85 | 2.19 |
| 25.97 | 3.54 | 2.02 |
| 39.68 | 8.43 | 4.07 |
| 36.3 | 6.98 | 2.29 |
| 37 | 6.83 | 3.07 |
| 34.19 | 6.25 | 2.1 |
| 34.67 | 6.79 | 2.97 |
| 39.1 | 7.58 | 3.36 |
| 30.95 | 6.42 | 2.55 |
| 39.51 | 7.26 | 2.01 |
| 43.37 | 8.04 | 3.51 |
| 35.71 | 7.2 | 2.93 |
| 33.52 | 7.08 | 1.91 |
| 38.26 | 8.24 | 3.69 |
| 30.3 | 5.96 | 1.79 |
| 15.5 | 3.93 | 0.4 |
| 16.22 | 4.8 | 0.72 |
| 14.13 | 3.24 | 0.66 |
| 12.16 | 2.67 | 0.64 |
| 14.46 | 2.79 | 0.52 |
| 12.52 | 2.53 | 0.37 |
| 13.81 | 3.5 | 0.12 |
| 15.6 | 4.17 | 1.02 |
| 13.68 | 3.38 | 0.91 |
| 15.27 | 3.55 | 0.81 |
| 14.86 | 3.69 | 0.91 |
| 17.34 | 4.24 | 1.17 |
| 18.18 | 4.73 | 1.28 |
| 20.62 | 5.14 | 1.96 |
| 18.6 | 5.15 | 1.71 |
| 15.23 | 4.4 | 0.64 |
| 13.36 | 5.06 | 0.48 |
| 15.72 | 4.86 | 0.35 |
| 14.73 | 4.89 | 0.56 |
| 16.29 | 5.82 | 0.28 |
| 18.58 | 6.14 | 1.11 |
| 18.13 | 5.68 | 0.77 |
| 16.48 | 5.19 | 0.37 |
| 17.48 | 5.78 | 0.67 |

*(continued)*

## Table C.1 (continued)

| Leg stiffness (f(x,y)) (nm/kg/rad) | Ankle stiffness (x) (nm/kg/rad) | Knee stiffness (y) (nm/kg/rad) |
|---|---|---|
| 17.81 | 6.01 | 1.25 |
| 17.96 | 6.22 | 0.69 |
| 17.6 | 5.77 | 0.65 |
| 17.27 | 5.84 | 0.38 |
| 19.02 | 5.88 | 0.77 |
| 18.38 | 5.89 | 0.57 |
| 23.3 | 6.77 | 1.88 |
| 18.97 | 6.22 | 0.97 |
| 21.53 | 6.89 | 1.88 |
| 22.48 | 6.57 | 2.09 |
| 22.6 | 6.94 | 2.81 |
| 22.93 | 8.53 | 3.07 |
| 20.26 | 5.6 | 0.56 |
| 22.22 | 6.82 | 2.64 |
| 19.48 | 5.8 | 1.42 |
| 22.26 | 8 | 2.89 |
| 19.56 | 6.51 | 1.39 |
| 19.66 | 5.67 | 1.05 |
| 23.56 | 9.23 | 2.67 |
| 20.74 | 7.09 | 2.21 |
| 16 | 5.58 | 1.1 |
| 29.91 | 6.85 | 2.19 |
| 25.97 | 3.54 | 2.02 |
| 39.68 | 8.43 | 4.07 |
| 36.3 | 6.98 | 2.29 |
| 37 | 6.83 | 3.07 |
| 34.19 | 6.25 | 2.1 |
| 34.67 | 6.79 | 2.97 |
| 39.1 | 7.58 | 3.36 |
| 30.95 | 6.42 | 2.55 |
| 39.51 | 7.26 | 2.01 |
| 43.37 | 8.04 | 3.51 |
| 35.71 | 7.2 | 2.93 |
| 33.52 | 7.08 | 1.91 |
| 38.26 | 8.24 | 3.69 |
| 30.3 | 5.96 | 1.79 |
| 15.5 | 3.93 | 0.4 |
| 16.22 | 4.8 | 0.72 |
| 14.13 | 3.24 | 0.66 |
| 12.16 | 2.67 | 0.64 |
| 14.46 | 2.79 | 0.52 |

| Leg stiffness (f(x,y)) (nm/kg/rad) | Ankle stiffness (x) (nm/kg/rad) | Knee stiffness (y) (nm/kg/rad) |
|---|---|---|
| 12.52 | 2.53 | 0.37 |
| 13.81 | 3.5 | 0.12 |
| 15.6 | 4.17 | 1.02 |
| 13.68 | 3.38 | 0.91 |
| 15.27 | 3.55 | 0.81 |
| 14.86 | 3.69 | 0.91 |
| 17.34 | 4.24 | 1.17 |
| 18.18 | 4.73 | 1.28 |
| 20.62 | 5.14 | 1.96 |
| 18.6 | 5.15 | 1.71 |
| 15.23 | 4.4 | 0.64 |
| 13.36 | 5.06 | 0.48 |
| 15.72 | 4.86 | 0.35 |
| 14.73 | 4.89 | 0.56 |
| 16.29 | 5.82 | 0.28 |
| 18.58 | 6.14 | 1.11 |
| 18.13 | 5.68 | 0.77 |
| 16.48 | 5.19 | 0.37 |
| 17.48 | 5.78 | 0.67 |
| 17.81 | 6.01 | 1.25 |
| 17.96 | 6.22 | 0.69 |
| 17.6 | 5.77 | 0.65 |
| 17.27 | 5.84 | 0.38 |
| 19.02 | 5.88 | 0.77 |
| 18.38 | 5.89 | 0.57 |
| 23.3 | 6.77 | 1.88 |
| 18.97 | 6.22 | 0.97 |
| 21.53 | 6.89 | 1.88 |
| 22.48 | 6.57 | 2.09 |
| 22.6 | 6.94 | 2.81 |
| 22.93 | 8.53 | 3.07 |
| 20.26 | 5.6 | 0.56 |
| 22.22 | 6.82 | 2.64 |
| 19.48 | 5.8 | 1.42 |
| 22.26 | 8 | 2.89 |
| 19.56 | 6.51 | 1.39 |
| 19.66 | 5.67 | 1.05 |
| 23.56 | 9.23 | 2.67 |
| 20.74 | 7.09 | 2.21 |
| 16 | 5.58 | 1.1 |
| 29.91 | 6.85 | 2.19 |

*(continued)*

| Table C.1 (continued) | | |
|---|---|---|
| **Leg stiffness (f(x,y))** <br> **(nm/kg/rad)** | **Ankle stiffness (x)** <br> **(nm/kg/rad)** | **Knee stiffness (y)** <br> **(nm/kg/rad)** |
| 25.97 | 3.54 | 2.02 |
| 39.68 | 8.43 | 4.07 |
| 36.3 | 6.98 | 2.29 |
| 37 | 6.83 | 3.07 |
| 34.19 | 6.25 | 2.1 |
| 34.67 | 6.79 | 2.97 |
| 39.1 | 7.58 | 3.36 |
| 30.95 | 6.42 | 2.55 |
| 39.51 | 7.26 | 2.01 |
| 43.37 | 8.04 | 3.51 |
| 35.71 | 7.2 | 2.93 |
| 33.52 | 7.08 | 1.91 |
| 38.26 | 8.24 | 3.69 |
| 30.3 | 5.96 | 1.79 |
| 15.5 | 3.93 | 0.4 |
| 16.22 | 4.8 | 0.72 |
| 14.13 | 3.24 | 0.66 |
| 12.16 | 2.67 | 0.64 |
| 14.46 | 2.79 | 0.52 |
| 12.52 | 2.53 | 0.37 |
| 13.81 | 3.5 | 0.12 |
| 15.6 | 4.17 | 1.02 |
| 13.68 | 3.38 | 0.91 |
| 15.27 | 3.55 | 0.81 |
| 14.86 | 3.69 | 0.91 |
| 17.34 | 4.24 | 1.17 |
| 18.18 | 4.73 | 1.28 |
| 20.62 | 5.14 | 1.96 |
| 18.6 | 5.15 | 1.71 |
| 15.23 | 4.4 | 0.64 |
| 13.36 | 5.06 | 0.48 |
| 15.72 | 4.86 | 0.35 |
| 14.73 | 4.89 | 0.56 |
| 16.29 | 5.82 | 0.28 |
| 18.58 | 6.14 | 1.11 |
| 18.13 | 5.68 | 0.77 |
| 16.48 | 5.19 | 0.37 |
| 17.48 | 5.78 | 0.67 |
| 17.81 | 6.01 | 1.25 |
| ›17.96 | 6.22 | 0.69 |
| 17.6 | 5.77 | 0.65 |

| Leg stiffness (f(x,y)) (nm/kg/rad) | Ankle stiffness (x) (nm/kg/rad) | Knee stiffness (y) (nm/kg/rad) |
| --- | --- | --- |
| 17.27 | 5.84 | 0.38 |
| 19.02 | 5.88 | 0.77 |
| 18.38 | 5.89 | 0.57 |
| 23.3 | 6.77 | 1.88 |
| 18.97 | 6.22 | 0.97 |
| 21.53 | 6.89 | 1.88 |
| 22.48 | 6.57 | 2.09 |
| 22.6 | 6.94 | 2.81 |
| 22.93 | 8.53 | 3.07 |
| 20.26 | 5.6 | 0.56 |
| 22.22 | 6.82 | 2.64 |
| 19.48 | 5.8 | 1.42 |
| 22.26 | 8 | 2.89 |
| 19.56 | 6.51 | 1.39 |
| 19.66 | 5.67 | 1.05 |
| 23.56 | 9.23 | 2.67 |
| 20.74 | 7.09 | 2.21 |
| 16 | 5.58 | 1.1 |
| 29.91 | 6.85 | 2.19 |
| 25.97 | 3.54 | 2.02 |
| 39.68 | 8.43 | 4.07 |
| 36.3 | 6.98 | 2.29 |
| 37 | 6.83 | 3.07 |
| 34.19 | 6.25 | 2.1 |
| 34.67 | 6.79 | 2.97 |
| 39.1 | 7.58 | 3.36 |
| 30.95 | 6.42 | 2.55 |
| 39.51 | 7.26 | 2.01 |
| 43.37 | 8.04 | 3.51 |
| 35.71 | 7.2 | 2.93 |
| 33.52 | 7.08 | 1.91 |
| 38.26 | 8.24 | 3.69 |
| 30.3 | 5.96 | 1.79 |
| 15.5 | 3.93 | 0.4 |
| 16.22 | 4.8 | 0.72 |
| 14.13 | 3.24 | 0.66 |
| 12.16 | 2.67 | 0.64 |
| 14.46 | 2.79 | 0.52 |
| 12.52 | 2.53 | 0.37 |
| 13.81 | 3.5 | 0.12 |
| 15.6 | 4.17 | 1.02 |

*(continued)*

Table C.1 (continued)

| Leg stiffness (f(x,y)) (nm/kg/rad) | Ankle stiffness (x) (nm/kg/rad) | Knee stiffness (y) (nm/kg/rad) |
|---|---|---|
| 13.68 | 3.38 | 0.91 |
| 15.27 | 3.55 | 0.81 |
| 14.86 | 3.69 | 0.91 |
| 17.34 | 4.24 | 1.17 |
| 18.18 | 4.73 | 1.28 |
| 20.62 | 5.14 | 1.96 |
| 18.6 | 5.15 | 1.71 |
| 15.23 | 4.4 | 0.64 |
| 13.36 | 5.06 | 0.48 |
| 15.72 | 4.86 | 0.35 |
| 14.73 | 4.89 | 0.56 |
| 16.29 | 5.82 | 0.28 |
| 18.58 | 6.14 | 1.11 |
| 18.13 | 5.68 | 0.77 |
| 16.48 | 5.19 | 0.37 |
| 17.48 | 5.78 | 0.67 |
| 17.81 | 6.01 | 1.25 |
| 17.96 | 6.22 | 0.69 |
| 17.6 | 5.77 | 0.65 |
| 17.27 | 5.84 | 0.38 |
| 19.02 | 5.88 | 0.77 |
| 18.38 | 5.89 | 0.57 |
| 23.3 | 6.77 | 1.88 |
| 18.97 | 6.22 | 0.97 |
| 21.53 | 6.89 | 1.88 |
| 22.48 | 6.57 | 2.09 |
| 22.6 | 6.94 | 2.81 |
| 22.93 | 8.53 | 3.07 |
| 20.26 | 5.6 | 0.56 |
| 22.22 | 6.82 | 2.64 |
| 19.48 | 5.8 | 1.42 |
| 22.26 | 8 | 2.89 |
| 19.56 | 6.51 | 1.39 |
| 19.66 | 5.67 | 1.05 |
| 23.56 | 9.23 | 2.67 |
| 20.74 | 7.09 | 2.21 |
| 16 | 5.58 | 1.1 |
| 29.91 | 6.85 | 2.19 |
| 25.97 | 3.54 | 2.02 |
| 39.68 | 8.43 | 4.07 |

| Leg stiffness (f(x,y)) (nm/kg/rad) | Ankle stiffness (x) (nm/kg/rad) | Knee stiffness (y) (nm/kg/rad) |
|---|---|---|
| 36.3 | 6.98 | 2.29 |
| 37 | 6.83 | 3.07 |
| 34.19 | 6.25 | 2.1 |
| 34.67 | 6.79 | 2.97 |
| 39.1 | 7.58 | 3.36 |
| 30.95 | 6.42 | 2.55 |
| 39.51 | 7.26 | 2.01 |

### Table C.2 Exemplar Mean Ensemble Curve Values for the Knee Joint During the Stance Period of the Running Gait Cycle (Values Presented in Degrees)

| Mean | SD+ | SD– |
|---|---|---|
| 44.54989 | 54.6262 | 34.47359 |
| 45.48623 | 55.36844 | 35.60402 |
| 46.44262 | 56.10429 | 36.78094 |
| 47.41158 | 56.82215 | 38.00101 |
| 48.38639 | 57.51371 | 39.25907 |
| 49.36268 | 58.17591 | 40.54944 |
| 50.33482 | 58.80692 | 41.86272 |
| 51.29644 | 59.40558 | 43.18729 |
| 52.24105 | 59.97135 | 44.51075 |
| 53.16236 | 60.50429 | 45.82043 |
| 54.05508 | 61.00499 | 47.10516 |
| 54.91487 | 61.47435 | 48.35539 |
| 55.73868 | 61.91342 | 49.56393 |
| 56.52435 | 62.32312 | 50.72557 |
| 57.27106 | 62.70449 | 51.83764 |
| 57.97899 | 63.05863 | 52.89936 |
| 58.64894 | 63.38625 | 53.91164 |
| 59.28225 | 63.68805 | 54.87645 |
| 59.88038 | 63.96489 | 55.79587 |
| 60.44447 | 64.21735 | 56.67159 |
| 60.97544 | 64.44614 | 57.50473 |
| 61.47385 | 64.65239 | 58.2953 |
| 61.93989 | 64.83727 | 59.04251 |
| 62.37341 | 65.0024 | 59.74442 |
| 62.77427 | 65.14966 | 60.39889 |
| 63.14236 | 65.2812 | 61.00352 |
| 63.47764 | 65.39915 | 61.55614 |
| 63.78045 | 65.50592 | 62.05498 |

*(continued)*

| Table C.2 | (continued) | |
|---|---|---|
| **Mean** | **SD+** | **SD–** |
| 64.0515 | 65.60378 | 62.49922 |
| 64.29165 | 65.69496 | 62.88834 |
| 64.5023 | 65.78152 | 63.22308 |
| 64.68507 | 65.86536 | 63.50479 |
| 64.84175 | 65.94789 | 63.73561 |
| 64.97401 | 66.02958 | 63.91845 |
| 65.08369 | 66.11047 | 64.05691 |
| 65.1721 | 66.18938 | 64.15481 |
| 65.24048 | 66.26469 | 64.21627 |
| 65.28968 | 66.33426 | 64.2451 |
| 65.32027 | 66.39583 | 64.24471 |
| 65.33278 | 66.44746 | 64.2181 |
| 65.32765 | 66.48741 | 64.16789 |
| 65.30521 | 66.51434 | 64.09608 |
| 65.26576 | 66.52708 | 64.00445 |
| 65.2095 | 66.52449 | 63.89452 |
| 65.13627 | 66.50525 | 63.76728 |
| 65.04575 | 66.46783 | 63.62367 |
| 64.93736 | 66.41048 | 63.46424 |
| 64.81059 | 66.33166 | 63.28952 |
| 64.66492 | 66.22992 | 63.09991 |
| 64.49992 | 66.10425 | 62.8956 |
| 64.31557 | 65.95415 | 62.67698 |
| 64.11185 | 65.77964 | 62.44405 |
| 63.88926 | 65.58133 | 62.19719 |
| 63.64837 | 65.3601 | 61.93663 |
| 63.3897 | 65.11694 | 61.66246 |
| 63.11404 | 64.85294 | 61.37514 |
| 62.8219 | 64.56885 | 61.07495 |
| 62.51386 | 64.26536 | 60.76236 |
| 62.19027 | 63.94284 | 60.43769 |
| 61.85171 | 63.60189 | 60.10152 |
| 61.49878 | 63.24313 | 59.75443 |
| 61.13238 | 62.86772 | 59.39704 |
| 60.75354 | 62.47709 | 59.02999 |
| 60.3634 | 62.07287 | 58.65392 |
| 59.96328 | 61.65713 | 58.26943 |
| 59.55442 | 61.23187 | 57.87697 |
| 59.13801 | 60.79929 | 57.47672 |
| 58.71524 | 60.36147 | 57.06902 |
| 58.28711 | 59.92038 | 56.65384 |

| Mean | SD+ | SD– |
| --- | --- | --- |
| 57.85452 | 59.4779 | 56.23113 |
| 57.41822 | 59.03554 | 55.8009 |
| 56.97871 | 58.59448 | 55.36294 |
| 56.5364 | 58.15565 | 54.91715 |
| 56.09156 | 57.71969 | 54.46343 |
| 55.64445 | 57.28699 | 54.00191 |
| 55.19516 | 56.8575 | 53.53282 |
| 54.74402 | 56.43127 | 53.05677 |
| 54.29121 | 56.00809 | 52.57433 |
| 53.83692 | 55.5876 | 52.08624 |
| 53.38158 | 55.16968 | 51.59349 |
| 52.92558 | 54.75422 | 51.09693 |
| 52.46934 | 54.341 | 50.59768 |
| 52.01326 | 53.93 | 50.09652 |
| 51.55779 | 53.52119 | 49.59439 |
| 51.10324 | 53.11444 | 49.09203 |
| 50.64991 | 52.70974 | 48.59009 |
| 50.19807 | 52.30707 | 48.08907 |
| 49.74823 | 51.90651 | 47.58995 |
| 49.3007 | 51.50787 | 47.09352 |
| 48.85615 | 51.1113 | 46.60099 |
| 48.41513 | 50.71676 | 46.1135 |
| 47.97819 | 50.32428 | 45.63209 |
| 47.54589 | 49.93387 | 45.15791 |
| 47.11878 | 49.54528 | 44.69228 |
| 46.69743 | 49.1585 | 44.23636 |
| 46.28248 | 48.77332 | 43.79163 |
| 45.87455 | 48.38949 | 43.35961 |
| 45.47436 | 48.0069 | 42.94182 |
| 45.08279 | 47.6255 | 42.54008 |
| 44.70058 | 47.24518 | 42.15598 |

# Glossary

$\alpha$—The significance level. This parameter, which is set before an experiment is conducted, represents the level at which the researcher will reject the null hypothesis.

**accommodating response**—A response to a perturbation intermediate between a Newtonian and a neuromuscular response, sometimes referred to as a biomechanical, overaccommodating, or underaccommodating response.

**analysis of variance (ANOVA)**—Statistical procedure that allows one to compare mean values obtained from more than two samples and to test the probability that the samples have been drawn from populations having the same mean.

**approximate entropy (ApEn)**—A measure that can quantify the regularity or predictability of a time series.

**arithmetic mean**—The average of a set of data.

**autocorrelation**—A function that expresses the degree of similarity between a series of data points and the same data points systematically shifted.

**bimodal**—Referring to data distributed such that they cluster at two distinct peaks.

**bootstrap**—A randomization procedure to address questions of temporal effects on an experimental outcome through evaluation of the mean differences for all combinations of empirical data.

**bootstrapping**—Invented by Bradley Efron, a statistician at Stanford University, bootstrapping is a computer-based resampling process in which samples are randomly drawn with replacement from a sample.

**bootstrap samples**—Subsamples randomly drawn, with replacement, from a sample.

**bootstrap sampling distribution**—Distribution of a bootstrap statistic, for example, the bootstrap sampling distribution of trimmed means.

**central tendency**—A number representing the "middle" of the data, or the number with the highest probability of occurring.

**computer-intensive statistics**—New statistics that are heavily based on modern high-speed computers. The payoff for such intensive computation is freedom from two major limiting factors that have dominated classical statistical theory since its beginning: the assumption that the data conform to a bell-shaped curve and the need to focus on statistical measures whose theoretical properties can be analyzed mathematically.

**concave down**—A function is concave down if the derivative of the function is decreasing and the second derivative is negative.

**concave up**—A function is concave up if the derivative of the function is increasing and the second derivative is positive.

**confidence interval**—The range of values in which an estimated value is expected to occur, given a level of probability.

**control parameter**—A variable to which the neuromuscular system is sensitive. A variable is a control parameter if variability in the relative phase increases prior to a bifurcation in the behavior and the variable promotes new relative phasing relationships (behaviors).

**correlation**—Measurement of the association or relationship between two samples. A significant correlation does not imply causation.

**correlation dimension (CoD)**—A measure that can approximate the fractal dimension of the region in state space occupied by the dynamical system.

**correlogram**—A plot of the cross-correlation or autocorrelation values at each lag.

**critical point**—The point on the response model surface where there is a local minimum or maximum. The critical point of the surface represents the preferred interaction of the modeled joints' behaviors.

**cross-correlation**—A function that expresses the degree of similarity between two series of data points.

**deterministic**—Referring to a signal in which future data points can be exactly predicted from past data points.

**deterministic origin**—The philosophical point of view, often associated with Albert Einstein, that nature is inherently deterministic and that any deviations from determinism are due to limitations on human knowledge.

**deterministic system**—A system whose evolving properties are completely determined by its current state and past history. Such a system has a predictable behavior at any future time.

**detrended fluctuation analysis (DFA)**—A type of analysis that can evaluate the presence of long-range, power-law correlations, as part of multifractal cascades that exist over a wide range of time scales.

**deviation phase**—The phase that quantifies variability in the relative phase relationship between two interacting segments. A low value indicates a more stable (less variable) relationship. A high value indicates a less stable relationship.

**differentiation**—The mathematical process by which one can calculate the higher-order derivatives of a signal, for example from displacement to calculate velocity and subsequently acceleration.

**discrete point analysis**—The process by which individual points on a curve are identified (usually local or global maximums and minimums).

**discrete relative phase**—Measure of relative phase at a specific local minimum or maximum of the relative phase curve.

**dispersion**—The spread in the data. Data that are highly clustered have low dispersion.

**dynamical diseases**—A class of diseases caused by variation of biological rhythms outside of normal limits. Variability or lack thereof is thought to reflect the nature and severity of the system's state of health or disease.

**electromyography**—The electrical activity in a contracting muscle composed of multiple motor unit action potentials.

**embedding dimension**—The minimum number of variables that is required to form a valid state space from a given time series.

**error variability ($V_e$)**—A component of total variability ($V_T$) caused by error originating within a biological system, during the measurement process, or from external or environmental variation. Traditional statistical techniques (e.g., standard deviation, coefficient of variation) can be used to quantify $V_e$.

**Fast Fourier Transformations**—A fast approximation to calculate the Fourier transform of a signal.

**filtering**—In human movement analysis, the general definition of filtering is the reduction or amplification of certain components of the signal.

**Fourier coefficients**—Coefficients applied to the sine and cosine functions of a Fourier series; they indicate the amount of the sine or cosine wave in a particular signal.

**Fourier transformation**—The linear function transforming a signal into frequency domain.

**frequency domain**—A representation of a signal as a function of frequency.

**gradient vector**—A vector that points to the critical point of a surface. It is based on the second partial derivative of the surface. The components of the gradient vector represent the behavior of the respective modeled joints.

**grid line curvature**—Measure of how sharply a grid line bends. Theoretically, the larger the curvature of the grid line, the less variability in the joint behavior. The less curvature of the grid line, the more variability in the joint behavior.

**grid lines**—Projections of the response surface model that provide a two-dimensional image of a surface shape. Grid lines isolate the behavior of the respective joints included in the model.

**homogeneity of variance**—Assumption that the variances of two groups are equal.

**independence**—The implication that the selection of any one element or subject from the population for inclusion in a sample does not alter the likelihood of drawing any other element of the population into the sample. In most cases, random sampling is used to accomplish independence.

**interindividual movement variability**—Variation in movement processes and outcomes that exist among individuals.

**intra-individual movement variability**—Variation in movement processes and outcomes that occur within an individual across multiple repetitions of movement execution.

**jackknife**—A resampling method in which each observation is omitted, in turn, to generate subsamples from a sample.

**joint time-frequency domain**—A two-dimensional space defined by the time and frequency axes.

**lag**—The number of data points by which a series is shifted when one is calculating a cross-correlation or autocorrelation.

**lag correlations**—Description of 1 to $n - 8$ ($n$ = total number of trials) for autocorrelation.

**local dynamic stability**—The sensitivity of a dynamical system to small perturbations (e.g., the natural stride-to-stride variations present during locomotion).

**Lyapunov exponent (LyE)**—A measure of the rate at which nearby trajectories in state space diverge.

**Mann-Whitney *U* Test**—A nonparametric statistical test that is used to evaluate the difference between two independent samples.

**matched filter**—A correlation-based algorithm used to detect a signal that is embedded within a more complex signal.

**mean absolute relative phase**—Quantifies the overall relative phase relationship of two segments. A value close to zero degrees indicates that the segments are in phase, while a value close to 180° indicates that the segments are out of phase.

**Model statistic**—Statistical procedure that takes advantage of the repeated-measures concept of the within-subject design to evaluate differences between conditions in a single-subject experiment.

**Monte Carlo simulation**—A type of computer-based simulation in which a hypothesized population is defined based on researchers' best knowledge of the phenomenon and random samples are then drawn from the population.

**motor unit action potential decomposition**—The process of identifying individual motor unit firings within an electromyographic signal.

**multicollinearity**—A condition in which there are high intercorrelations among predictor variables in a multiple regression analysis.

**multiple regression**—A statistical technique for predicting a dependent variable from a set of predictor variables. Simple regression predicts a dependent variable from a single predictor variable.

**noise**—That part of a system description that is not deterministic. For simplicity, it is usually assumed to have a simple form such as white noise.

**nonparametric**—Referring to data for which there is no underlying probability distribution.

**nonparametric tests**—Statistical tests that do not test hypotheses about specific parameters and do not require the assumptions of parametric tests; also called distribution-free tests since they do not require that the scores be normally distributed.

**normal distribution**—A set of data in which the probability of a value occurring increases the closer one gets to the mean. This is the classic symmetrical bell-shaped curve.

**normality**—The assumption that sample scores or characteristics are normally distributed.

**null hypothesis**—A statement of no difference between samples of data. For example, a null hypothesis could read, "The mean value of sample A will not be different than that of sample B." If the $p$-value is less than $\alpha$, then the null hypothesis is rejected, which suggests that there is a difference between the samples.

**order parameter**—A single variable that defines the dynamic state of the neuromuscular system and compresses the multiple degrees of freedom contained in the movement pattern into one value. Relative phase of the lower extremity segments has been shown to be an order parameter for gait.

**paired**—Referring to two samples of data that are not independent of one another. Samples are usually paired when two measurements are taken from the same subject.

**parametric**—Describes situation in which there is an underlying distribution of the data that can be characterized or modeled by functions having defined parameters. For example, data that are normally distributed are characterized by the parameters mean and standard deviation.

**phase angle**—The angle formed between the x-axis of the phase plot and the vector r of the phase plot trajectory. This angle quantifies where the trajectory is located in the phase plot as time progresses. It quantifies the behavior of the lower extremity segment and is used to calculate relative phase.

**phase portrait**—Provides a qualitative picture of the organization of the neuromuscular system. Constructed by plotting a segment's angular position versus its angular velocity.

**phase shift**—The amount by which one signal is temporally shifted relative to a second signal.

**phase space (or phase plane)**—A representation of the behavior of the dynamic system in state space. Typically, it takes the form of a two-dimensional plot of the position X of the time series (on the horizontal axis) versus the first derivative X' (on the vertical axis).

$P_n$—Represents the number of polynomial terms in a given equation.

**point estimate relative phase**—Provides a measure of the relationship between two segments based on their relative times to reach a local maximum or minimum; is a measure of the phase lag between two segments.

**power spectral density**—The square amplitude of the Fourier transform of the autocorrelation function of a stochastic signal.

**power spectrum**—The power of individual frequency components that are present in a signal.

**pseudo-population**—A "population" defined based on a representative sample drawn from the targeted population.

**p-value**—The probability that a result is due to chance. A $p$-value of 0.01 for a result indicates that this result would occur by chance only 1 time in 100.

**randomization**—A process by which all elements of a population have an equal probability of being selected.

**rearfoot angle**—The angle formed between the calcaneus and the tibia in a frontal plane.

**relative phase**—A measure of the interaction or coordination of two segments. Calculated by subtracting the phase angle of the proximal segment from that of the distal segment for each $i$th data point of the movement.

$R^n$—Represents the number of dimensions required to express a given set of values. For example, $R^2$ would indicate that the values are two-dimensional.

**sampling distribution**—Distribution of a sample statistic, for example the sampling distribution of means.

**single-subject (SS) analysis**—An experimental technique that involves an in-depth examination of individuals in order to better understand what unique movement characteristics, if any, they have in common. It does not imply "case study" investigation.

**stance phase**—The portion of a walking or running stride in which the foot is on the ground.

**state space**—A vector space where a dynamical system (e.g., a swaying body during posture) can be defined at any point.

**stationarity**—The statistical similarity of successive parts of a time series.

**stochastic**—Referring to a signal in which future data points are only partly determined by past values.

**strategy**—A unique neuromusculoskeletal solution for the performance of a motor task.

**surrogation**—A technique that can accurately determine if the source of a given time series is actually deterministic in nature by comparing the original time series with a random equivalent one.

**time domain**—A representation of a signal as a function of time. In biomechanics all signals are collected as a function of time (e.g., force plate data).

**time series**—A list of numbers assumed to measure some process sequentially in time.

**total variability ($V_T$)**—Total variability that exists within a system and is observable during movement. $V_T$ is a function of both variability due to nonlinear dynamical processes within the system ($V_n$) and variability due to error ($V_e$).

**trimmed mean**—The mean of a set of "trimmed" data, in which certain percentages of the data have been excluded to eliminate the effect of outliers. It is a useful central tendency measure when sample data come from a long-tailed population.

**uniform distribution**—A set of data in which any value is equally likely to occur. Random number generators produce uniform distributions.

**unimodal**—Describing data distributed such that they cluster around one peak, usually the mean value.

**variability and overuse injury hypothesis**—Hypothesis (model) predicting that some types of overuse injuries (e.g., stress fractures) might be caused by too little variability in tissue loading characteristics, leading to accumulation of trauma or an insufficient or delayed adaptive response. Conversely, the hypothesis predicts that some variability provides protection of the tissue from injury resulting from invariant repetitive loading.

**variance**—An index that reflects the degree of variability in a group of scores.

# Index

*Note:* The italicized *f* and *t* following page numbers refer to figures and tables, respectively.